Elements of Concave Analysis and Applications

Elements of Concave Analysis and Applications

Prem K. Kythe

CRC Press
Taylor & Francis Group
Boca Raton London New York

CRC Press is an imprint of the
Taylor & Francis Group, an **informa** business

CRC Press
Taylor & Francis Group
6000 Broken Sound Parkway NW, Suite 300
Boca Raton, FL 33487-2742

Printed on acid-free paper
Version Date: 20180414

International Standard Book Number-13: 978-1-138-70528-9 (Hardback)

Library of Congress Cataloging-in-Publication Data

Names: Kythe, Prem K., author.
Title: Elements of concave analysis and applications / Prem K. Kythe.
Description: Boca Raton, Florida : CRC Press, [2018] | Includes
bibliographical references and index.
Identifiers: LCCN 2018006905| ISBN 9781138705289 (hardback : alk. paper) |
ISBN 9781315202259 (ebook : alk. paper)
Subjects: LCSH: Concave functions--Textbooks. | Convex functions--Textbooks.
| Functions of real variables--Textbooks. | Matrices--Textbooks.
Classification: LCC QA353.C64 K96 2018 | DDC 515/.88--dc23
LC record available at https://lccn.loc.gov/2018006905

Visit the Taylor & Francis Web site at
http://www.taylorandfrancis.com

and the CRC Press Web site at
http://www.crcpress.com

To the memory of

Dr. James H. & Mrs. Mickey Abbott

with reverence and love

Contents

Preface

This textbook on concave analysis aims at two goals. Firstly, it provides simple yet comprehensive subject matter to the readers who are undergraduate seniors and beginning graduate students in mathematical economics and business mathematics. For most readers the only prerequisites are courses in matrix algebra and differential calculus including partial differentiation; however, for the last chapter a thorough working knowledge of linear partial differential equations and the Laplace transforms is required. The readers can omit this chapter if not required. The subject of the book centers mostly around concave and convex optimization; other related topics are also included. The details are provided below in the overview section.

Although there are many excellent books on the market, almost all of them are at times difficult to understand. They are very heavy on theoretical aspects, and generally fail to provide ample worked-out examples to give readers easy understanding and workability.

The second goal is elucidated below in the section 'To Readers'.

Motivation

The subject of convexity and quasi-convexity has been a model for economic theorists to make decisions about cost minimization and revenue maximization. This has resulted in a lot of publications in convex optimization. So why is there keen interest in concave and quasi-concave functions? Firstly, economic theory dictates that all utility functions are quasi-convex and that all cost functions are concave in input prices. Therefore, a cost function that is not concave in input prices is not a cost function. Secondly, the standard model in economic theory consists in a set of alternatives and an ordering of these alternatives, according to different priorities and interests. The process that a decision maker follows is to choose a favorite alternative with the property that no other alternative exceeds the ordering. In such a situation the decision maker often uses a function that 'represents' this ordering. Thus, for example, suppose there are four alternatives, say, a, b, c and d, and suppose that the decision maker prefers a to b and treats both c and d as equally desirable. Any function, like f, with $f(a) > f(b) > f(c) = f(d)$ may represent the ordering, irrespective of whatever numerical values (level curves) such ordering has. However, the situation changes when the decision maker is a consumer who is choosing between different goods and prices. In this case the consumer's ordering, based on the level curves (or indifference curves) of chosen alternatives, can look quite different from that of the businessperson. It so happens that the consumer's ordering becomes concave (less expense, more goods). But this situation implies that any function that represents the consumer's interest is, at best, quasi-concave. This natural phenomenon is, of course, based on a deep but simple result that concave and convex functions are diametrically opposite to each other in behavior and intent.

However, the subject of concave analysis, with emphasis on concave, quasi-concave and log-concave functions, has appeal to both the consumer and business organizations.

Overview

A general description of the topics covered in the book is as follows: Chapter 1 introduces a review of matrix algebra that includes definitions, matrix inversion, solutions of systems of linear algebraic equations, definite and semi-definite matrices, Jacobian, two types of Hessian matrices, and the Hessian test. Chapter 2 is a review of calculus, with topics dealing with limits, derivative, global and local extrema, first and second derivative tests, vector-valued functions, optimization, multivariate functions, and basic concepts of mathematical economics.

Concave and convex functions are introduced in Chapter 3, starting with the notion of convex sets, Jensen's inequalities for both concave and convex functions, and unconstrained optimization. Chapter 4 deals with concave programming; it is devoted to optimization problems on maximization mostly with inequality constraints, and using the Lagrange method of multipliers and the KKT necessary and sufficient conditions. Applications to mathematical economics include the topic of peak price loading, and comparative statics is discussed. Optimization problems focusing on minimization are introduced in Chapter 5 on convex programming, in order to compare it with concave optimization. Nonlinear programming is discussed; the Fritz John and Slater conditions are presented, and the topic of Lagrangian duality is discussed.

Chapters 6 and 7 deal with quasi-concave and quasi-convex functions. Both topics are important in their own applications. The single-function bordered Hessian test on quasi-concavity and quasi-convexity is presented, and optimization problems with types of functions and the minmax theorem are provided. Chapter 8 deals with log-concave functions; general results on log-concavity are presented, with application on mean residual life; and the Asplund sum is introduced, with its algebra, derivatives, and area measure. Log-concavity of nonnegative sequences is discussed, and all log-concave probability distributions with their density functions and cumulative density functions are presented in detail.

Chapter 9 deals with the quadratic programming to optimization problems, and presents the following numerical methods: (i) Hildreth-D'Esopo method; (ii) Beale's method; and (iii) Wolfe's method. The optimal control theory is discussed in Chapter 10, using the Hamiltonian and different types of optimization problems. Three types of demands, namely Marshallian, Hicksian, and Walrasian, are introduced in Chapter 11, using the Shephard's lemma and the Slusky equation with applications to cost minimization, and the Giffen and Veblen goods. Chapter 12 is exclusively devoted to the Black-Scholes differential equation, its solution, log-normal distribution, Black-Scholes call price, Feynman-Kac theorem, capital asset pricing model, dividends, stocks, bonds, and discounted stock prices.

There are six appendices: (A) Some useful topics on probability; (B) differentiation of operators involving the Gateaux differential, Fréchet derivative, the concept of the gradient, and Taylor's series first- and second-order approximations; (C) a list of probability distributions; (D) a self-contained detailed discussion of the Laplace transforms; (E) implicit function theorem; and (F) locally nonsatiated function. The bibliography toward the end of the book contains the references cited in the book, followed by the Index.

The book contains over 330 examples and exercises; most exercises are provided with solutions, simpler ones with hints and answers. Since the book uses and discusses vectors and matrices, care is taken, in order to avoid confusion, to set all (row/column) vectors and matrices in bold lowercase and bold uppercase, respectively.

This is an introductory textbook that provides a good combination of methodology, applications, and hands-on projects for students with diverse interests from mathematical economics, business mathematics, engineering, and other related applied mathematics courses.

To Readers

The second goal concerns specifically the abuse and misuse of a couple of standard mathematical notations in this field of scientific study. They are the gradient ∇f and the Laplacian $\nabla^2 f$ of a function $f(\mathbf{x})$ in $\mathbb{R}^n$. Somehow, and somewhere, a tradition started to replace the first-order partials of the function f by its gradient ∇f. It seems that this tradition started without any rigorous mathematical argument in its support. This book has provided a result (Theorem 2.18) that establishes that only under a specific *necessary condition* the column vector $[\partial f/\partial x_1 \;\cdots\; \partial f/\partial x_n]$ can replace the gradient vector ∇f, and these two quantities, although isomorphic to each other, are not equal. Moreover, it is shown that any indiscriminate replacement between these two quantities leads to certain incorrect results (§3.5).

The other misuse deals with the Laplacian $\nabla^2 f$, which has been used to represent the Hessian matrix (§1.6.2), without realizing that $\nabla^2 f$ is the trace (i.e., sum of the diagonal elements) of the Hessian matrix itself. This abuse makes a part equal to the whole. Moreover, ∇^2 is the well-known linear partial differential operator of the elliptic type known as the Laplacian.

It appears that this misuse perhaps happened because of the term 'vector', which is used (i) as a scalar quantity, having only magnitude, as in the row or column vectors (in the sense of a matrix), and (ii) as a physical quantity, such as force, velocity, acceleration, and momentum, having both magnitude and direction. The other factor for the abuse in the case of the gradient is the above-mentioned linear isomorphic mapping between the gradient vector ∇f and the (scalar) column vector $[\partial f/\partial x_1 \ \cdots \ \partial f/\partial x_n]^T$. This isomorphism has been then literally used as 'equality' between these two quantities. Once the case for ∇f became the tradition, the next choice $\nabla^2 f$ for the Hessian matrix became another obvious, but incorrect, tradition.

As readers, you will find an attention symbol, **!!!**, at different parts of the book. It is used to point out the significance of the statements found there.

The other less important notations are the $\prec$, the $\oplus$ and the $\odot$ symbols. Although borrowed from physics and astronomy, these symbols are acceptable with a different but almost similar meaning provided that they are properly defined as given in the section on Notations. Moreover, the $\oplus$ and the $\odot$ symbols have now become so common due to the advancement in cell phones and related electronic technology that they are probably losing their rigorous mathematical significance.

Acknowledgments

I take this opportunity to thank Mr. Sarfraz Khan, Executive Editor, Taylor & Francis, for his support, and Mr. Callum Fraser for coordinating the book project. I also thank the Project Editor Michele A. Dimont for doing a great job of editing the text. Thanks are due to the reviewers and to some of my colleagues who made some very valuable suggestions to improve the book. Lastly, I thank my friend Michael R. Schäferkotter for help and advice freely given whenever needed.

Prem K. Kythe

Notations, Definitions, and Acronyms

A list of the notations, definitions, abbreviations, and acronyms used in this book is given below.

a.s., almost surely
$\mathbf{A}$, matrix
$\mathbf{A}^T$, transpose of the matrix $\mathbf{A}$
$|\mathbf{A}|$, determinant of a matrix $\mathbf{A}$
adj $\mathbf{A}$, adjoint of a matrix $\mathbf{A}$
$\det(\mathbf{A})$ or $|\mathbf{A}|$, determinant of a matrix $\mathbf{A}$
$\mathbf{A}^{-1}$, inverse of a matrix $\mathbf{A}$
a_{ij}, element of a matrix $\mathbf{A}$ in the ith row and jth column
$B(c, r)$, ball with center c and radius r
$B(X, Y)$, class of all bounded linear operators from X into Y
$\mathbf{B}$, bordered Hessian matrix: one function
$C^0(D)$, class of functions continuous on a region D
$C^k(D)$, class of continuous functions with kth continuous derivative on a region D, $0 \le k < \infty$
$C^\infty(D)$, class of continuous functions infinitely differentiable on a region D
C-function, same as a C^0-function; continuous function
$\mathbf{C}$, cofactor matrix
C_{ij}, cofactor of the element in the ith row and jth column
C^T, transpose of $C = \text{adj}\,\mathbf{A}$
C_i, eigenvectors
C, cost; consumption
$C(Q)$, production cost
CAPM, capital asset pricing model
CES, constant elasticity of substitution
CLPD, constant positive linear dependence constraint qualification
CLT, central limit theorem
CQ, constraint qualification
CRCQ, constant rank constraint qualification
CU, concave upward
CD, concave downward

Cov, covariance
c.d.f., cumulative distribution function
$c(w, y)$, cost function
D_t, derivative with respect to t
$Df(\mathbf{x})$, derivative of $f(\mathbf{x})$ in $\mathbb{R}^n$
D, aggregated demand
D, domain, usually in the z-plane
dist(A, B), distance between points (or sets) A and B
dom(f), domain of a function f
DRS, decreasing return to scale
e, expenditure function
E, amount allocated for expenditure
$E[X]$, expected value of a random vector X
$E(f)$, entropy of f
Eq(s)., Equation(s) (when followed by an equation number)
$\mathbf{e}_i$, ith unit vector, $i = 1, \dots, n$
$[\mathbf{e}]$, set of the unit vectors $\mathbf{e}_i$ in $\mathbb{R}^n$
epi(f), epigraph of f
$e(p, u)$, expenditure function
F, field
$f : X \mapsto Y$, function f maps the set X into (onto) the set Y
$f \circ g$, composite function of f and g: $(f \circ g)(\cdot) = f(g(\cdot))$
f', first derivative of f
f'', second derivative of f
$f^{(n)}$, nth derivative of f
$\dfrac{\partial f(\mathbf{x})}{\partial x_i}$, first-order partials of f in $\mathbb{R}^n$, also written f_i, for $i = 1, \dots, n$; also written as f_x, f_y, f_x for $\dfrac{\partial f}{\partial x}, \dfrac{\partial f}{\partial y}, \dfrac{\partial f}{\partial z}$ in $\mathbb{R}^3$
$\dfrac{\partial^2 f(\mathbf{x})}{\partial x_i \partial x_j}$, second-order partials of f in $\mathbb{R}^n$, also written as f_{ij} for $i, j = 1, \dots, n$; also written as f_{xx}, f_{yy}, f_{zx} for $\dfrac{\partial^2 f}{\partial x^2}, \dfrac{\partial^2 f}{\partial y^2}, \dfrac{\partial^2 f}{\partial z^2}$ in $\mathbb{R}^3$
$(f \circ g)(x), = f(g(x))$, composition of functions f and g
$f \star g$, convolution of $f(t)$ and $g(t)$ ($= \int_0^t f(t-u) g(u)\, du = \int_0^t f(u_g(t-u)\, du = \mathcal{L}^{-1}\{G(s)F(s)\}$)
FJ, Fritz John conditions
$F(s)$, Laplace transform of $f(t)$ ($= \int_0^\infty e^{st} f(t)\, dt$)
G, government expenditure; constrained set
$G_{\min}$, positive minimal accepted level of profit
$\mathcal{G}(\cdot; \cdot)$, Green's function

Geom(p), geometric distribution with probability distribution
H, Hamiltonian
hyp, hypograph of f
$h_j(p, u)$, Hicksian demand for good j
h, hours of work, $h = T - \tau$
$\mathbf{H}$, Hessian matrix
$\bar{\mathbf{H}}$, bordered Hessian matrix: two functions, f and the constraint g
iff, if and only if
i, income
I, income
$I(p)$, e-beam intensity at position p
$I_K(x) = 0$ if $x \in K; \infty$ if $x \notin K$
IS, commodity equilibrium
Int (D), interior of a domain D
IRS, increasing return to scale
$\mathbf{I}$, identity matrix
j , goods number; input factor
$|\mathbf{J}|$, or $\mathbf{J}$, or simply J, Jacobian determinant
K, capital
KKT, Karush-Kuhn-Tucker conditions
L, Lagrangian function; also, labor
$L_x,\ L_y, = \dfrac{\partial L}{\partial x}, \dfrac{\partial L}{\partial y}$ first-order partials of L
$L_{xx},\ L_{yy},\ L_{xy},\ L_{yx}, = \dfrac{\partial^2 L}{\partial x^2}, \dfrac{\partial^2 L}{\partial y^2}, \dfrac{\partial^2 L}{\partial x \partial y}, \dfrac{\partial^2 L}{\partial y \partial x}$, second-order partials of L
$\mathcal{L}\{f(t)\}$, Laplace transform of $f(t)$, also denoted by $F(s)$
$\mathcal{L}^{-1}\{f(t)\}$, inverse Laplace transform, also denoted by $f(t)$
$\mathcal{L}_2\{f(t); s\}$, Laplace$_2$ transform (also known as $\mathcal{L}_2$-transform) of $f(t; s)$
LICQ, linear independence constraint qualification
LM, monetary equilibrium
LM, monetary equilibrium
LQC, linearity constrained qualifications
mrl(x), mean residual lifetime function
M_{ij}, minor of the element a_{ij} of a matrix $\mathbf{A}$
M_d, demand for money
M_s, supply of money
M_t, transition-precautionary demand for money
M_z, speculative demand for money
MC, marginal cost
MFCQ, Mangasarian-Fromovitz constraint qualification
ME, marginal expenditure

MPC, marginal propensity to consume
MR, marginal revenue
MRTS, marginal rate of technical substitution
$\mathbb{N}$, set of natural numbers (positive integers)
$\mathfrak{N}(\mathbf{A})$, null space of matrix $\mathbf{A}$
$\mathcal{N}$, numéraire
ND, negative definite
NSD, negative semidefinite
p.d.f., probability density function, or density function
p, prices (a vector with component p_j)
P, profit; supply; nonlabor income
P_V, probability function for a random variable V
P_K price for K
P_L, price for L
$\mathfrak{P}$, probability
PD, positive definite
PSD, positive semidefinite
q, Cobb-Douglas production function; also, amount of consumed good
Q, quantity of output produced; risk neutral measure
Q_s, supply
Q_d, demand
Q, level of output
QNCQ, quasi-normality constraint qualification
QP, quadratic programming
r, eigenvalue of a matrix $\mathbf{A}$
$\mathbf{r}(t)$, vector-valued function $(= f(t)\mathbf{i} + g(t)\mathbf{j} + h(t)\mathbf{k})$
R, revenue
$R(Q)$, sales revenue
$\mathcal{R}(f)$, range of function f
$\mathbb{R}$, the real line; real plane
$\mathbb{R}^3$, three-dimensional real space
$\mathbb{R}^n$, n-dimensional real space
$\mathbb{R}_+$, nonnegative real numbers
$\mathbb{R}_{++}$, positive real numbers
s, variable of Laplace transform
S, slackness variable; savings
$S(p, w)$, Slutsky matrix
SC, Slater condition
SDE, stochastic differential equation
SOSC, second-order sufficient condition
tr$(\mathbf{A})$, trace of a matrix $\mathbf{A}$ (sum of the diagonal elements)

T, total time available
TC, total cost
TE, total expenditure
TP, total product
TP$_k$, totally positive of order k
TR, total revenue
u, utility
$\dot{u}$, time derivative of a function u
$U(y \mid f)$, upper contour set of f at y
U_α, upper-level set
$v(p, m)$, indirect utility function
V, value function
w, wage (a vector with components w_j);
wh, labor income(wage/hour)
x, quantity (vector)
$x_j(w, y)$, conditional factor demand for each input factor or good j
$\lfloor x \rfloor$, greatest integer function, floor of x
$\mathbf{x}$, eigenvector of a matrix $\mathbf{A}$; a point in $\mathbb{R}$
$\mathbf{x}^*$, critical number, critical point (maximizer or minimizer)
$\mathbf{x} \succeq \mathbf{y}$, componentwise inequality between vectors $\mathbf{x}$ and $\mathbf{y}$
$\mathbf{x} \succ \mathbf{y}$, strict componentwise inequality between vectors $\mathbf{x}$ and $\mathbf{y}$
X, real normed space
X, exports
y, output;
Y, income
Y_c, convex hull of $Y \subset X$
Z, imports
$\mathbb{Z}$, set of integers
$\mathbb{Z}_+$, set of nonnegative integers
$\delta f(x, h)$, Gateaux differential (or G-differential) of f at x and h real
λ, $\boldsymbol{\lambda}$ Lagrange multiplier(s)
μ, $\boldsymbol{\mu}$ Lagrange multiplier(s); also measure, Lebesgue measure
π, profit
ρ, rank of a matrix; $\rho(\mathbf{A})$, rank of the matrix $\mathbf{A}$; correlation coefficient
Σ, σ-algebra
τ, labor; also, time spent for leisure
$\chi_K(x) = 1$ if $x \in K$; 0 if $x \notin K$, characteristic function defined on a convex set K
$\|\cdot\|$, norm
$\mathbf{0}$, null (or zero) vector, $= [0\ 0\ \dots\ 0]$ in $\mathbb{R}^n$

∇, 'del' operator, $\nabla = \mathbf{i}\frac{\partial}{\partial x} + \mathbf{j}\frac{\partial}{\partial y} + \mathbf{k}\frac{\partial}{\partial z}$, $((x, y, z) \in \mathbb{R}^3)$; an operator defined in $\mathbb{R}^n$ as $\nabla = \mathbf{e}_1\frac{\partial}{\partial x_1} + \cdots + \mathbf{e}_n\frac{\partial}{\partial x_n}$

∇f, gradient of a function f, a vector in $\mathbb{R}^3$, defined by $\nabla = \mathbf{i}\frac{\partial f}{\partial x} + \mathbf{j}\frac{\partial f}{\partial y} + \mathbf{k}\frac{\partial f}{\partial z}$; a vector in $\mathbb{R}^n$ defined by $\nabla = \frac{\partial f}{\partial x_1}\mathbf{e}_1 + \cdots + \frac{\partial f}{\partial x_n}\mathbf{e}_n$ for $\mathbf{x} = (x_1, \dots, x_n) \in \mathbb{R}^n$ (dimension $1 \times n$), or by $\nabla = \mathbf{e}_1\frac{\partial f}{\partial x_1} + \cdots + \mathbf{e}_n\frac{\partial f}{\partial x_n}$ for $\mathbf{x} = (x_1, \dots, x_n) \in \mathbb{R}^n$ (dimension $n \times 1$)

∇^2, Laplacian operator defined on $\mathbb{R}^n$ as $\frac{\partial^2}{\partial x_1^2} + \cdots + \frac{\partial^2}{\partial x_n^2}$; it is a linear elliptic partial differential operator

$\nabla^2 f(\mathbf{x}) = \frac{\partial^2 f}{\partial x_1^2} + \cdots + \frac{\partial^2 f}{\partial x_n^2}$, $\mathbf{x} = (x_1, \dots, x_n) \in \mathbb{R}^n$, Laplacian of $f(\mathbf{x})$; also the trace of the Hessian matrix $\mathbf{H}$

$\|\mathbf{x}\|_1$, l_1-norm of a vector $\mathbf{x}$

$\|\mathbf{x}\|_2$, l_2-norm, or Euclidean norm, of a vector $\mathbf{x}$

$\|\mathbf{x}\|_\infty$, l_∞-norm of a vector $\mathbf{x}$

$\succ$, $\succeq$, (subordination (predecessor): $\mathbf{A} \succeq \mathbf{B}$, matrix inequality between matrices $\mathbf{A}$ and $\mathbf{B}$; $\mathbf{A} \succ \mathbf{B}$, strict matrix inequality between matrices $\mathbf{A}$ and $\mathbf{B}$

$\prec$, $\preceq$, subordination (successor) , e.g., $f \prec g$ is equivalent to $f(0) = g(0)$ and $f(E) \subset g(E)$, where E is the open disks ; but here $\mathbf{x} \prec \mathbf{y}$ is used for componentwise strict inequality, and $\mathbf{x} \preceq \mathbf{y}$ for componentwise inequality between vectors $\mathbf{x}$ and $\mathbf{y}$

$(f \oplus g)(z)$, $= \sup_{x+y}\{f(x)g(y)\}$, where f and g are log-concave functions

$(s \odot f)(x)$, $= sf\big(\frac{x}{s}\big)$, where f is a log-concave function, and $s > 0$

$\binom{n}{k}$, binomial coefficient $= \frac{n!}{k!\,(n-k)!} = \binom{n}{n-k}$

$\overset{\text{iso}}{=}$, isomorphic to; for example, $A \overset{\text{iso}}{=} B$ means A is isomorphic to B, and conversely

■ end of a proof, or an example

!!! attention symbol

1

Matrix Algebra

Some basic concepts and results from linear and matrix algebra, and from finite-dimensional vector spaces are presented. Proofs for most of the results can be found in many books, for example, Bellman [1970], Halmos [1958], Hoffman and Kunze [1961] Lipschutz [1968], and Michel and Herget [2007].

1.1 Definitions

A matrix $\mathbf{A}$ is a rectangular array of *elements* (numbers, parameters, or variables), where the elements in a horizontal line are called *rows*, and those in a vertical line *columns*. The *dimension* of a matrix is defined by the number of rows m and the number of columns n, and we say that such a matrix has dimension $m \times n$, or simply that the matrix is $m \times n$. If $m = n$, then we have a *square* matrix. If the matrix is $1 \times n$, we call it a *row vector*, and if the matrix is $m \times 1$, then it is called a *column vector*. A matrix that converts the rows of a matrix $\mathbf{A}$ to columns and the columns of $\mathbf{A}$ to rows is called the *transpose* of $\mathbf{A}$ and is denoted by $\mathbf{A}^T$.

Let two 3×3 matrices A and B be defined as

$$\mathbf{A} = \begin{bmatrix} a_{11} & a_{12} & a_{13} \\ a_{21} & a_{22} & a_{23} \\ a_{31} & a_{32} & a_{33} \end{bmatrix}, \quad \mathbf{B} = \begin{bmatrix} b_{11} & b_{12} & b_{13} \\ b_{21} & b_{22} & b_{23} \\ b_{31} & b_{32} & b_{33} \end{bmatrix}. \tag{1.1.1}$$

Then, for example,

$$\mathbf{A}^T = \begin{bmatrix} a_{11} & a_{21} & a_{31} \\ a_{12} & a_{22} & a_{32} \\ a_{13} & a_{23} & a_{33} \end{bmatrix},$$

and similarly for $\mathbf{B}^T$. For addition (or subtraction) of two matrices ($\mathbf{A} + \mathbf{B}$, or $\mathbf{A} - \mathbf{B}$) the two matrices $\mathbf{A}$ and $\mathbf{B}$ must be of equal dimension. Each element of one matrix ($\mathbf{B}$) is added to (or subtracted from) the corresponding element of the other matrix ($\mathbf{A}$). Thus, the element b_{11} in $\mathbf{B}$ is added to (or subtracted from) a_{11} in **A**; b_{12} to (or from) a_{12}, and so on. Multiplication of a matrix by a number or scalar involves multiplication of each element of the matrix by the scalar, and it is called *scalar multiplication*, since it scales the matrix up or down by the size of the scalar.

A row vector **A** and a column vector **B** are written, respectively, as

$$\mathbf{A} = [\,a_{11} \quad a_{12} \quad a_{13}\,]_{1\times 3}, \quad \mathbf{B} = \begin{bmatrix} b_{11} \\ b_{21} \\ b_{31} \end{bmatrix}_{3\times 1}.$$

However, to save space, **B** is often written as its transpose, i.e.,

$$\mathbf{B} = [\,b_{11} \quad b_{21} \quad b_{31}\,]^T.$$

Multiplication of a row vector **A** by a column vector **B** requires that each vector has the same number of elements. This multiplication is then carried out by multiplying each element of the row vector by its corresponding element in the column vector and summing the product. Thus,

$$\begin{aligned}\mathbf{AB} &= [\,a_{11} \quad a_{12} \quad a_{13}\,]_{1\times 3} \times [\,b_{11} \quad b_{21} \quad b_{31}\,]^T_{3\times 1} \\ &= [a_{11}b_{11} + a_{12}b_{21} + a_{13}b_{31}]_{1\times 1}.\end{aligned}$$

The above technique of multiplication of a row and a column vector is used to obtain the multiplication of any two vectors with a precondition that the number of rows and columns of one matrix must be the same as the number of columns and rows of the other matrix. Then the two matrices are said to be *conformable* for multiplication. Thus, an $m \times n$ matrix can be multiplied by another $n \times m$ matrix, and the resulting matrix will be $m \times m$.

Example 1.1. Given

$$\mathbf{A} = \begin{bmatrix} 3 & 6 & 11 \\ 12 & 8 & 5 \end{bmatrix}_{2\times 3}, \quad \mathbf{B} = \begin{bmatrix} 5 & 13 \\ 7 & 8 \\ 9 & 10 \end{bmatrix}_{3\times 2}, \quad \mathbf{E} = \begin{bmatrix} 1 & 4 & 7 \\ 2 & 4 & 9 \end{bmatrix}_{2\times 3},$$

the matrices A and B, and B and E are conformable for multiplication, but A and C are not conformable. Thus,

$$\mathbf{AB} = \begin{bmatrix} 3\cdot 5 + 6\cdot 7 + 11\cdot 9 & 3\cdot 13 + 6\cdot 8 + 11\cdot 10 \\ 12\cdot 5 + 8\cdot 7 + 5\cdot 9 & 12\cdot 13 + 8\cdot 8 + 5\cdot 10 \end{bmatrix} = \begin{bmatrix} 156 & 97 \\ 161 & 270 \end{bmatrix}_{2\times 2},$$

$$\mathbf{BE} = \begin{bmatrix} 5\cdot 1 + 13\cdot 2 & 5\cdot 4 + 13\cdot 4 & 5\cdot 7 + 13\times 9 \\ 7\cdot 1 + 8\cdot 2 & 7\cdot 4 + 8\cdot 4 & 7\cdot 7 + 8\times 9 \\ 9\cdot 1 + 10\cdot 2 & 9\cdot 4 + 10\cdot 4 & 9\cdot 7 + 10\times 9 \end{bmatrix} = \begin{bmatrix} 31 & 72 & 152 \\ 23 & 60 & 121 \\ 29 & 76 & 153 \end{bmatrix}_{3\times 3}.$$

Example 1.2. In business, one method of keeping track of sales of different types of products at different outlets is to keep the inventory in the form of a matrix. Thus, suppose that a construction company has four different outlets selling (a) bricks, (b) lumber, (c) cement, and (d) roof shingles. The inventory and the price of each item are expressed as matrices **A** and **P**:

$$\mathbf{A} = \begin{array}{c} \begin{matrix} (a) & (b) & (c) & (d) \end{matrix} \\ \begin{bmatrix} 100 & 110 & 80 & 115 \\ 210 & 230 & 150 & 400 \\ 165 & 95 & 68 & 145 \\ 150 & 190 & 130 & 300 \end{bmatrix}_{4\times 4} \end{array}, \quad \mathbf{P} = \begin{bmatrix} 220 \\ 65 \\ 114 \\ 168 \end{bmatrix}_{4\times 1}.$$

Since the two matrices are conformable, their product **AP** will give the values **V** (in dollars) of the stock at each outlet:

$$\mathbf{V} = \mathbf{AP} = \begin{bmatrix} 100\cdot 220 + 110\cdot 65 + 80\cdot 114 + 115\cdot 168 \\ 210\cdot 220 + 230\cdot 65 + 150\cdot 114 + 400\cdot 168 \\ 165\cdot 220 + 95\cdot 65 + 68\cdot 114 + 145\cdot 168 \\ 150\cdot 220 + 190\cdot 65 + 130\cdot 114 + 300\cdot 168 \end{bmatrix} = \begin{bmatrix} 57,590 \\ 145,450 \\ 74,587 \\ 110,520 \end{bmatrix}. \blacksquare$$

1.2 Properties

The following properties of matrices are useful.

1. Matrix addition is commutative and associative i.e., $\mathbf{A}+\mathbf{B} = \mathbf{B}+\mathbf{A}$, and $(\mathbf{A}+\mathbf{B})+\mathbf{C} = \mathbf{A}+(\mathbf{B}+\mathbf{C})$. These properties also hold for matrix subtraction, since $\mathbf{A}-\mathbf{B} = \mathbf{A}+(-\mathbf{B})$.

2. Matrix multiplication, with a few exceptions, is not commutative, i.e., $\mathbf{AB} \neq \mathbf{BA}$. Scalar multiplication is commutative, i.e., $c\mathbf{A} = \mathbf{A}c$. If three or more matrices are conformable, i.e., if $\mathbf{A}_{j\times k}, \mathbf{B}_{m\times n}, \mathbf{C}_{p\times q}$, where $k = m$ and $n = p$, the associative law applies as long as matrices are multiplied in their order of conformability. Thus, $(\mathbf{AB})\mathbf{C} = \mathbf{A}(\mathbf{BC})$. Under the same conditions the matrix multiplication is also distributive, i.e., $\mathbf{A}(\mathbf{B}+\mathbf{C}) = \mathbf{AB}+\mathbf{BC}$.

Example 1.3. Given

$$\mathbf{A} = \begin{bmatrix} 7 & 4 \\ 1 & 5 \\ 8 & 9 \end{bmatrix}_{3\times 2}, \quad \mathbf{B} = \begin{bmatrix} 4 & 3 & 10 \\ 3 & 5 & 6 \end{bmatrix}_{2\times 3}, \quad \mathbf{C} = \begin{bmatrix} 7 \\ 8 \\ 9 \end{bmatrix}_{3\times 1},$$

we get

$$\mathbf{AB} = \begin{bmatrix} 7\cdot 4 + 4\cdot 3 & 7\cdot 3 + 4\cdot 5 & 7\cdot 10 + 4\cdot 6 \\ 1\cdot 4 + 5\cdot 3 & 1\cdot 3 + 5\cdot 5 & 1\cdot 10 + 5\cdot 6 \\ 8\cdot 4 + 9\cdot 3 & 8\cdot 3 + 9\cdot 5 & 8\cdot 10 + 9\cdot 6 \end{bmatrix} = \begin{bmatrix} 40 & 41 & 94 \\ 19 & 28 & 40 \\ 59 & 69 & 134 \end{bmatrix}_{3\times 3};$$

$$(\mathbf{AB})\mathbf{C} = \begin{bmatrix} 40 & 41 & 94 \\ 19 & 28 & 40 \\ 59 & 69 & 134 \end{bmatrix}_{3\times 3} \begin{bmatrix} 7 \\ 8 \\ 9 \end{bmatrix}_{3\times 1} = \begin{bmatrix} 40\cdot 7 + 41\cdot 8 + 94\cdot 9 \\ 19\cdot 7 + 28\cdot 8 + 40\cdot 9 \\ 59\cdot 7 + 69\cdot 8 + 134\cdot 9 \end{bmatrix} = \begin{bmatrix} 1454 \\ 717 \\ 2171 \end{bmatrix}_{3\times 1};$$

$$\mathbf{BC} = \begin{bmatrix} 4 & 3 & 10 \\ 3 & 5 & 6 \end{bmatrix}_{2\times 3} \begin{bmatrix} 7 \\ 8 \\ 9 \end{bmatrix}_{3\times 1} = \begin{bmatrix} 4\cdot 7 + 3\cdot 8 + 10\cdot 9 \\ 3\cdot 7 + 5\cdot 8 + 6\cdot 9 \end{bmatrix} = \begin{bmatrix} 142 \\ 115 \end{bmatrix}_{2\times 1};$$

$$\mathbf{A}(\mathbf{BC}) = \begin{bmatrix} 7 & 4 \\ 1 & 5 \\ 8 & 9 \end{bmatrix}_{3\times 2} \begin{bmatrix} 142 \\ 115 \end{bmatrix} = \begin{bmatrix} 7\cdot 142 + 4\cdot 115 \\ 1\cdot 142 + 5\cdot 115 \\ 8\cdot 142 + 9\cdot 115 \end{bmatrix}_{3\times 1} = \begin{bmatrix} 1454 \\ 717 \\ 2171 \end{bmatrix}_{3\times 1}. \blacksquare$$

3. An *identity matrix* $\mathbf{I}$ is a square matrix whose diagonal elements are all 1 and all remaining elements are 0. An $n \times n$ identity matrix is sometimes denoted by $\mathbf{I}_n$. The identity matrix $\mathbf{I}$ is the unity in matrix algebra just as the numeral 1 is the unity in algebra. Thus, the multiplication of a matrix by an identity matrix leaves the original matrix unchanged; so also the multiplication of an identity matrix by itself leaves the identity matrix unchanged. Hence, $\mathbf{AI} = \mathbf{IA} = \mathbf{A}$, and $\mathbf{I} \times \mathbf{I} = \mathbf{I}^2 = \mathbf{I}$.

4. A matrix $\mathbf{A}$ for which $\mathbf{A} = \mathbf{A}^T$ is called a *symmetric matrix*. A symmetric matrix $\mathbf{A}$ for which $\mathbf{A} \times \mathbf{A} = \mathbf{A}$ is an *idempotent matrix*. The identity matrix $\mathbf{I}$ is both symmetric and idempotent.

5. A *null matrix* $\mathbf{0}$ is composed of all 0s and can have any dimension; it is not necessarily square. Obviously, addition or subtraction of null matrices leaves the original matrix unchanged; multiplication by a null matrix yields a null matrix. A scalar zero 0 has dimension 1×1.

6. A matrix with zero elements everywhere below (or above) the principal diagonal is called *upper* (or *lower*) *triangular matrix*, also known as *upper* or *lower echelon form*. Thus,

$$\begin{bmatrix} a_{11} & a_{12} & \cdots & a_{1,n-1} & a_{1,n} \\ 0 & a_{22} & \cdots & a_{2,n-1} & a_{2n} \\ \cdots & \cdots & \cdots & \cdots & \cdots \\ 0 & 0 & \cdots & 0 & a_{nn} \end{bmatrix}, \quad \text{or} \quad \begin{bmatrix} a_{11} & 0 & \cdots & 0 & 0 \\ a_{21} & a_{22} & \cdots & 0 & 0 \\ \cdots & \cdots & \cdots & \cdots & \cdots \\ a_{n1} & a_{n2} & \cdots & a_{n,n-1} & a_{nn} \end{bmatrix}$$

represent the upper or lower triangular matrix.

7. The sum of all the elements on the principal diagonal of an $n \times n$ matrix $\mathbf{A}$ is called the *trace* of the matrix $\mathbf{A}$ and is denoted by $\text{tr}(\mathbf{A})$.

See Exercise 1.6 for the dimensions of some most used expressions in $\mathbb{R}^n$. Remember that dimensions are always listed row by column, and a vector $\mathbf{a}$ is a column vector, and scalars are of dimension 1×1.

1.3 Matrix Inversion

The *determinant* of a 2×2 matrix $\mathbf{A}$ is called a *second-order determinant*, and it is obtained by taking the product of the two elements on the principal diagonal and subtracting it from the product of the two elements off the principal diagonal. Thus, for the 2×2 matrix

$$\mathbf{A} = \begin{bmatrix} a_{11} & a_{12} \\ a_{21} & a_{22} \end{bmatrix},$$

the determinant is

$$|\mathbf{A}| = \begin{vmatrix} a_{11} & a_{12} \\ a_{21} & a_{22} \end{vmatrix} = a_{11}a_{22} - a_{12}a_{21}. \tag{1.3.1}$$

The determinant $|A|$ is sometimes denoted by $\det(\mathbf{A})$. It is a number or a scalar and is obtained only for square matrices. If $|\mathbf{A}| = 0$, then the determinant is said to *vanish* and the matrix $\mathbf{A}$ is said to be *singular*. A singular matrix is one in which there exists a linear dependence between at least two rows or columns. If $|\mathbf{A}| \neq 0$, then the matrix $\mathbf{A}$ is *nonsingular* and all its rows and columns are linearly independent.

Generally, in the case of systems of linear equations, if there is linear dependence, the system may have no solution or an infinite number of possible solutions, and a unique solution cannot be determined. Thus, the determinant is an important test that points to specific problems before the system should be solved. Given a system of equations with the coefficient matrix $\mathbf{A}$,

(i) if $|\mathbf{A}| = 0$, the matrix is singular, indicating that there is a linear dependence among the equations, and so no unique solution exists.

(ii) if $|\mathbf{A}| \neq 0$, the matrix is nonsingular and there is linear independence among the equations, and so a unique solution exists and can be determined.

The *rank* ρ of a matrix is defined as the maximum number of linearly independent rows or columns in the matrix. The rank of a matrix is used for a simple test of linear dependence, as follows: Assume that $\mathbf{A}$ is a square matrix of order n. Then

(i) if $\rho(\mathbf{A}) = n$, the $\mathbf{A}$ is nonsingular and there is linear independence;

(ii) if $\rho(\mathbf{A}) < n$, then $\mathbf{A}$ is singular and there is linear dependence.

Example 1.4. Consider $\mathbf{A} = \begin{bmatrix} 7 & 4 \\ 8 & 9 \end{bmatrix}$ and $\mathbf{B} = \begin{bmatrix} 6 & 9 \\ 8 & 12 \end{bmatrix}$. Then $|A| = 7(9) - 4(8) = 63 - 32 = 31 \neq 0$, and so the matrix $\mathbf{A}$ is nonsingular, i.e., there is linear independence between any rows or columns, and $\rho(\mathbf{A}) = 2$. On the other hand, $|\mathbf{B}| = 6(12) - 9(8) = 72 - 72 = 0$, and so the matrix $\mathbf{B}$ is singular and linear dependence exists between its rows and columns (casual inspection reveals that row 2 is 4/3 times row 1, and column 2 is 3/2 times column 1), and thus $\rho(\mathbf{B}) = 1$. ■

The determinant of a 3×3 matrix $\mathbf{A}$

$$\mathbf{A} = \begin{bmatrix} a_{11} & a_{12} & a_{13} \\ a_{21} & a_{22} & a_{23} \\ a_{31} & a_{32} & a_{32} \end{bmatrix} \tag{1.3.2}$$

is called a *third-order determinant*, and it is a sum of three products, which are derived as follows:

1. Take the first element a_{11} of the first row and (mentally) delete the row and the column in which it appears. Then multiply a_{11} by the determinant of the remaining second-order matrix.
2. Take the second element a_{12} of the first row and (mentally) delete the row and the column in which it appears. Then multiply a_{12} by (-1) and the determinant of the remaining second-order matrix.
3. Take the first element a_{13} of the first row and (mentally) delete the row and the column in which it appears. Then multiply a_{13} by the determinant of the remaining second-order matrix.

This process yields

$$\begin{aligned} |\mathbf{A}| &= a_{11}\begin{vmatrix} a_{22} & a_{23} \\ a_{32} & a_{33} \end{vmatrix} + a_{12}(-1)\begin{vmatrix} a_{21} & a_{23} \\ a_{31} & a_{33} \end{vmatrix} + a_{13}\begin{vmatrix} a_{21} & a_{22} \\ a_{31} & a_{32} \end{vmatrix} \\ &= a_{11}(a_{22}a_{33} - a_{23}a_{32}) - a_{12}(a_{21}a_{32} - a_{23}a_{31}) + a_{13}(a_{21}a_{32} - a_{22}a_{31}), \end{aligned} \tag{1.3.3}$$

which is a scalar quantity.

The determinant of a 4×4 matrix can be determined similarly as the sum of four products, and so on.

The first line of Eq (1.3.3) can also be written as

$$|\mathbf{A}| = a_{11}|M_{11}| + a_{12}(-1)|M_{12}| + a_{13}|M_{13}|, \tag{1.3.4}$$

where

$$|M_{11}| = \begin{vmatrix} a_{22} & a_{23} \\ a_{32} & a_{33} \end{vmatrix}, \quad |M_{12}| = \begin{vmatrix} a_{21} & a_{23} \\ a_{31} & a_{33} \end{vmatrix}, \quad |M_{13}| = \begin{vmatrix} a_{21} & a_{22} \\ a_{31} & a_{32} \end{vmatrix},$$

where $|M_{11}|$ is the *minor* of a_{11}, $|M_{12}|$ the *minor* of a_{12}, and $|M_{13}|$ the *minor* of a_{13}. A *cofactor* $|C_{ij}|$ is a minor with a prescribed sign, which follows the rule

$$|C_{ij}| = (-1)^{i+j}|M_{ij}|. \tag{1.3.5}$$

Thus, depending on an even or odd power of (-1) we have

$$\begin{cases} \text{if } i+j \text{ is an even number, then} & |C_{ij}| = |M_{ij}|, \\ \text{if } i+j \text{ is an odd number, then} & |C_{ij}| = -|M_{ij}|. \end{cases}$$

Then the expansion (1.3.4) can be expressed in terms of the cofactors as

$$|\mathbf{A}| = a_{11}|C_{11}| + a_{12}|C_{12}| + a_{13}|C_{13}|, \tag{1.3.6}$$

which is known as the third-order *Laplace expansion* of the determinant $|A|$. This expansion can be extended to an nth-order determinant. The Laplace expansion allows us to evaluate a determinant along any row or column. In practice, however, selection of a row or column with more zeros than others simplifies the evaluation by eliminating terms.

Let $\mathbf{A}$ and $\mathbf{B}$ be $(n \times n)$ matrices. Some properties of determinants are stated below; proofs can be found in Michel and Herget [2007].

1. $|\mathbf{A}^T| = |\mathbf{A}|$; and $|\mathbf{A}B| = |\mathbf{A}||\mathbf{B}|$.
2. If all elements of a column (or row) of $\mathbf{A}$ are zero, then $|\mathbf{A}| = 0$.
3. If $\mathbf{B}$ is the matrix obtained by multiplying every element in a column (or row) of $\mathbf{A}$ by a constant c, while all other columns of $\mathbf{B}$ are the same as those in $\mathbf{A}$, then $|\mathbf{B}| = c\,|\mathbf{A}|$.
4. If $\mathbf{B}$ is the same except that two columns (or rows) are interchanged, then $|\mathbf{B}| = -|\mathbf{A}|$.
5. If two columns (or rows) of $\mathbf{A}$ are identical, or proportional, or linearly dependent, then $|\mathbf{A}| = 0$.
6. Addition or subtraction of any nonzero multiple of one row (or column) to (or from) another row (or column) of $\mathbf{A}$ has no effect on the value of a determinant.
7. Interchanging any two rows or columns of $\mathbf{A}$ reverses the sign of the determinant, but this cannot change the singularity of the matrix.
8. The determinant of a triangular matrix is equal to the product of the elements on the principal diagonal.
9. Let $\mathbf{I}$ be the $(n \times n)$ identity matrix, and let $\mathbf{0}$ be the $(n \times n)$ zero matrix. Then $|\mathbf{I}) = 1$ and $|\mathbf{0}) = 0$.

1.3.1 Cofactor Matrix. A *cofactor matrix* $\mathbf{C}$ is a matrix in which every element a_{ij} is replaced by its cofactor $|C_{ij}|$. An *adjoint matrix* is the transpose of a cofactor matrix. Thus, for a 3×3 matrix $\mathbf{A}$ defined in (1.1.1),

$$\mathbf{C} = \begin{bmatrix} |C_{11}| & |C_{12}| & |C_{13}| \\ |C_{21}| & |C_{22}| & |C_{23}| \\ |C_{31}| & |C_{32}| & |C_{33}| \end{bmatrix}, \quad \operatorname{adj}\mathbf{A} = \mathbf{C}^T = \begin{bmatrix} |C_{11}| & |C_{21}| & |C_{31}| \\ |C_{12}| & |C_{22}| & |C_{32}| \\ |C_{13}| & |C_{23}| & |C_{33}| \end{bmatrix}. \tag{1.3.7}$$

Let $\mathbf{A}$ be a square matrix. Then its *inverse matrix* $\mathbf{A}^{-1}$ satisfies the relation

$$\mathbf{A}\mathbf{A}^{-1} = \mathbf{I} = \mathbf{A}^{-1}\mathbf{A}. \tag{1.3.8}$$

The *inverse matrix* $\mathbf{A}^{-1}$ exists only if $\mathbf{A}$ is a square matrix. Thus, multiplying a matrix by its inverse reduces to the identity matrix. This property is similar to the property of the reciprocal in algebra. The inverse matrix can be obtained using the formula

$$\mathbf{A}^{-1} = \frac{1}{|\mathbf{A}|} \operatorname{adj} \mathbf{A}, \quad |\mathbf{A}| \neq 0. \tag{1.3.9}$$

Example 1.5. Consider the matrix $\mathbf{A} = \begin{bmatrix} 3 & 2 & -4 \\ -2 & 5 & 1 \\ 3 & -2 & 7 \end{bmatrix}$. Then

$$|\mathbf{A}| = 3[5(7) - 1(-2)] - 2[(-2)(7) - (1)(3)] - 4[(-2)(-2) - 5(3)] = 127 \neq 0.$$

So the matrix $\mathbf{A}$ is nonsingular, and $\rho(\mathbf{A}) = 3$. The cofactor matrix of $\mathbf{A}$ is

$$\mathbf{C} = \begin{bmatrix} \begin{vmatrix} 5 & 1 \\ -2 & 7 \end{vmatrix} & -\begin{vmatrix} -2 & 1 \\ 3 & 7 \end{vmatrix} & \begin{vmatrix} -2 & -2 \\ 3 & -2 \end{vmatrix} \\ -\begin{vmatrix} 2 & -4 \\ -2 & 7 \end{vmatrix} & \begin{vmatrix} 3 & -4 \\ 3 & 7 \end{vmatrix} & \begin{vmatrix} 3 & 2 \\ 3 & -2 \end{vmatrix} \\ \begin{vmatrix} 3 & 2 \\ -2 & 5 \end{vmatrix} & -\begin{vmatrix} 3 & -4 \\ -2 & 1 \end{vmatrix} & \begin{vmatrix} 3 & 2 \\ -2 & 5 \end{vmatrix} \end{bmatrix} = \begin{bmatrix} 37 & 17 & 10 \\ -22 & 33 & -12 \\ 14 & -11 & 14 \end{bmatrix}.$$

Then the transpose of the cofactor matrix C gives the adjoint matrix:

$$\operatorname{adj} \mathbf{A} = \mathbf{C}^T = \begin{bmatrix} 37 & -22 & 14 \\ 17 & 33 & -11 \\ 10 & -12 & 14 \end{bmatrix}.$$

Thus, by formula (1.3.9),

$$\begin{aligned} \mathbf{A}^{-1} &= \frac{1}{127} \begin{bmatrix} 37 & -22 & 14 \\ 17 & 33 & -11 \\ 10 & -12 & 14 \end{bmatrix} = \begin{bmatrix} \frac{37}{127} & -\frac{22}{127} & \frac{14}{127} \\ \frac{17}{127} & \frac{33}{127} & -\frac{11}{127} \\ \frac{10}{127} & -\frac{12}{127} & \frac{14}{127} \end{bmatrix} \\ &= \begin{bmatrix} 0.29 & 0.17 & 0.11 \\ 0.13 & 0.26 & -0.09 \\ 0.08 & -0.09 & 0.11 \end{bmatrix}. \end{aligned}$$

To check the answer, evaluate both $\mathbf{AA}^{-1}$ and $\mathbf{A}^{-1}\mathbf{A}$; both of these products should be equal to $\mathbf{I}$. ■

Example 1.6. This is a very useful time-saving result. Given $\mathbf{A} = \begin{bmatrix} a & b \\ c & d \end{bmatrix}$, find $\operatorname{adj} \mathbf{A}$ and $\mathbf{A}^{-1}$. Assuming $|\mathbf{A}| = ad - bc \neq 0$, the cofactor matrix of $\mathbf{A}$ is

$$\mathbf{C} = \begin{bmatrix} d & -c \\ -b & a \end{bmatrix},$$

and thus,

$$\operatorname{adj}\mathbf{A} = \mathbf{C}^T = \begin{bmatrix} d & -b \\ -c & a \end{bmatrix}.$$

Hence,

$$\mathbf{A}^{-1} = \frac{1}{|\mathbf{A}|}\operatorname{adj}\mathbf{A} = \frac{1}{ad-bc}\begin{bmatrix} d & -b \\ -c & a \end{bmatrix}. \blacksquare \tag{1.3.10}$$

1.4 Systems of Linear Equations

A system of linear equations can be expressed in the form of a matrix equation. For example, consider the following system of n linear equations:

$$\begin{aligned} a_{11}x_1 + a_{12}x_2 + \cdots + a_{1n}x_n &= b_1, \\ a_{21}x_1 + a_{22}x_2 + \cdots + a_{2n}x_n &= b_2, \\ \cdots\cdots \\ a_{n1}x_1 + a_{n2}x_2 + \cdots + a_{nn}x_n &= b_n. \end{aligned} \tag{1.4.1}$$

Then this system can be expressed in the matrix form as

$$\mathbf{A}\mathbf{x} = \mathbf{b}, \tag{1.4.2}$$

where

$$\mathbf{A} = \begin{bmatrix} a_{11} & a_{12} & \cdots & a_{1n} \\ a_{21} & a_{22} & \cdots & a_{2n} \\ \cdots & \cdots & \cdots & \\ a_{n1} & a_{n2} & \cdots & a_{nn} \end{bmatrix}, \quad \mathbf{x} = \begin{bmatrix} x_1 \\ x_2 \\ \vdots \\ x_n \end{bmatrix} \quad \mathbf{b} = \begin{bmatrix} b_1 \\ b_2 \\ \vdots \\ b_n \end{bmatrix}, \tag{1.4.3}$$

where $\mathbf{A}$ is called the *coefficient matrix*, $\mathbf{x}$ the *solution vector*, and $\mathbf{b}$ the *vector of constant terms*. Note that $\mathbf{x}$ and $\mathbf{b}$ are always column vectors.

Example 1.6. Consider

$$\begin{aligned} 5x_1 + 8x_2 &= 42 \\ 4x_1 + 9x_2 &= 39. \end{aligned}$$

Expressing this system as $\mathbf{A}\mathbf{x} = \mathbf{b}$, we have

$$\mathbf{A} = \begin{bmatrix} 5 & 8 \\ 4 & 9 \end{bmatrix}, \quad \mathbf{x} = \begin{bmatrix} x_1 \\ x_2 \end{bmatrix}, \quad \mathbf{b} = \begin{bmatrix} 42 \\ 39 \end{bmatrix}.$$

Thus,

$$\mathbf{A}\mathbf{x} = \begin{bmatrix} 5 & 8 \\ 4 & 9 \end{bmatrix}\begin{bmatrix} x_1 \\ x_2 \end{bmatrix} = \begin{bmatrix} 5x_1 + 8x_2 \\ 4x_1 + 9x_2 \end{bmatrix}_{2\times 1},$$

and so we get the matrix form

$$\mathbf{Ax} = \mathbf{b}: \quad \begin{bmatrix} 5x_1 + 8x_2 \\ 4x_1 + 9x_2 \end{bmatrix} = \begin{bmatrix} 42 \\ 39 \end{bmatrix}. \blacksquare$$

Systems of linear equations can be solved by the following methods.

1.4.1 Solution with the Inverse. Any system of linear equations can be solved with the inverse. Thus, to solve the equation

$$\mathbf{A}_{n\times n}\mathbf{x}_{n\times 1} = \mathbf{b}_{n\times 1},$$

provided that $\mathbf{A}^{-1}$ exists, we multiply both sides of this equation by $\mathbf{A}^{-1}$, following the laws of conformability:

$$\mathbf{A}^{-1}_{n\times n}\mathbf{A}_{n\times n}\mathbf{x}_{n\times 1} = \mathbf{A}^{-1}_{n\times n}\mathbf{b}_{n\times 1},$$

which, using $\mathbf{A}^{-1}\mathbf{A} = \mathbf{I}$, gives

$$\mathbf{I}_{n\times n}\mathbf{x}_{n\times 1} = \mathbf{A}^{-1}_{n\times n}\mathbf{b}_{n\times 1}.$$

Since $\mathbf{Ix} = \mathbf{x}$, we have the formula:

$$\mathbf{x}_{n\times 1} = \mathbf{A}^{-1}_{n\times n}\mathbf{b}_{n\times 1}. \tag{1.4.4}$$

Example 1.8. To solve the system of equations in Example 1.7, we have $|\mathbf{A}| = 5(9) - 8(4) = 13 \neq 0$. The cofactor matrix of $\mathbf{A}$ is

$$\mathbf{C} = \begin{bmatrix} 9 & -4 \\ -8 & 5 \end{bmatrix}.$$

Then

$$\operatorname{adj}\mathbf{A} = \mathbf{C}^T = \begin{bmatrix} 9 & -8 \\ -4 & 5 \end{bmatrix}, \quad \mathbf{A}^{-1} = \frac{1}{13}\begin{bmatrix} 9 & -8 \\ -4 & 5 \end{bmatrix} = \begin{bmatrix} \frac{9}{13} & -\frac{8}{13} \\ -\frac{4}{13} & \frac{5}{13} \end{bmatrix}.$$

Hence, by formula (1.4.4), we get

$$\mathbf{x} = \begin{bmatrix} \frac{9}{13} & -\frac{8}{13} \\ -\frac{4}{13} & \frac{5}{13} \end{bmatrix}\begin{bmatrix} 42 \\ 39 \end{bmatrix} = \begin{bmatrix} \frac{66}{13} \\ \frac{27}{13} \end{bmatrix}$$

which gives $x_1 = \dfrac{66}{13} \approx 5.07$ and $x_2 = \dfrac{27}{13} \approx 2.07$. $\blacksquare$

1.4.2 Cramer's Rule. To solve the system $\mathbf{Ax} = \mathbf{b}$, Cramer's rule is as follows: Let $\mathbf{D}$ denote the coefficient matrix $\mathbf{A}$, and $\mathbf{D}_i$ the matrix obtained by replacing the coefficients x_i in $\mathbf{D}$ by the elements of the column matrix $\mathbf{b}$. Then, if $|\mathbf{D}| \neq 0$, the solution of the system is given by

$$x_i = \frac{|\mathbf{D}_i|}{|\mathbf{D}|}, \quad i = 1, 2, \dots, n. \tag{1.4.5}$$

Example 1.9. Solve

$$\begin{aligned} x - 2z &= 3 \\ -y + 3z &= 1 \\ 2x + 5z &= 0. \end{aligned}$$

We have $|\mathbf{D}| = \begin{vmatrix} 1 & 0 & -2 \\ 0 & -1 & 3 \\ 2 & 0 & 5 \end{vmatrix} = -9 \neq 0$; $|\mathbf{D}_x| = \begin{vmatrix} 3 & 0 & -2 \\ 1 & -1 & 3 \\ 0 & 0 & 5 \end{vmatrix} = -15$;
$|\mathbf{D}_y| = \begin{vmatrix} 1 & 3 & -2 \\ 0 & 1 & 3 \\ 2 & 0 & 5 \end{vmatrix} = 27$; $|\mathbf{D}_z| = \begin{vmatrix} 1 & 0 & 3 \\ 0 & -1 & 1 \\ 2 & 0 & 0 \end{vmatrix} = 6$. Thus,

$$x = \frac{|\mathbf{D}_x|}{|\mathbf{D}|} = \frac{-15}{-9} = \frac{5}{3}; \quad y = \frac{|\mathbf{D}_y|}{|\mathbf{D}|} = \frac{27}{-9} = -3; \quad z = \frac{|\mathbf{D}_z|}{|\mathbf{D}|} = \frac{6}{-9} = -\frac{2}{3}. \blacksquare$$

1.4.3 Gaussian Elimination Method. To solve the system (1.4.1), this method involves replacing equations by combinations of equations in such a way that a triangular system in the upper echelon form is obtained:

$$\begin{aligned} x_1 + a_{12}x_2 + \cdots + a_{1,n-1}x_{n-1} + a_{1n}x_n &= b'_1 \\ x_2 + \cdots + a_{2,n-1}x_{n-1} + a_{2n}x_n &= b'_2 \\ \cdots\cdots \\ x_{n-1} + a_{n-1,n}x_n &= b'_{n-1} \\ x_n &= b'_n. \end{aligned}$$

The last equation determines x_n. Then the process of backward substitution is used to find $x_{n-1}, \dots, x_2, x_1$.

Example 1.10. Use the Gaussian elimination method to solve the system

$$\begin{aligned} x - 2y + 3z &= 4 \\ 2x + y - 4z &= 3 \\ -3x + 4y - z &= -2. \end{aligned}$$

In order to reduce the matrix to the upper triangular form, first, any nonzero coefficient (usually the largest) in absolute value is located, and brought to the upper left corner by interchanges of equations and columns. In this example this is already in place. Keeping the first equation intact, the correct multiple of the second equation (which is 1/2) and that of the third equation (which is 1/3) are subtracted, respectively, from the first, which gives

$$\begin{aligned} x - 2y + 3z &= 4 \\ y - 2z &= -1 \\ 4z &= 8. \end{aligned}$$

Thus, the solution is $x = 4, y = 3, z = 2$. ■

Example 1.11. Use the Gaussian elimination method to solve the system

$$\begin{aligned} x - 2z + 2w &= 1 \\ -2x + 3y + 4z &= 1 \\ y + z - w &= 0 \\ 3x + y - 2z - w &= 3. \end{aligned}$$

By following the procedure, this system reduced to the triangular form

$$\begin{aligned} x - 2z + 2w &= 1 \\ y + z - w &= 0 \\ -3z + 7w &= 1 \\ w &= 1, \end{aligned}$$

which gives the solution as $x = 3, y = -1, z = 2, w = 1$. ■

1.5 Definite and Semidefinite Matrices

Let $\mathbf{A}$ be an $n \times n$ matrix. If there exists a scalar r and a non-zero vector $\mathbf{x}$ such that

$$\mathbf{Ax} = r\mathbf{x}, \tag{1.5.1}$$

then r is called an *eigenvalue*[1] of $\mathbf{A}$ and $\mathbf{x}$ is called an *eigenvector* of $\mathbf{A}$ corresponding to the eigenvalue r. An important result is: r is an eigenvalue of $\mathbf{A}$ iff

$$|\mathbf{A} - r\mathbf{I}| = 0. \tag{1.5.2}$$

The left side of this equation can be expanded in terms of the elements of $\mathbf{A}$ as

$$|\mathbf{A} - r\mathbf{I}| = \begin{vmatrix} a_{11} - r & a_{12} & \cdots & a_{1n} \\ a_{21} & a_{22} - r & \cdots & a_{2n} \\ \vdots & \vdots & \cdots & \vdots \\ a_{n1} & a_{n2} & \cdots & a_{nn} - r \end{vmatrix}, \tag{1.5.3}$$

[1] Although λ is generally used to denote an eigenvalue of a matrix, we are using r instead, so that it does not conflict with the Lagrange multiplier λ.

which yields a polynomial of degree n. In order for r to be an eigenvalue of $\mathbf{A}$ it must satisfy Eq (1.5.2).

For example, the eigenvalues r_i, $i = 1, 2$, of a 2×2 matrix $\mathbf{A}$ can be found by solving the quadratic equation

$$r_i = \frac{\text{tr}(\mathbf{A}) \pm \sqrt{\text{tr}(\mathbf{A})} - 4|\mathbf{A}|}{2}, \tag{1.5.4}$$

where $\text{tr}(\mathbf{A}) = a_{11} + a_{22}$ is the trace of $\mathbf{A}$. In this case, Eq (1.5.3) is $|\mathbf{A} - r_i\mathbf{I}| = 0$, or

$$\begin{vmatrix} a_{11} - r & a_{12} \\ a_{21} & a_{22} - r \end{vmatrix} = 0,$$

solving which we have $r^2 - (a_{11} + a_{22})r + (a_{11}a_{22} - a_{12}a_{21}) = 0$, or in matrix notation,

$$r^2 - \text{tr}(\mathbf{A})r + |\mathbf{A}| = 0,$$

which leads to the quadratic formula (1.5.4).

Eq (1.5.2) is called the *characteristic equation* or *characteristic polynomial* for the matrix $\mathbf{A}$. For a non-trivial solution, $(\mathbf{A} - r\mathbf{I})$ must be singular.

The eigenvalues of any real symmetric matrix $\mathbf{A}$ in a complex vector space are all real. A $(n \times n)$ matrix $\mathbf{A}$ is said to be *positive semidefinite* (PSD) (or *negative semidefinite* (NDS)) if the eigenvalues r_i are nonnegative or nonpositive; or, alternatively, if for all $\mathbf{z} \in \mathbb{R}^n$,

$$\mathbf{z} \cdot \mathbf{A}\mathbf{z} \geq (\leq) 0.$$

If equality is strict for all $\mathbf{z} \neq 0$, then $\mathbf{A}$ is *positive definite* (PD)(or *negative definite* (ND)).

Example 1.12. (a) The identity matrix $\mathbf{I} = \begin{bmatrix} 1 & 0 \\ 0 & 1 \end{bmatrix}$ is positive (semi) definite for all $\mathbf{z} = \begin{bmatrix} x \\ y \end{bmatrix}$, since $\mathbf{z} \cdot \mathbf{I}\mathbf{z} = \begin{bmatrix} x \\ y \end{bmatrix} \cdot \begin{bmatrix} 1 & 0 \\ 0 & 1 \end{bmatrix} \begin{bmatrix} x \\ y \end{bmatrix} = \begin{bmatrix} x \\ y \end{bmatrix} \begin{bmatrix} x \\ y \end{bmatrix} = x^2 + y^2 \geq 0$. Note that $-\mathbf{I}$ is negative (semi) definite.

(b) The matrix $\mathbf{M} = \begin{bmatrix} -1 & -1 \\ -1 & -1 \end{bmatrix}$ is negative semidefinite, since for all $\mathbf{z} = \begin{bmatrix} x \\ y \end{bmatrix}$,

$$\mathbf{z} \cdot \mathbf{M}\mathbf{z} = \begin{bmatrix} x \\ y \end{bmatrix} \begin{bmatrix} -1 & -1 \\ -1 & -1 \end{bmatrix} \begin{bmatrix} x \\ y \end{bmatrix} = \begin{bmatrix} x \\ y \end{bmatrix} \begin{bmatrix} -x - y \\ -x - y \end{bmatrix} = -(x + y)^2 \leq 0,$$

but it is not negative definite at $x = -y$, since $\mathbf{z} \cdot \mathbf{Mz} = 0$. However, not all matrices are either positive or negative definite, as for example, the matrix $\mathbf{D} = \begin{bmatrix} 1 & 0 \\ 0 & -1 \end{bmatrix}$ shows that

$$\mathbf{z} \cdot \mathbf{Dz} = \begin{bmatrix} x \\ y \end{bmatrix} \begin{bmatrix} 1 & 0 \\ 0 & -1 \end{bmatrix} \begin{bmatrix} x \\ y \end{bmatrix} = x^2 - y^2. \blacksquare$$

Some properties of positive (negative) definite and positive (negative) semidefinite matrices are:

(i) $\mathbf{M}$ is PSD [PD] $\Leftrightarrow$ $-\mathbf{M}$ is NSD [ND].
(ii) $\mathbf{M}$ is ND [PD] $\Leftrightarrow$ $\mathbf{M}^{-1}$ is ND [PD].
(iii) $\mathbf{M}$ is ND [PD] $\Leftrightarrow$ $\mathbf{M}$ NSD [PSD], but $\mathbf{M}$ is NSD [PSD] $\not\Rightarrow$ $\mathbf{M}$ is ND [PD].
(iv) $\mathbf{M}$ is ND [PD] $\Rightarrow$ $\mathbf{M} + \mathbf{M}'$ is ND [PD].

PROOF. (i) $\mathbf{z} \cdot (-\mathbf{M})\mathbf{Az} = -(\mathbf{z} \cdot \mathbf{Mz})$. (ii) $\mathbf{z} \cdot \mathbf{Mz} = (\mathbf{z} \cdot \mathbf{Mz})' = \mathbf{z} \cdot \mathbf{M}'\mathbf{z} = \mathbf{z} \cdot \mathbf{MM}^{-1}\mathbf{M}'\mathbf{z} = \mathbf{M}'\mathbf{z} \cdot \mathbf{M}^{-1}\mathbf{z} = \mathbf{M}'\mathbf{z} \cdot \mathbf{M}^{-1}\mathbf{M}'\mathbf{z}$. (iii) For all $\mathbf{z} \neq 0$, $\mathbf{z} \cdot \mathbf{Mz} > 0$ and for all $\mathbf{z} = 0$, $\mathbf{z} \cdot \mathbf{Mz} = 0$, then for all $\mathbf{z}$ we have $\mathbf{z} \cdot \mathbf{Mz} \geq 0$; and (iv) $\mathbf{z} \cdot (\mathbf{M} + \mathbf{M}') = 2\mathbf{z} \cdot \mathbf{Mz}$. ■

1.6 Special Determinants

We will discuss the following determinants in detail: the Jacobian, the Hessian, and the two bordered Hessians, one involving two functions f and (the constraint) g, and the other involving one function f only.

1.6.1 Jacobian. The *Jacobian determinant* $|\mathbf{J}|$ is used to test functional dependence, both linear and nonlinear. It is composed of all the first-order partial derivatives of a system of equations, arranged in an ordered sequence. For a system of nonlinear equations

$$\begin{aligned} f_1 &= f_1(x_1, x_2, \dots, x_m) \\ f_2 &= f_2(x_1, x_2, \dots, x_m) \\ &\vdots \\ f_n &= f_n(x_1, x_2, \dots, x_m), \end{aligned} \tag{1.6.1}$$

the Jacobian determinant is given by

$$|\mathbf{J}| = \left| \frac{\partial(f_1, f_2, \dots, f_n)}{\partial(x_1, x_2, \dots, x_m)} \right| = \begin{vmatrix} \frac{\partial f_1}{\partial x_1} & \frac{\partial f_1}{\partial x_2} & \cdots & \frac{\partial f_1}{\partial x_m} \\ \frac{\partial f_2}{\partial x_1} & \frac{\partial f_2}{\partial x_2} & \cdots & \frac{\partial f_2}{\partial x_m} \\ \vdots & \vdots & \cdots & \vdots \\ \frac{\partial f_n}{\partial x_1} & \frac{\partial f_n}{\partial x_2} & \cdots & \frac{\partial f_n}{\partial x_m} \end{vmatrix}. \tag{1.6.2}$$

If $|\mathbf{J}| = 0$, the equations are functionally dependent; if $|\mathbf{J}| \neq 0$, they are functionally independent. Note that the Jacobian $|\mathbf{J}|$ is also denoted by J or $\mathbf{J}$.

Example 1.13. Given the system

$$\begin{aligned} y_1 &= 5x_1 - 4x_2 \\ y_2 &= 25x_1^2 - 40x_1x_2 + 16x_2^2, \end{aligned}$$

the Jacobian is

$$\begin{aligned} |\mathbf{J}| &= \begin{vmatrix} 5 & -4 \\ 50x_1 - 40x_2 & -40x_1 + 32x_2 \end{vmatrix} \\ &= -200x_1 + 160x_2 + 200x_1 - 160x_2 = 0. \end{aligned}$$

So there is functional dependence, which is $(5x_1 - 4x_2)^2 = 25x_1^2 - 40x_1x_2 + 16x_2^2$. ■

1.6.2 Hessian. The *Hessian* matrix $\mathbf{H}$ of a twice-differentiable function $f(\mathbf{x})$ at a point $\mathbf{x} \in \mathbb{R}^n$ is defined by the square $n \times n$ matrix by

$$\mathbf{H} = \begin{bmatrix} \frac{\partial^2 f}{\partial x_1^2} & \frac{\partial^2 f}{\partial x_1 \partial x_2} & \cdots & \frac{\partial^2 f}{\partial x_1 \partial x_n} \\ \frac{\partial^2 f}{\partial x_2 \partial x_1} & \frac{\partial^2 f}{\partial x_2^2} & \cdots & \frac{\partial^2 f}{\partial x_2 \partial x_n} \\ \vdots & \vdots & \cdots & \vdots \\ \frac{\partial^2 f}{\partial x_n \partial x_1} & \frac{\partial^2 f}{\partial x_n \partial x_2} & \cdots & \frac{\partial^2 f}{\partial x_n^2} \end{bmatrix}, \tag{1.6.3}$$

or, componentwise by $\mathbf{H}_{i,j} = \dfrac{\partial^2 f}{\partial x_i \partial x_j}$. The determinant $\det \mathbf{H}$ is denoted by $|\mathbf{H}|$. Generally, the function for which the Hessian is used is obvious from the context.

Like any matrix, the Hessian additions are commutative, associative and distributive, and multiplication of a Hessian by a scalar and the product of two Hessians are defined in the usual way.

!!! *Some authors denote the Hessian* $\mathbf{H}$ *of a twice-differentiable function* $f = f(x_1, \dots, x_n)$ *incorrectly as* $\nabla^2 f(x)$. *This notation is in conflict with the established notation for the Laplacian of a twice-differentiable function* $f = f(x_1, \dots, x_n)$, *denoted by* $\nabla^2 f$ *and defined by*

$$\nabla^2 f = \nabla \cdot \nabla f = \frac{\partial^2 f}{\partial x_1^2} + \frac{\partial^2 f}{\partial x_2^2} + \cdots + \frac{\partial^2 f}{\partial x_n^2},$$

which is obviously not a matrix, but a second-order elliptic partial differential operator. Moreover, $\nabla^2 f$ is the trace of the matrix (1.6.3).

As seen from the definition (1.6.3), the mixed partial derivatives occupy the entries that are off-diagonal. By Clairaut's theorem, so long as they are continuous, the order of differentiation does not matter. Thus,

$$\frac{\partial}{\partial x_i}\Big(\frac{\partial f}{\partial x_j}\Big) = \frac{\partial}{\partial x_j}\Big(\frac{\partial f}{\partial x_i}\Big).$$

Hence, if the second derivatives of f are all continuous in a neighborhood D, then the Hessian of f is a symmetric matrix throughout D.

The column vector of the first-order partial derivatives of a function f is denoted by

$$\left[\frac{\partial f}{\partial x_1}\ \frac{\partial f}{\partial x_2}\ \cdots\ \frac{\partial f}{\partial x_m}\right]^T.$$

If this column vector is zero at a point $\mathbf{x}$, then f has a *critical point* at $\mathbf{x}$, which is denoted by $\mathbf{x}^*$.

The main application of the Hessian is found in large-scale optimization problems that use Newton-type methods. The second derivative test (§2.4) checks if the point $\mathbf{x}$ is a local extremum. For example, a function f attains a local maximum, or a local minimum, at $\mathbf{x}$ according as the Hessian is positive definite, or negative definite, at $\mathbf{x}$. The function f has a saddle point at $\mathbf{x}$, if the Hessian has both positive and negative eigenvalues. In other cases the test is inconclusive. However, if the Hessian is semidefinite in a neighborhood of $\mathbf{x}$, we can definitely conclude whether f is locally convex or concave, respectively.

The second derivative test for a single variable function f is explained in §2.3. In the case of a bivariate function, the discriminant $|\mathbf{H}|$ can be used since it is the product of the eigenvalues. If this product is positive, then the eigenvalues are both positive, or both negative; if it is negative, then the two eigenvalues have different signs; if it is zero, then the second derivative test fails.

For functions of several variables, the second-order conditions that are sufficient for a local maximum or minimum can be expressed in terms of the sequence of principal (i.e., the upper-leftmost) minors (determinants of sub-matrices) of the Hessian. These conditions are a special case of bordered Hessian, defined below, which is used for the second-derivative test in certain constrained optimization problems.

For example, in an optimization problem with a function f of two variables (x, y), if the first-order conditions $f_x = f_y = 0$ are satisfied, a sufficient condition for a function $z = f(x, y)$ to be an optimum is:

$$(1) \quad \begin{cases} f_{xx}, f_{yy} > 0 & \text{for a minimum,} \\ f_{xx}, f_{yy} < 0 & \text{for a maximum;} \end{cases}$$

$$(2) \quad f_{xx}f_{yy} > (f_{xy})^2.$$

This is the *Hessian test* for second-order derivatives, and for a symmetric (2×2) positive definite matrix it is defined by

$$|\mathbf{H}| = \begin{vmatrix} f_{xx} & f_{xy} \\ f_{yx} & f_{yy} \end{vmatrix}. \tag{1.6.4}$$

If the first element on the principal diagonal, known as the *first principal minor* and denoted by $|H_1| = f_{xx}$ is positive and the *second principal minor*

$$|H_2| = |\mathbf{H}| = f_{xx}f_{yy} - (f_{xy})^2 > 0,$$

then the second-order conditions for a minimum are met. When $|H_1| > 0$ and $|H_2| > 0$, the Hessian $|\mathbf{H}|$ is called *positive definite*. A positive definite Hessian satisfies the second-order conditions for a minimum.

If the first principal minor $|H_1| = f_{xx} < 0$ and $|H_2| > 0$, the Hessian $|\mathbf{H}|$ is called *negative definite*. A negative definite Hessian satisfies the second-order conditions for a maximum.

Example 1.14. Consider $f = 3x^2 - xy + 2y^2 - 4x - 7y + 12$. Then $f_{xx} = 6, f_{xy} = -1, f_{yy} = 4$. Thus,

$$|\mathbf{H}| = \begin{bmatrix} 6 & -1 \\ -1 & 4 \end{bmatrix},$$

which gives $|H_1| = 6 > 0$ and $|H_2| = |\mathbf{H}| = 24 - 1 = 23 > 0$. Hence, the Hessian is positive definite, and f is minimized at the critical values, which are given by the solution of $f_x = 6x - y - 4 = 0, f_y = -x + 4y - 7 = 0$, i.e., at $x^* = 1, y^* = 2$. ■

1.6.3 Bordered Hessian: Two Functions. Given a function f with a constraint function g such that $g(\mathbf{x}) = c$, the bordered Hessian is defined by

$$\bar{\mathbf{H}}(f,g) = \begin{bmatrix} 0 & \frac{\partial g}{\partial x_1} & \frac{\partial g}{\partial x_2} & \cdots & \frac{\partial g}{\partial x_n} \\ \frac{\partial g}{\partial x_1} & \frac{\partial^2 f}{\partial x_1^2} & \frac{\partial^2 f}{\partial x_1 \partial x_2} & \cdots & \frac{\partial^2 f}{\partial x_1 \partial x_n} \\ \frac{\partial g}{\partial x_2} & \frac{\partial^2 f}{\partial x_2 \partial x_1} & \frac{\partial^2 f}{\partial x_2^2} & \cdots & \frac{\partial^2 f}{\partial x_2 \partial x_n} \\ \vdots & \vdots & \vdots & \cdots & \vdots \\ \frac{\partial g}{\partial x_n} & \frac{\partial^2 f}{\partial x_n \partial x_1} & \frac{\partial^2 f}{\partial x_n \partial x_2} & \cdots & \frac{\partial^2 f}{\partial x_n^2} \end{bmatrix}. \tag{1.6.5}$$

In the case of m constraints, the zero in the top-left first element is to be replaced by an $m \times m$ square block of zeros, and there will be m border rows at the top and m border columns at the left.

Note that since a bordered Hessian can neither be positive definite nor negative definite, the rules stated above about extrema being characterized by a positive definite or negative definite Hessian do not apply. The second derivative test in such a case consists of signs of restrictions of the determinants of a certain set of $(n - m)$ sub-matrices of the bordered Hessian.

In the case of a function $f(x, y)$ subject to a constraint $g(x, y)$ the *bordered Hessian* for constrained optimization is defined as follows: form a new function $F(x, y) = f(x, y) + r[k - g(x, y)]$, where the first-order conditions are $F_x = F_y = F_r = 0$. Then the second-order conditions can be expressed in terms of a bordered Hessian $|\bar{\mathbf{H}}|$ in one of the following two ways:

$$|\bar{\mathbf{H}}| = \begin{vmatrix} F_{xx} & F_{xy} & g_x \\ F_{yx} & F_{yy} & g_y \\ g_x & g_y & 0 \end{vmatrix}, \quad \text{or} \quad \begin{vmatrix} 0 & g_x & g_y \\ g_x & F_{xx} & F_{xy} \\ g_y & F_{yx} & F_{yy} \end{vmatrix}, \tag{1.6.6}$$

which is simply the usual Hessian (1.6.4) (with f replaced by F), i.e.,

$$|\mathbf{H}| = \begin{vmatrix} F_{xx} & F_{xy} \\ F_{yx} & F_{yy} \end{vmatrix},$$

bordered by the first derivatives of the constraint with zero on the principal diagonal. The order of a bordered Hessian is determined by the order of the principal minor being bordered. Thus, $|\bar{\mathbf{H}}|$ in (1.6.6) represents a second bordered principal minor $|\bar{H}_2|$, because the principal minor being bordered is 2×2.

In general, for a function $F(x_1, x_2, \dots, x_n)$ in n variables subject to the constraint $g(x_1, x_2, \dots, x_n)$, the bordered Hessian, defined by (1.6.5), can also be expressed as

$$|\bar{\mathbf{H}}| = \begin{vmatrix} F_{11} & F_{12} & \cdots & F_{1n} & g_1 \\ F_{21} & F_{22} & \cdots & F_{2n} & g_2 \\ \cdots & \cdots & \cdots & \cdots & \\ F_{n1} & F_{n2} & \cdots & F_{nn} & g_n \\ g_1 & g_2 & \cdots & g_n & 0 \end{vmatrix}, \quad \text{or} \quad \begin{vmatrix} 0 & g_1 & g_2 & \cdots & g_n \\ g_1 & F_{11} & F_{12} & \cdots & F_{1n} \\ g_2 & F_{21} & F_{22} & \cdots & F_{2n} \\ \cdots & \cdots & \cdots & \cdots & \\ g_n & F_{n1} & F_{n2} & \cdots & F_{nn} \end{vmatrix}, \tag{1.6.7}$$

where the $n \times n$ principal minor is being bordered. If all the principal minors are negative, i.e., if $|\bar{H}_2|, |\bar{H}_3|, \dots, |\bar{H}_n| < 0$, then the bordered Hessian is positive definite, and a positive definite Hessian always satisfies the sufficient condition for a relative (local) minimum. Similarly, if the principal minors alternate in sign from positive to negative, i.e., if $|\bar{H}_2| > 0$, $|\bar{H}_3| < 0$, $|\bar{H}_4| > 0$, and so on, then the bordered Hessian is negative definite, and a negative definite Hessian always satisfies the sufficient condition for F to be concave and have a relative (local) maximum.

If $|\mathbf{H}| = 0$ and $|H_1| = 0 = |H_2|$, then $\mathbf{H}$ is not negative definite, but it is negative semidefinite with $|H_1| \le 0$ and $|H_2| = |\mathbf{H}| \ge 0$. However, for the semidefinite test, we must check the signs of these discriminants with adj $\mathbf{H}$. Then if $|H_1| < 0$ and $|H_2| = |\mathbf{H}| = 0$, both discriminants are negative semidefinite, and it satisfies the sufficient condition for F to be concave and have a relative (local) maximum.

1.6.4 Bordered Hessian: Single Function. If a function $f(\mathbf{x})$, $\mathbf{x} \in \mathbb{R}^n$, is twice continuously differentiable, then the bordered Hessian $|\bar{\mathbf{B}}|$ for a single function is defined by

$$|\bar{\mathbf{B}}| = \begin{vmatrix} 0 & f_1 & f_2 & \cdots & f_n \\ f_1 & f_{11} & f_{12} & \cdots & f_{1n} \\ f_2 & f_{21} & f_{22} & \cdots & f_{2n} \\ \cdots & \cdots & \cdots & \cdots & \cdots \\ f_n & f_{n1} & f_{n2} & \cdots & f_{nn} \end{vmatrix}. \tag{1.6.13}$$

Note that the bordered Hessian $|\mathbf{B}|$ is composed of the first derivatives of the function f rather than an extraneous constraint g. The leading principal minors are

$$|B_1| = \begin{vmatrix} 0 & f_1 \\ f_1 & f_{11} \end{vmatrix}, \ |B_2| = \begin{vmatrix} 0 & f_1 & f_2 \\ f_1 & f_{11} & f_{12} \\ f_2 & f_{21} & f_{22} \end{vmatrix}, \ |B_n| = |\mathbf{B}|. \tag{1.6.14}$$

This Hessian is used to check quasi-concavity and quasi-convexity for a function $z = f(\mathbf{x})$, which are discussed in Chapters 6 and 7. We will state two conditions, one of which is necessary and the other is sufficient, and both relate to quasi-concavity on a domain consisting only of the nonnegative orthant (the n-dimensional analogue of the nonnegative quadrant) which is defined by $x_1, \ldots, x_n \ge 0$. These conditions are as follows:

The necessary condition for a function $z = f(\mathbf{x})$ to be quasi-concave on the nonnegative orthant is (see §6.2; also Arrow and Enthoven [1961])

$$|\bar{B}_1| \le 0, \quad |\bar{B}_2| \ge 0, \quad \text{and } |\bar{B}_n| \begin{Bmatrix} \le \\ \ge \end{Bmatrix} 0 \ \text{ if } n \ \text{ is } \begin{Bmatrix} \text{odd} \\ \text{even,} \end{Bmatrix} \tag{1.6.15}$$

where the partial derivatives are evaluated in the nonnegative orthant. Note that the first condition in (1.6.15) is automatically satisfied since $|\bar{B}_1| = -f_1^2 = -\left(\dfrac{\partial f}{\partial x_1}\right)^2$.

The sufficient condition for f to be strictly quasi-concave on the nonnegative orthant is that

$$|\bar{B}_1| < 0, \quad \{\bar{B}_2| > 0, \quad |\bar{B}_n| \begin{cases} \leq \\ \geq \end{cases} 0 \ \text{ if } n \ \text{ is } \begin{cases} \text{odd} \\ \text{even}, \end{cases} \tag{1.6.16}$$

where the partial derivatives are evaluated in the nonnegative orthant.

Notice that unlike the bordered Hessian for a function f that is to be optimized under a constraint g, the bordered Hessian $|\mathbf{B}|$ is used to check whether a given function is quasi-concave or quasi-convex. Examples where this Hessian is used are provided in Chapters 6 and 7.

1.7 Exercises

1.1. Use any method to solve each system of equations. If the system has no solution, mark it as inconsistent.

$$\text{(i)} \begin{cases} x - 2y + 3z = 7 \\ 2x + y + z = 4 \\ -3x + 2y - 2z = -10. \end{cases} \qquad \text{(ii)} \begin{cases} x + y - z = 6 \\ 3x - 2y + z = -5 \\ x + 3y - 2z = 14. \end{cases}$$

$$\text{(iii)} \begin{cases} 2x - 2y - 2z = 2 \\ 2x + 3y + z = 2 \\ 3x + 2y = 0. \end{cases} \qquad \text{(iv)} \begin{cases} x + 2y - z = -3 \\ 2x - 4y + z = -7 \\ -2x + 2y - 3z = 4. \end{cases}$$

ANS. (i) $x = 2, y = -1, z = 1$; (ii) $x = 1, y = 3, z = -2$; (iii) inconsistent; (iv) $x = -3, y = \frac{1}{2}, z = 1$.

1.2. A store sells almonds for \$6.00 per pound and peanuts for \$1.50 per pound. The manager decides to mix 40 pounds of peanuts and some almonds and sell the mixture for \$3.00 per pound. How many pounds of almonds are mixed with the peanuts so that the mixture will generate the same revenue as each kind of nut sold separately. ANS. 20 pounds.

1.3. Let $\quad \mathbf{A} = \begin{bmatrix} a_{11} & a_{12} & a_{13} \\ a_{21} & a_{22} & a_{23} \\ a_{31} & a_{32} & a_{33} \end{bmatrix}$

be a 3×3 matrix. Show that by interchanging columns 1 and 3 the value of the new determinant is -1 times the value of the original determinant.

SOLUTION.

$|\mathbf{A}| = \begin{vmatrix} a_{13} & a_{12} & a_{11} \\ a_{23} & a_{22} & a_{21} \\ a_{33} & a_{32} & a_{31} \end{vmatrix} = a_{13}(a_{22}a_{31} - a_{32}a_{21}) - a_{12}(a_{21}a_{33} - a_{33}a_{21}) +$
$a_{11}(a_{23}a_{32} - a_{33}a_{22})$

$$= -[a_{11}(a_{22}a_{33} - a_{32}a_{23}) - a_{12}(a_{21}a_{33} - a_{31}a_{23}) + a_{13}(a_{21}a_{32} - a_{31}a_{22})]$$

$$= -\begin{vmatrix} a_{11} & a_{12} & a_{13} \\ a_{21} & a_{22} & a_{23} \\ a_{31} & a_{32} & a_{33} \end{vmatrix}.$$

1.4. Given the column vector **u** and the row vector **v**, find the matrix **uv**: (a) $\mathbf{u} = [3 \quad 2 \quad 5 \quad 4]^T_{4\times1}$, and $\mathbf{v} = [2 \quad 7 \quad 4 \quad 8]_{1\times4}$.

ANS. The matrix $\mathbf{uv} = \begin{bmatrix} 6 & 21 & 12 & 24 \\ 4 & 14 & 8 & 16 \\ 10 & 35 & 20 & 40 \\ 8 & 28 & 16 & 32 \end{bmatrix}_{4\times4}$.

(b) $\mathbf{u} = [4 \quad 7 \quad 8 \quad 1]_{1\times4}$, and $\mathbf{v} = [2 \quad 9 \quad 7]^T_{3\times1}$. ANS. The matrix **uv** is not defined and cannot be multiplied.

1.5. Consider the system of equations $\begin{cases} a_1x + b_1y = c_1, \\ a_2x + b_2y = c_2. \end{cases}$

If $D = a_1b_2 - a_2b_1 \neq 0$, use the matrix method to show that the solution is

$$x = \frac{c_1b_2 - c_2b_1}{D}, \; y = \frac{a_1c_2 - a_2c_1}{D}.$$

SOLUTION. If $a_1 \neq 0$, then

$$\left[\begin{array}{cc|c} a_1 & b_1 & c_1 \\ a_2 & b_2 & c_2 \end{array}\right] \to \left[\begin{array}{cc|c} 1 & \dfrac{b_1}{a_1} & \dfrac{c_1}{a_1} \\ a_2 & b_2 & c_2 \end{array}\right] \to$$

$$\left[\begin{array}{cc|c} 1 & \dfrac{b_1}{a_1} & \dfrac{c_1}{a_1} \\ 0 & \dfrac{-a_2b_1}{a_1} + b_2 & \dfrac{-a_1c_1}{a_1} + c_2 \end{array}\right] \to \left[\begin{array}{cc|c} 1 & \dfrac{b_1}{a_1} & \dfrac{c_1}{a_1} \\ 0 & \dfrac{-a_2b_1 + b_2a_1}{a_1} & \dfrac{-a_2c_1 + c_2a_1}{a_1} \end{array}\right]$$

$$\to \left[\begin{array}{cc|c} 1 & \dfrac{b_1}{a_1} & \dfrac{c_1}{a_1} \\ 0 & 1 & \dfrac{-a_2c_1 + c_2a_1}{a_1} \cdot \dfrac{a_1}{-a_2b_1 + b_2a_1} \end{array}\right]$$

$$\to \left[\begin{array}{cc|c} 1 & \dfrac{b_1}{a_1} & \dfrac{c_1}{a_1} \\ 0 & 1 & \dfrac{-a_2c_1 + c_2a_1}{-a_2b_1 + b_2a_1} \end{array}\right] \to \left[\begin{array}{cc|c} 1 & 0 & \dfrac{-b_1c_2 + b_2c_1}{-a_2b_1 + b_2a_1} \\ 0 & 1 & \dfrac{-a_2c_1 + c_2a_1}{-a_2b_1 + b_2a_1} \end{array}\right]$$

which yields

$$x = \frac{c_1b_2 - c_2b_1}{a_1b_2 - a_2b_1} = \frac{1}{D}(c_1b_2 - c_2b_1), \; y = \frac{a_1c_2 - c_2b_1}{a_1b_2 - a_2c_1} = \frac{1}{D}(a_1c_2 - a_2c_1).$$

If $a_1 = 0$, then

$$\begin{bmatrix} 0 & b_1 & | & c_1 \\ a_2 & b_2 & | & c_2 \end{bmatrix} \to \begin{bmatrix} a_2 & b_2 & | & c_2 \\ 0 & b_1 & | & c_1 \end{bmatrix} \to \begin{bmatrix} 1 & \dfrac{b_2}{a_2} & | & \dfrac{c_2}{a_2} \\ 0 & b_1 & | & c_1 \end{bmatrix}$$

$$\to \begin{bmatrix} 1 & \dfrac{b_2}{a_2} & | & \dfrac{c_2}{a_2} \\ 0 & 1 & | & \dfrac{c_1}{b_1} \end{bmatrix} \to \begin{bmatrix} 1 & 0 & | & \dfrac{c_2}{a_2} - \dfrac{b_2 c_1}{a_2 b_1} = \dfrac{c_1 b_2 - c_2 b_1}{-a_2 b_1} \\ 0 & 1 & | & \dfrac{c_1}{b_1} = \dfrac{-a_2 c_1}{-a_2 b_1} \end{bmatrix},$$

which yields $x = \dfrac{c_1 b_2 - c_2 b_1}{D}$ and $y = \dfrac{-a_2 c_1}{D}$.

1.6. Given $\mathbf{a}$ and $\mathbf{x}$ as vectors, $\mathbf{e}_i$ as the unit column vector with ith element 1 and all other elements zeros, $[\mathbf{e}] = [\mathbf{e} - 1, \dots, \mathbf{e}_n]$ as the column vector with ith element $\mathbf{e}_i$ for all $i = 1, \dots, n$, and $\mathbf{A}$ and $\mathbf{H}$ as square matrices in $\mathbb{R}^n$, verify the dimensions in the following table.

Expression	Dimension	Expression	Dimension
$\mathbf{x}$	$n \times 1$	$\mathbf{a}$	$n \times 1$
$f(\mathbf{x})$	1×1	$f(\mathbf{a})$	1×1
$\mathbf{e}_1$	$n \times 1$	$[\mathbf{e}]$	$n \times 1$
$\dfrac{\partial f(\mathbf{x})}{\partial \mathbf{x}}$	$n \times 1$	$\left(\dfrac{\partial f(\mathbf{x})}{\partial \mathbf{x}}\right)^T$	$1 \times n$
$\nabla f(\mathbf{x})$	$n \times 1$, or $1 \times n^\dagger$	$\nabla^2 f = \operatorname{tr} \mathbf{H}$	1×1
$\mathbf{A}$	$n \times n$	$\mathbf{H}$	$n \times n$

† depending on whether ∇f is written as $\mathbf{e}_1 \dfrac{\partial f}{\partial x_1} + \cdots + \mathbf{e}_n \dfrac{\partial f}{\partial x_n}$, or as $\dfrac{\partial f}{\partial x_1}\mathbf{e}_1 + \cdots + \dfrac{\partial f}{\partial x_n}\mathbf{e}_n$.

1.7. Determine the rank of the following matrices: (a) $\mathbf{A} = \begin{bmatrix} -2 & 7 & 3 \\ 1 & 6 & 4 \\ 3 & -8 & 5 \end{bmatrix}$. ANS. $|\mathbf{A}| = -153 \neq 0$. Thus, the matrix $\mathbf{A}$ is nonsingular and the three rows and three columns are linearly independent. Hence, $\rho(\mathbf{A}) = 3$.

(b) $\mathbf{B} = \begin{bmatrix} 7 & -2 & 5 \\ 2 & 10 & -4 \\ -3 & -15 & 6 \end{bmatrix}$. ANS. $|\mathbf{B}| = 0$. Hence the matrix $\mathbf{B}$ is singular, and the three rows and three columns are not linearly independent. Hence, $\rho(\mathbf{B}) \neq 3$. However, the 2×2 submatrix in the upper left corner gives $\begin{vmatrix} 7 & -2 \\ 2 & 10 \end{vmatrix} = 74 \neq 0$. Thus, $\rho(\mathbf{B}) = 2$. Note that row 3 is -1.5 times row 2, and column 3 is $-2/5$ times column 2.

1.8. If the two columns of a 3×3 determinant are equal, then the value of the determinant is zero.

SOLUTION. $\begin{vmatrix} a_{11} & a_{12} & a_{13} \\ a_{21} & a_{22} & a_{23} \\ a_{11} & a_{12} & a_{13} \end{vmatrix} = a_{11}(a_{22}a_{31} - a_{32}a_{21}) - a_{12}(a_{21}a_{31} - a_{31}a_{21})$

$+a_{11}(a_{21}a_{32} - a_{31}a_{22})$

$= a_{11}a_{22}a_{31} - a_{11}a_{32}a_{21} - a_{12}(0) + a_{11}a_{21}a_{22} - a_{11}a_{31}a_{22}$

$= a_{31}(a_{11}a_{22} - a_{11}a_{22}) + a_{32}(a_{11}a_{21} - a_{11}a_{21}) = 0.$

1.9. Show that the matrix $\mathbf{A} = \begin{bmatrix} 4 & 6 \\ 2 & 3 \end{bmatrix}$ has no inverse.

SOLUTION 1. $\mathbf{A}\mathbf{I} = \left[\begin{array}{cc|cc} 4 & 6 & 1 & 0 \\ 2 & 3 & 0 & 1 \end{array}\right] \to \left[\begin{array}{cc|cc} 1 & \frac{3}{2} & \frac{1}{4} & 0 \\ 2 & 3 & 0 & 1 \end{array}\right]$

$\to \left[\begin{array}{cc|cc} 1 & \frac{3}{2} & \frac{1}{4} & 0 \\ 0 & 1 & -\frac{1}{2} & 1 \end{array}\right].$

Since this reduced form of the matrix $\mathbf{A}\mathbf{I}$ shows that the identity matrix $\mathbf{I}$ cannot appear on the left of the vertical bar, the matrix $\mathbf{A}$ has no inverse.

SOLUTION 2. Since $|\mathbf{A}| = 0$, the matrix A has no inverse, in view of formula (1.3.10).

1.10. Consider the system

$$\begin{bmatrix} a_{11} & a_{12} \\ a_{21} & a_{22} \end{bmatrix} \begin{bmatrix} x_t \\ y_t \end{bmatrix} = \begin{bmatrix} a_{13} & a_{14} \\ a_{23} & a_{24} \end{bmatrix} \begin{bmatrix} x_{t-1} \\ y_{t-1} \end{bmatrix} + \begin{bmatrix} b_1 \\ b_2 \end{bmatrix},$$

or in matrix form

$$\mathbf{A}\mathbf{x}_t = \mathbf{E}\,\mathbf{x}_{t-1} + \mathbf{b}, \tag{1.6.11}$$

where $\mathbf{x} = (x, y)$. Prove that $(\mathbf{E} - r_i\mathbf{A})\mathbf{C}_i = 0$.

PROOF. Without loss of generality assume that $\mathbf{b}$ is a null matrix, and also assume distinct real roots. Then we have

$$\mathbf{x}_t - k_i\mathbf{C}_i(r_i)^t, \quad \mathbf{x}_{t-1} - k_i\mathbf{C}_i(r_i)^{t-1}. \tag{1.6.12}$$

Substituting these into the homogeneous form of the given equation, we get

$$\mathbf{A}k_i\mathbf{C}_i(r_i)^t = \mathbf{E}k_i\mathbf{C}_i(r_i)^{t-1},$$

or

$$\mathbf{E}k_i\mathbf{C}_i(r_i)^{t-1} - \mathbf{A}k_i\mathbf{C}_i(r_i)^t = 0,$$

which, when evaluated at $t = 1$, gives $(\mathbf{E} - r_i\mathbf{A})\mathbf{C}_i = 0$.

1.11. Use any method to solve the following system:

(a) $\begin{cases} \ln x = 4\ln y \\ \log_3 x = 2 + 2\log_3 y. \end{cases}$; (b) $\begin{cases} \log_x y = 3\ln y \\ \log_x(4y) = 5. \end{cases}$

ANS. (a) $x = 81, y = 3$; (b) $x = 2, y = 8$.

1.12. Use the Gaussian elimination method to solve the system

$$10x + 7y + 8z + 7w = 32, \quad 5x + 6y + 5z + 7w = 23,$$
$$6x + 10y + 9z + 8w = 33, \quad 5x + 9y + 10z + 7w = 31.$$

SOLUTION. In order to reduce the matrix to the upper triangular form, first the largest coefficient in absolute value is located, and brought to the upper left corner by interchanges of equations and columns. This is called the first pivot. In this example this is already in place. Keeping the first equation intact, multiply the second and fourth equations by 2, third equation by 5/3, and then subtract all these three equations from the first, to give

$$\begin{aligned} 10x + 7y + 8z + 7w &= 32 \\ -5y - 2z - 7w &= -14 \\ -\tfrac{29}{3}y - 7z - \tfrac{19}{3}w &= -23 \\ -11y - 12z - 7w &= -30. \end{aligned}$$

Now eliminate y from the second and third equations, and also from the third and fourth equations, and then z from the third and fourth equations. This gives the echelon form, from which we first find w, then z, then y, and finally x. ANS. $x = y = z = w = 1$.

1.13. Use the Hessian to determine whether the function $F(x, y, z) = 2x^2 - 7x - xy + 5y^2 - 3y + 4yz + 6z^2 + 3z - 4xz$ is minimized or maximized at the critical points. SOLUTION. The first-order criterion gives

$$F_x = 4x - 7 - y - 4z = 0;\ F_y = -x + 10y - 3 + 4z = 0;\ F_z = 4y + 12z + 3 - 4x = 0,$$

which in the matrix form $\mathbf{Ax} = \mathbf{b}$ is

$$\begin{bmatrix} 4 & -1 & -4 \\ -1 & 10 & 4 \\ -4 & 4 & 12 \end{bmatrix} \begin{bmatrix} x \\ y \\ z \end{bmatrix} = \begin{bmatrix} 7 \\ 3 \\ -3 \end{bmatrix}.$$

Using Cramer's method, we get $|\mathbf{A}| = (120-16)+1(-12+16)-4(-4+40) = 276$, and

$$|\mathbf{A}_1| = \begin{vmatrix} 7 & -1 & -4 \\ 3 & 10 & 4 \\ -3 & 4 & 12 \end{vmatrix}, \; |\mathbf{A}_2| = \begin{vmatrix} 4 & 7 & -4 \\ -1 & 3 & 4 \\ -4 & -3 & 12 \end{vmatrix}, \; |\mathbf{A}_3| = \begin{vmatrix} 4 & -1 & 7 \\ -1 & 10 & 3 \\ 4 & 4 & -3 \end{vmatrix},$$

yielding $\mathbf{A}_1 = 608, \mathbf{A}_2 = 104, \mathbf{A}_3 = -485$; thus, the critical points are $x^* = (608)/(276), y^* = (104)/(276)$, and $z^* = -(485)/(276)$. Next, taking the second-order derivatives we have $F_{xx} = 4, F_{yy} = 10, F_{zz} = 12, F_{xy} = -1 = F_{yx}, F_{yz} = 4 = F_{zy}, F_{zx} = -4 = F_{xz}$, which yields the Hessian

$$|\mathbf{H}| = \begin{vmatrix} 4 & -1 & -4 \\ -1 & 10 & 4 \\ -4 & 4 & 12 \end{vmatrix}.$$

Since $|H_1| = 4 > 0$, $|H_2| = \begin{vmatrix} 4 & -1 \\ -1 & 10 \end{vmatrix} = 42 > 0$, and $|H_3| = |\mathbf{H}| = 276 > 0$, we find that $|\mathbf{H}|$ is positive definite, which means that $F(x, y, z)$ is minimized at the critical points.

1.14. Find the characteristic root of the matrix $\mathbf{A} = \begin{bmatrix} -8 & 4 \\ 4 & -8 \end{bmatrix}$. Ans. Solving $|\mathbf{A} - r\,\mathbf{I}| = 0$, we get

$$|\mathbf{A} - r\,\mathbf{I}| = \begin{vmatrix} -8-r & 4 \\ 4 & -8-r \end{vmatrix} = r^2 + 16r + 48 = 0,$$

which gives $r = -4, -12$. Also note that since both characteristic roots are negative, the matrix $\mathbf{A}$ is negative definite. Also, check that the trace (sum of the principal diagonal) of the matrix $\mathbf{A}$ must be equal to the sum of the two characteristic roots.

1.15. Find the inverse of the matrix $\mathbf{A} = \begin{bmatrix} 25 & 61 & -12 \\ 18 & -2 & 4 \\ 8 & 35 & 21 \end{bmatrix}$.

Ans. $\mathbf{A}^{-1} = \begin{bmatrix} 0.01 & 0.05 & -0.01 \\ 0.01 & -0.02 & 0.01 \\ -0.02 & 0.01 & 0.03 \end{bmatrix}$.

1.16. Use matrix inversion to solve the following system of equations: (a) $5x + 3y = 29$, $2x + 7y = 58$; and (b) $6x + 9y = 48$, $2x + 5y = 36$.

SOLUTION. Write the system in the matrix form $\mathbf{Ax} = \mathbf{b}$: $\begin{bmatrix} 5 & 3 \\ 2 & 7 \end{bmatrix} \begin{bmatrix} x \\ y \end{bmatrix} = \begin{bmatrix} 29 \\ 58 \end{bmatrix}$. Then $|\mathbf{A}| = 29$; $\operatorname{adj} \mathbf{A} = \begin{bmatrix} 7 & -3 \\ -2 & 5 \end{bmatrix}$; thus, $\mathbf{A}^{-1} = \dfrac{1}{29} \begin{bmatrix} 7 & -3 \\ -2 & 5 \end{bmatrix}$, giving $\mathbf{x} = \mathbf{A}^{-1}\mathbf{x} = \begin{bmatrix} 1 \\ 8 \end{bmatrix}$. (b) Write the system in the matrix form $\mathbf{Ax} = \mathbf{b}$: $\begin{bmatrix} 6 & 9 \\ 2 & 5 \end{bmatrix} \begin{bmatrix} x \\ y \end{bmatrix} = \begin{bmatrix} 48 \\ 36 \end{bmatrix}$. Then $|\mathbf{A}| = 12$; $\operatorname{adj} \mathbf{A} = \begin{bmatrix} 5 & -9 \\ -2 & 6 \end{bmatrix}$; thus, $\mathbf{A}^{-1} = \dfrac{1}{12} \begin{bmatrix} 5 & -9 \\ -2 & 6 \end{bmatrix}$, giving $\mathbf{x} = \mathbf{A}^{-1}\mathbf{x} = \begin{bmatrix} -7 \\ 10 \end{bmatrix}$.

1.17. The equilibrium conditions for two related goods are given by $6P_1 - 9P_2 = 42$, $-P_1 + 7P_2 = 66$, where P_1 and P_2 are respective prices. Find the equilibrium prices P_1 and P_2.

SOLUTION. The system of equations in the matrix form is $\mathbf{Ax} = \mathbf{b} = \begin{bmatrix} 6 & -9 \\ -1 & 7 \end{bmatrix} \begin{bmatrix} P_1 \\ P_2 \end{bmatrix} = \begin{bmatrix} 42 \\ 66 \end{bmatrix}$. Then $|\mathbf{A}| = 33$; $\operatorname{adj} \mathbf{A} = \begin{bmatrix} 7 & 9 \\ 1 & 6 \end{bmatrix}$. Then $\mathbf{A}^{-1} = \dfrac{1}{|\mathbf{A}|} \operatorname{adj} \mathbf{A} = \dfrac{1}{33} \begin{bmatrix} 7 & 9 \\ 1 & 6 \end{bmatrix}$. This yields $\begin{bmatrix} P_1 \\ P_2 \end{bmatrix} = \mathbf{A}^{-1}\mathbf{b} = \begin{bmatrix} 26.96 \\ 13.28 \end{bmatrix}$.

1.18. The Bayou Steel Company produces stainless steel and aluminum containers. On a typical day, they manufactured 750 steel containers with 10-gallon capacity, 500 with 5-gallon capacity, and 600 with 1-gallon capacity. On the same day they manufactured 900 aluminum containers with 10-gallon capacity, 700 with 5-gallon capacity, and 1100 with 1-gallon capacity. (a) Represent the above data as two different matrices. (b) If the amount of the material used in the 10-gallon capacity containers is 20 pounds, that used in 5-gallon containers is 12 pounds, and that in 1-gallon containers is 5 pounds, find the matrix representing the amount of material. (c) If the stainless steel costs \$0.25 per pound and aluminum costs \$0.10 per pound, find the matrix representing cost. (d) Find the total cost of the day's production.

SOLUTION. (a) The data can be represented as

$$\begin{bmatrix} 750 & 500 & 600 \\ 900 & 700 & 1100 \end{bmatrix}_{2\times 3}, \text{ or } \begin{bmatrix} 750 & 900 \\ 500 & 700 \\ 600 & 1100 \end{bmatrix}_{3\times 2};$$

(b) the amount of material: $\begin{bmatrix} 750 & 500 & 600 \\ 900 & 700 & 1100 \end{bmatrix}_{2\times 3} \begin{bmatrix} 20 \\ 12 \\ 5 \end{bmatrix}_{3\times 1} = \begin{bmatrix} 24000 \\ 31900 \end{bmatrix}_{2\times 1}$;

(c) the cost is equal to $[0.25 \quad 0.10]_{1\times 2}$; and (d) total cost of production $[0.25 \quad 0.10]_{1\times 2}\begin{bmatrix}124000\\3900\end{bmatrix}_{2\times 1} = \$9191.20.$

1.19. Use the Hessian to determine whether the function $F(x,y,z) = -4x^2 = 9x + xz - 2y^2 + 3y + 2yz - 4z^3$ is minimized or maximized at the critical points. SOLUTION. The first-order criterion gives

$$F_x = -8x + 9 + z = 0;\ F_y = -4y + 3 + 2z = 0;\ F_z = x + 2y - 12z = 0,$$

which in the matrix form $\mathbf{Ax} = \mathbf{b}$ is

$$\begin{bmatrix}-8 & 0 & 1\\ 0 & -4 & 2\\ 1 & 2 & -12\end{bmatrix}\begin{bmatrix}x\\y\\z\end{bmatrix} = \begin{bmatrix}-9\\-3\\0\end{bmatrix}.$$

Using Cramer's rule, we get $|\mathbf{A}| = -8(4804) + 1(4) = -348$, and

$$|\mathbf{A}_1| = \begin{vmatrix}-9 & 0 & 1\\ -3 & -4 & 2\\ 0 & 2 & 12\end{vmatrix},\ |\mathbf{A}_2| = \begin{vmatrix}-8 & -9 & 1\\ 0 & -3 & 2\\ 1 & 0 & -12\end{vmatrix},\ |\mathbf{A}_3| = \begin{vmatrix}8 & 0 & -9\\ 0 & -4 & -3\\ 1 & 2 & 0\end{vmatrix},$$

yielding $\mathbf{A}_1 = 462, \mathbf{A}_2 = -303, \mathbf{A}_3 = -84$; thus, the critical points are $x^* = -(462)/(348) = -1.33, y^* = (303)/(348) = 0.87, z^* = (84)/(348) = 0.24$. Next, taking the second-order derivatives we have $F_{xx} = -8, F_{yy} = -4, F_{zz} = -12, F_{xy} = 0 = F_{yx}, F_{yz} = 2 = F_{zy}, F_{zx} = 1 = F_{xz}$, which yields the Hessian

$$|\mathbf{H}| = \begin{vmatrix}-8 & 0 & 1\\ 0 & -4 & 2\\ 1 & 2 & -12\end{vmatrix}.$$

Since $|H_1| = -8 < 0$, $|H_2| = \begin{vmatrix}-8 & 0\\ 0 & -4\end{vmatrix} = 32 > 0$, and $|H_3| = |\mathbf{H}| = -348 < 0$; also $F_{xx}\cdot F_{yy} = 32 < (F_{xy})^2 = 144$. Thus, we find that $|\mathbf{H}|$ is negative definite, which means that $F(x,y,z)$ is maximized at the critical points.

1.20. Maximize the total profit function P for a manufacturing firm producing two related goods, in quantities x and y, so that the demand functions are defined by $P_1 = 60 - 4x - 2y$, and $P_2 = 40 - x - 4y$, and the total cost function TC$= 4x^2 + xy + 2y^2$. SOLUTION. Let $P =$TR -TC, where TR$= P_1x + P_2y$. Thus,

$$\begin{aligned}P &= \text{TR} - \text{TC} = (60 - 4x - 2y)x + (40 - x - 4y)y - (4x^2 + xy + 2y^2)\\ &= 60x - 8x^2 - 4xy + 40y - 6y^2.\end{aligned}$$

Thus, using the first-order criterion, we have $P_x = 60 - 16x - 4y = 0, P_y = -4x + 40 = 12y$, which in matrix form $\mathbf{Ax} = \mathbf{b}$ is

$$\begin{bmatrix} -16 & -4 \\ -4 & -12 \end{bmatrix} \begin{bmatrix} x \\ y \end{bmatrix} = \begin{bmatrix} -60 \\ -40 \end{bmatrix}.$$

Using Cramer's Rule, $|\mathbf{A}| = 176, |A_1| = 560, |A_2| = 400$, thus giving $x^8 = 3.18, y^* = 2.27$. The second-order derivatives are $P_{xx} = -16, P_{yy} = -12, P+xy - 4 = P_{yx}$, and the Hessian is

$$|\mathbf{H}| = \begin{vmatrix} -16 & -4 \\ -4 & -12 \end{vmatrix} = |\mathbf{A}| = 176 > 0,$$

and $|H_1| = -16 < 0$. Thus, $|\mathbf{H}|$ is negative definite, and P is maximized at (x^*, y^*).

2

Differential Calculus

Some basic concepts and results from real analysis are presented. The topics include limit theorems, differentiation, criterion for concavity and related theorems, and vector-valued functions. Proofs for almost all of the results can be found in many textbooks on calculus, e.g., Boas [1996], Hardy [1967], Royden [1968], and Rudin [1976].

2.1 Definitions

A *function* f is a *rule* which assigns to each value of a variable x, called the *argument* of the function, one and only one value $y = f(x)$ known as the *value of the function* at x. The *domain* of a function f, denoted dom(f), is the set of all possible values of x; the *range* of f, denoted by $\mathcal{R}(f)$, is the set of all possible values for $f(x)$. Examples of functions are:

Linear function: $f(x) = mx + b$,

Quadratic function: $f(x) = ax^2 + bx + c = 0,\ a \neq 0$.

Polynomial function of degree n: $f(x) = a_n x^n + a_{n-1}x^{n-1} + \cdots + a_1 x + a_0,\ a_n \neq 0$; n nonnegative integer.

Rational function: $f(x) = \dfrac{g(x)}{h(x)}$, where $g(x)$ and $h(x)$ are both polynomials and $h(x) \neq 0$.

Power function: $f(x) = ax^n$, where n is any real number.

2.1.1 Limit of a Function at a Point. Let a function f be defined throughout an open interval containing a, except possibly at a itself. Then the *limit* of $f(x)$ as x approaches a is L, i.e.,

$$\lim_{x \to a} f(x) = L, \tag{2.1.1}$$

if for every $\varepsilon > 0$ there corresponds a $\delta > 0$ such that $|f(x) - L| < \varepsilon$ whenever $0 < |x - a| < \delta$. In other words, $\lim_{x \to a} f(x) = L$ means that for every $\varepsilon > 0$

there corresponds a $\delta > 0$, such that $f(x)$ is in the interval $(L-\varepsilon, L+\varepsilon)$ whenever x is in the interval $(x-a, x+a)$ and $x \neq a$. Moreover, if $f(x)$ has a limit L as $x \to a$, then L is unique and finite.

Example 2.1. (a) $\lim\limits_{x\to 2}(5x-7) = 3$; (b) $\lim\limits_{x\to 3} c = c$, where c is a constant; (c) $\lim\limits_{x\to 0} \dfrac{|x|}{x}$ does not exist, because in every interval $(-\delta, \delta)$ there are numbers such that $|x|/x = 1$ and other numbers such that $|x|/x = -1$, so the limit is not unique; (d) if $f(x)$ is the greatest integer function defined as $f(x) = \lfloor x \rfloor$, i.e., f denotes the largest integer z such that $z \leq x$ and is known as the *floor of* f, and if n is any integer, then the limit from the right as $x \to n+$ is $\lim\limits_{x\to n+} \lfloor x \rfloor = n$, while the limit from the left as $x \to n-$ is $\lim\limits_{x\to n-} \lfloor x \rfloor = n-1$, and hence the limit $\lim\limits_{x\to n} \lfloor x \rfloor$ does not exist; (e) $\lim\limits_{x\to 0} \dfrac{\sin x}{x} = 1$ ■

Example 2.2. Let f be defined as follows: $f(x) = \begin{cases} 0 & \text{if } x \text{ is rational,} \\ 1 & \text{if } x \text{ is irrational.} \end{cases}$ Then for every real number a, $\lim\limits_{x\to a} f(x)$ does not exist. ■

2.2. Theorems on Limits

Some useful theorems on limits are given below, without proof.

Theorem 2.1. *If* $\lim\limits_{x\to a} f(x) = L$ *and* $\lim\limits_{x\to a} g(x) = M$, *then*

$$(i)\ \lim_{x\to a}[f(x) \pm g(x)] = L \pm M;$$
$$(ii)\ \lim_{x\to a}[f(x)\cdot g(x)] = L\cdot M,$$
$$(iii)\ \lim_{x\to a}[f(x)/g(x)] = L/M, \textit{provided } M \neq 0.$$

Example 2.3. (a) $\lim\limits_{x\to 2}[x^3(x+4)] = \lim\limits_{x\to 2} x^2 \cdot \lim\limits_{x\to 2}(x+4) = (2)^3\cdot(2+4) = 48$.

(b) $$\lim_{x\to 4}\frac{3x^2-5x}{x+6} = \frac{\lim_{x\to 4}(3x^2-5x)}{\lim_{x\to 4}(x+6)} = \frac{3(4)^2-5(4)}{4+6} = 2.8;$$

(c) $$\lim_{x\to 2}\sqrt{6x^3+1} = \Big[\lim_{x\to 2}(6x^3+1)\Big]^{1/2} = [6(3)^3+1]^{1/2} = (49)^{1/2} = 7.$$

(d) $$\lim_{x\to 8}\frac{x^{2/3}+3\sqrt{x}}{4-(16/x)} = \frac{\lim\limits_{x\to 8}[x^{2/3}+3\sqrt{x}]}{\lim\limits_{x\to 8}[4-(16/x)]} = \frac{\lim\limits_{x\to 8} x^{2/3} + \lim\limits_{x\to 8} 3\sqrt{x}}{\lim\limits_{x\to 8} 4 - \lim\limits_{x\to 8}(16/x)} = \frac{4+6\sqrt{2}}{4-2} = 2+3\sqrt{2};$$

(e) $$\lim_{x\to 7}\frac{x-7}{x^2-49} = \lim_{x\to 7}\frac{x-7}{(x+7)} = \lim_{x\to 7}\frac{1}{x+7} = \frac{1}{13};$$

(f) $\lim\limits_{x\to 0} \frac{2}{x}$, $x \neq 0$. Note that $\lim\limits_{x\to 0^+} \frac{2}{x} = \infty$, and $\lim\limits_{x\to 0^-} \frac{2}{x} = -\infty$. Since the value of the limit is not unique as $x \to 0$ from either right or left, this limit does not exist. ■

Theorem 2.2. *If $a > 0$ and n positive integer, or if $a < 0$ and n odd positive integer, then $\lim\limits_{x\to a} \sqrt[n]{x} = \sqrt[n]{a}$.*

Theorem 2.3. (Sandwich theorem) *If $f(x) \leq h(x)$ for all x in an open interval containing a, except possibly at a, and if $\lim\limits_{x\to a} f(x) = L = \lim\limits_{x\to a} g(x)$, then $\lim\limits_{x\to a} h(x) = L$.*

A function f is *continuous* at a point a if the following three conditions are satisfied: (i) a is in the domain of f, (ii) $\lim\limits_{x\to a} f(x)$ exists, and (iii) $\lim\limits_{x\to a} f(x) = f(a)$.

Theorem 2.4. (Intermediate value theorem) *If a function f is continuous on a closed interval $[a, b]$ and if $f(a) \neq f(b)$, then f takes on every value between $f(a)$ and $f(b)$.*

2.2.1 Limit at Infinity. If f is defined on an interval (c, ∞), then $\lim\limits_{x\to\infty} f(x) = L$ means that for every $\varepsilon > 0$ there corresponds an $N > 0$ such that $|f(x) - L| < \varepsilon$ whenever $x > N$.

Theorem 2.5. *If k is a positive rational number and c is any nonzero real number, then*

$$\lim_{x\to\infty} \frac{c}{x^k} = 0, \quad \textit{and} \quad \lim_{x\to-\infty} \frac{c}{x^k} = 0.$$

2.2.2 Infinite Limits $\lim\limits_{x\to a} f(x) = \pm\infty$. If $f(x)$ exists on an open interval containing a, except possibly at $x = a$, then $f(x)$ becomes *infinite* (or *increases without bound*), written as $\lim\limits_{x\to a} f(x) = \infty$, if, for every positive number N, there corresponds a $\delta > 0$ such that $f(x) > N$ whenever $0 < |x - a| < \delta$. Sometimes we say that $f(x)$ becomes *positively* infinite as x approaches a. A similar definition for $\lim\limits_{x\to a} f(x) = -\infty$ is: If $f(x)$ exists on an open interval containing a, except possibly at $x = a$, then $f(x)$ becomes *negatively infinite* (or *decreases without bound*), written as $\lim\limits_{x\to a} f(x) = -\infty$, if, for every positive number M, there corresponds a $\delta > 0$ such that $f(x) < M$ whenever $0 < |x - a| < \delta$.

Example 2.4. (a) $\lim\limits_{x\to a} \frac{1}{(x-a)^n} = \infty$ if n is an even positive number; (b) $\lim\limits_{x\to a+} \frac{1}{(x-a)^n} = \infty$, and $\lim\limits_{x\to a-} \frac{1}{(x-a)^n} = -\infty$ if n is an odd positive number.

Theorem 2.6. If $\lim_{x\to a} f(x) = \infty$ and $\lim_{x\to a} g(x) = c \neq 0$, then

(i) $\lim_{x\to a} [g(x) + f(x)] = \infty$;

(ii) If $c > 0$, then $\lim_{x\to a} [g(x)f(x)] = \infty$ and $\lim_{x\to a} \dfrac{f(x)}{g(x)} = \infty$;

(iii) If $c < 0$, then $\lim_{x\to a} [g(x)f(x)] = -\infty$ and $\lim_{x\to a} \dfrac{f(x)}{g(x)} = -\infty$;

(iv) $\lim_{x\to a} \dfrac{g(x)}{f(x)} = 0$.

2.3 Global and Local Extrema of Functions

A function f is *increasing* on an interval I if $f(x_1) < f(x_2)$ whenever $x_1 < x_2$ for $x_1, x_2 \in I$. A function f is *decreasing* on an interval I if $f(x_1) > f(x_2)$ whenever $x_1 < x_2$, $x_1, x_2 \in I$. Let a function f be defined on an interval I, and let u and v be numbers in I. If $f(x) \leq f(v)$ for all $x \in I$, then $f(v)$ is called the *maximum value* of f on I. Similarly, if $f(x) \geq f(u)$ for all $x \in I$, then $f(u)$ is called the *minimum value* of f on I.

Theorem 2.7. *If a function f is continuous on a closed interval $[a, b]$, then f takes on a minimum value $f(u)$ and a maximum value $f(v)$ at some numbers u and v in $[a, b]$.*

These extrema are called the *absolute minimum* and the *absolute maximum* for f on an interval. However, the *local extrema* of f are defined as follows: Let c be a number in the domain of a function f. If there exists an open interval (a, b) containing c such that $f(x) \leq f(c)$ for all $x \in (a, b)$, then $f(c)$ is a *local maximum* of f. If there exists an open interval (a, b) containing c such that $f(x) \geq f(c)$ for all $x \in (a, b)$, then $f(c)$ is a *local minimum* of f. Sometimes, the term *extremum* is used to mean either a maximum or a minimum.

Theorem 2.8. *If a function f has a local extremum at a number c, then either $f'(c) = 0$ or $f'(c)$ does not exist.*

Theorem 2.9. *If a function f is continuous on a closed interval $[a, b]$ and has its maximum (or minimum) value at a number c in the open interval (a, b), then either $f'(c) = 0$ or $f'(c)$ does not exist.*

The number c in the domain of a function f is known as a *critical number* of f if either $f'(c) = 0$ or $f'(c)$ does not exist.

Theorem 2.10. (Rolle's theorem) *If a function f is continuous on a closed interval $[a, b]$, differentiable on the open interval (a, b), and if $f(a) = f(b)$, then $f'(c) = 0$ for at least one number $c \in (a, b)$.*

Corollary 2.1. *If f has a derivative at $c \in (a, b)$ and $f'(c) \neq 0$, then $f(c)$*

is not a local extremum of f.

This result is a particular case of *Fermat's theorem*, which follows.

Theorem 2.11. (Fermat's theorem) *If* f *has a relative extremum at a point* $c \in (a, b)$, *and if* $f'(c)$ *exists, then* $f'(c) = 0$.

PROOF. Assume that f has a relative maximum at $c \in (a, b)$ and that $f'(c)$ exists. The existence of $f'(c)$ implies that the limit $\lim_{x \to c} \dfrac{f(x) - f(c)}{x - c}$ exists. This being a two-sided limit, we will show that both one-sided limits exist and are equal to $f'(c)$:

$$f'(c) = \lim_{x \to c^-} \frac{f(x) - f(c)}{x - c} \geq 0 \quad \text{since } x - c < 0 \text{ as } x \to c \text{ from left,}$$

$$f'(c) = \lim_{x \to c^+} \frac{f(x) - f(c)}{x - c} \leq 0 \quad \text{since } x - c > 0 \text{ as } x \to c \text{ from right.}$$

Since zero is the only number which is both non-negative and non-positive, so $f'(c) = 0$.

In the case of a relative minimum, assume in the above proof that f has a relative minimum at c and note that $f(x) - f(c) \geq 0$ for all x sufficiently close to c, and reverse the sign in the two inequalities above. ■

Corollary 2.2. *If a function* f *is continuous on a closed interval* $[a, b]$ *and if* $f(a) = f(b)$, *then* f *has at least one critical number in the open interval* (a, b).

Theorem 2.12. (Mean-value theorem) *If a function* f *is continuous on a closed interval* $[a, b]$ *and differentiable on the open interval* (a, b), *then there exists a number* $c \in (a, b)$ *such that* $f(b) - f(a) = f'(c)(b - a)$.

A function f is said to have an *absolute maximum* or (a *global maximum*) on the domain D at the point $c \in D$ if $f(c) \geq f(x)$ for all $x \in D$, where $f(c)$ is called the *maximum value* of f on D. Similarly, if $f(c) \leq f(x)$ for all $x \in D$, then we say that f has an *absolute minimum* or (a *global minimum*) on the domain D at the point $c \in D$, where $f(c)$ is called the *minimum value* of f on D. These extreme values are termed *absolute* or *global* because they are the largest and the smallest value, respectively, of the function f on D.

A function f is said to have a *relative maximum* or (a *local maximum*) at the point c if $f(c) \geq f(x)$ for all x in an open interval containing c. Similarly, if $f(c) \leq f(x)$ for all x in an open interval containing c, then we say that f has an *relative minimum* or (a *local minimum*) at c. This definition can be extended to include the endpoints of the interval $[a, b]$ by saying that f has a relative extremum at an endpoint of $[a, b]$ if f attains its maximum or minimum value at that endpoint in the half-open interval containing it.

2.4 First and Second Derivative Tests

The following theorem is useful in determining the intervals on which a function is increasing or decreasing.

Theorem 2.13. *Let a function f be continuous on a closed interval $[a, b]$ and differentiable on the open interval (a, b). If $f'(x) > 0$ for all $x \in (a, b)$, then f is increasing on $[a, b]$. If $f'(x) < 0$ for all $x \in (a, b)$, then f is decreasing on $[a, b]$.*

Theorem 2.14. (First derivative test) *Let $c \in (a, b)$ be a critical number of a function f, and let f be continuous on $[a, b]$ and differentiable on (a, b), except possibly at c. If $f'(x) > 0$ for $a < c < b$ and $f'(x) < 0$ for $c < x < b$, then $f(c)$ is a local maximum for f; if $f'(x) < 0$ for $a < c < b$ and $f'(x) > 0$ for $c < x < b$, then $f(c)$ is a local minimum for f; and if $f'(x) > 0$ or if $f'(x) < 0$ for all $x \in (a, b)$ except $x = c$, then $f(c)$ is not a local extremum of the function f.*

2.4.1 Definition of Concavity. Let a function f be differentiable at c. The graph of f is *concave upward*, or *concave up* (CU), at the point $P(c, f(c))$ if there exists an open interval (a, b) containing c such that the graph of f is above the tangent line through the point P. The graph is *concave downward*, or *concave down* (CD), at $P : (c, f(c))$ if there exists an open interval (a, b) containing c such that on (a, b) the graph of f is below the tangent through the point P.

Concavity refers to the shape of a curve rather than its direction. Although we have used the conventional terms 'concave up' (CU) and 'concave down' (CD), we will henceforth use the terms 'convex' and 'concave', respectively, when discussing concavity of a function f on an interval (a, b).

This definition is related to the above Theorems 2.13 and 2.14, which are used to determine whether a function is decreasing or increasing depending on whether f' is negative or positive (see Figure 2.1).

If the function f' has a derivative, then these theorems can also be applied to f''. In other words, if f has a second derivative on some interval $I = (a, b)$, then on that interval the following theorem holds:

Theorem 2.15. (Test for concavity) *If f is differentiable on an open interval I containing c, then at the point $P : (c, f(c))$ the graph of f is (i) convex (i.e., CU) iff $f''(c) > 0$, and (ii) concave (i.e., CD) iff $f''(c) < 0$.*

Hence, we have:

A function f is convex (i.e., CU) iff $f'' > 0$, and concave (i.e., CD) iff $f'' < 0$.

A point $P(c, f(c))$ on the graph of a function f is a *point of inflection* (or an *inflection point*) if there exists an open interval (a, b) containing c such that (i) $f''(x) > 0$ if $a < x < c$ and $f''(x) < 0$ if $c < x < b$; or (ii) $f''(x) < 0$

if $a < x < c$ and $f''(x) > 0$ if $c < x < b$ (see Figure 2.2.).

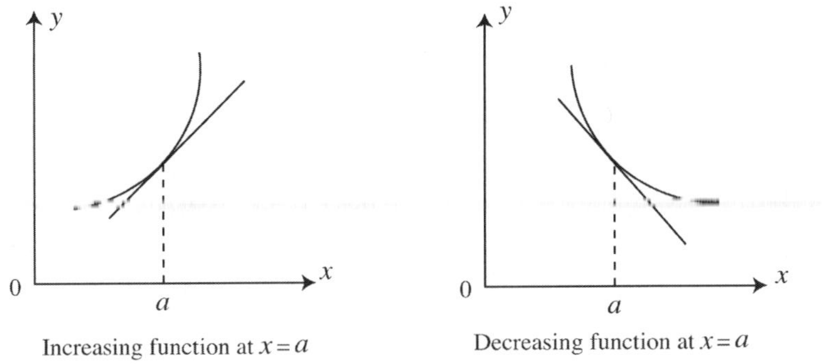

Figure 2.1 Increasing and decreasing functions at $x = a$.

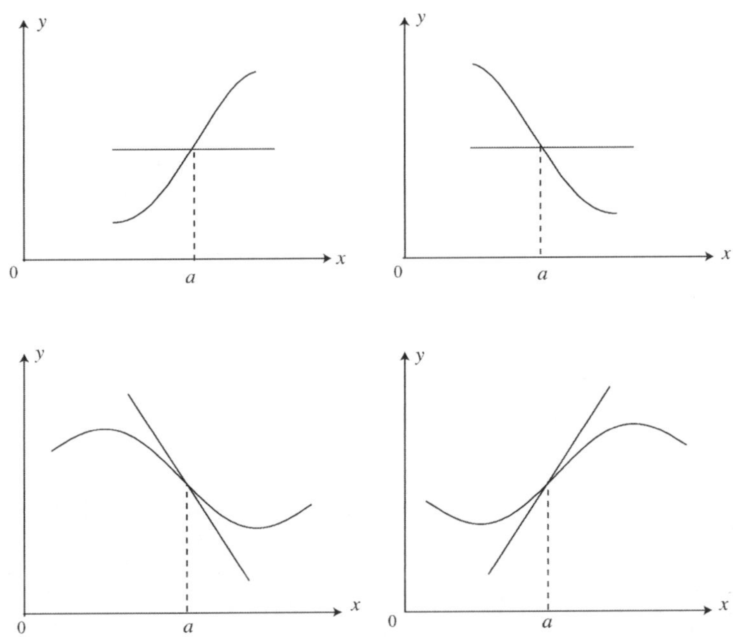

Figure 2.2 Points of inflection at $x = a$.

Theorem 2.16. (Second derivative test) *Let a function f be differentiable on an open interval containing c and let $f'(c) = 0$. If $f''(c) < 0$, then f has a local maximum at c; and if $f''(c) > 0$, then f has a local minimum at c.*

The functions that are convex or concave at a point are presented graphically in Figure 2.3.

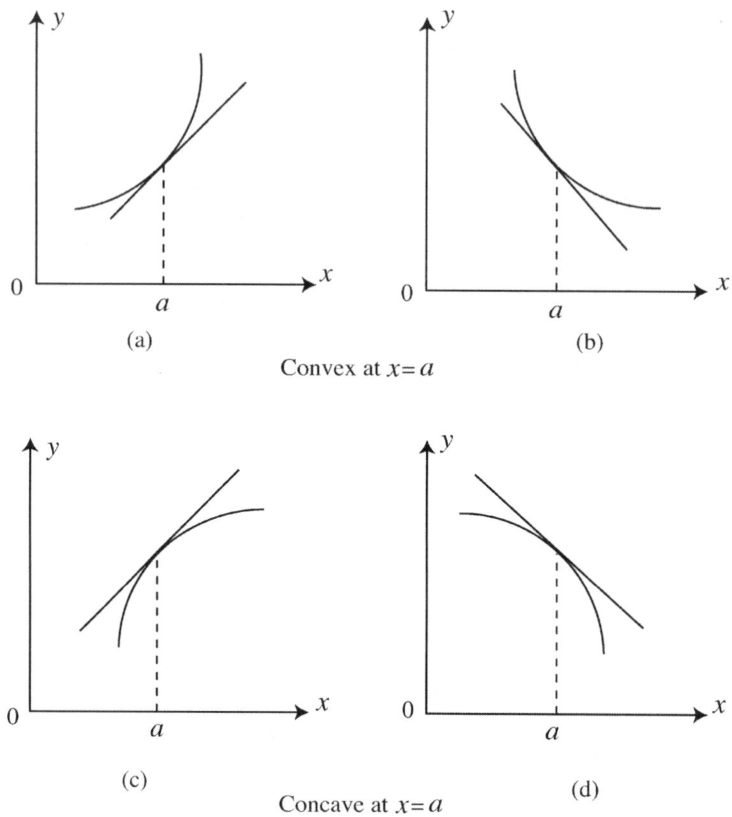

Figure 2.3 Convex and concave functions at a point.

Figure 2.3(a): Slope at $x = a$ is positive; the function $f(x)$ is increasing at $x = a$; $f'(a) > 0$, and $f''(a) > 0$, so the function is convex (CU) at $x = a$.

Figure 2.3(b): Slope at $x = a$ is negative; the function $f(x)$ is decreasing at $x = a$; $f'(a) < 0$, and $f''(a) > 0$, so the function is convex (CU) at $x = a$.

Figure 2.3(c): Slope at $x = a$ is positive; the function $f(x)$ is increasing at $x = a$; $f'(a) > 0$, and $f''(a) < 0$, so the function is concave (CD) at $x = a$.

Figure 2.3(d): Slope at $x = a$ is negative; the function $f(x)$ is decreasing at $x = a$; $f'(a) < 0$, and $f''(a) < 0$, so the function is concave (CD) at $x = a$.

Example 2.5. Let $f(x) = x^5 - 5x^3$. Then $f'(x) = 5x^4 - 15x^2 = 5x^2(x^2 - 3)$ and $f''(x) = 20x^3 - 30x = 10x(2x^2 - 3)$. The critical numbers are $0, \pm\sqrt{3}$, and the values are $f''(0) = 0, f''(-\sqrt{3}) = 30\sqrt{3}.0, f''(-\sqrt{3}) = -30\sqrt{3} < 0$. Thus, by the second derivative test, f has a local minimum at $x = \sqrt{3}$ and a local maximum at $x = -\sqrt{3}$, given by $f(\sqrt{3}) = -5\sqrt{3}$ and $f(-\sqrt{3}) = 6\sqrt{3}$. Since $f''(0) = 0$, the second derivative test does not apply at this critical number. Then using the first derivative test we find that $f'(x) < 0$ in $-\sqrt{3} < x < 0$ and $f'(x) < 0$ in $0 < x < \sqrt{3}$, i.e., $f'(x)$ does not change sign, so there is no local extremum at $x = 0$. Solving $f''(x) = 0$, i.e., $10x(2x^3 - 3) = 0$, we get the solution set $\{-\sqrt{6}/2, 0, \sqrt{6}/2\}$ which are the points of inflection. The results are given in the following table.

Interval	$f''(x)$	Concavity
$(-\infty, -\sqrt{6}/2)$	$-$	concave (CD)
$(-\sqrt{6}/2, 0)$	$+$	convex (CU)
$(0, \sqrt{6}/2)$	$-$	concave (CD)
$(\sqrt{6}/2, \infty)$	$+$	convex (CU) ■

Example 2.6. Let $f(x) = 12 + 2x^2 - x^4$. Then $f'(x) = 4x - 4x^3 = 4x(1 - x^2)$, and $f''(x) = 4 - 12x^2 = 4(1 - 3x^2)$. From $f'(x) = 0$ we get the critical numbers $0, 1$, and -1, giving $f''(0) = 4 > 0, f''(1) = -8 < 0$ and $f''(-1) = -8 < 0$, with the corresponding functional values $f(0) = 12, f(1) = 13 = f(-1)$. Hence, by the second derivative test there is a local minimum at $x = 0$ and a local maxima at $x = \pm 1$. The possible points of inflection are obtained by solving $f''(x) = 0$, which has the solution $\{-1/\sqrt{3}, 1/\sqrt{3}\}$. Hence, we find that

(i) f is concave on the interval $(-\infty, -1/\sqrt{3})$ since $f''(x) < 0$ there;

(ii) f is convex on the interval $-1/\sqrt{3}, 1/\sqrt{3})$ since $f''(x) > 0$ there; and

(iii) f is concave on the interval $(1/\sqrt{3}, \infty)$ since $f''(x) < 0$ there. ■

2.5 Vector-Valued Functions

A function f from a set $X \subseteq \mathbb{R}$ to a set $Y \subseteq \mathbb{R}^3$ is a correspondence defined by a *vector-valued function*, denoted by $\mathbf{r}(t)$, which defines a unique vector $\mathbf{r}(t) = x\,\mathbf{i} + y\,\mathbf{j} + z\,\mathbf{k} \equiv \langle x, y, z \rangle$, where $t \in \mathbb{R}$. Let the components x, y, z be defined by $x = f(t), y = g(t), z = h(t)$, respectively. Then

$$\mathbf{r}(t) = \langle f(t), g(t), h(t) \rangle = f(t)\mathbf{i} + g(t)\mathbf{j} + h(t)\mathbf{k} \tag{2.5.1}$$

for all numbers $t \in \mathbb{R}$, where $\mathbf{i}, \mathbf{j}, \mathbf{k}$ are the unit vectors along the coordinate axes. Conversely, if f, g, h are functions from X to $\mathbb{R}$, then a vector-valued function $\mathbf{r}$ may be defined by Eq (2.5.1). Thus, $\mathbf{r}$ is a vector-valued function iff $\mathbf{r}(t)$ is defined by (2.5.1). The domain of $\mathbf{r}$ is assumed to be the intersection of the domains of f, g, and h. A geometrical interpretation of Eq (2.5.1) is presented in Figure 2.4(a): If OP is the position vector corresponding to $\mathbf{r}(t)$, then as x varies through X, and the endpoint $P(f(t), g(t), h(t))$ traces the curve with parametric equations $x = f(t), y = g(t), z = h(t)$. For example, if $\mathbf{r}(t) = a\cos t\,\mathbf{i} + a\sin t\,\mathbf{j} + bt\,\mathbf{k}$, then the endpoint of the position vector corresponds to t tracing the circular helix shown in Figure 2.4(b).

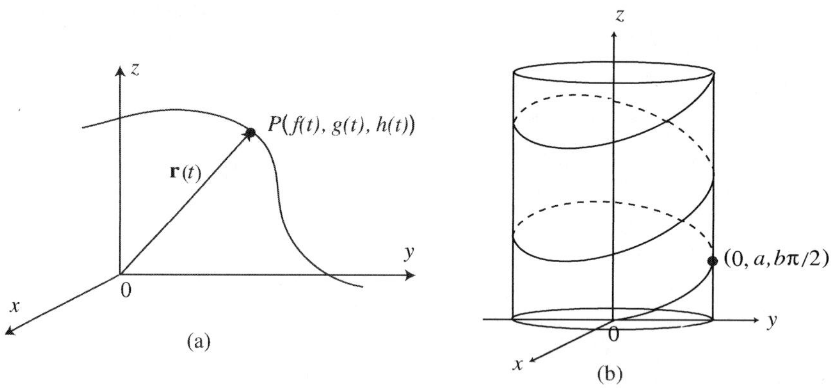

Figure 2.4 (a) Eq (1.5.1). (b) Circular helix.

If $\mathbf{r}(t)$ is defined by (2.5.1), then

$$\lim_{t\to a}\mathbf{r}(t) = \big\langle \lim_{t\to a} f(t),\ \lim_{t\to a} g(t),\ \lim_{t\to a} h(t)\big\rangle, \tag{2.5.2}$$

provided f, g, and h have limits as $t \to a$. If we denote $\lim_{t\to a} f(t) = a_1$, $\lim_{t\to a} g(t) = a_2$, $\lim_{t\to a} h(t) = a_3$, then $\lim_{t\to a}\mathbf{r}(t) = \big\langle a_1, a_2, a_3\big\rangle = a_1\,\mathbf{i} + a_2\,\mathbf{j} + a_3\,\mathbf{k} = \mathbf{a}$.

A vector-valued function $\mathbf{r}(t)$ is *continuous* at a if $\lim_{t\to a}\mathbf{r}(t) = \mathbf{r}(a)$. The *derivative* $\mathbf{r}'(t)$ of a vector-valued function $\mathbf{r}(t)$ is defined by

$$\mathbf{r}'(t) = \lim_{t_0\to 0}\frac{\mathbf{r}(t+t_0) - \mathbf{r}(t)}{t_0}, \tag{2.5.3}$$

for all t such that the limit exists.

Example 2.7. Given $\mathbf{r}(t) = (\ln t)\mathbf{i} + e^{-3t}\mathbf{j} + t^2\mathbf{k}$, the domain of $\mathbf{r}$ is the set of positive real numbers where $\mathbf{r}$ is continuous, and $\mathbf{r}'(t) = (1/t)\mathbf{i} - 3e^{-3t}\mathbf{j} + 2t\mathbf{k}$, and $\mathbf{r}''(t) = (-1/t^2)\mathbf{i} + 9e^{-3t}\mathbf{j} + 2\mathbf{k}$. ■

Example 2.8. To prove that if $\mathbf{r}(t)$ is constant, then $\mathbf{r}'(t)$ is orthogonal to $\mathbf{r}(t)$ for every t, note that $\mathbf{r}(t)\cdot\mathbf{r}(t) = |\mathbf{r}(t)|^2 = c$ for some scalar c. Since

$\mathbf{r}(t) = \langle f(t), g(t), h(t)\rangle = f(t)\,\mathbf{i} + g(t)\,\mathbf{j} + h(t)\,\mathbf{k}$, the above equation becomes $[f(t)]^2 + [g(t)]^2 + [h(t)]^2 = c$, which when differentiated implicitly gives

$$2f(t)f'(t) + 2g(t)g'(t) + 2h(t)h'(t) = 0,$$

i.e., $2\mathbf{r}(t) \cdot \mathbf{r}'(t) = 0$. ■

Let $D_t\mathbf{u}(t)$ denote the derivative of $\mathbf{u}(t)$ with respect to t. Then

Theorem 2.17. *If* $\mathbf{u}$ *and* $\mathbf{v}$ *are differentiable vector-valued functions and* c *is a scalar, then the following relations hold:*

$$\begin{aligned}
&D_t[\mathbf{u}(t) \pm \mathbf{v}(t)] = \mathbf{u}'(t) \pm \mathbf{v}'(t);\\
&D_t[c\,\mathbf{u}(t)] = c\,\mathbf{u}(t);\\
&D_t[\mathbf{u}(t) \cdot \mathbf{v}(t)] = \mathbf{u}(t) \cdot \mathbf{v}'(t) + \mathbf{u}'(t) \cdot \mathbf{v}(t);\\
&D_t[\mathbf{u}(t) \times \mathbf{v}(t)] = \mathbf{u}(t) \times \mathbf{v}'(t) + \mathbf{u}'(t) \times \mathbf{v}(t).
\end{aligned}$$

Definite integrals of vector-valued functions are defined as follows: If $\mathbf{r}(t) = f(t)\mathbf{i} + g(t)\mathbf{j} + h(t)\mathbf{k}$, where the functions f, g, and h are integrable on an interval $[a, b]$, then by definition

$$\int_a^b \mathbf{r}(t)\,dt = \Big(\int_a^b f(t)\,dt\Big)\mathbf{i} + \Big(\int_a^b g(t)\,dt\Big)\mathbf{j}\Big(\int_a^b h(t)\,dt\Big)\mathbf{k}, \tag{2.5.4}$$

and we say that $\mathbf{r}(t)$ is *integrable* on $[a, b]$. Moreover, if $\mathbf{R}(t)$ is an antiderivative of $\mathbf{r}(t)$ in the sense that $\mathbf{R}'(t) = \mathbf{r}(t)$ for all $t \in [a, b]$, then

$$\int_a^b \mathbf{r}(t)\,dt = \mathbf{R}(t)\Big|_a^b = \mathbf{R}(b) - \mathbf{R}(a). \tag{2.5.5}$$

If $\mathbf{R}(t)$ is an antiderivative of $\mathbf{r}(t)$, then every antiderivative has the form $\mathbf{R}(t) + \mathbf{c}$ for some constant vector $\mathbf{c}$, and we write

$$\int \mathbf{r}(t)\,dt = \mathbf{R}(t) + \mathbf{c}.$$

Example 2.9. Given $\mathbf{u}(t) = t\mathbf{i}+t^2\mathbf{j}+t^3\mathbf{k}$, and $\mathbf{v}(t) = \sin t\mathbf{i}+\cos t\mathbf{j}+2\sin t\mathbf{k}$, we have

$$\begin{aligned}
D_t[\mathbf{u}(t) \cdot \mathbf{v}(t)] &= (1 + 5t^2)\sin t + (3t + 2t^2)\cos t,\\
D_t[\mathbf{u}(t) \times \mathbf{v}(t)] &= [(t^2 + 4t)\sin t - t^2\cos t]\mathbf{i} + [(3t^2 - 2)\sin t + (t^2 - 2t)\cos t]\mathbf{j}\\
&\quad + [-3t\sin t + (1 - t^2)\cos t]\mathbf{k}. ■
\end{aligned}$$

Example 2.10. If $\mathbf{u}'(t) = t^2\mathbf{i} + (6t + 1)\mathbf{j} + 8t^3\mathbf{k}$ and $\mathbf{u}(0) = 2\mathbf{i} - 3\mathbf{j} + \mathbf{k}$, then

$$\mathbf{u}(t) = \Big(\frac{t^3}{3} + 2\Big)\mathbf{i} + (3t^2 + t - 3)\mathbf{j} + (2t^4 + 1)\mathbf{k}. ■$$

2.5.1 Geometric Meaning of the Inflection Point. An inflection point (IP) on the graph of a function f is a point where the second derivative $f'' = 0$. What is the meaning of this? The relation of the inflection point to intervals where the curve is concave upward (CU) or concave downward (CD) is exactly the same as the relation of critical points to the intervals where the function is increasing or decreasing. Thus, the inflection points mark the boundaries of the two different kinds of behavior. Also, only one sample value of f' is needed between each pair of consecutive IPs in order to see whether the curve is CU or CD along that interval.

2.6 Optimization

An application of the study of global and local extrema of a function leads to certain optimization problems. Recall that an optimal solution corresponds to the point or points where a given function attains an absolute maximum or absolute minimum value. Certain useful guidelines to solve such optimization problems are as follows:

(i) Relative to a given problem, define a function to be optimized, then plot its graph and label the relevant quantities, if possible; (ii) label the quantity that needs to be optimized, and signify the appropriate domain, also known as the *feasible domain*, for the problem; and (iii) using the methods of §2.3, solve the problem.

Example 2.11. To find the rectangle of largest possible area that can be inscribed in a semi-circle of radius r, let the rectangle be of height h, length w, and area $A = hw = h\sqrt{r^2 - h^2}$, where $w = 2\sqrt{r^2 - h^2}$, $0 < h < r$ (see Figure 2.5).

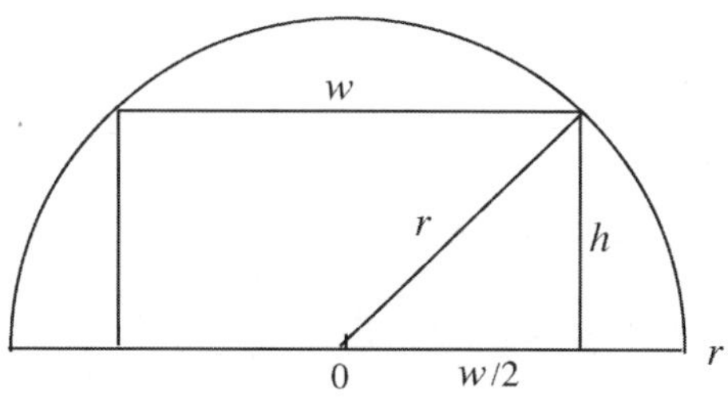

Figure 2.5 Example 2.11.

Since the domain A is closed and bounded on the interval $[0, r]$, we find, in view of Rolle's theorem, that the absolute maximum area of A is guaranteed to exist. First, we find the critical point(s):

$$A'(h) = 2\sqrt{r^2 - h^2} - \frac{2h^2}{\sqrt{r^2 - h^2}} = \frac{2r^2 - 4h^2}{\sqrt{r^2 - h^2}} = 0,$$

which gives $h = \pm r/\sqrt{2}$. Note that A' is undetermined when $h = r$ (obvious from the geometry of the problem). Also, we can discard the negative solution for h. Then at the two endpoints and one critical point in $[0, r]$, we have $A(0) = 0 = A(r)$, and $A(h) = A(r/\sqrt{2}) = r^2$. Hence, the maximum possible area of the rectangle is $A(h) = r^2$, and it occurs when $h = r/\sqrt{2}$ and $w = r\sqrt{2}$. Also check that this maximum area A is smaller that the area $\pi r^2/2$ of the semi-circle. ■

Example 2.12. A 15" × 24" piece of sheet metal is formed into an open box by cutting out a square from each of the four corners and folding up the remaining piece. How large should each square be to obtain a box of maximum volume?

Let V denote the volume of the open box. From Figure 2.6, we find that $V(x) = x(15 - 2x)(24 - 2x)$, and the critical points are given by $V'(x) = 0$.

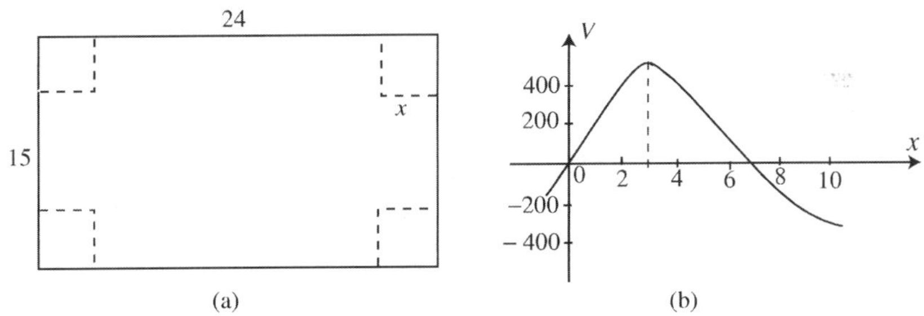

Figure 2.6 (a) Example 2.12. (b) Graph of V.

Now,

$$V'(x) = x(15-2x)(-2)+x(-2)(24-2x)+(15-2x)(24-2x) = 12(x-3)(x-10).$$

Thus, $V'(x) = 0$ at $x = 3$ or $x = 10$. But since the only critical point in $[0, 15/2]$ is at $x = 3$, we find that $V(0) = 0 = V(15/2)$, and thus, the maximum possible volume is $V(3) = 486$ in^2 (see Figure 2.6(b)). ■

Example 2.13. The profit $P(x)$ by selling x units of a given product is related to the cost $C(x)$ of its production and the revenue $R(x)$ by the formula $P(x) = R(x) - C(x)$. In order to maximize profit, we will assume that C and R are differentiable functions, so we can compute the critical points of P from $P'(x) = 0$, which are also the solutions of $C'(x) = R'(x)$. A typical cost-revenue graph is presented in Figure 2.7, which shows that cost initially exceeds revenue and then falls below revenue as bulk manufacturing and transportation costs are realized, and finally exceeds revenue after the production capacity and market saturation are reached. The profit zone is bounded by two positive break-even points, at which $C(x) = R(x)$; and the points of maximum profit and loss happen at the points where $C'(x) = R'(x)$.

Using a specific example, suppose that for a certain book printing company the cost and revenue functions in a particular year are defined, in thousands of dollars, by $C(x) = 2x^3 - 12x^2 + 30x$ and $R(x) = -x^3 + 9x^2$, where x represents the units of 1000 books. It is assumed that this model is accurate up to approximately $x = 6$. We are required to find out what the company's profit zone is and what level of production will maximize the company's profit.

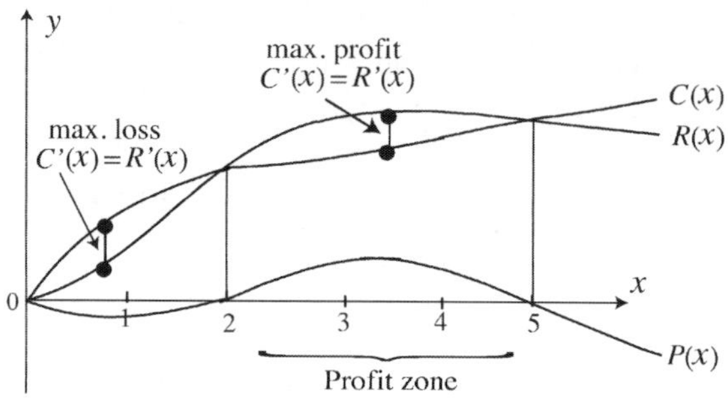

Figure 2.7 Example 2.13, graph of $P(x)$.

The profit function is $P(x) = R(x) - C(x) = -3x^3 + 21x^2 - 30x = -3x(x-2)(x-5)$. The solution set of the equation $P(x) = 0$ is $x = \{0, 2, 5\}$. Neglecting $x = 0$, we know that the positive break-even points are $x = 2$ and $x = 5$, i.e., 2000 and 5000 books. Again, solving $P'(x) = 0$, or equivalently $C'(x) = R'(x)$, we get $3x^2 - 14x + 30 = 0$, which gives $x = \dfrac{7 \pm \sqrt{19}}{3} \approx 3.786$ or 0.880. Using the first derivative test we find that P has a relative maximum at 3.786 and a relative minimum at 0.880, i.e., the relative maximum at 3786 and relative minimum at 880 books, respectively. Thus, the maximum profit is $P(3.786) = 24.626$, or \$24,626. This is presented in Figure 2.7(b), in which the correspondence between the relative extrema of P and those of C and R are marked by vertical segments between the graphs of $C(x)$ and $R(x)$ at the points 0.88 and 3.786. ■

2.7 Multivariate Functions. The following three conditions must be met for a multivariate function to be a relative maximum or minimum at a critical point:

(a) All first-order partial derivatives must be zero simultaneously. When solved, they yield the critical point $\mathbf{x}^*$ where the function is neither increasing nor decreasing with respect to the coordinate axes, although there is a relative plateau at this point.

(b) The second-order direct partial derivatives at the point $\mathbf{x}^*$ must all be negative for a relative maximum and positive for a relative minimum. Geometrically, it means that the function is concave and moving downward from the relative plateau (i.e., at the point $\mathbf{x}^*$) in relation to the coordinate axes to be a relative maximum and moving upward in relation to the coordinate axes to be a minimum.

(c) In the case of a function of two variables, the product of the second-order direct partial derivatives evaluated at the point $\mathbf{x}^*$ must be greater than the product of the cross partial derivatives also evaluated at the critical point $\mathbf{x}^*$. This condition is needed to exclude the cases of an inflection point or a saddle point at $\mathbf{x}^*$.

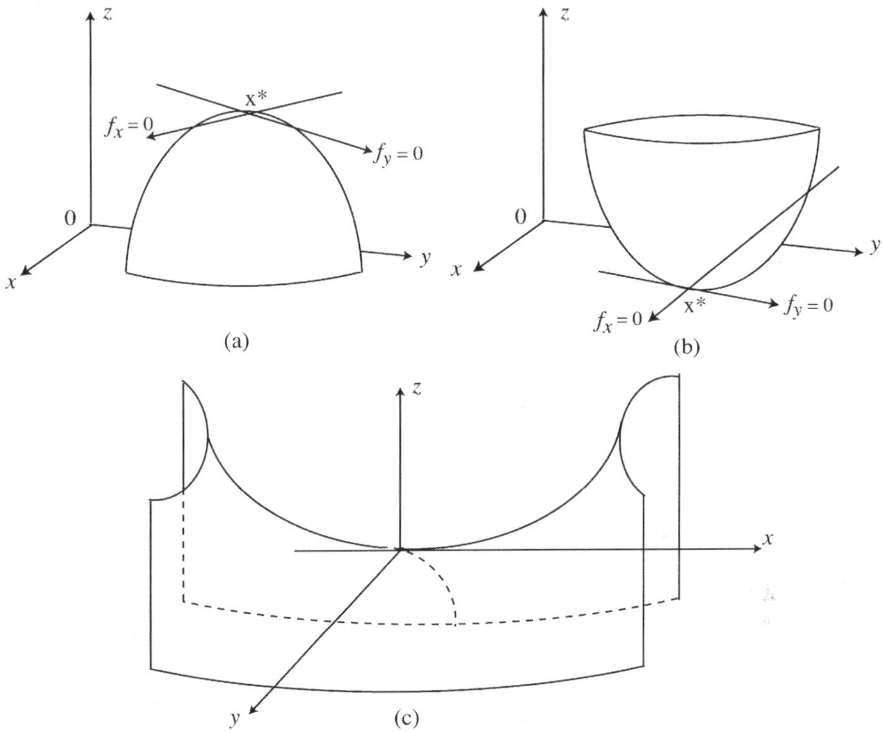

Figure 2.8 Relative extrema for a multivariate function.

The above conditions are presented in Figure 2.8 for a function of two variables $z = f(x, y)$, where we have a relative maximum (Figure 2.8(a)) and a relative minimum (Figure 2.8(b)).

The conditions satisfied in each case are as follows:

For a relative maximum: $f_x, f_y = 0$; $f_{xx}, f_{yy} < 0$; and $f_{xx} \cdot f_{yy} > (f_{xy})^2$;

For a relative minimum: $f_x, f_y = 0$; $f_{xx}, f_{yy} > 0$; and $f_{xx} \cdot f_{yy} > (f_{xy})^2$.

The last condition can also be written as $f_{xx} \cdot f_{yy} - (f_{xy})^2 > 0$.

If $f_{xx} \cdot f_{yy} < (f_{xy})^2$, we get

(i) an *inflection point* if f_{xx} and f_{yy} have the same signs;

(ii) a *saddle point* if f_{xx} and f_{yy} have different signs where the function has a maximum when viewed from one axis and a minimum when viewed from the other axis (see Figure 2.8(c));

(iii) if $f_{xx} \cdot f_{yy} = (f_{xy})^2$, the test fails; and

(iv) if the function is strictly concave up (CU) (convex down (CD)) in x and y, as in Figures 2.8(a)-(b), there is only one extremum, called the global maximum (minimum). If the function is simply concave (convex) in x and y on an interval, there will be a relative maximum (minimum) at the critical point.

Example 2.14. Consider $f(x, y) = y^3 - 2x^3 + 294x - 27y + 72$. The first-order partial derivatives are:

$f_x = -6x^2 + 294 = 0$, or $x^2 = 49$, which gives $x = \pm 7$;

$f_y = 3y^2 - 27 = 0$, or $y^2 = 9$, which gives $y = \pm 3$.

Hence, there are four critical points: $(7, 3), (7, -3), (-7, 3)$, and $(-7, -3)$. Next, using the second-order partial derivatives at each critical point, we check the signs: $f_{xx} = -12x$, $f_{yy} = 6y$, $f_{xy} = 0 = f_{yx}$:

$$
\begin{array}{ll}
(1)\ f_{xx}(7,3) = -84 < 0, & f_{yy}(7,3) = 18 > 0, \\
(2)\ f_{xx}(7,-3) = -84 < 0, & f_{yy}(7,-3) = -18 < 0, \\
(3)\ f_{xx}(-7,3) = 84 > 0, & f_{yy}(-7,3) = 18 > 0, \\
(4)\ f_{xx}(-7,-3) = 84 > 0, & f_{yy}(-7,-3) = -18 < 0.
\end{array}
$$

Since there are different signs in each of the second-order derivative values in the cases (1) and (4), the function f cannot have a relative extremum at the critical points $(7, 3)$ and $(-7, -3)$. In the case when f_{xx} and f_{yy} are of different signs, the product $f_{xx} f_{yy}$ cannot be greater than $(f_{xy})^2$, and the function f is at a saddle point. Next we check the sign of $f_{xx} \cdot f_{yy} > (f_{xy})^2$ at the remaining two critical points $(7, -3)$ and $(-7, 3)$:

At the point $(7, -3)$ we have $(-84) \cdot (-18) > (0)^2$; thus, we have a relative maximum at $(7, -3)$. Also, at the point $(-7, 3)$ we have $(84) \cdot (18) > (0)^2$; thus, we have a relative minimum at $(-7, 3)$.

As an alternative method, we can use the Hessian (see §1.6.2), which for this example is defined by

$$
|\mathbf{H}| = \begin{vmatrix} f_{xx} & f_{xy} \\ f_{yx} & f_{yy} \end{vmatrix}.
$$

Then at each of the critical points

$$\text{at } (7,3): \quad |\mathbf{H}| = \begin{vmatrix} -84 & 0 \\ 0 & 18 \end{vmatrix}; \quad \text{at } (7,-3): \quad |\mathbf{H}| = \begin{vmatrix} -84 & 0 \\ 0 & -18 \end{vmatrix};$$

$$\text{at } (-7,3): \quad |\mathbf{H}| = \begin{vmatrix} 84 & 0 \\ 0 & 18 \end{vmatrix}; \quad \text{at } (-7,-3): \quad |\mathbf{H}| = \begin{vmatrix} 84 & 0 \\ 0 & -18 \end{vmatrix};$$

which leads to the same answer as above. ■

Thus, we have three ways to present the optimization analysis: (i) by tabular form, (ii) by using Hessian, or (iii) by simple explanation.

2.7.1 Geometrical Interpretation. The geometrical concepts of minima, saddle points, and maxima can be visualized from the following illustration: Consider a mountainous terrain M. If a function $f \mapsto \mathbb{R}$ sends each point to its elevation, then the inverse image of a point $a \in \mathbb{R}$ is simply a contour line, such that each connected component of a contour line is either a point, a simple closed curve, or a closed curve with a double point. Contour lines may have points of higher order, like triple points, but they, being unstable, may be removed by a slight deformation of M. The double points in contour lines occur at saddle points, or passes, since at these points the surrounding terrain curves up in one direction and down in the other.

Just imagine there is flooding in the terrain M. Then the region covered by water when it reaches an elevation of the point a is given by $f^{-1}(-\infty, a]$, i.e., water reaches the points with at most elevation a. When water passes the height of the point a, a critical point where the gradient $\nabla f = 0$, then the water either (i) starts filling the terrain basin, (ii) covers a saddle point (a mountain pass), or (iii) submerges a peak. In each of these three types of critical points (basins, passes, and peaks), we have the case of minima, saddle points, and maxima, respectively. The safest place to escape flooding is either the front or back high elevation at a saddle point (known as the horn and the cantle of a saddle), or the highest peak (maximum elevation), and the worst places are the basins (minima) and the inflection points.

2.7.2 Gradient at a Point. The gradient of a function f at a point $\mathbf{x} \in \mathbb{R}^n$ is a multiplication of vectors by scalars. For example, in $\mathbb{R}^3$, the gradient of f at a point $\mathbf{x} = (x, y, z)$ is denoted by grad f, or ∇f, and defined by

$$\nabla f(\mathbf{x}) = \frac{\partial f}{\partial x}\mathbf{i} + \frac{\partial f}{\partial y}\mathbf{j} + \frac{\partial f}{\partial z}\mathbf{k}, \tag{2.7.1}$$

where $\mathbf{i}, \mathbf{j}, \mathbf{k}$ are the unit vectors along the coordinate axes. In $\mathbb{R}^n$, we have

$$\nabla f(\mathbf{x}) = \frac{\partial f}{\partial x_1}\mathbf{e}_1 + \frac{\partial f}{\partial x_2}\mathbf{e}_2 + \cdots + \frac{\partial f}{\partial x_n}\mathbf{e}_n, \tag{2.7.2}$$

where $\mathbf{e}_i$ is the unit vector along the ith coordinate axis, $i = 1, 2, \dots, n$, and all these vectors form a linearly independent set in $\mathbb{R}^n$.

Theorem 2.18. (Michel and Herget [2007: 88]) *Let $\{\mathbf{e}_1, \mathbf{e}_2, \dots, \mathbf{e}_n\}$ be the linearly independent set of unit vectors along the coordinate axes in a vector space $\mathbb{R}^n$. If $\sum_{i=1}^{n} \frac{\partial f}{\partial x_i} \mathbf{e}_i = \sum_{i=1}^{n} \frac{\partial g}{\partial x_i} \mathbf{e}_i$, then $\frac{\partial f}{\partial x_i} = \frac{\partial g}{\partial x_i}$ for all $i = 1, 2, \dots, n$.*

PROOF. If $\sum_{i=1}^{n} \frac{\partial f}{\partial x_i} \mathbf{e}_i = \sum_{i=1}^{n} \frac{\partial g}{\partial x_i} \mathbf{e}_i$, then, by matrix multiplication, we have

$$\left[\frac{\partial f}{\partial x_i} - \frac{\partial g}{\partial x_i}\right]_{1\times n} [\mathbf{e}_i]^T_{n\times 1} = [\mathbf{0}]_{1\times 1} = [\mathbf{0}]_{1\times n} [\mathbf{e}_i]^T_{n\times 1}, \tag{2.7.4}$$

which is a relation in the sense that it is an equation, expressed in matrix form as the matrix product of scalar terms $\left(\frac{\partial f}{\partial x_i} - \frac{\partial g}{\partial x_i}\right)$ and unit vectors $[\mathbf{e}_i]^T$ for $i = 1, 2, \dots, n$, and equated to the matrix product of the zero vector and unit vectors $[\mathbf{e}_i]^T$. Since the set $\{\mathbf{e}_1, \mathbf{e}_2, \dots, \mathbf{e}_n\}$ is linearly independent, we get $\frac{\partial f}{\partial x_i} - \frac{\partial g}{\partial x_i} = 0$ for all $i = 1, 2, \dots, n$, which proves the theorem. ■

This leads to the following result:

Condition A. Let $f(\mathbf{x}) \in \mathbb{R}^n$ be a continuously differentiable function. If

$$\begin{aligned} \nabla f(\mathbf{x}) &= \frac{\partial f}{\partial x_1} \mathbf{e}_1 + \frac{\partial f}{\partial x_2} \mathbf{e}_2 + \cdots + \frac{\partial f}{\partial x_n} \mathbf{e}_n \\ &= \left[\frac{\partial f}{\partial x_1} \; \frac{\partial f}{\partial x_2} \; \cdots \; \frac{\partial f}{\partial x_n}\right] [\mathbf{e}]^T = \mathbf{0}, \end{aligned} \tag{2.7.5}$$

where $[\mathbf{e}] = [\mathbf{e}_1 \; \mathbf{e}_2 \; \dots, \mathbf{e}_n]$, then

$$\frac{\partial f}{\partial x_1} = 0, \; \frac{\partial f}{\partial x_2} = 0, \; \dots, \; \frac{\partial f}{\partial x_n} = 0. \tag{2.7.6}$$

This is the *necessary condition* that not only establishes an isomorphism

$$\nabla f(\mathbf{x}) \overset{\text{iso}}{=} \left[\frac{\partial f}{\partial x_1} \quad \frac{\partial f}{\partial x_2} \quad \cdots \quad \frac{\partial f}{\partial x_n}\right] \tag{2.7.7}$$

between the gradient of a linear mapping, ∇f, and the first partial derivatives $\frac{\partial f}{\partial x_i}$ for $i = 1, \dots, n$, but also imposes the restriction that the equations (2.7.6) will hold only when condition (2.7.5), $\nabla f(\mathbf{x}) = \mathbf{0}$, is satisfied.

This condition is used in the Lagrange multiplier method, the KKT conditions, and the Fritz John condition in optimization problems. However, there are a couple of cases involving the first- and higher-order Taylor series approximations, where the above isomorphism is misused (see §3.5).

2.8 Mathematical Economics

Some terms, notations, and definitions, useful in the study of elementary mathematical economics, are introduced.

2.8.1 Isocost Lines. In economics, an isocost line represents the different combinations of two inputs or factors of production that can be purchased with a given sum of money. It is represented, in general, by the formula

$$P_K K + P_L L = E, \tag{2.8.1}$$

where K and L denote capital and labor, P_K and P_L their respective prices, and E the amount allotted for expenditures. In isocost analysis the prices and the expenditures for individual items are initially held constant; only different inputs are allowed to vary. Solve the above formula for K, and show that a change in P_L and P_K will affect the slope and the vertical intercept.

Solving Eq (2.8.1) for an isocost line we get

$$K = \frac{E - P_L L}{P_K}, \quad \text{or } K = \frac{E}{P_K} - \Big(\frac{P_L}{P_K}\Big) L,$$

which is a straight line of the form $y = mx + b$, where the slope $m = -P_L/P_K$ and the vertical intercept, also called the y-intercept, $b = E/P_K$ (see Figure 2.9).

The effect of a change in any one of the parameters can be easily seen from Figure 2.9. For example, an increase in expenditure from E to E' will increase the vertical intercept and the isocost line (dashed line) will shift out to the right parallel to the previous line; however the slope remains unaffected because it depends on the ratio of the prices $-P_L/P_K$ and prices are not affected by change in expenditures. A change in P_L will change the slope of the line but not the vertical intercept, but a change in P_K will change both the slope and the vertical intercept.

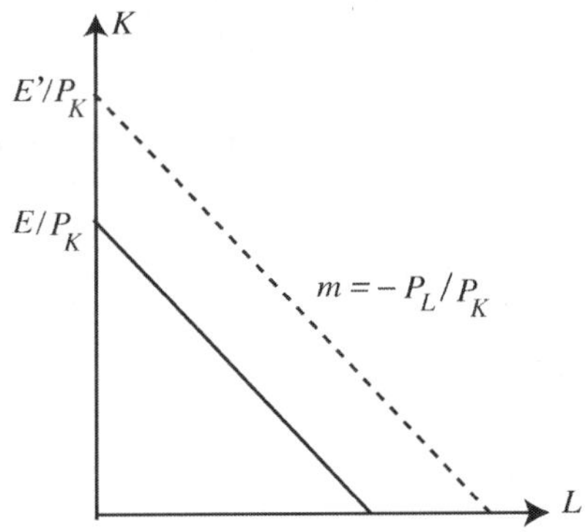

Figure 2.9 Isocost line.

2.8.2 Supply and Demand. Let Q_s and Q_d denote the supply and demand functions, respectively. Equilibrium in supply and demand occurs when $Q_s = Q_d$. For example, the equilibrium prices and quantity are determined in the following situation: Given $Q_s = 4P - 7$, $Q_d = 14 - 3P$, in equilibrium we have $Q_s = Q_d$, or $4P - 7 = 14 - 3P$, or $P = 3$. Then substituting $P = 3$ in either equation we get $Q_s = 4P - 7 = 12 - 7 = 5 = Q_d$.

The equilibrium equation is $Y = C + I + G + (X - Z)$, where Y is income, C consumption, I investment, G government expenditures, X exports, and Z imports.

Example 2.15. In the case of a two-sector economy, $Y = C + I$, $C = C_0 + bY$, and $I = I_0$. Given $C_0 = 95, b = 0.8$, and $I_0 = 65$, calculate the equilibrium level of income in terms of (a) the general parameters, and (b) the specific values assigned to these parameters. In the case (a), the equilibrium equation is $Y = C+I = C_0+bY+I_0$, which on solving for Y gives $Y = \dfrac{C_0 + I_0}{1 - b}$, known as the *reduced form* which expresses the endogenous variable Y as an explicit function of the exogenous variables C_0, I_0 and the parameter b. In case (b), $Y = \dfrac{95 + 65}{1 - 0.8} = \dfrac{160}{0.2} = 800$. Note that the term $1/(1 - b)$ is called the *autonomous expenditure multiplier*, which measures the multiple effect each dollar of autonomous spending has on the equilibrium level of income. ■

The parameter b is called the *marginal propensity to consume* (MPC) in an income determination model; it is the proportion of an aggregate raise in pay that a consumer spends on the consumption of goods and services, as opposed to saving it. In this sense, the above autonomous expenditure multiplier can be expressed as $1/(1 - \text{MPC})$.

2.8.3 IS-LM Equation. The *IS-schedule* is the locus of points representing all possible combinations of interest rates and income levels consistent with equilibrium in the commodity market and the *LM-schedule* is the locus of similar points in the money market. Thus, the IS-LM analysis is used to find the level of income and the rate of interest at which both the commodity (goods) market and the money market will be in equilibrium. The commodity market for a single two-sector economy is in equilibrium when $Y = C + I$, while the money market is in equilibrium when the supply of money M_s is equal to the demand for money M_d, with $M_d = M_t + M_z$, where M_t is the transition-precautionary demand for money, and M_z the speculative demand for money.

Example 2.16. Assuming a two-sector economy where $C = 56 + 0.8Y$, $I = 94 - 70i, M_s = 210, M_t = 0.3Y$, and $M_z = 55 - 140i$, compute the IS and LM. The IS (commodity equilibrium) exists when $Y = C + I$. With the above data,

$$Y = 56 + 0.8Y + 94 - 70i, \quad \text{or} \quad 0.2Y + 70i - 150 = 0. \tag{2.8.2}$$

The LM (monetary equilibrium) exists when $M_s = M_t + M_z$, which with the above data is

$$210 = 0.3Y + 55 - 140i, \quad \text{or} \quad 0.3Y - 140i - 155 = 0. \tag{2.8.3}$$

Solving Eqs (2.8.2) and (2.8.3), we find the condition of simultaneous equilibrium in both markets. Thus, multiplying Eq (2.8.2) by 2 and adding to Eq (2.8.3) gives $0.7Y = 455$, or $Y = 650$. Then substituting this value of Y into Eq (2.8.2) we get $130 + 70i = 150$, $i = 2/7 \approx 0.29$. For these values of Y and i, the equilibrium values of C, M_t, and M_z are: $C = 56 + 0.8Y = 56 + (0.8)(650) = 576$, $M_t = 0.3Y = (0.3)(650) = 195$, $M_z = 55 - 140i = 55 - 140(2/7) = 15$. To check, $C + I = 56 + 0.8Y + 94 = 70i = 56 + (0.8)(650) + 94 - 70(2/7) = 650$, and $M_t + M_z = 0.3Y + 55 - 140i = (0.3)(650) + 55 - 140(27) = 195 + 55 - 40 = 210 = M_s$.

2.8.4 Marginal of an Economic Function. Let MC = marginal cost; MR= marginal revenue; TC = total cost; TR = total revenue; Q = level of output; P = demand; TR=PQ; R = revenue; C = cost; π = profit = R − C; and TP = total product. Then the *marginal cost* (MC) in economics is defined as the change in total cost incurred from the production of an additional unit. The *marginal revenue* (MR) is defined as the change in total revenue brought about by the sale of an extra good. Since total cost (TC) and total revenue (TR) are both functions of the level of output (Q), both MC and MR are expressed as derivatives of their total functions, respectively, i.e.,

$$\text{if TC= TC(Q), then MC=}\frac{d(\text{TC})}{dQ}; \quad \text{if TR= TR(Q), then MR=}\frac{d(\text{TR})}{dQ}.$$

For example, (i) let TR= 62 Q-5 Q^2. Then MR=62-10 Q; and (ii) TC = Q^3-19 Q+34. Then MC = $3Q^2-19$.

In general, the marginal of any economic function is the derivative of its total function.

The marginal propensity to consume (MCP) is defined as MCP=$\dfrac{dC}{dY}$; and average cost function is denoted by AC.

Theorem 2.19. *MC = MR at the profit maximizing state.*

PROOF. Since π = TR − TC, to maximize we have

$$\frac{d\pi}{dQ} = \frac{d(\text{TR})}{dQ} - \frac{d(\text{TC})}{dQ} = 0,$$

which implies that $\dfrac{d(\text{TR})}{dQ} = \dfrac{d(\text{TC})}{dQ}$, or MR = MC. ■

2.8.5 Marginal Rate of Technical Substitution (MRTS). An *isoquant* determines the different levels of inputs K and L that can be used to produce a specific level of output Q. One such isoquant for the output level $Q = k$, k constant, is defined by $aK^{1/4}L^{3/4} = k$, and its slope dK/dL is known as MRTS. The general form of an isoquant is $aK^pL^{1-p} = k$, where a is real, $0 < p < 1$, and k is a constant.

Example 2.17. Let an isoquant at the output level 2016 be defined by $Q = 24K^{1/6}L^{5/6} = 2016$. (a) Determine the slope dK/dL, and (b) evaluate the MRTS at $K = 308$ and $L = 115$. First, (a) since the given isoquant is $K^{1/6}L^{5/6} = 84$, using the implicit differentiation we get

$$\frac{1}{6}K^{-5/6}L^{5/6}\frac{dK}{dL} + K^{1/6}\frac{5}{6}L^{-5/6} = 0,$$

which simplifies to give

$$\frac{dK}{dL} = \frac{-5K^{1/6}L^{-5/6}}{K^{-5/6}L^{5/6}} = -\frac{5K}{L}. \tag{2.8.4}$$

Thus, MRTS $= -\dfrac{5K}{L}$. (b) Substituting $K = 308$ and $L = 115$ into (2.8.4), we find that MRTS $= -\dfrac{(5)(308)}{115} = -13.39$. Thus, in the case of a constant production level, if L is increased by a meager 1 unit, K must be decreased by 13.39 units in order to remain on the production isoquant. ■

2.9 Exercises

2.1. Prove that if f is a linear function, then f satisfies the hypotheses of the mean value theorem on every closed interval $[a, b]$, and that every number c satisfies the conclusion of the theorem.

2.2. If f is a quadratic function and $[a, b]$ is any closed interval, prove that there is precisely one number $c \in (a, b)$ which satisfies the conclusion of the mean value theorem. HINT. Consider $f(x) = ax^2 + bx + c$.

2.3. Find the intervals where the function $f(x) = x^2 - 2x + 1$ is decreasing or increasing. HINT. $f'(x) = 2x - 2 = 2(x - 1)$; $f' > 0$ when $x > 1$ and negative when $x < 1$. ANS. $f(x)$ is decreasing in the interval $(-\infty, 1)$ and increasing in the interval $(1, \infty)$.

2.4. For the function $f(x) = x^4 = 8x^2 + 16$, find the intervals where it is increasing or decreasing. HINT. $f'(x) = 4x^3 - 16x = 4x(2)(x + 2)$. Three cases: (i) $f' < 0$ if $x < -2$, (ii) $f' > 0$ if $0 < x < 2$, and (iii) $f' > 0$ if $x > 2$. ANS. f is increasing in the intervals $(-2, 0)$ and $(2, \infty)$ and decreasing in the intervals $(-\infty, -2)$ and $0, 2)$.

2.5. If f is a polynomial function of degree 3 and $[a, b]$ is any closed interval, prove that there are at most two numbers in (a, b) which satisfy the conclusion of the mean value theorem. Generalize this result to a polynomial function of degree n, where n is a positive integer. HINT. If f has degree 3, then $f'(x)$ is a polynomial of degree 2. Then $f(b) - f(a) = f'(x)(b - a)$ has at most two solutions x_1 and x_2. If f has degree n, then there are at most $n - 1$ solutions.

2.6. For the function $f(x) = x^3 + x^2 - 2x - 1$, find the intervals where it is concave downward or upward, and find the point of inflection. HINT. $f'(x) = 3x^2 + 2x - 2$; $f''(x) = 6x + 2 = 2(3x + 1)$. ANS. f is concave downward if $x < -1/3$ and concave upward if $x > -1/3$. Since the graph of f changes the direction of its concavity at $x = -1/3$, the point $(-1/3, f(-1/3) = (-1/3, -59/27)$ is the inflection point.

2.7. Find the local extrema of f', and describe the intervals in which f' is increasing or decreasing, given (i) $f(x) = x^4 - 6x^2$; (ii) $f(x) = x^{4/3} + 4x^{1/3}$.

ANS. (i) maximum $f'(-1) = 8$; minimum $f'(1) = -8$; f' increasing on $(-\infty, -1]$ and $1, \infty)$, decreasing on $[-1, 1]$. (ii) minimum $f(-1) = -3$; increasing on $[-1, \infty)$, decreasing on $(-\infty, -1]$.

2.8. Use the second derivative test, whenever applicable, to find the local extrema of f, and the intervals of concavity of f: (i) $f(x) = 3x^4 - 4x^3 + 6$; (ii) $f(x) = 2x^6 - 6x^4$; and (iii) $x^2 - 27/x^2$.

ANS. (i) minimum: $f(1) = 5$; CU (concave upward) on $(-\infty, 0)$ and $(2/3, 0)$; CD (concave downward) on $(0, 2/3)$; abscissas of point of inflection are 0 and 2/3. (ii) maximum $f(0) = 0$ (by first derivative test); minimum $f(-\sqrt{2}) = f(\sqrt{2}) = -8$; CU on $(-\infty, -\sqrt{6/5})$ and $(\sqrt{6/5}, \infty)$; CD on $(-\sqrt{6/5}, \sqrt{6/5})$; abscissas of points of inflection are $\pm\sqrt{6/5}$. (iii) No maximum or minimum; CU on $(-\infty, -3)$ and $(3, \infty)$; CD on $(-3, 0)$ and $(0, 3)$; abscissas of points of inflection are ± 3.

2.9. Find the intervals where the functions (i) $f(x) = xe^x$, and (ii) $g(x) = \cos x$ are concave upward or downward, and the points of inflection, if any. ANS. (i) $f''(x) = (2 + x)\, e^x$; $f(x)$ is concave upward if $x > -2$ and downward if $x < -2$; -2 is an inflection point. (ii) $g''(x) = -\cos x$; so g is concave upward where it is negative and downward where it is positive, with the point of inflection at 0.

2.10. For the $f(x) = 3x^2 - 9x + 6$, determine the points of inflection, if any, and intervals of concavity. ANS. $f''(x) > 0$, so no points of inflection, and the curve is entirely concave upward.

2.11. Find the parametric equation of the tangent line to the curve C defined by $C = \{(e^t, te^t, t+4) : t \in \mathbb{R}\}$ at the point $P = (1, 0, 4)$. ANS. $x = 1 + t, y = t, z = 4$.

2.12. Find two different unit tangent vectors to the curve $C = \{x = e^{2t}, y = e^{-t}, x = t^2\}$ at the point $P = (1, 1, 4)$. ANS. $\pm(1/\sqrt{5})\langle 2, -1, 0\rangle$. ■

2.13. If a function f and a vector-valued function $\mathbf{u}$ have limits as $t \to a$, then prove that $\lim_{t\to a} f(t)\mathbf{u}(t) = \left[\lim_{t\to a} f(t)\right]\left[\lim_{t\to a} \mathbf{u}(t)\right]$. ■

2.14. If $\mathbf{u}$ and $\mathbf{v}$ are vector-valued functions which have limits as $t \to a$, then prove that $\lim_{t\to a}[\mathbf{u}(t) \times \mathbf{v}(t)] = \lim_{t\to a} \mathbf{u}(t) \times \lim_{t\to a} \mathbf{v}(t)$. ■

2.15. Let $\mathbf{r}(t)$ be the position vector of a particle, and s denote the arc length along a curve C traced by the motion of the particle. Since the magnitude of the velocity $\mathbf{r}'(t) = ds/dt$, and the direction of $\mathbf{r}'(t)$ is the same as that of the unit vector $\mathbf{T}(s)$ defined by $\mathbf{T}(s) = \mathbf{r}'(t)/|\mathbf{r}'(t)|$, we may write $\mathbf{r}'(t) = \dfrac{ds}{dt}\mathbf{T}(s)$, which after differentiating with respect to t gives

$$\mathbf{r}''(t) = \frac{d^2 s}{dt^2}\mathbf{T}(s) + \frac{ds}{dt}\frac{d}{dt}\mathbf{T}(s) + \frac{d^2 s}{dt^2}\mathbf{T}(s) + \frac{ds}{dt}\frac{ds}{dt}\mathbf{T}'(s).$$

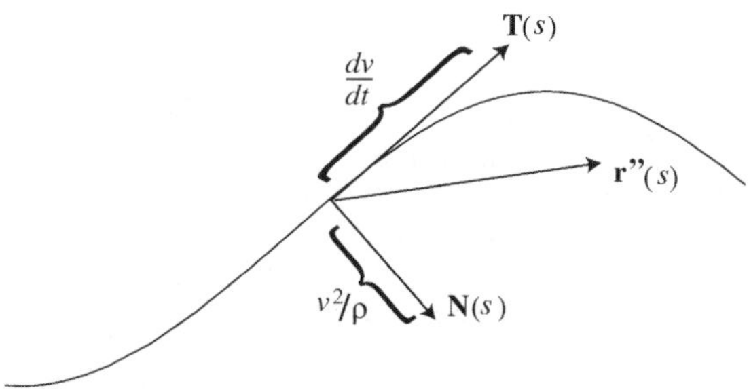

Figure 2.10 Acceleration.

We know that $\mathbf{T}'(s) = K\mathbf{N}(s)$, where $K = |\mathbf{T}'(s)|$ and $\mathbf{N}(s) = \mathbf{T}'(s)/|\mathbf{T}(s)|$ is a unit vector orthogonal to $\mathbf{T}(s)$ (called the *principal unit normal vector* to C at the point $P(s)$). If we denote the speed ds/dt by v and write $K = 1/\rho$, where ρ is the radius of curvature of C, then

$$\mathbf{r}''(t) = \frac{dv}{dt}\mathbf{T}(s) + \frac{v^2}{\rho}\mathbf{N}(s).$$

The result is known as the acceleration in terms of a tangential component dv/dt (the rate of change of speed with respect to time) and a normal component v^2/ρ (see Figure 2.10).

2.16. An electrical power station (A) is being built on the bank of a river. Cables to the substation need to be laid out underground and underwater from another substation (B) 3 km upstream and on the opposite bank of the river. The river follows a straight line course through this stretch and has a fairly constant width of 1 km. Given that the cost of laying cable underground is \$30,000 per km and the cost of laying cables underwater is \$50,000 per km, in what proportion should the cable be laid in order to minimize cost?

SOLUTION. A simple solution is to connect substations A and B by a straight line because that will be the shortest distance of $\sqrt{10}$ between them. But will it be cost effective since the cable must be laid completely underwater, at a cost of $50000 \times \sqrt{10} \approx \$\,158,114$, which may not be the minimum cost.

Another approach, to minimize the underwater installation cost, would be to cross the river along the shortest possible path and then run the underground cable along the bank of the river. Any combination of these two paths may lead to the absolute minimum cost. Thus, to find the optimal solution, we consider the path as shown in Figure 2.11.

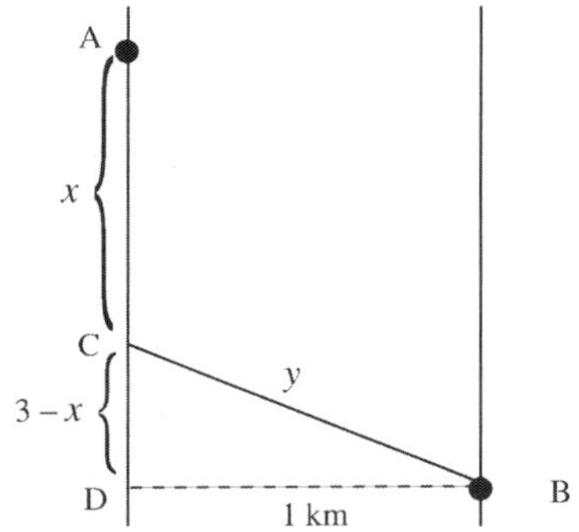

Figure 2.11 Possible path of the cable.

Since the vertical distance between stations A and B is 3 km, we take a point C at a distance x from A and join CB. Denoting CB by y, we note x is the length of the underground cable and y the length of underwater cable. Note that $x = 0$ corresponds to the case in which all the cable is underwater, and $x = 3$ corresponds to the case in which the underwater cable will be the shortest (along the line segment DB). From the right triangle CDB, we have $y^2 = 1^2 + (3 - x)^2$, or $y = \sqrt{1 + (3 - x)^2}$. Then choosing the path as A to C to B, the cost of laying the cable is

$$C(x) = 30000\,x + 50000\,y = 30000\,x + 50000\sqrt{1+(3-x)^2}.$$

Since we want to find the absolute minimum of the function $C(x)$ on the interval $[0,3]$, we first find the critical points:

$$\begin{aligned} C'(x) &= 30000 + 50000 \cdot \frac{1}{2}\big(1 = (3-x)^2\big)^{-1/2}\big(2(3-x)\big)\cdot(-1) \\ &= 30000 + \frac{50000(x-3)}{\sqrt{1+(3-x)^2}}. \end{aligned}$$

Although $C'(x)$ exists for all $x \in [0,3]$, there are two values of x for which $C'(x) = 0$, solving which we find that

$$30000 + \frac{50000(x-3)}{\sqrt{1+(3-x)^2}} = 0,$$

which simplifies to $(x-3)^2 = \dfrac{9}{16}$, thus leading to two solutions: $x = 3 \pm \frac{3}{4}$. Since the solution with the plus sign is outside the feasible interval $[0,3]$, we have only one solution at $x = 9/4$. Thus, we have

$$C(0) = \$\,158,114 \text{ (as above)};\ C(9/4) = \$\,130,000; \text{and } C(3) = \$\,140,000.$$

Hence, the minimum cost of running the cable from station A to B is achieved by laying a vertical underground cable for a distance of 2.25 km (A to C) and a diagonal underwater cable of length $1 + \sqrt{(3/4)^2} = 1.25$ km (along CB) at a cost of \$ 130,000.

2.17. An oil company wants to manufacture a half-liter can for a brand of motor oil in the shape of a right circular cylinder. Determine the dimensions for such a can of radius r and height h that will minimize the amount of metal used to form it.

HINT. Using the fact that 1 liter corresponds to a volume of 1000 cm^3, we have $V = \pi r^2 h = 500$, and using the area of base and the top plus the area of the circular part, the total area of the metal used for a can is $A(r) = 2\pi r^2 + 2\pi r h = 2\pi r \dfrac{500}{\pi r^2} = 2\pi r^2 + \dfrac{1000}{r}$. ANS. Relative minimum at $r = 10/(4\pi)^{1/3} \approx 4.3$ cm.

2.18. Let the investment be nonautonomous but a function of income. Find (a) the equilibrium level of income Y, given that $Y = C+I$, $C = C_0 + bY$, $I = I_0 + aY$, and (b) determine the change in the multiplier (MCP) b in this case.

SOLUTION. (a) $Y = C + I = C_0 + bY + I_0 + aY$, or $(1-b-a)Y = C_0 + I_0$, which gives $Y = \dfrac{C_0 + I_0}{1-b-a}$. Assume that $C_0 = 75, I_0 = 80, b = 0.6, a = 0.2$. Then $Y = \dfrac{75+80}{1-0.6-0.2} = \dfrac{155}{0.2} = 775$.

(b) $\dfrac{1}{1-b} = \dfrac{1}{1-0.6} = \dfrac{1}{0.4} = 2.5$, whereas $\dfrac{1}{1-b-a} = \dfrac{1}{1-0.6-0.2} = \dfrac{1}{0.2} = 5$, thus the multiplier MCP increases while changing from $\dfrac{1}{1-b}$ to $\dfrac{1}{1-b-a}$. With the above data, we get $\dfrac{C_0+I_0}{1-b} = \dfrac{155}{0.6} = \dfrac{155}{0.4} = 387.5$, which is half of the value of 775 determined above in the case of nonautonomous investment.

2.19. Supply and demand problems generally involve more than one market. Determine the equilibrium price P and quantity Q for the following three goods:

$$Q_{d1} = -5P_1 + P_2 + P_3 + 23, \quad Q_{s1} = 6P_1 - 8,$$
$$Q_{d2} = P_1 - 3P_2 + 2P_3 - 15, \quad Q_{s2} = 3P_2 - 11,$$
$$Q_{d3} = P_1 + 2P_2 - 4P_3 + 19, \quad Q_{s3} = 3P_3 - 5,$$

where $s1, s2, s3$ and $d1, c2, d3$ denote the three supply and demand indices.

SOLUTION. For the equilibrium we have in each market the following equations:

$$\text{Market 1: } Q_{d1} = Q_{s1} \quad \text{gives} \quad \text{(a) } 11P_1 - P_2 - P_3 = 31,$$
$$\text{Market 2: } Q_{d2} = Q_{s2} \quad \text{gives} \quad \text{(b) } P_1 - 6P_2 + 2P_3 = -26,$$
$$\text{Market 3: } Q_{d3} = Q_{s3} \quad \text{gives} \quad \text{(c) } P_1 - 2P_2 - 7P_3 = -24.$$

METHOD 1. We will use Cramer's rule (A.15) to solve this system of three equations $\mathbf{AP} = \mathbf{b}$. Thus,

$$|\mathbf{A}| = \begin{vmatrix} 11 & -1 & -1 \\ 1 & -6 & 2 \\ 1 & 2 & -7 \end{vmatrix} = 11[42-4] + [-7-2] - [2+6] = 401,$$

$$|\mathbf{A}_1| = \begin{vmatrix} 31 & -1 & -1 \\ -26 & -6 & 2 \\ -24 & 2 & -7 \end{vmatrix} = 1604, \Longrightarrow P_1 = \frac{|\mathbf{A}_1|}{|\mathbf{A}|} = \frac{1604}{401} = 4;$$

$$|\mathbf{A}_2| = \begin{vmatrix} 11 & 31 & -1 \\ 1 & -26 & 2 \\ 1 & -24 & -7 \end{vmatrix} = 2807, \Longrightarrow P_2 = \frac{|\mathbf{A}_2|}{|\mathbf{A}|} = \frac{2807}{401} = 7;$$

$$|\mathbf{A}_3| = \begin{vmatrix} 11 & -1 & 31 \\ 1 & 1 & -26 \\ 1 & 2 & -24 \end{vmatrix} = 2406, \Longrightarrow P_3 = \frac{|\mathbf{A}_3|}{|\mathbf{A}|} = \frac{2406}{401} = 6.$$

METHOD 2. This problem can also be solved by the *Gauss elimination method* (§1.4.3), as follows: Keeping Eq (a) fixed, eliminate P_1 between Eqs (a) and (b), i.e., multiply Eq (b) by 11 and subtract from Eq (a), to get

$$\text{(d)} \quad 65P_2 - 23P_3 = 317.$$

Again, eliminate P_1 between Eqs (b) and (c), to get

$$\text{(e)} \quad -8P_2 + 9P_3 = -2.$$

Next, eliminate P_2 between Eqs (d) and (e), i.e., multiply Eq (d) by 23 and Eq (e) by 65, and add, to get: $401P_3 = 2406$, or $P_3 = 6$. Hence, the system reduces to the triangularized system

$$\begin{aligned} 11P_1 - P_2 - P_3 &= 31, \\ 65P_2 - 23P_3 &= 317, \\ P_3 &= 6. \end{aligned}$$

The values of P_1, P_2 are now found by back substitution, i.e., substituting $P_3 = 6$ in the second equation we get $P_2 = 7$, and then substituting these values of P_2 and P_3 into the first equations gives $P_1 = 4$.

METHOD 3. We can use the formula $[\mathbf{P}] = [\mathbf{A}]^{-1}[\mathbf{b}]$ (see §1.4), where $[\mathbf{A}]^{-1}$ is obtained using formulas (A.8) and (A.10). Since $|\mathbf{A}| = 401 \neq 0$, we have the cofactor matrix $\mathbf{C}$ as

$$\mathbf{C} = \begin{bmatrix} \begin{vmatrix} -6 & 2 \\ 2 & -7 \end{vmatrix} & -\begin{vmatrix} 1 & 2 \\ 1 & -7 \end{vmatrix} & \begin{vmatrix} 1 & -6 \\ 1 & 2 \end{vmatrix} \\ \begin{vmatrix} -1 & -1 \\ 2 & -7 \end{vmatrix} & -\begin{vmatrix} 11 & -1 \\ 1 & -7 \end{vmatrix} & \begin{vmatrix} 11 & -1 \\ 1 & 2 \end{vmatrix} \\ -\begin{vmatrix} -1 & -1 \\ -6 & 2 \end{vmatrix} & -\begin{vmatrix} 11 & -1 \\ 1 & 2 \end{vmatrix} & -\begin{vmatrix} 11 & -1 \\ 1 & -6 \end{vmatrix} \end{bmatrix} = \begin{bmatrix} 38 & -9 & -8 \\ -9 & -76 & -23 \\ -8 & -23 & -65 \end{bmatrix},$$

which yields

$$\text{adj}(\mathbf{A}) = \mathbf{C}^T = \begin{bmatrix} 38 & 9 & 8 \\ -9 & -76 & -23 \\ -8 & -23 & -65 \end{bmatrix},$$

where $[\mathbf{A}]^{-1} = \frac{1}{\{\mathbf{A}}|\,\text{adj}(\mathbf{A})$, and $[\mathbf{b}] = [31 \ -26 \ -24]^T$. Hence, the vector $\mathbf{P} = [P_1 \ P_2 \ P_3]^T$ is given by

$$[\,P_1 \quad P_2 \quad P_3\,]^T = [\mathbf{A}]^{-1} = \frac{1}{|\mathbf{A}|} = [\mathbf{A}]^{-1}[\mathbf{b}]^T,$$

or

$$\begin{bmatrix} P_1 \\ P_2 \\ P_3 \end{bmatrix} = \frac{1}{401} \begin{bmatrix} 38 & -9 & -8 \\ 9 & 8 & -23 \\ 8 & -23 & -65 \end{bmatrix} \begin{bmatrix} 31 \\ -26 \\ -24 \end{bmatrix} = \frac{1}{401} \begin{bmatrix} 1604 \\ 2807 \\ 2306 \end{bmatrix} = \begin{bmatrix} 4 \\ 7 \\ 6 \end{bmatrix}.$$

METHOD 4. Using the simple elimination method, multiply Eq (a) by 2 to get $22P_1 - 2P_2 - 2P_3 = 62$, and subtract it from Eq (c) to get

$$\text{(d')} \quad 23P_1 - 19P_3 = 38.$$

Multiply Eq (b) by 3 and add to Eq (b), to get

$$\text{(e')} \quad 4P_1 - 19P_3 = -98.$$

Next, multiply Eq (d') by 19 and Eq (e') by 9, and subtract; this gives $401P_1 = 1604$, or $P_1 = 4$. Substitute this value of P_1 into Eq (e') to get $P_3 = 6$. Finally, substitute these values of P_1 and P_3 into any one of Eqs (a), (b) or (c), to get $P_2 = 7$.

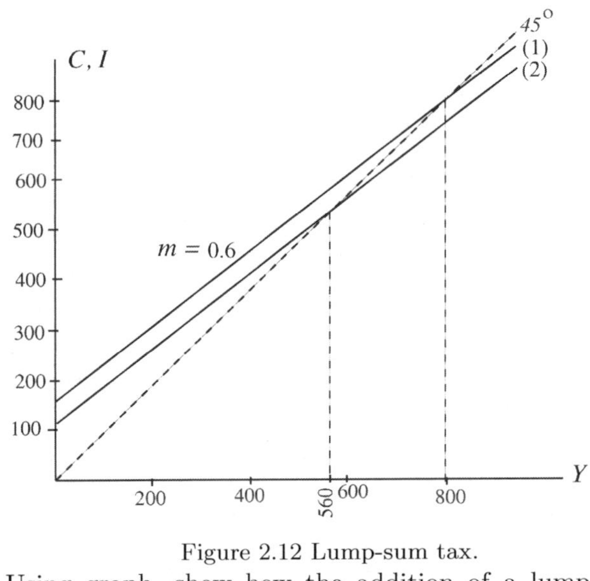

Figure 2.12 Lump-sum tax.

2.20. Using graph, show how the addition of a lump-sum tax (a tax independent of income) affects the parameters of the income determination model. Plot two systems: (1) $Y = C + I$, $C = 120 + 0.8Y$, $I_0 = 40$ (solid line marked (1) in Figure 2.12); and (2) $Y = C + I$, $C = 120 + 0.8Y$, $I_0 = 40$, $Y_d = Y - T$, $T = 60$ (sold line marked (2) in Figure 2.12).

SOLUTION. System (1): Aggregated demand function $D = C + I = 120 + 0.8Y + 40 = 160 + 0.8Y$; slope$= 0.8$; solving $160 + 0.8Y = y$ yields the income of 800.

System (2): Aggregated demand function $D = C + I = 120 + 0.8Y_d + 40 = 120 + 0.8(Y - 60) + 40 = 112 + 0.8Y$; slope $= 0.8$; solving $112 + 0.8Y = y$ yields the income of 560. Thus, the lump-sum tax has a negative effect on the vertical intercept of D equal to $-MPC(T) = -0.8(60) = -48$. As a result of this tax, income falls from 800 to 560.

2.21. (Optimization of economic functions) Maximize the profit π, given $R = 3500Q - 64Q^2$, and $C = 4Q^3 - 10Q^2 - 1300Q + 6000$. Thus,

$$\pi = R - C = -4Q^3 - 54Q^2 + 4800Q - 6000.$$

Then

$$\pi' = \frac{d\pi}{dQ} = -12Q^2 - 108Q + 4800 = -12(Q^2 + 9Q - 400) = -12(Q - 16)(Q + 25).$$

Equating π' to zero, we get the critical points as $Q = 16, -25$. Next,

$$\pi'' = -24Q - 108 \Longrightarrow \pi''(16) = -492 < 0,$$

which is concave and π has a relative maximum at $Q = 16$. Since $\pi''(-25) = 492 > 0$, this critical point is rejected. Also, $\pi(16) = -4(16)^3 - 54(16)^2 + 4800(16) - 6000 = 40492$.

2.22. The marginal expenditure function (ME) is associated with the supply function $P = 1 + 3Q + 4Q^2$. Find ME when $P = 3$ and $P = 5$.

SOLUTION. ME $= \dfrac{d(\text{TE})}{dQ}$. Here TE $= PQ = (1 + 3Q + 4Q^2)Q = Q + 3Q^2 + 4Q^3$. Then ME $= 1 + 6Q + 12Q^2$. At $Q = 2$, ME $= 1 + 6(2) + 12(4) = 61$; and at $Q = 5$, ME $= 1 + 6(5) + 12(25) = 331$.

2.23. Using graphs, show how the addition of a proportional tax (a tax depending on income, also known as *super tax*) affects the parameters of the income determination model. Plot two systems: (1) $Y = C + I$, $C = 90 + 0.8Y$, $I_0 = 40$ (solid line marked (1) in Figure 2.13); and (2) $Y = C + I$, $C = 90 + 0.8Y_d$, $I_0 = 40$, $Y_d = Y - T$, $T = 25 + 0.25Y$, where 25 is the lump-sum tax (sold line marked (2) in Figure 2.13).

SOLUTION. System (1): Aggregated demand function $D = C + I = 90 + 0.8Y + 40 = 130 + 0.8Y$; slope$= 0.8$; this yields the income of 650.

System (2): Aggregated demand function $D = C + I = 90 + 0.8Y_d + 40 = 130 + 0.8(Y - 25 - 0.25Y) = 110 + 0.6Y$; slope $= 0.6$; this yields the income of 275. Thus, the proportional tax affects not only the slope of the line, or the MCP, from $m = 0.8$ to $m = 0.6$, but also the vertical intercept that is lowered since the tax includes a lump-sum tax of 25. As a result of this tax, income falls from 650 to 275.

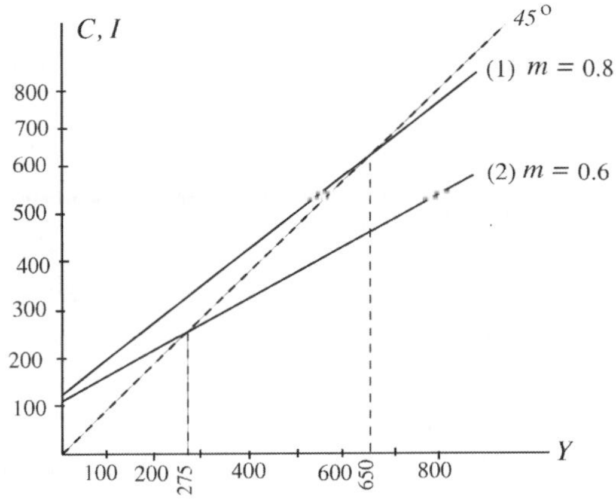

Figure 2.13 Proportional tax.

2.24. (a) Given $C = C_0 + bY$, we get MCP$= \dfrac{dC}{dY} = b$.

(b) Given $C = 1100 + 0.75Y_d$, where $Y_d = Y - T$, $T = 80$, we have $C = 1100 + 0.75(Y - 80) = 1160 + 0.75Y$. Then MCP $= \dfrac{dC}{dY} = 0.75$.

2.25. The average cost function is given by AC $= 1.6Q + 4 + \dfrac{44}{Q}$. Find the marginal cost MC. HINT. MC is determined by first finding TC $=$ (AC)Q, and then using the formula MC$= \dfrac{d(\text{TC})}{dQ}$. ANS. MC $= 3.2Q + 4$.

2.26. Optimize the following functions: (a) $y = \frac{1}{2}x^4 - 10x^3 + 100x^2 + 16$; (b) $y = -3x^4 - 28x^2 + 108x + 37$; (c) $y = -(x + 11)^4$; and (d) $y = (7 - 5x)^4$.

ANS. (a) Critical values $-5, 0, 20$; convex, and relative minimum at $x = -5, 20$; concave and relative maximum at $x = 0$. (b) Critical values $-9, 0, 2$; concave, and relative minimum at $x = -9, 2$; convex and relative minimum at $x = 0$. (c) Critical value -11; concave and relative maximum at $x = -11$. (d) Critical values $\frac{7}{5}$; convex, and relative minimum at $x = \frac{7}{5}$. HINT. In (c) the test fails at $y'(-11), y''(-11), y'''(-11)$, and in (d) the test fails at $y'(7/5), y''(7/5), y'''(7/5)$.

2.27. Maximize the following total revenue function TR and total profit function π by finding the critical value(s), by testing the second-order conditions, and by calculating the maximum TR or π: (a) $\pi = -Q^3 - 75Q^2 +$

$1800Q - 310$; (b) TC $= \frac{1}{2}Q^3 - 18Q^2 + 420Q$; (c) TR $= 1200Q - 8Q^2$, and TC $= 1400 + 80Q$; and (d) TR $= 6400Q + 15Q^2$, and TC $= Q^3 + 75Q^2 + 100Q + 754$.

ANS. (a) Critical values $Q = -60, 10$; $\pi''(-60) < 0$, $\pi''(10) > 0$, so concave and relative maximum is at $Q = -60$, and convex and relative minimum at $Q = 10$. (b) AC $= \left(\frac{1}{2}Q^3 - 18Q^2 + 420Q\right)/Q = \frac{1}{2}Q^2 - 18Q + 420$. Then AC$' = Q - 18$, so the critical value is $Q = 18$. Since AC$'' = 1$, the convex and relative minimum is at $Q = 18$. (c) $\pi =$ TR $-$ TC; the critical value is at $Q = 70$; since $\pi'' < 0$, we have concave and relative maximum at $Q = 70$. (d) $\pi = 6400Q + 15Q^2 - Q^3 - 75Q^2 - 100Q - 754 = -Q^3 - 60Q^2 + 6300Q - 754$, giving $\pi' = -3Q^2 - 120Q + 6300 = -3(Q^2 + 40Q - 2100)$. The critical values are $Q = 70, -30$, and $\pi''(70) < 0$, so the concave and relative maximum is at $Q = 70$; $\pi''(-30) > 0$, so convex and relative minimum is at $Q = -30$.

2.28. Let an isoquant be defined by $24K^{1/4}L^{3/4} = 2414$. Use implicit differentiation with respect to L to find the slope of the isoquant dK/dL, or the MRTS for given values of $K = 260$ and $L = 120$, and interpret the result.

SOLUTION. By implicit differentiation, we get $dK/dL = -3K/L$. Then MRTS $= -3(260)/120 = -6.5$. This means that if L is increased by 1 unit, K must be decreased by 6.5 units in order to retain the production isoquant when the production level is constant.

2.29. Let an isoquant be defined by $50K^{3/5}L^{2/5} = 5000$. Use implicit differentiation with respect to L to find the slope of the isoquant dK/dL, or the MRTS for given values of $K = 360$ and $L = 160$, and interpret the result.

ANS. $\dfrac{dK}{dL} = -\dfrac{2L}{3K}$, MRTS$= -1.5$

2.30. Consider the function $f(x, y) = 4x^2 - xy + y^2 - 4x - 7y + 10$. Find the extrema at the critical points for this function.

SOLUTION. We have $f_x = 8x - y - 4 = 0$, $f_y = -x + 2y - 7 = 0$, solving which we get the critical point $(1, 4)$. Next, $f_{xx} = 8$, $f_{yy} = 2$, $f_{xy} = -1 = f_{yx}$, so $f_{xx}(1, 4) \cdot f_{yy}(1, 4) = 16 > [f_{xy}(1, 4)]^2 = 1$. Hence, the function f has a global minimum at $(1, 4)$. The same results are obtained by using the Hessian at each critical point:

$$\text{At } (1,4): \quad |\mathbf{H}| = \begin{vmatrix} 8 & -1 \\ -1 & 2 \end{vmatrix}.$$

2.31. Consider the function $f(x, y) = 36y - 4x^2 - 8xy - 2y^2 + 72x$. Find the extrema at the critical points for this function.

SOLUTION. We have $f_x = -8x - 8y + 72 = 0$, $f_y = 36 - 8x - 4y = 0$, solving which we get the critical point $(0, 9)$. Next, $f_{xx} = -8$, $f_{yy} = -4$, $f_{xy} =$

$-8 = f_{yx}$, so $f_{xx}(0,9) \cdot f_{yy}(0,9) = 32 < [f_{xy}(1,4)]^2 = 64$. Hence, the function f has an inflection point at $(0,9)$.

2.32 Consider the function $f(x,y) = 6x^2 - 3y^2 - 24x + 6y + 6xy$. Find the extrema at the critical points for this function.

SOLUTION. We have $f_x = 12x - 24 + 6y = 0$, $f_y = -6y + 6 + 6x = 0$, solving which we get the critical point $(1,2)$. Next, $f_{xx} = 12$, $f_{yy} = -6$, $f_{xy} = 6 = f_{yx}$. Since f_{xx} and f_{yy} are of different signs, we get $f_{xx} \cdot f_{yy} = -72 < (f_{xy})^2 = 36$. Hence, the function f has a saddle point at $(1,2)$. The same results are obtained by using the Hessian at the critical point:

$$\text{At } (1,2): \quad |\mathbf{H}| = \begin{vmatrix} 12 & 6 \\ 6 & -6 \end{vmatrix}.$$

2.33. Consider the function $f(x,y) = 2x^3 + 3y^3 + 18y^2 - 90x - 189y$. Find the extrema at the critical points for this function.

SOLUTION. We have $f_x = 6x^2 - 12x - 90 = 0$, $f_y = 9y^2 + 36y - 189 = 0$, solving which we get the critical points $(-3,3), (-3,-7), (5,3), (5,-7)$. Next, $f_{xx} = 12x - 12$, $f_{yy} = 18y + 36$, $f_{xy} = 0 = f_{yx}$. Then

$$\begin{aligned}
&(1)\ f_{xx}(-3,3) = -48 < 0, && f_{yy}(-3,3) = 90 > 0,\\
&(2)\ f_{xx}(-3,-7) = -48 < 0, && f_{yy}(-3,-7) = -18 < 0,\\
&(3)\ f_{xx}(5,3) = 48 > 0, && f_{yy}(5,3) = -18 > 0,\\
&(4)\ f_{xx}(5,-7) = 48 > 0, && f_{yy}(5,-7) = -90 < 0.
\end{aligned}$$

Since f_{xx} and f_{yy} are of different signs in cases (1) and (4), we have saddle points at $(-3,3)$ and $(5,-7)$. Next, in case (2) we have $f_{xx} \cdot f_{yy} = (-48)(-18) > (0)^2$, so we have a relative maximum at $(-3,-7)$. Again, in case (3) we have $f_{xx} \cdot f_{yy} = (48)(18) > (0)^2$, so we have a relative minimum at $(5,3)$. The same results are obtained by using the Hessian at each critical point:

$$\text{At } (-3,3): \quad |\mathbf{H}| = \begin{vmatrix} -48 & 0 \\ 0 & 90 \end{vmatrix}; \quad \text{At } (-3,-7): \quad |\mathbf{H}| = \begin{vmatrix} -48 & 0 \\ 0 & -18 \end{vmatrix};$$

$$\text{At } (5,3): \quad |\mathbf{H}| = \begin{vmatrix} 48 & 0 \\ 0 & -18 \end{vmatrix}; \quad \text{At } (5,-7): \quad |\mathbf{H}| = \begin{vmatrix} 48 & 0 \\ 0 & -90 \end{vmatrix}.$$

2.34. The equation of a production isoquant is given as $8K^{1/4}L^{3/4} = 1008$. (a) Use implicit differentiation to find the MRTS $\dfrac{dK}{dL}$; and (b) evaluate the MRTS at $K = 238$ and $L = 183$.

ANS. (a) $dK/dL = -3K/L$; (b) MRTS $= -(3)(238)/183 = -3.9$. Hence, an increase in L by 1 unit will result in a decrease of 3.9 units in K, in order to remain on the production isoquant.

2.35. The elasticity of substitution σ, $0 \leq \sigma \leq \infty$, measures the percentage change in the least-cost (K/L) input ratio resulting from a small percentage change in the input-price ratio (P_L/P_K). If $\sigma = 0$, there is no substitution. A Cobb-Douglas function has a constant elasticity of substitution (CES), defined by $q = A\left[\alpha K^{-\beta} + (1-\alpha)L^{-\beta}\right]^{-1/\beta}$, where $A > 0$ is the efficiency parameter, α $(0 < \alpha < 1)$ is the distribution parameter denoting relative factor shares, and $\beta > -1$ is the substitution parameter that determines the value of the elasticity of substitution. Prove that the CES production function $f(kK, kL)$ is a homogeneous function of degree 1.

SOLUTION. Multiply inputs K and L in the expression for q by a constant $k > 0$, to get

$$\begin{aligned}
f(kK, kL) &= A[\alpha(kK)^{-\beta} + (1-\alpha)(kL)^{-\beta}]^{-1/\beta} \\
&= A[k^{-\beta}(\alpha K^{-\beta} + (1-\alpha)L^{-\beta})]^{-1/\beta} \\
&= A(k^{-\beta})^{-1/\beta}[\alpha K^{-\beta} + (1-\alpha)L^{-\beta}]^{-1/\beta} \\
&= kA[\alpha K^{-\beta} + (1-\alpha)L^{-\beta}]^{-1/\beta} = kq.
\end{aligned}$$

3

Concave and Convex Functions

The concept of a convex set is used to define concave and convex functions. Although the names appear similar, a convex set should not be confused with a convex function. However, a concave and a convex function are defined in terms of a convex set.

3.1 Convex Sets

Let X denote a vector space over a field F. Let $\|\cdot\|$ denote a norm on X, which is a mapping from X into $\mathbb{R}$ such that for every $x, y \in X$ and every $t \in F$, the following conditions are satisfied: (i) $\|x\| \geq 0$; (ii) $\|x\| = 0$ iff $x = 0$; (iii) $\|tx\| = |t| \cdot \|x\|$; and $\|x + y\| \leq \|x\| + \|y\|$. Then X is called a *real normed linear space*.

Let $x, y \in X$, and let a set $\overline{xy}$ be defined by

$$\overline{xy} = \{z \in X : z = tx + (1-t)y \text{ for all } t \in \mathbb{R},\ 0 \leq t \leq 1\}. \tag{3.1.1}$$

The set $\overline{xy}$ is the *line segment* joining the points x and y in X. Then we have the following definition: The set $Y \subset X$ is said to be a *convex set* if Y contains the line segment $\overline{xy}$ whenever x and y are two arbitrary points in Y (see Figure 3.1). A convex set is called a *convex body* if it contains at least one interior point, i.e., if it completely contains some sphere.

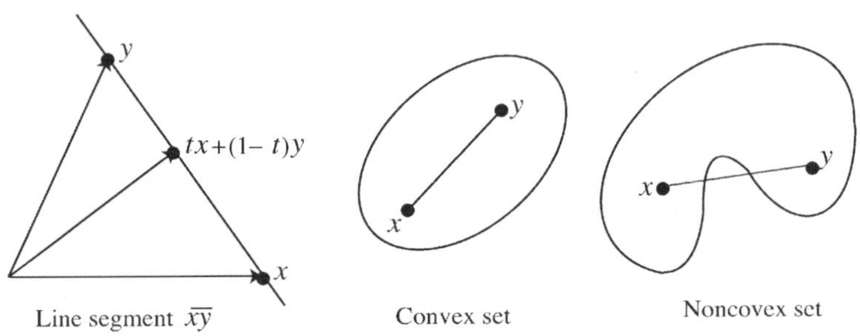

Figure 3.1 Convex and non-convex sets.

Example 3.1. The following sets are convex: (i) the empty set; (ii) a set containing one point; a line segment and a plane in $\mathbb{R}^3$; and (iv) any linear subspace of X, and (v) a cube and a sphere in $\mathbb{R}^3$. ■

Example 3.2. Let Y and Z be convex sets in X, let $\alpha, \beta \in \mathbb{R}$, and let $\alpha Y = \{x \in X : x = \alpha y, y \in Y\}$, and $\beta Z = \{x \in X : x = \beta Z, z \in Z\}$. Then the set $\alpha Y + \beta Z = \{x \in X : x = \alpha y + \beta z, y \in Y, z \in Z\}$ is a convex set in X. ■

Theorem 3.1. *Let Y be a convex set in X, and let $\alpha, \beta \in \mathbb{R}$ be positive scalars. Then $(\alpha+\beta)Y = \alpha Y + \beta Y$.*

PROOF. If $x \in (\alpha+\beta)Y$, then $x = (\alpha+\beta)y = \alpha y + \beta y \in \alpha Y + \beta Y$. Hence, $(\alpha+\beta)Y \subset \alpha Y + \beta Y$. Let Y be convex, and let $x = \alpha y + \beta z$, where $y, z \in Y$. Then since $\dfrac{\alpha}{\alpha+\beta} + \dfrac{\beta}{\alpha+\beta} = \dfrac{\alpha+\beta}{\alpha+\beta} = 1$, we get

$$\frac{1}{\alpha+\beta}x = \frac{\alpha}{\alpha+\beta}y + \frac{\beta}{\alpha+\beta}z \in Y.$$

Hence, $x \in (\alpha+\beta)Y$, which gives $\alpha Y + \beta Y \subset (\alpha+\beta)Y$. ■

This theorem implies that the intersection of an arbitrary collection of convex sets is also a convex set.

Let Y be a subset of X. Then the *convex hull* of Y, denoted by Y_c, is the intersection of all convex sets which contain Y. This convex hull is also called the *convex cover* of Y, and therefore, it is always a convex set. Some convex hulls are presented in Figure 3.2. The convex hull is also known as *convex envelope*.

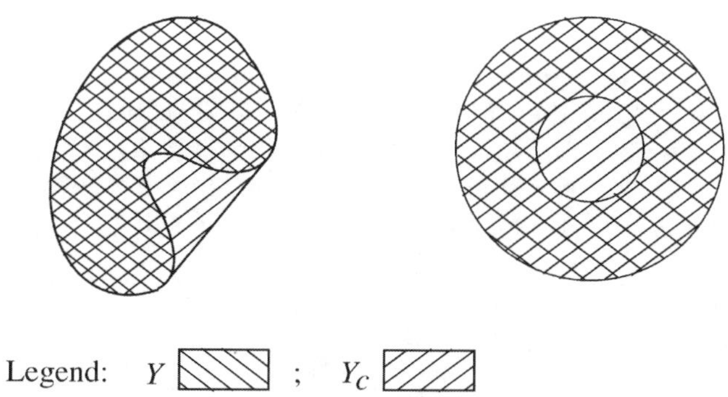

Figure 3.2 Convex hulls.

Theorem 3.2. *Let Y be a subset of X. The convex hull of Y is the set of points $\alpha_1 y_1 + \alpha_2 y_2 + \cdots + \alpha_k y_k$, where $y_1, \dots, y_k \in Y$, and $\alpha_i > 0$, $i = 1, \dots, k$, such that $\sum_{i=1}^{k} \alpha_i = 1$, where k is not fixed.*

PROOF. If Z is the set of points as described in the theorem, then obviously Z is convex, and also $Y \subset Z$. Hence, $Y_c \subset Z$. Now, we will show that $Z \subset Y_c$, i.e., we show that Z is contained in every convex set which contains Y. We will use the method of induction on the number of elements of Y that appear in the representation of an element of Z. Suppose U is a convex set such that $Y \subset U$. If $z = \alpha_1 z_1 \in Z$ for $n = 1$, then $\alpha_1 = 1$ and $z \in U$. Now assume that an element of Z is in U if it is represented in terms of $(n-1)$ elements of Y. Then, let $z = \alpha_1 z_1 + \cdots + \alpha_n z_n$ be in Z, and let $\beta = \alpha_1 + \cdots + \alpha_{n-1}$, and take $\beta_i = \alpha_i/\beta$, $i = 1, \dots, n-1$. Further, let $u = \beta_1 z_1 + \cdots + \beta_{n-1} z_{n-1}$. Then $u \in U$, by induction. However, $z_n \in U$, $\alpha_n = 1-\beta$, and $z = \beta u + (1-\beta) z_n \in U$, since U is convex. Thus, by induction, $Z \subset U$, which implies that $Z \subset Y_c$. ■

Corollary 3.1. (i) Let Y be a convex set in X. Then the closure $\bar{Y}$ of Y is also a convex set; and (ii) since the intersection of finitely many closed sets is always closed, the intersection of finitely many closed convex sets is also a closed convex set.

A set Y in X is called a *cone* with vertex at the origin if $y \in Y$ implies $\alpha y \in Y$ for all $\alpha \geq 0$. Hence, if Y is a cone with vertex at the origin, then the set $x_0 + Y$, where $x_0 \in X$, is called a *cone with vertex* x_0. A convex cone is a set which is both convex and a cone. Some examples of cones are shown in Figure 3.3.

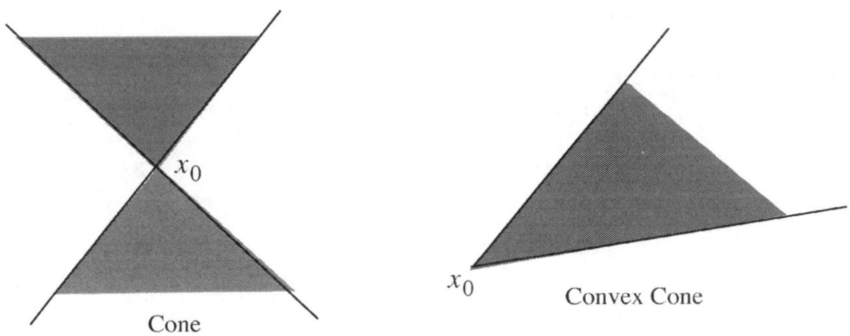

Figure 3.3 (a) Cone. (b) Convex cone.

Theorem 3.3. *Any sphere in X is a convex set.*

PROOF. Without loss of generality, consider the unit sphere $Y = \{x \in X : \|x\| < 1\}$. If $x_0, y_0 \in Y$, then $\|x_0\| < 1$ and $\|y_0\| < 1$. Now if $\alpha \geq 0$ and $\beta \geq 0$, where $\alpha + \beta = 1$, then $\|\alpha x_0 + \beta y_0\| \leq \|\alpha x_0\| + \|\beta y_0\| = \alpha\|x_0\| + \beta\|y_0\| < \alpha + \beta = 1$, and therefore, $\alpha x_0 + \beta y_0 \in Y$. ■

3.2 Concave Functions

Note that a function f is convex iff $-f$ is concave. Concave and convex functions can be defined in terms of convex sets known as hypographs and epigraphs, respectively. The *hypograph*[1] hyp(f) of a real-valued function $f : \mathbb{R}^n \mapsto \mathbb{R}$ is defined as the *area* below f. It is a set in $\mathbb{R}^{n+1}$ given by

$$\text{hyp}(f) = \{(x, y) \,|\, y \leq f(x), \tag{3.2.1}$$

Thus, a function f is said to be *concave* if hyp(f) is a convex set (Figure 3.4). This definition implies that a function f is concave if for any $x, x' \in \text{dom}(f)$,

$$f(tx + (1-t)x') \geq tf(x) + (1-t)f(x') \quad \text{for } 0 \leq t \leq 1. \tag{3.2.2}$$

The left-hand side of (3.2.2) denotes the functional value of a convex function, which exceeds the combination of functional values on the right-hand side (see Figure 3.5).

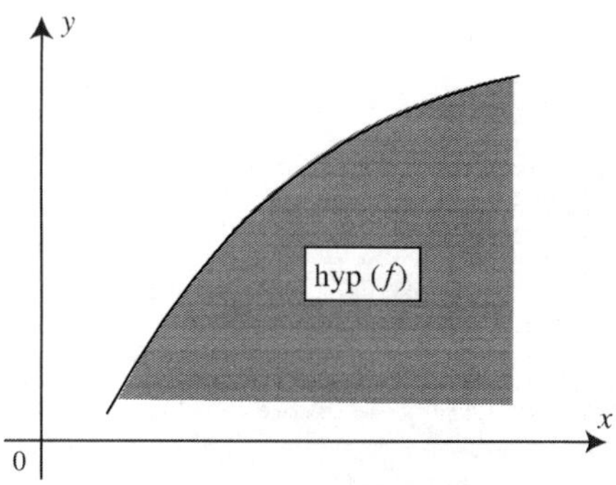

Figure 3.4 Concave function.

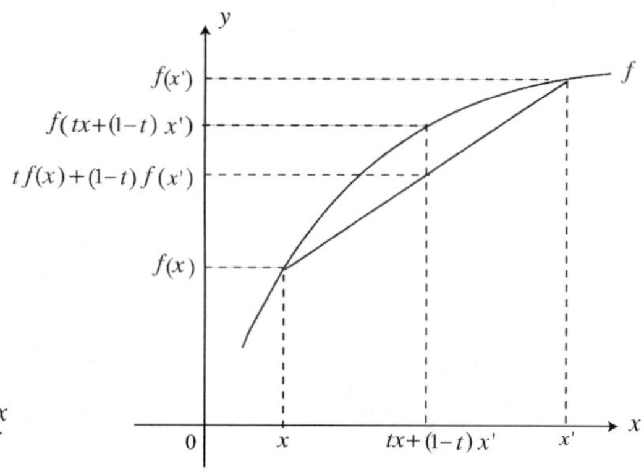

Figure 3.5 Definition (3.2.2).

Theorem 3.4. *The definitions (3.2.1) and (3.2.2) of a concave function are equivalent.*

PROOF. Let (x, x') be in hyp(f), where $y = f(x)$ and $y' = f(x')$. If hyp(f) is a convex set, then for any $0 \leq t \leq 1$ the convex combination $(tx + (1-t)x', ty + (1-t)y')$ is also in hyp(f). Thus,

$$tf(x) = (1-t)f(x') = ty + (1-t)y' \leq f(tx + (1-t)x'). \tag{3.2.3}$$

The left-side of this inequality, which is equal to $tf(x) + (1-t)f(x')$, implies (3.2.2). Hence, the inequality (3.2.1) implies (3.2.2). Conversely, assume that $tf(x) + (1-t)f(x') \leq f(tx + (1-t)x')$. Choose y and y' such that $y \leq f(x)$ and $y' \leq f(x')$. Obviously, (x, y) and (x', y') are both in hyp(f). Thus, $ty \leq tf(x)$, and $(1-t)y' \leq (1-t)f(x')$ for any $0 \leq t \leq 1$. But this implies that $ty + (1-t)y' \leq tf(x) + (1-t)f(x')$. Since the right side of this inequality

[1] The prefix *hypo*- or *hyp*– (from Greek and Latin) mean 'under' or 'beneath'.

is assumed to be not greater than $tf(x) + (1-t)f(x')$, we have the inequality (3.2.3).

$$ty + (1-t)y' \le f(tx + (1-t)x').$$

Hence, $(tx + (1-t)x', ty + (1-t)y') \in \text{hyp}(f)$, i.e., hyp$(f)$ is a convex set, and the inequality (3.2.2) implies (3.2.1), which completes the proof. ■

If the inequality in (3.2.3) is strict, then the function f is called a *strictly concave* function; i.e., for any $x, x' \in \text{dom}(f)$,

$$f(tx + (1-t)x') > tf(x) + (1-t)f(x'), \tag{3.2.4}$$

where $t \in (0,1)$. Thus, a function is strictly concave if its hypograph is a strictly convex set.

Theorem 3.5. (Combinations of concave functions) (i) *Let f and g be concave functions defined on a convex subset of $\mathbb{R}^n$. Then their sum $f+g$ is a concave function; also if one of them is strictly concave, then the sum $f+g$ is strictly concave.*

(ii) Let f be a (strictly) concave function on a convex subset of $\mathbb{R}^n$, and α be a positive scalar. Then, αf is a (strictly) concave function.

(iii) An affine combination of concave functions is a concave function, i.e., $\alpha f + \beta g$ is a concave function on a subset of $\mathbb{R}^n$, where $\alpha, \beta \ge 0$.

(iv) Let f be a (strictly) concave function on a convex subset of $\mathbb{R}^n$, and let g be a strictly increasing concave function defined on $\mathcal{R}(f) \in \mathbb{R}$. Then, the composite function $f \circ g$ is a (strictly) concave function.

(v) Let f and g be concave functions on a convex subset of $\mathbb{R}^n$, and bounded from below. Then, the pointwise infimum function $\min\{f(x), g(x)\}$ *is a concave function.*

(vi) Let f and g be concave functions on a convex subset of $\mathbb{R}^n$, and let S be a subset of the domain space dom(f). *Then, f is continuous on the interior of S, except possibly a point of discontinuity (singularity) only on the boundary ∂S.*

(vii) Let f be a function defined on a convex subset of $\mathbb{R}^n$. Then f is concave iff its restriction to every chord in the convex domain is a concave function.

Geometrically, part (vii) is about a convex slice that a vertical hyperplane cuts out of the hypograph. Recall that a function is concave iff its hypograph is a convex set, and a hypograph is convex if every hyperplane intersecting it produces a slice that is a convex set.

3.2.1 Properties of Concave Functions

1. A differentiable function f is concave on an interval I if its derivative f' is monotone decreasing on I; i.e., a concave function has a non-increasing slope (i.e., not decreasing but allowing zero slope).

2. Points where concavity changes (between concave and convex) are inflection points.

3. Near a local maximum in the interior of the domain of a function, the function must be concave; as a partial converse, if the derivative of a strictly concave function is zero at some point, then the point is a local maximum.

4. If f is twice-differentiable, then f is concave iff f'' is nonpositive (or, if the acceleration is nonpositive). If f'' is negative, then it is strictly concave, but the converse is not true as shown by $f(x) = -x^4$.

5. Any local maximum of a concave function is also a global maximum. A strictly concave function will have at most one global maximum.

6. If f is concave and differentiable, then it is bounded above by its first-order Taylor approximation

$$f(y) \leq f(x) + f'(x)(y - x).$$

7. A continuous function on C is concave iff for any $x, y \in C$

$$f\Big(\frac{x+y}{2}\Big) \leq \frac{f(x) + f(y)}{2}.$$

8. If a function f is concave, and $f(0) = 0$, then f is subadditive.
PROOF. Since f is concave, let $y = 0$, then $f(tx) = f(tx + (1-t)\cdot 0) \geq tf(x) + (1-t)\cdot 0 \geq tf(x)$. Also,

$$\begin{aligned} f(a) + f(b) &= f\Big((a+b)\frac{a}{a+b}\Big) + f\Big((a+b)\frac{b}{a+b}\Big) \\ &\geq \frac{a}{a+b} f(a+b) + \frac{b}{a+b} f(a+b) = f(a+b). \ \blacksquare \end{aligned}$$

Example 3.3. Some examples of concave and convex functions are

1. The functions $f(x) = -x^2$ and $g(x) = \sqrt{x}$ are concave on their domains, as their second derivatives $f''(x) = -2$ and $g''(x) = -\frac{3}{4}x^{-5/4}$ are always negative.

2. The logarithm function $f(x) = \log x$ is concave on its domain $(0, \infty)$, as its derivative $f'(x) = 1/x$ is strictly decreasing function.

3. Any affine function $f(x) = ax + b$ is both concave and convex, but not strictly concave nor strictly convex.

4. The sine function $f(x) = \sin x$ is concave on the interval $[0, \pi]$.

3.3 Jensen's Inequality for Concave Functions

Jensen's inequality is a general form of the definition (3.2.2) of a concave function; it uses general convex combinations. It states that a real-valued function $f : \mathbb{R}^n \mapsto \mathbb{R}$ is concave iff the function value of a convex combination is at least as large as the convex combination of functional values. Thus, a real-valued function $f : \mathbb{R}^n \mapsto \mathbb{R}$ is concave iff the function value of a concave combination is larger than the concave combination of the functional value, i.e., iff for any k vectors $\mathbf{x}_1, \dots, \mathbf{x}_k \in \text{dom}(f)$ and $t_1 \geq 0, \dots, t_k \geq 0$, $t_1 + \cdots + t_k = 1$, we have

$$f(t_1\mathbf{x}_1 + \cdots + t_k\mathbf{x}_k) \geq t_1 f(\mathbf{x}_1) + \cdots + t_k f(\mathbf{x}_k). \tag{3.3.1}$$

This inequality can also be written as $f\left(\sum_i^k t_i\mathbf{x}_i\right) \geq \sum_i^k t_i f(\mathbf{x}_i)$ for $\sum_i^k t_1 = 1$ and $t_i \geq 0$ for $i = 1, \dots, k$.

Besides the above general definition of Jensen's inequality, there are other forms for this inequality for concave functions. Let $f : \mathbb{R}^n \mapsto \mathbb{R}$ be a concave function. Then

(i) For two points $x, x' \in \mathbb{R}$, Jensen's inequality is

$$f(tx + (1-t)x') \geq tf(x) + (1-t)f(x'), \tag{3.3.2}$$

where $t \in [0, 1]$. This is inequality (3.2.2) valid for concave functions defined on $\mathbb{R}$.

(ii) For more than two points $x_i \in \mathbb{R}$, $i = 1, \dots, k$, Jensen's inequality is

$$f\Big(\sum_i^k t_i x_i\Big) \geq \sum_i^k t_i f(x_i), \tag{3.3.3}$$

where $\sum_i^k t_i = 1$, $t_i \geq 0$.

(iii) Let $p(x) \geq 0$ such that $\int p(x)\,dx = 1$. Then the continuous form of Jensen's inequality is

$$f\Big(\int x\,p(x)\,dx\Big) \geq \int f(x)p(x)\,dx. \tag{3.3.4}$$

(iv) For any probability distribution on x, Jensen's inequality is

$$f(\mathbf{E}x) \geq \mathbf{E}f(x). \tag{3.3.5}$$

Jensen's inequality (Figure 3.6) can be interpreted as follows: The (zero mean) randomization decreases the average value of a concave function.

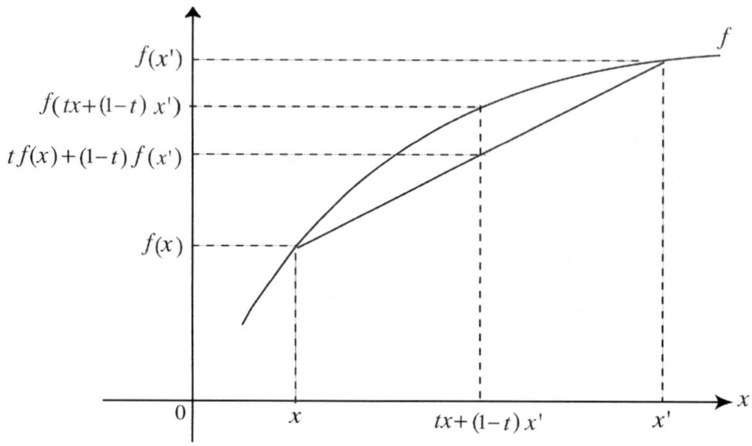

Figure 3.6 Jensen's inequality.

This inequality is a basic result which has produced many other useful inequalities. For example, we have the arithmetic-geometric mean inequality:

$$\frac{a+b}{2} \geq \sqrt{ab} \quad \text{for } a, b \geq 0, \tag{3.3.6}$$

which can be easily proved by using the inequality $\log\left(\frac{a+b}{2}\right) \geq \frac{1}{2}(\log a + \log b)$ for the function $f(x) = \log x$, which is concave for all $x > 0$.

3.4 Convex Functions

Let $f : \mathbb{R}^n \mapsto \mathbb{R}$. Then the function f is said to be *convex* if for all $x, y \in \text{dom}(f)$,

$$f(tx + (1-t)y) \leq tf(x) + (1-t)f(y), \tag{3.4.1}$$

where $0 \leq t \leq 1$. The definition (3.4.1) simply says that a convex combination of the functional values is greater than or equal to their convex combinations (see Figure 3.7).

Convex functions can be defined in terms of the convex set called *epigraphs*. The epigraph[2] epi(f) of a real-valued function $f : \mathbb{R}^n \mapsto \mathbb{R}$ is defined as the area above f (see Figure 3.8). It is a set in $\mathbb{R}^n$, defined by

$$\text{epi}(f) = \{(x, y) : y \geq f(x)\}. \tag{3.4.2}$$

[2] The prefix *epi-* from Greek and Latin means 'over' or 'upon.'

Thus, a function f is said to be *convex* if epi(f) is a convex set. This definition implies that a function f satisfies the inequality (3.4.1). Obviously,

A function f is convex if and only if $-f$ is concave.

An extended-value extension of a convex function f is defined as

$$\tilde{f}(x) = \begin{cases} f(x) & \text{if } x \in \text{dom } (f), \\ +\infty & \text{if } x \notin \text{dom}(f). \end{cases} \tag{3.4.3}$$

Then the inequality

$$\tilde{f}(tx + (1-t)y) \leq t\tilde{f}(x) + (1-t)\tilde{f}(y) \tag{3.4.4}$$

holds for all $x, y \in \mathbb{R}^n$, $t \in [0,1]$. The inequality (3.4.4) is an extension of (3.4.1) defined on $\mathbb{R} \cup \{+\infty\}$.

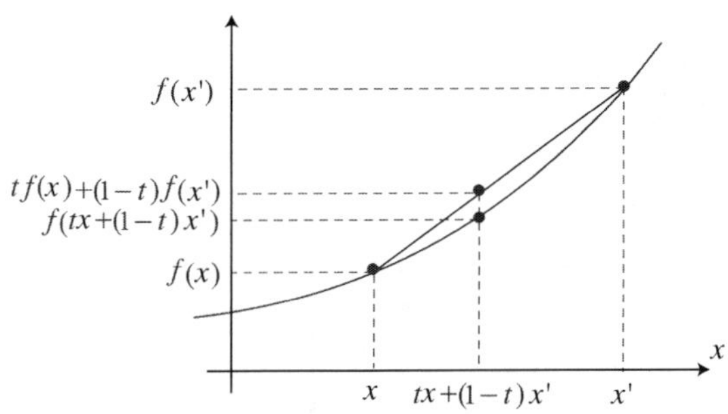

Figure 3.7 Definition (3.4.1).

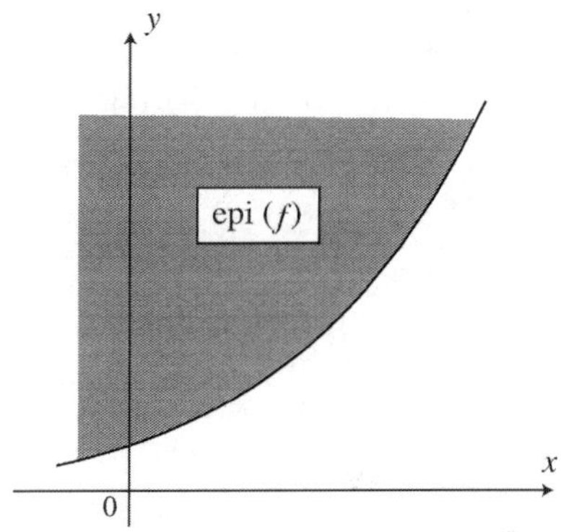

Figure 3.8 Convex set epi(f).

Note that the property (3.4.1) is, in many cases, weakened by requiring that

$$f(tx + (1-t)x') \leq \max\{f(x), f(x')\} \tag{3.4.5}$$

for all $x, x' \in X$ and $t \in [0,1]$. The inequality (3.4.5) is known as the *modified Jensen's inequality*.

We will assume that all convex functions are extendable, and hence, use the same notation f for a convex function as well as its extension $\tilde{f}$.

Let $f : X \mapsto Y$ denote the mapping of a set $X \in \mathbb{R}^n$ into another set $Y \in \mathbb{R}^n$. Then a mapping f_1 of a subset $X_1 \subset X$ into Y is called the *restriction* of f to the set X_1. This definition leads to the following result:

Theorem 3.6. *Let f be a convex function defined on a convex subset of $\mathbb{R}^n$. Then f is convex iff its restriction to every chord in the convex domain set is a convex function.*

Geometrically, a function is convex iff its epigraph is a convex set. An epigraph is convex if every hyperplane intersecting it produces a convex shaped slice. This theorem is about a convex slice which is cut out of the epigraph by a vertical hyperplane.

A function f is *strictly convex* if the strict inequality holds in the definition (3.4.1).

3.4.1 Properties of Convex Functions. Some useful properties of convex functions are as follows.

(i) Given two convex functions f and g defined on a convex subset of $\mathbb{R}^n$, their sum $f + g$ is a convex function. Moreover, if at least one of them is strictly convex, the sum $f + g$ is strictly convex. This property extends to infinite sums and integrals, i.e., if $f_1, \dots f_n, \dots$ are convex functions, then $\sum\limits_{n=1}^{\infty} f_n$ is convex; and if $g(x, y)$ is convex in x, then $\int g(x, y)\, dy$ is convex.

(ii) Given f as a (strictly) convex function on a convex subset of $\mathbb{R}^n$, and a positive scalar α, the function αf is (strictly) convex.

(iii) An affine combination of convex functions is again a convex function, i.e., if f and g are both convex functions on a convex subset of $\mathbb{R}^n$, then (a) $\alpha f + \beta g$, where $\alpha, \beta \geq 0$, is a convex function; and (b) if f is convex, then $f(Ax + b)$ is convex. This is called *affine transformation of the domain.*

(iv) Let f be a (strictly) convex function on a convex subset of $\mathbb{R}^n$, and g be a strictly increasing convex function defined on $\mathcal{R}(f)$ in $\mathbb{R}$. Then the composite function $f \circ g$ is a (strictly) convex function.

(v) Let f_1 and f_2 be convex functions on a convex subset of $\mathbb{R}^n$, and bounded from above. Then the pointwise supremum function $\max\{f_1(x), f_2(x)\}$ is a convex function. This property corresponds to the intersection of epigraphs (see Exercise 3.5).

(vi) A function f is convex iff it is convex on all lines, i.e., f is convex at x_0 iff $f(x_0 + th)$ is convex in t for all x_0 and h.

(vii) A positive multiple of a convex function is convex, i.e., f is convex iff αf is convex for $\alpha \geq 0$.

(viii) If $\{f_a\}_{\alpha \in \mathcal{A}}$ is convex, then $\sup_{\alpha \in \mathcal{A}}\{f_a\}$ is convex. This is known as *pointwise supremum*; see property (v).

Example 3.4. (i) The piecewise-linear function $f(x) = \max\limits_{i}\{A_i^T x_i + b_i\}$ is convex, and its epi (f) is a polyhedron;
(ii) $\sup_{s \in S} \|x - s\|$, which is the maximum distance to any set, is convex in x;
(iii) $f(x) = x_{\lfloor 1 \rfloor} + x_{\lfloor 2 \rfloor} + x_{\lfloor 3 \rfloor}$ is convex on $\mathbb{R}^n$, where $x_{\lfloor i \rfloor}$ is the ith largest integer x; and

(iv) $f(x) = \sum_{i=1}^{m} \log(b_i - a_i^T x)^{-1}$ is convex, where dom $(f) = \{x \mid a_i^T x < b_i,\ i = 1, \dots, m\}$. ■

3.4.2 Jensen's Inequality for Convex Functions. Let $f : \mathbb{R}^n \mapsto \mathbb{R}$ be a convex function. Then Jensen's inequality for convex functions is defined in the four cases discussed in §3.3, except that the $\geq$ sign in inequalities (3.3.1) through (3.3.5) is replaced by the $\leq$ sign.

Jensen's inequality for convex functions is presented in Figure 3.9. It can be interpreted as follows: The (zero mean) randomization increases the average value of a convex function.

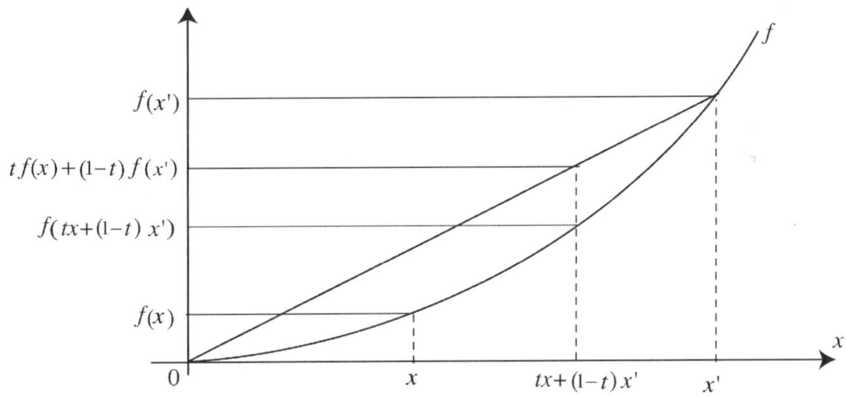

Figure 3.9 Jensen's inequality.

As in the case of concave functions, a general definition of convex functions, based on Jensen's inequality, uses general convex combinations. Thus, a real-valued function $f : \mathbb{R}^n \mapsto \mathbb{R}$ is convex iff the function value of a convex combination is not larger than the convex combination of the functional value, i.e., iff for any k vectors $\mathbf{x}_1, \dots, \mathbf{x}_k \in \text{dom}(f)$ and $t_1 \geq 0, \dots, t_k \geq 0$, $t_1 + \dots + t_k = 1$,

$$f(t_1\mathbf{x}_1 + \dots + t_k\mathbf{x}_k) \leq t_1 f(\mathbf{x}_1) + \dots + t_k f(\mathbf{x}_k). \tag{3.4.6}$$

Example 3.5. In $\mathbb{R}$, (i) x^α is convex on $\mathbb{R}_+$ for $\alpha \geq 1$ and $a \leq 0$, and concave for $0 \leq \alpha \leq 1$.

(ii) $\log x$ is concave on $\mathbb{R}_+$, but $x \log x$ is concave on $\mathbb{R}_{++}$, because $\lim_{x \to 0} x \log x = 0$.

(iii) e^{ax} is convex.

(iv) $|x|$, and $\max(0, \pm x)$ are convex, but $\log \int_{-x}^{x} e^{-t^2}\, dt$ is concave.

(v) $f(x) = e^{g(x)}$ is convex if g is convex; $f(x) = 1/g(x)$ is convex if g is concave and positive; $f(x) = g(x)^p$, $p \geq 1$, is convex if $g(x)$ is convex and positive; and $f(x) = -\sum_i \log(-f_i(x))$ is convex on $\{x|f_i(x) < 0\}$ if f_i are convex.

(vi) Since f is *concave* if $-f$ is convex, the function $f(x) = x^2$ is convex for all $x \in \mathbb{R}$; the function $f(x) = 1/x$ is convex for $x \in \mathbb{R}_+$; and the function $f(x) = \log x$ is concave for $x \in \mathbb{R}_+$.

(vi) If $f(x) = (h \circ g)(x) = h(g(x))$, where $g : \mathbb{R} \mapsto \mathbb{R}$ and $h : \mathbb{R} \mapsto \mathbb{R}$, is a composition of two functions g and h in $\mathbb{R}$, then (a) if g and h are both convex, then f is nondecreasing, and (b) if g and h are both concave, then f is nonincreasing.

PROOF. For differentiable functions f, g, h, if $f(x) = h(g(x))$, then $f''(x) = h'(g(x))g''(x) + h''(x)(g'(x))^2$. ■

For the composition of three functions defined by $f(x) = h(g_1(g_2))(x)$, where f, h, g_1 and g_2 are differentiable, we have the following result.

Theorem 3.7. Let $f(x) = h(g_1(g_2))(x)$, where $h : \mathbb{R}^2 \mapsto \mathbb{R}$, and $g_i : \mathbb{R} \mapsto \mathbb{R}$. Then f is convex if h is a univariate convex and nondecreasing function in each argument, and g_1, g_2 are convex.

PROOF. For the composition $f(x) = h(g_1(g_2))(x)$,

$$f''(x) = h''(g_1(g_2))g_1'(g_2)g_2' + h'(g_1(g_2))\big[g_1'(g_2)g_2'' + g_1''(g_1')^2\big]. \blacksquare \tag{3.4.7}$$

This result can be extended to the k-composition of the form $f(x) = h(g_1(g_2 \cdots (g_k)))$.

Example 3.6. If each $g_i(x)$ is convex, then (i) $f(x) = \max_i g_i(x)$ is convex, and (ii) $f(x) = \log \sum_i e^{g_i(x)}$ is convex. ■

3.5 Differentiable Functions

Let $f : \mathbb{R}^n \mapsto \mathbb{R}$ be a concave or convex function. We say that a function f is differentiable at the point $\mathbf{x} = (x_1, x_2, \dots, x_n)$ if the first-order partial derivatives $\dfrac{\partial f}{\partial x_i}$, $i = 1, 2, \dots, n$, exist. Then the second-order partial derivatives of f are $\dfrac{\partial^2 f}{\partial x_i \partial x_j}$ for $i, j = 1, 2, \dots, n$.

The *first-order Taylor's series approximation* of f at $\mathbf{a} \in \mathbb{R}^n$ is defined as

$$f(\mathbf{x}) = f(\mathbf{a}) + \frac{\partial f(\mathbf{a})}{\partial \mathbf{x}}(\mathbf{x} - \mathbf{a}), \tag{3.5.1}$$

which can be written in matrix form as

$$[f(\mathbf{x})]_{1\times 1} = [f(\mathbf{a})]_{1\times 1} + \left[\frac{\partial f(\mathbf{a})}{\partial \mathbf{x}}\right]^T_{1\times n} [\mathbf{x}-\mathbf{a}]_{n\times 1}. \tag{3.5.2}$$

The *second-order Taylor's series approximation* of f at $\mathbf{a} \in \mathbb{R}^2$ in matrix form is

$$\begin{aligned} [f(\mathbf{x})]_{1\times 1} &= [f(\mathbf{a})]_{1\times 1} + [\mathbf{x}-\mathbf{a}]^T_{1\times n} \left[\frac{\partial f(\mathbf{a})}{\partial \mathbf{x}}\right]_{n\times 1} \\ &+ \frac{1}{2}[\mathbf{x}-\mathbf{a}]^T_{1\times n} \left[\frac{\partial^2 f}{\partial x_i\, \partial x_j}\right]_{n\times n} [(\mathbf{x}-\mathbf{a})]_{n\times 1}, \end{aligned} \tag{3.5.3}$$

for $i, j = 1, 2, \dots, n$. The term $\left[\dfrac{\partial^2 f}{\partial x_i\, \partial x_j}\right]$ in (3.5.3) represents the (i,j)th element of the Hessian matrix $\mathbf{H}$ for the function $f(\mathbf{a})$, and the vector $[\mathbf{x}-\mathbf{a}]^T$ is for index i while the second vector $[\mathbf{x}-\mathbf{a}]$ is for index j. In particular, in $\mathbb{R}^2$ with $\mathbf{x} = (x, y)$, the second-order Taylor approximation at a point $\mathbf{a} = (a, b)$ is given by

$$\begin{aligned} f(x,y) &= f(a,b) + \frac{\partial f(a,b)}{\partial x}(x-a) + \frac{\partial f(a,b)}{\partial y}(y-b) + \frac{1}{2}\Big[\frac{\partial^2 f(a,b)}{\partial x^2}(x-a)^2 \\ &+ 2\frac{\partial f(a,b)}{\partial x \partial y}(x-a)(y-b) + \frac{\partial^2 f(a,b)}{\partial y^2}(y-b)^2\Big]. \end{aligned} \tag{3.5.4}$$

The matrix form (3.5.3) can be compared with the summation form of the second-order Taylor's series approximation given in §B.3.

!!! *As a tradition, some authors define the first-order derivatives of f as $\nabla f(\mathbf{x})$, and the first-order Taylor's series approximation of $f(\mathbf{x})$ at a point $\mathbf{a}$ as*

$$f(\mathbf{x}) = f(\mathbf{a}) + \nabla f(\mathbf{x})(\mathbf{x}-\mathbf{a}), \tag{3.5.5}$$

which is misleading because it is based on the isomorphic equality (2.7.7). But isomorphism and equality are two different operations. If we rewrite this approximation as

$$f(\mathbf{x}) - f(\mathbf{a}) = \nabla f(\mathbf{x})(\mathbf{x}-\mathbf{a}),$$

we see that the left-hand side is a scalar, while the right-hand side is a vector; thus, this equality cannot be justified by any argument. Similarly, the second-order Taylor's series approximation is defined as

$$f(\mathbf{x}) = f(\mathbf{a}) + \nabla f(\mathbf{x})^T(\mathbf{x}-\mathbf{a}) + \frac{1}{2}(\mathbf{x}-\mathbf{a})^T \nabla^2 f(\mathbf{x})(\mathbf{x}-\mathbf{a}), \tag{3.5.6}$$

which is again abused on two counts: (i) the second term on the right side is already shown to be misleading, and (ii) the last term on the right involves $\nabla^2 f$*, which is the Laplacian of* f*, same as the trace of the Hessian matrix* **H***, and hence, it does not represent all second-order derivatives* $\dfrac{\partial^2 f}{\partial x_i \partial x_j}$ *for* $i, j = 1, 2, \dots, n$ *(for more details, refer to §1.6.2). In fact, the second-order Taylor approximation (3.5.6) does not reduce to the second-order Taylor approximation (3.5.4) for a function of two variables.*

3.6 Unconstrained Optimization

In view of §1.3-1.6, a multivariate function f is a relative minimum or maximum if the following three conditions are met:

(i) The first-order direct partial derivatives must be zero simultaneously, which means that at a critical point the function is neither increasing nor decreasing with respect to the principal axes.

(ii) The second-order partial derivatives, when calculated at the critical point, must be negative for a relative maximum and positive for a relative minimum. Thus, in a relative way with respect to the critical point the function is concave and moving downward in relation to the principal axes for a maximum, and convex and moving upward relative to the principal axes for a minimum.

(iii) The product of the second-order direct partial derivatives evaluated at the critical point must exceed the product of the cross partial derivatives also evaluated at the critical point. This condition is used to check for an inflection point or a saddle point.

Example 3.7. Consider $f(x, y) = 2y^3 - x^3 + 12x - 54y + 12$. Equating the first-order partial derivatives of f to zero, we get $f_x = -3x^2 + 12 = 0$, $f_y = 6y^2 - 54 = 0$, i.e., $x = \pm 2$ and $y = \pm 3$. Thus, the critical numbers are $(2, 3), (2, -3), (-2, 3), (-2, -3)$. Next, take the second-order direct partial derivatives and evaluate them at each of the four critical points to check for their signs:

$$f_{xx} = -6x, \quad f_{yy} = 12y.$$

Then

$$\begin{aligned}
&(1)\ f_{xx}(2, 3) = -6(2) = -12 < 0, && f_{yy}(2, 3) = 12(3) = 36 > 0, \\
&(2)\ f_{xx}(2, -3) = -6(2) = -12 < 0, && f_{yy}(2, -3) = 12(-3) = -36 < 0, \\
&(3)\ f_{xx}(-2, 3) = -6(-2) = 12 > 0, && f_{yy}(-2, 3) = 12(3) = 36 > 0, \\
&(4)\ f_{xx}(-2, -3) = -6(-2) = 12 > 0, && f_{yy}(-2, -3) = 12(-3) = -36 < 0.
\end{aligned}$$

Since there are different signs for each of the second direct partials in (1) and (4), the function f cannot be at a relative extremum at $(2, 3)$ and $(-2, -3)$.

However, since the signs of second partials are both negative in (2) and positive in (3) above, the function f may have a relative maximum at $(2,-3)$ and a relative minimum at $(-2,3)$. Since f_{xx} and f_{yy} are of different signs, the product of f_{xx} and f_{yy} cannot be greater than $(f_{xy})^2$.

Since $f_{xy} = 0 = f_{yx}$, we check $f_{xx} \cdot f_{yy} > (f_{xy})^2$ at the critical points $(2,-3)$ and $(-2,3)$:

$$f_{xx}(2,-3)\cdot f_{yy}(2,-3)=(-12)(-36) > 0, \; f_{xx}(-2,3)\cdot f_{yy}(-2,3)=(12)(36) > 0.$$

Thus, f has a relative maximum at $(2,-3)$ and a relative minimum at $(-2,3)$. ■

Example 3.8 Consider $f(x,y) = 3x^2 - xy + 2y^2 - 4x - 7y + 8$. The first-order partial derivatives, equated to zero, give: $f_x = 6x - y - 4 = 0$, $f_y = -x + 4y - 7 = 0$, solving which we get $x = 1, y = 2$. Thus, the critical number is $(1,2)$. The second-order partial derivatives are: $f_{xx} = 6, f_{xy} = f_{yx} = -1, f_{yy} = 4$. Checking the condition $f_{xx} \cdot f_{yy} > (f_{xy})^2$, we have $6 \cdot 4 > (-1)^2$. Since both f_{xx} and $_{yy}$ are positive, we have a global minimum at $(1,2)$. ■

Example 3.9. Consider $f(x,y) = 52x + 36y - 4xy - 6x^2 - 3y^2 + 5$. The first-order partial derivatives, equated to zero, give $f_x = 52 - 4y - 12x = 0$, $f_y = 36 - 4x - 6y = 0$, so the critical point is $(3,4)$. The second-order partial derivatives are: $f_{xx} = -12, f_{xy} = -4, f_{yy} = -6$. Since both $f_{xx} < 0$, $F_{yy} < 0$, and $(f_{xy})^2 = (-4)^2 > 0$, and $f_{xx} \cdot f_{yy} > (f_{xy})^2$ at the point $(3,4)$, the function f has a global maximum at $(3,4)$. ■

Example 3.10. Consider $f(x,y) = 48y - 3x^2 - 6xy - 2y^2 + 60x$. The first-order partial derivatives, equated to zero, give $f_x = -6x - 6y + 60 = 0$, $f_y = -6x - 4y + 48 = 0$, so the critical point is $(4,6)$, The second-order partial derivatives are: $f_{xx} = -6 < 0, f_{xy} = -6 = f_{yx} < 0, f_{yy} = -4 < 0$, and $f_{xx} \cdot f_{xy} = (-6)(-4) = 24 < (f_{xY})^2 = 36$. The function f has an inflection point at $(4,6)$. ■

Example 3.11. Optimize the following total profit function π:
(a) $\pi = -Q^2 + 15Q = 36$; and (b) $\pi = Q63 - 25Q^2 + 1200Q - 316$, where Q is the total output function.

ANS. (a) Critical point $Q = 7.5$; $\pi''(Q) = -2$, so $\pi''(7.5) = -2 < 0$, convex, relative minimum at $Q = 7.5$.

(b) $\pi = 3Q^2 - 50Q + 1200 = (3Q + 40)(Q - 30) = 0$, so critical points are $Q = -(40)/3, 30$; $\pi'' = 6Q - 30$, so $Q(-40/3) < 0$, concave, relative maximum; $Q(30) > 0$, convex, relative minimum.

3.7 Exercises

3.1. Some graphs of functions and domains are presented below in Figure 3.10. What can you say about each one of these graphs?

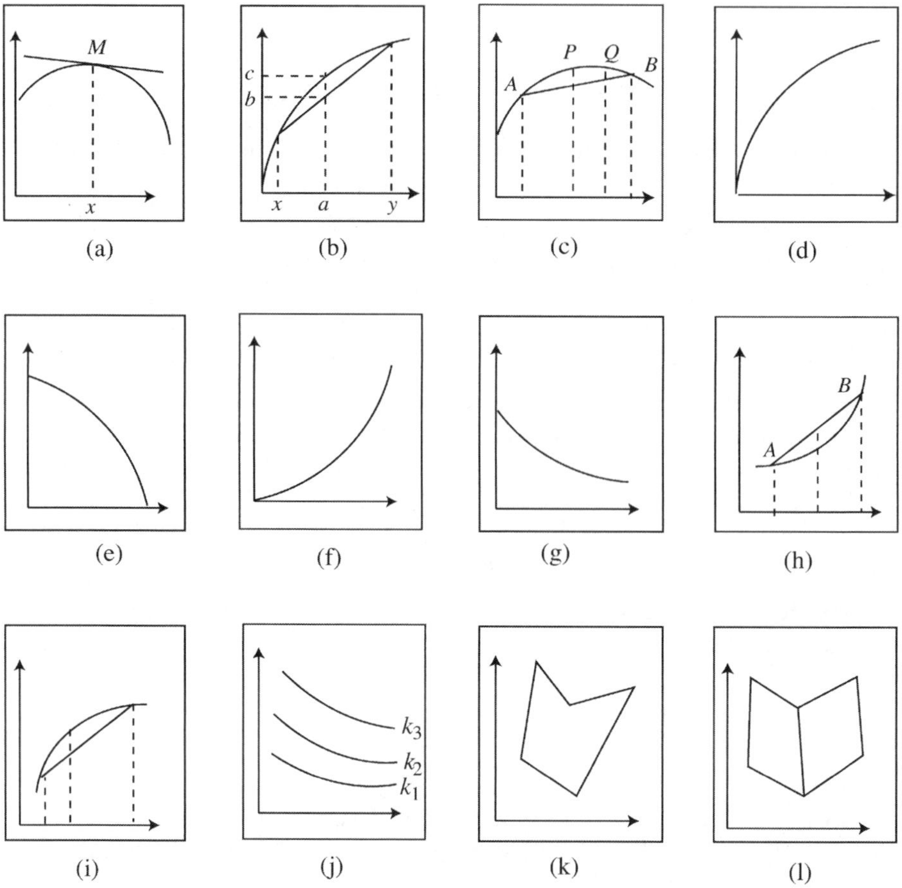

Figure 3.10 Convex and concave functions.

Ans. (a)-(e) concave; (f)-(i): convex; (j): indifference curve; (k)-(l): non-convex sets.

An *indifference curve* is the set of all (x, y) where a utility function $u(x, y)$ has a constant value. In Figure 3.10(j) the values k_1, k_2, k_3 represent the indifference curves, each one obeying a different level of utility; for example, $k_1 = 4, k_2 = 12, k_3 = 16$.

3.2. Given the following plots of polygons (Figure 3.11), determine which polygons are convex. HINT. First row convex; justify.

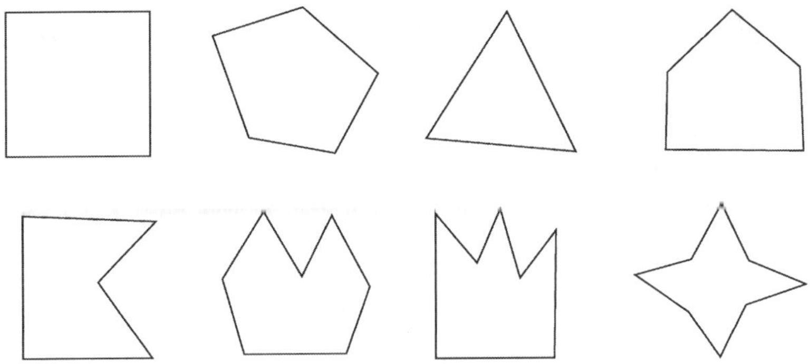

Figure 3.11 Polygons.

3.3. Prove that if f_1, f_2 are convex, then the pointwise supremum function $\max\{f_1(x), f_2(x)\}$ is convex. PROOF. Note that epi $\max\{f_1(x), f_2(x)\}$ corresponds to the intersection of the two epigraphs (Figure 3.12).

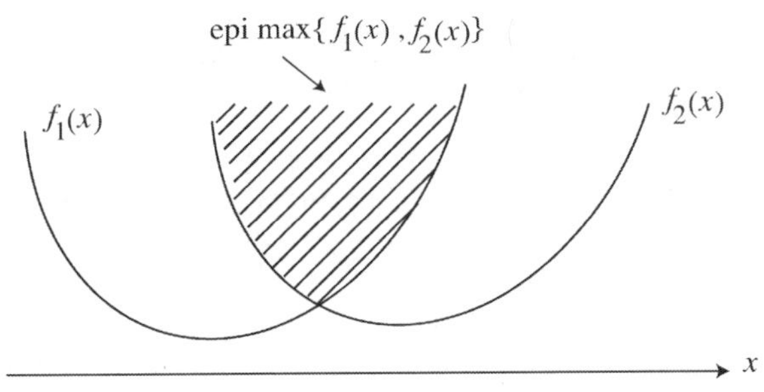

Figure 3.12 epi $\max\{f_1(x), f_2(x)\}$.

3.4. Let Y be a convex set in X. Prove that the closure $\bar{Y}$ is a convex set. HINT. Use definition (3.1.1) from a line segment from a boundary point to another.

3.5. Prove that if f is convex on the interval $[a, b]$ iff for any $a < x_1 < x_2 < x_3 < b$, we have

$$\frac{f(x_2) - f(x_1)}{x_2 - x_1} \leq \frac{f(x_3) - f(x_2)}{x_3 - x_2}. \tag{3.7.1}$$

PROOF. Let f be convex and twice differentiable on $[a, b]$, and choose $a <$

$x_1 < x_2 < x_3 < b$. Then by definition of a convex function, we have

$$f(x_2) = f\Big(\frac{x_3 - x_2}{x_3 - x_1}\, x_1 + \frac{x_2 - x_1}{x_3 - x_1}\, x_3\Big) \le \frac{x_3 - x_2}{x_3 - x_1} f(x_1) + \frac{x_2 - x_2}{x_3 - x_1}\, f(x_3).$$

Thus,

$$x_3 - x - 1)f(x_2) \le (x_3 - x - 2)f(x_1) + (x_2 - x_1)f(x_3).$$

or

$$\begin{aligned}(x_3 - x_1)[f(x_3) - f(x_1)] &\le (x - 1 - x_2)f(x_1) + (x_2 - x_1)f(x_3)\\ &= ((x - 2 - x_1)[f(x_3) - f(x_1)],\end{aligned}$$

which yields (3.10.1). ■

3.6. Prove that a function f which is twice-differentiable on the interval $[a, b] \in \mathbb{R}$ is convex iff $f''(x) \ge 0$ for $a \le x \le b$. PROOF. Let f be convex and twice-differentiable on $[a, b]$. Choose $a < x_1 < x_2 < x_3 < x_4 < b$. Then, by definition,

$$\frac{f(x_2) - f(x_1)}{x_2 - x_1} \le \frac{f(x_4) - f(x_3)}{x_4 - x_3}.$$

Letting $x_2 \to x_1^+$ and $x_3 \to x_4^-$, we find that $f'(x_1) \le f(x_4)$. Since these points are arbitrary, f' is increasing on (a, b). Hence, $f''(x) \ge 0$ for all $x \in (a, b)$. ■

3.7. Choose k points $(x_1, f(x_1)), \dots , (x_k, f(x_k))$ on the graph of the function $y = f(x)$, and assign these k points normalized masses (weights) $p_1, \dots , p_k \in [0, 1]$ such that $p_1 + \cdots + p_k = 1$. Then the *center of gravity* is defined at the point (x_g, y_g) with $x_g = \sum_{i=1}^{k}, y_g = \sum_{i=1}^{k} pif(x_i)$. Prove that if f is a convex function on the interval $[a, b]$, then for any choice of $\{x_1, \dots , x_k\} \in [a, b]$ and associated weights $p_1, \dots p_k \in [0, 1]$ with $\sum_i p_i^k = 1$, there holds the inequality $f(x_g) \le y_g$. HINT. Use induction on k.

3.8. Let f and g be real-valued concave functions with the same domain D. Define a function h so that $h(x) = f(x)+g(x)$ for all $x \in D$. Is h a concave function? If it is, prove it; otherwise provide a counterexample. ANS. Since f and g are concave functions with domain D, then for $x \in D$ and $y \in D$, $x \ne y$, and for all $t \in [0, 1]$, we have

$$f(tx + (1 - t)y) \ge tf(x) + (1 - t)f(y), \quad g(tx + (1 - t)y) \ge tg(x) + (1 - t)g(y).$$

Using these two inequalities we find that

$$\begin{aligned} h(tx+(1-t)y) &= f(tx+(1-t)y) + g(tx+(1-t)y) \\ &\ge tf(x) + (1-t)f(y) + tg(x) + (1-t)g(y) \\ &= t[f(x)+g(x)] + (1-t)[f(y)+g(y)] \\ &= t\,h(x) + (1-t)h(y), \end{aligned}$$

which means that h is a concave function. A similar result holds for convex functions; just replace the word 'concave' by 'convex' and the operation '$\ge$' by '$\le$' in the above statement and proof.

3.9. Let f be a convex function on $[0,1]$. Prove that for any two sets $p_1,\dots,p_k \in [0,1]$ with $\sum_{i=1}^k p_i = 1$, and $q_1,\dots,q_k \in [0,1]$ with $\sum_{i=1}^k q_i = 1$, there holds the inequality $\sum\limits_{i=1}^k p_i \log q_i \le \sum\limits_{i=1}^k p_i \log p_i$. PROOF. Let $x_i = q_i/p_i$. Using Exercise 3.6, we find that for a convex function,

$$f\Big(\sum_{i=1}^k p_i x_i\Big) \le \sum_{i=1}^k p_i f(x_i).$$

Since $f(x) = \log(1/x)$ is a convex function, we have $0 = \log 1 = \log(1/1) = \log\left(\frac{1}{\sum_i q_i}\right) = \log\left(\frac{1}{\sum_i p_i x_i}\right) \le p_i \log\frac{1}{x_i} = \sum p_i \log\frac{1}{\sum_i p_i x_i}$. In particular, using $0 \le \sum_i p_i \log\frac{p_i}{q_i} = \sum_i p_i[\log p_i - \log q_i]$, whence we get $\sum_i p_i \log q_i \le \sum_i p_i \log p_i$. ■

3.10. Show that the function $f(x) = \log x$ is concave for $x > 0$.

PROOF. Take $t = \frac{1}{2}$, and use the definition (3.4.2); we should show that $f\Big(\frac{x+y}{2}\Big) \ge \frac{f(x)+f(y)}{2}$, or that $\log\left(\frac{x+y}{2}\right) \ge \frac{\log x + \log y}{2}$, or that $\log\left(\frac{x+y}{2}\right) \ge \frac{\log(xy)}{2}$ for all $x, y > 0$. This leads to $\log\left(\frac{x+y}{2}\right) \ge \log(xy)^{1/2}$, which after exponentiating both sides gives $\frac{x+y}{2} \ge \sqrt{xy}$ for all $x, y > 0$ (see inequality (3.3.6)).

3.11. Show that the function $f(x) = x^\alpha$ is concave for $0 \le \alpha \le 1$.

HINT. Concavity or convexity can be verified using the inequality (3.2.2) and (3.4.1), respectively, or by checking that the second derivative is nonpositive or nonnegative.

3.12. Show that the following functions are convex on the specified domain: (a) $f(x) = x^\alpha$ on $\mathbb{R}_{++}$ for $\alpha \ge 1$ or $\alpha \le 0$; (b) $f(x) = e^{ax}$ on $\mathbb{R}$ for any $a \in \mathbb{R}$; (c) $f(x) = |x|^p$, $p \geq 1$ on $\mathbb{R}$; (d) the negative entropy function $f(x) = x \log x$, defined on $\mathbb{R}_{++}$, or on $\mathbb{R}_+$, and defined 0 for $x = 0$; and (e) $f(x, y) = x^2/y$, $(x, y) \in \mathbb{R}^2, y > 0$; (f) $f(x) = \max_i \{x_i\}$.

ANS. (a)-(c) are simple; (d) $f'''(x) = 1/x > 0$ for $x > 0$; (e) the Hessian for $y > 0$ Is

$$|\mathbf{H}| = \begin{vmatrix} 2/y & -2x/y^2 \\ -2x/y^2 & 2x^2/y^3 \end{vmatrix} = \frac{2}{y^3} \cdot 2x^2y^2 = \frac{4x^2}{y} \geq 0 \text{ for } y > 0,$$

which shows that the Hessian is nonnegative, and hence, $f(x, y)$ is convex.

3.13. Let $x_{\lfloor i \rfloor}$ denote the ith largest integral component of x (i.e., largest integer not exceeding x), which means that the terms $x_{\lfloor 1 \rfloor} \geq x_{\lfloor 2 \rfloor} \geq \cdots \geq x_{\lfloor n \rfloor}$ are in nondecreasing order. Then the function $f(x) = \sum_{i=1}^{m} x_{\lfloor i \rfloor}$ is a convex function.

HINT. The result follows by writing $f(x) = \sum_{i=1}^{m} x_{\lfloor i \rfloor} = \max \big\{x_{i_1} + x_{i_2} + \cdots x_{i_m} \,|\, 1 \leq i_1 \leq i_2 \leq i_m \leq n\big\}$, which is the maximum of all possible sums of m different largest integral components of x, and it is a convex function since it is the pointwise maximum of $\dfrac{n!}{m!(n-m)!}$ of linear functions.

3.14. Use the inequality (3.3.6) to derive the *Hölder inequality*: For $p, q > 1$, $\frac{1}{p} + \frac{1}{q} = 1$, and $x, y \in \mathbb{R}^n$,

$$\sum_{i=1}^{n} x_i y_i = \Big(\sum_{i=1}^{n} |x_i|^p\Big)^{1/p} \Big(\sum_{i=1}^{n} |y_i|^p\Big)^{1/q}. \tag{3.7.2}$$

SOLUTION. Using (3.2.2) or (3.4.1), the general form of inequality (3.3.6) is

$$a^t b^{1-t} \leq ta + (1-t)b,\ a, b \geq 0,\ t \in [0, 1]. \tag{3.7.3}$$

Take $a = \dfrac{|x_i|^p}{\sum_{j=1}^{n} |x_j|^p}$, $b = \dfrac{|y_i|^q}{\sum_{j=1}^{n} |y_j|^q}$, and set $t = \frac{1}{p}$ and $1 - t = \frac{1}{q}$. Then the inequality (3.7.3) becomes

$$\left(\frac{|x_i|^p}{\sum_{j=1}^{n} |x_j|^p}\right)^{1/p} \left(\frac{|y_i|^q}{\sum_{j=1}^{n} |y_j|^q}\right)^{1/q} \leq \frac{1}{p} \frac{|x_i|^p}{\sum_{j=1}^{n} |x_j|^p} + \frac{1}{q} \frac{|y_i|^q}{\sum_{j=1}^{n} |y_j|^q}.$$

Then summing over i we get the Hölder inequality. The equality in (2.7.2) holds iff $\dfrac{|x_1|^{p-1}}{|y_1|} = \dfrac{|x_2|^{p-1}}{|y_2|} = \cdots = \dfrac{|x_n|^{p-1}}{|y_n|}$. Note that for $p = 2$ this inequality reduces to *Cauchy-Schwarz inequality*:

$$\Big(\sum_{i=1}^{n} |x|_i y_i|\Big)^2 \le \sum_{i=1}^{n} |x_i|^2 \sum_{j=1}^{n} |y_j|^2, \tag{3.7.4}$$

and the *triangle inequality* is:

$$|\sum_{i=1}^{n} x_i| \le \sum_{i=1}^{n} |x_i|. \tag{3.7.5}$$

3.15. Consider $f(x, y) = x^3 + 2y^3 - 3x^2 + 9y^2 - 45x - 60y$. The first-order partial derivatives are: $f_x = 3x^2 - 6x - 45$, $f_y = 6y^2 + 18y - 60$, so the solutions of $f_x = 0$ are $x = -3, 5$ and of $f_y = 0$ are $y = 2, -5$, giving four critical points: $(-3, 2), (-3, -5), (5, 2), (5, -5)$, The second-order partial derivatives are: $f_{xx} = 6x - 6, f_{xy} = 0 = f_{yx}, f_{yy} = 12y + 18$. Then

$$\begin{aligned}
&(1)\ f_{xx}(-3, 2) = -24 < 0, && f_{yy}(-3, 2) = 42 > 0,\\
&(2)\ f_{xx}(-3, -5) = -24 < 0, && f_{yy}(-3, -5) = -42 < 0,\\
&(3)\ f_{xx}(5, 2) = 24 > 0, && f_{yy}(7, 2) = 42 > 0,\\
&(4)\ f_{xx}(5, -5) = 24 > 0, && f_{yy}(7, -5) = -42 < 0.
\end{aligned}$$

Since the signs in (1) and (4) are different, they are ignored; they can be saddle points. Since $f_{xy} = 0 = f_{yx}$, we get from (2): $(-24)(-42) > (0)^2$, and from (3): $(24)(42) > (0)^2$. Hence, the function f has a relative maximum at $(-3, -5)$ and a relative minimum at $(5, 2)$, and a saddle point at $(-3, 2)$ and $(5, -5)$. ■

3.16. For the following function, find the critical points and determine if at these points the function is a relative maximum, relative minimum, inflection point, or saddle point.

(a) $f(x, y) = 3x^2 - xy + 2y^2 - 4x - 7y + 10$.
(b) $f(x, y) = 48y - 3x^2 - 6xy - 2y^2 + 72x$.
(c) $f(x, y) = 5x^2 - 3y^2 - 30x + 7y + 4xy$.
(d) $f(x, y) = 3x^3 - 5y^2 - 225x + 70y + 20$.

ANS. (a) Critical point $(1, 2)$; f has a global minimum at $(1, 2)$.
(b) Critical point $(0, 12)$; inflection point at $(0, 12)$.
(c) Critical point $(2, 5/2)$; saddle point.

(d) Critical points $(5,7), (-5,7)$; relative maximum at $(-5,7)$, saddle point at $(5,7)$.

3.17. Find (a) the critical points, and (b) test whether the following function is at a relative maximum or minimum: $z = 3y^3 - x^3 + 108x - 81y + 32$.

ANS. (a) Equate the first-order partial derivatives to zero, and solve for x and y: $z_x = -3x^2 + 108 = 0$, which gives $x = \pm 6$. Similarly, $z_y = 9y^2 - 81 = 0$, which gives $y = \pm 3$. Thus, there are four distinct critical points at $(6,3), (6,-3), (-6,3), (-6,-3)$. The second partials are $z_{xy} = z_{yx} = 0$.

(b) Take the second-order direct partial derivatives, evaluate them at each critical point, and check the signs:

$$
\begin{array}{ll}
z_{xx} = -6x, & z_{yy} = 18y, \\
(1)\quad z_{xx}(6,3) = -6(6) = -36 < 0 & z_{yy}(6,3) = 18(3) = 54 > 0, \\
(2)\quad z_{xx}(6,-3) = -6(-6) = -36 < 0 & z_{yy}(6,-3) = 18(-3) = -54 > 0, \\
(3)\quad z_{xx}(-6,3) = -6(-6) = 36 > 0 & z_{yy}(-6,3) = 18(3) = 54 < 0, \\
(4)\quad z_{xx}(-6,-3) = -6(-6) = 36 > 0 & z_{yy}(-6,-3) = 18(-3) = -54 < 0.
\end{array}
$$

Since there are different signs for each second partial in (1) and (4), the function cannot be a relative maximum or minimum at $(6,3)$ or $(-6,-3)$. Also, since $z_{xx}z_{yy} - (z_x y)^2$ is negative at $(6,3)$ and positive at $(-6,-3)$, there is a saddle point at each of these two points. Now, since z_{xx} and z_{yy} are of different signs in (2) and (3), the function may have a relative maximum at $(6,-3)$ and a relative minimum at $(-6,3)$. But since z_{xx} and z_{yy} are of the same sign in (2) and (3), while $z_{xx}z_{yy} < (z_x y)^2$ in (3), there is an inflection point at $(-6,3)$; and thus, there is a relative maximum at $(6,-3)$.

3.18. Test for relative maxima and minima of the function $f(x,y) = x^3 + 3xy^2 - 3x^2 - 3y^2 - 40$.

ANS. $f_x = 3x^2 + 3y^2 - 6x = 0$, $f_y = 6xy - 6y = 0$. Solving these equations simultaneously, we get the critical points as $(0,0), (2,0), (1,1), (1,-1)$. Also, $f_{xx} = 6x - 6$, $f_{yy} = 6x - 6$, $f_{xy} = f_{yx} = 6y$. Then

at $(0,0)$: $f_{xx}f_{yy} - (f_{xy})^2 > 0$ and $f_{xx} < 0 \Rightarrow$ relative maximum;
at $(2,0)$: $f_{xx}f_{yy} - (f_{xy})^2 > 0$ and $f_{xx} > 0 \Rightarrow$ relative minimum;
at $(1,1)$: $f_{xx}f_{yy} - (f_{xy})^2 < 0 \Rightarrow$ saddle point;
at $(1,-1)$: $f_{xx}f_{yy} - (f_{xy})^2 < 0 \Rightarrow$ saddle point.

3.19. Find the minimum distance between the origin and the surface $z^2 = x^2y + 4$.

ANS. Let $P(z, y, z)$ be any point on the surface. Then the square of the distance $0P$ is $d^2 = x^2 + y^2 + z^2$. Thus, we find the coordinates of P such that d^2 is minimum. Let $d^2 \equiv f(x,y) = x^2 + y^2 + x^2y + 4$. Then $f_x = 2x + 2xy = 0$, $f_y = 2y + x^2 = 0$; solving these equations we get the critical points as $(,0), \pm\sqrt{2}, -1)$. Also, $f_{xx} = 2+2y$, $f_{yy} = 2$, $f_{xy} = f_{yx} = 2x$. Then at $(0,0)$ we have $f_{xx}f_{yy} - (f_{xy})^2 = 4 > 0$. Hence, we have the relative minimum at $x = 4$, and $d^2 = 4$, so that $d = 2$.

3.20. A manufacturer determines that his cost function $C = \frac{1}{3}Q^2 + 3Q + 300$, where Q is the number of units produced. At what level of output will the average cost per unit be a minimum, and what is the minimum?

ANS. Average cost $\tilde{C} = C/Q = \dfrac{Q^2/3 + 3Q + 300}{Q} = \dfrac{Q}{3} + 3 + \dfrac{300}{Q}$, so that $\dfrac{d\tilde{C}}{dQ} = \dfrac{1}{3} - \dfrac{300}{Q^2} = 0$ gives the critical point $Q = \pm 30$. Since $\dfrac{d^2\tilde{C}}{dQ^2} > 0$ at $Q = 30$ and $\tilde{C}$ has relative minimum at $Q = 30$, so $Q = 30$ is the absolute minimum.

3.21. A manufacturer produces and sells 30,000 units of products throughout the year. He would like to determine the number of units to be manufactured in each production run so that he can minimize annual set-up costs and carrying costs. The production cost of each unit is \$20 and carrying costs (insurance, storage, etc.), estimated to be 10% of the value of the average set-up costs per production run, are \$30. Find the economic order quantity (or economic lot size) which is the size of each production run.

ANS. Let q be the number of units in a production factory. Since the sales are distributed at a uniform rate, assume that the inventory varies uniformly from q to 0 between production runs. Thus, we take the average inventory to be $q/2$ units. The production costs are \$ 20 per unit, so the value of the average inventory is $20(q/2) = 10q$. The carrying costs are 10% of this value, i.e., they are equal to $(0.1)(10q)$. The number of production runs per year is $30,000/q$. Thus, the total set-up costs are $= 30(30000/q)$. Hence, the total annual carrying costs plus the set-up costs are $C = (0.1(10q) + 30\dfrac{30000}{q} = q + \dfrac{900000}{q}$, whence $\dfrac{dC}{dq} = 1 - \dfrac{900000}{q^2} = 0$ gives the critical points as $\pm 300\sqrt{10}$. Since $\dfrac{d^2C}{dq^2} = \dfrac{1800000}{q^3} > 0$ at $q = 300\sqrt{10} \approx 949$, the absolute minimum is at $q \approx 949$.

3.22. Show that the density function $f(x) = \dfrac{\lambda}{2}e^{-\lambda|x|}$, $\lambda > 0$, of the Laplace distribution $\ln f(x) = \ln|x|$ is a concave function. What can you say about $f'(x)$? ANS. $f'(x) < 0$ for $x \neq 0$; $f'(x)$ does not exist at $x = 0$. Also see Appendix C.

3.23. Prove that a (concave, convex, or any) function $f : \mathbb{R}^n \mapsto \mathbb{R}$ is differentiable at a point $x \in \text{dom}(f)$ iff the gradient ∇f exists. PROOF. The gradient of f is defined at $x \in \mathbb{R}^n$ by

$$\nabla f(x) = \mathbf{e}_1 \frac{\partial f}{\partial x_1} + \mathbf{e}_2 \frac{\partial f}{\partial x_2} + \cdots + \mathbf{e}_n \frac{\partial f}{\partial x_n} = [\mathbf{e}]^T \begin{bmatrix} \frac{\partial f}{\partial x_1} & \frac{\partial f}{\partial x_2} & \cdots & \frac{\partial f}{\partial x_n} \end{bmatrix},$$

where $[\mathbf{e}]$ is the $1 \times n$ matrix of unit vectors $\mathbf{e}_i$, each in the direction of the respective axis for $i = 1, 2, \dots, n$. If f is differentiable at x, then all $\dfrac{\partial f}{\partial x_i}$ exist, which implies that ∇f exists. Conversely, if ∇f exists, then all first-order partial derivatives $\dfrac{\partial f}{\partial x_i}$ exist for each $i = 1, 2, \dots, n$. However, note that using ∇f is taking the question too far; simply the existence of $\dfrac{\partial f}{\partial x_i}$, $i = 1, 2, \dots, n$, should suffice.

3.24. Prove that the definitions (3.2.1) and (3.2.2) are equivalent. HINT. Follow the proof of Theorem 3.4 by using proper inequality signs.

3.25. Let $f : \mathbb{R}^n \mapsto \mathbb{R}$ be defined by $f(x) = p\,x$. Consider the problem of minimizing $p \cdot x$ by choosing x subject to the condition that x belongs to a constraint set G. Prove that the minimum value of $C(p) = \min\{p\,x \mid x \in G\}$ is a linear homogeneous and concave function of p.

3.26. Optimize the following functions by (i) finding the critical values at which the function is optimized, and (ii) testing the second-order condition to determine if it is a relative maximum or minimum.

(a) $f(x) = -x^3 + 6x^2 + 135x - 26$; (b) $f(x) = x^4 - 4x^3 - 80x^2 + 108$; and (c) $f(x) = (11 - 5x)^4$.

ANS. (a) Critical points $x = -5, 9$; $f''(-5) > 0$, convex, relative minimum at $x = -5$; $f''(9) < 0$, concave, relative maximum at $x = 9$.

(b) Critical points $-5, 0, 8$; $f''(-5) > 0$, convex, relative minimum at $x = -5$; $f''(0) < 0$, concave, relative maximum at $x = 0$; $f''(8) > 0$, convex, relative minimum at $x = 8$.

(c) Critical points $x = \frac{11}{5}$; $f''(11/5) = 0$, test fails; $f'''(11/5) = 0$, test inconclusive; $f^{iv)}(11/5) > 0$, convex, relative minimum at $x = \frac{11}{5}$.

4

Concave Programming

The subject of concave programming deals with constrained optimization problems, in the sense of maximizing an objective function subject to constrained equality, and inequality constraints. In this chapter we will introduce the method of Lagrange multipliers to solve constrained optimization problems with equality and inequality constraints, in areas of both mathematics and mathematical finance.

4.1 Optimization

As we have seen, optimization problems deal with finding the maximum or minimum of a function (i.e., optimizing the objective function) subject to none or certain prescribed constraints.

We will consider the following four cases of necessary and sufficient conditions for (local) optimality: (1) no constraints; (2) only equality constraints; (3) equality and inequality constraints; and (4) only inequality constraints.

4.1.1 Unconstrained Optimization. Assume that the function $f : D \mapsto \mathbb{R}^n$ is a continuously differentiable function.

FOR UNCONSTRAINED MAXIMIZATION, the necessary and sufficient conditions for a local maximum of $f(\mathbf{x})$ at $\mathbf{x}^* = (x_1, x_2, \dots, x_n)$ are:

(i) the first partial derivatives of f with respect to each x_i, $i = 1, 2, \dots, n$, is zero, i.e.,

$$\frac{\partial f}{\partial x_i}(\mathbf{x}) = 0, \quad i = 1, 2, \dots, n, \tag{4.1.1}$$

where the critical point $\mathbf{x}^*$ is obtained by solving equations (4.1.1) simultaneously; and

(ii) the Hessian $|\mathbf{H}|$ of f at $\mathbf{x}^*$ is negative semidefinite (NSD), i.e.,

$$|\mathbf{H}|(f) \leq 0 \text{ for all } \mathbf{x}^*, \tag{4.1.2}$$

where the Hessian is defined in §1.6.2, and definite and semidefinite matrices in §1.5.

FOR UNCONSTRAINED MINIMIZATION, the necessary and sufficient conditions for a local minimum of $f(\mathbf{x})$ at $\mathbf{x}^*$ are:

(i) the first partial derivative of f with respect to each x_i, $i = 1, 2, \dots, n$, is zero, i.e.,

$$\frac{\partial f}{\partial x_i}(\mathbf{x}) = 0, \quad i = 1, 2, \dots, n, \tag{4.1.3}$$

where the critical point $\mathbf{x}^*$ is obtained by solving equations (4.1.3) simultaneously; and

(ii) the Hessian $|\mathbf{H}|$ of f at $\mathbf{x}^*$ is positive semidefinite (PSD), i.e.,

$$|\mathbf{H}|(f) \geq 0 \quad \text{for all } \mathbf{x}^*. \tag{4.1.4}$$

Many examples of minimization for the unconstrained case have been considered in the previous chapters, and some examples on minimization for this case will be presented in the next chapter.

!!! *The necessary and sufficient conditions (4.1.1) and (4.1.3) are sometimes defined by stating that $\nabla f(\mathbf{x}^*) = \mathbf{0}$, which reduces to (4.1.1) and (4.1.3).*

Example 4.1. Consider $f(x, y) = 2y^3 - x^3 + 12x - 54y + 12$. Equating the first-order partial derivatives of f to zero, we get $f_x = -3x^2 + 12 = 0$, $f_y = 6y^2 - 54 = 0$, i.e., $x = \pm 2$ and $y = \pm 3$. Thus, the critical numbers are $(2, 3), (2, -3), (-2, 3), (-2, -3)$. Next, take the second-order direct partial derivatives and evaluate them at each of the four critical points to check for their signs:

$$f_{xx} = -6x, \quad f_{yy} = 12y.$$

Then

$$\begin{aligned}
&(1)\ f_{xx}(2,3) = -6(2) = -12 < 0, && f_{yy}(2,3) = 12(3) = 36 > 0,\\
&(2)\ f_{xx}(2,-3) = -6(2) = -12 < 0, && f_{yy}(2,-3) = 12(-3) = -36 < 0,\\
&(3)\ f_{xx}(-2,3) = -6(-2) = 12 > 0, && f_{yy}(-2,3) = 12(3) = 36 > 0,\\
&(4)\ f_{xx}(-2,-3) = -6(-2) = 12 > 0, && f_{yy}(-2,-3) = 12(-3) = -36 < 0.
\end{aligned}$$

Since there are different signs for each of the second direct partials in (1) and (4), the function f cannot have a relative extremum at $(2, 3)$ and $(-2, -3)$. However, since the signs of second partials are both negative in (2) and both positive in (3) above, the function f may have a relative maximum at $(2, -3)$ and a relative minimum at $(-2, 3)$. Since f_{xx} and f_{yy} are of different signs, the product of f_{xx} and f_{yy} cannot be greater than $(f_{xy})^2$.

Since $f_{xy} = 0 = f_{yx}$, we check $f_{xx} \cdot f_{yy} > (f_{xy})^2$ at the critical points $(2, -3)$ and $(-2, 3)$:

$f_{xx}(2,-3)\cdot f_{yy}(2,-3) = (-12)(-36) > 0,\ f_{xx}(-2,3)\cdot f_{yy}(-2,3) = (12)(36) > 0.$

Hence, f has a relative maximum at $(2, -3)$ and a relative minimum at $(-2, 3)$. ■

Example 4.2. Maximize the utility function $u(x, y) = 3xy$ subject to the constraint $g(x) = 3x + 4y - 60$. The Lagrangian is $L(x, y, \lambda) = 3xy + \lambda(60 - 3x - 4y)$, and the first-order partial derivatives equated to zero give the system of equations $L_x = 3y - 3\lambda = 0$, $L_y = 3x - 4\lambda = 0$, $L_\lambda = 60 - 3x - 4y = 0$. Writing this system in the matrix form $\mathbf{Ax} = \mathbf{b}$:

$$\begin{bmatrix} 0 & 3 & -3 \\ 3 & 0 & -4 \\ -3 & -4 & 0 \end{bmatrix} \begin{bmatrix} x \\ y \\ \lambda \end{bmatrix} = \begin{bmatrix} 0 \\ 0 \\ -60 \end{bmatrix}.$$

Using Cramer's rule to solve this system, we have $|\mathbf{A}| = 72$, $|\mathbf{A}_1| = 720$, $|\mathbf{A}_2| = 540$, $|\mathbf{A}_3| = 540$, which give the critical point $x^*, y^* = 7.5, \lambda^* = 7.5$. Next, taking the second-order partial derivatives of U with respect to x, y, $(L_{xx} = 0, L_{xy} = 3 = L_{yx}, L_{yy} = 0)$, the first-order partial derivatives of g are: $g_x = 3, g_y = 4$, and writing in the left-side form of (1.6.7) we get

$$|\bar{\mathbf{H}}| = \begin{bmatrix} 0 & 3 & 4 \\ 3 & 0 & 3 \\ 4 & 3 & 0 \end{bmatrix}, \quad \text{or} \quad |\bar{\mathbf{H}}| = \begin{bmatrix} 0 & 3 & 3 \\ 3 & 0 & 4 \\ 3 & 4 & 0 \end{bmatrix}.$$

Thus, we find the value of $|\bar{H}_2| = |\bar{\mathbf{H}}| = 72$ from the bordered Hessian on the left, and that of $|\bar{H}_2| = |\bar{\mathbf{H}} = 72$ from the bordered Hessian on the right. Note that there is no need to use both forms of the bordered Hessian; the left-hand form works as well. Since $|\bar{\mathbf{H}}| = |\mathbf{A}| > 0$, the bordered Hessian $|\bar{\mathbf{H}}|$ is negative definite. Also, $u_{xx} \cdot u_{yy} = 0 < (u_{xy})^2 = 9$, the function u is maximized at the above critical point $(x^*, y^*) = (7.5, 7.5)$.

4.2 Method of Lagrange Multipliers

Given functions $f, g_1, \dots, g_m$ and $h_1, \dots, h_k$ defined on some domain $D \subset \mathbb{R}^n$, the maximization problem is stated as follows.

Determine $\max\limits_{\mathbf{x} \in D} f(\mathbf{x})$ subject to the constraints $g_i(\mathbf{x}) \le 0$ for all $i = 1, \dots, m$ and all $h_j(\mathbf{x}) = 0$ for all $j = 1, \dots, k$.

4.2.1. Constrained Optimization with Equality Constraint.

The method of Lagrange multipliers is used in this and other cases. Thus, given a function $f(x, y)$ subject to a constraint $g(x, y) = k$ (constant), a new function can be defined by either of the following two equivalent forms:

$$F(x, y, \lambda) = f(x, y) - \lambda(g(x, y) - k), \tag{4.2.1}$$

or

$$F(x, y, \lambda) = f(x, y) + \lambda(k - g(x, y)), \tag{4.2.2}$$

where $\lambda > 0$ is known as the *Lagrange multiplier*, $f(x, y)$ as the *original function* or the *objective function*, $g(x, y)$ as the *constraint*, and $F(x, y, \lambda)$ as the *Lagrangian function* or simply, the *Lagrangian.* Since the constraint is always set equal to zero, the product $\lambda(g(x, y) - k)$, or $\lambda(k - g(x, y))$ is also equal to zero, and thus, the addition of this term does not change the value of the objective function $f(x, y)$. In view of Theorem 2.18, if $\nabla F(x, y, \lambda) = \mathbf{0}$, then

$$F_x(x, y, \lambda) = 0, \quad F_y(x, y, \lambda) = 0, \quad F_\lambda(x, y, \lambda) = 0. \tag{4.2.3}$$

The critical values at which the objective function is optimized are denoted by x^*, y^*, λ^*, and are determined by solving equations (4.2.3) simultaneously.

The second-order conditions will obviously be different from those for the unconstrained optimization considered in §4.1; they are discussed in the sequel.

Example 4.3. Optimize $f(x, y) = 2x^2 + 12xy - 5y^2$, subject to the constraint $x + y = 30$. The Lagrangian in the form (4.2.2) is

$$F(x, y, \lambda) = 2x^2 + 12xy - 5y^2 + \lambda(30 - x - y).$$

The first-order partial derivatives are: $F_x = 4x + 12y - \lambda = 0$, $F_y = 12x - 10y - \lambda = 0$, $F_\lambda = 30 - x - y = 0$, which when solved simultaneously give the critical values: $x^* = 22, y^* = 8, \lambda^* = 184$. Substituting these values in $F(x, y, \lambda)$ we get $F(22, 8) = 2760 = f(22, 8)$. Notice that both functions $f(x, y)$ and $F(x, y)$ are equal at the critical values, since the constraint is equal to zero at these values.

The second-order derivatives are $F_{xx} = 4, F_{yy} = -10, F_{xy} = F_{yx} = 12$. Also, from the constraint $g(x, y) = x + y - 30$, we have $g_x = 1, g_y = 1$. Then the bordered Hessian (see Eq (1.6.7)) is

$$|\mathbf{H}| = \begin{vmatrix} 4 & 12 & 1 \\ 12 & -10 & 1 \\ 1 & 1 & 0 \end{vmatrix},$$

and its second principal minor is $|H_2| = |\mathbf{H}| = 4(-1) - 12(1) + 1(12 + 10) = 6 > 0$. Since $|\bar{H}_2| > 0$, $|\mathbf{H}|$ is negative definite, and so $F(x, y)$ has a local maximum at $(22, 8)$. ■

Note that the Lagrange multiplier λ approximates the marginal impact on the objective function caused by a small change in the constant of the constraint. Thus, in the above example, with the value of $\lambda^* = 184$, a 1-unit increase (or decrease) in the constant of constraint would result in an increase (or decrease) in F by approximately 184 units, as the following example shows.

Example 4.4. Suppose there is a 1-unit change (decrease) in the constant of the constraint. We want to determine what change it will make in F and λ^* in the above example. Then with the new constraint $g(x, y) = x + y - 29$, we have

$$F(x, y, \lambda) = 2x^2 + 12xy - 5y^2 + \lambda(29 - x - y),$$

which gives $F_x = 4x + 12y - \lambda = 0, F_y = 12x - 10y - \lambda = 0, F_\lambda = 29 - x - y = 0$. Solving these equations simultaneously we get $x^* = 21.266$, $y^* = 7.733$, $\lambda^* = 177.86$; thus, $F(21.266, 7.733) = 2578.887 = f(21.266, 7.733)$, which is approximately 181.1 (i.e., about 6.6%) smaller than the previous value of $F = 2760$, while the new value of λ^* is about 3.3% smaller than the previous value of $\lambda^* = 184$. ■

This is the reason why Lagrange multipliers are called *shadow prices*. Also, in utility optimization subject to a budget constraint, the value of λ will estimate the marginal utility of an extra dollar of income.

!!! *The function $F(x, y)$, defined by (4.2.1), can have the term $\lambda(k - g(x, y)) = 0$ either added to or subtracted from the objective function $f(x, y)$ without changing the value of x^* and y^*, except that only the sign of λ^* will be affected.*

4.3 Karush-Kuhn-Tucker Conditions

We will consider maximization problems subject to equality and inequality constraints. This case is also known as nonlinear programming, and the *Karush-Kuhn-Tucker (KKT) conditions* (also known as *Kuhn-Tucker conditions*) are used in such optimization problems. These conditions provide the first-order necessary conditions for an optimal solution in nonlinear programming, provided that certain regularity conditions are satisfied. Although the KKT conditions allow inequality constraints, this is a generalization of the method of Lagrange multipliers, which allows only equality constraints. The system of equations under the KKT conditions is generally not solved directly, except in a few special cases where a closed-form solution can be derived analytically. In general, many nonlinear optimization algorithms can be interpreted by numerical methods for solving the KKT system of equations.

4.3.1 Equality and Inequality Constraints. Consider the nonlinear programming problem

$$\text{Optimize } f(\mathbf{x}) \text{ subject to } g_i(\mathbf{x}) \le 0,\ h_j(\mathbf{x}) = 0, \tag{4.3.1}$$

where $\mathbf{x}$ is the optimization variable, f the objective (or utility) function, $g_i, i = 1, \dots, m$, are the inequality constraint functions, and $h_j, j = 1, \dots, k$, are the equality constraint functions.

4.3.2 Necessary Conditions. Suppose that the objective function $f : \mathbb{R}^n \mapsto \mathbb{R}$, and the constraint functions $g_i : \mathbb{R}^n \mapsto \mathbb{R}$ and $h_j : \mathbb{R}^n \mapsto \mathbb{R}$, are continuously differentiable at a point $\mathbf{x}^*$. If this point $\mathbf{x}^*$ is a local optimum that satisfies the following four regularity conditions (known as the *KKT conditions*), then there exist constants μ_i, $(i = 1, \dots, m)$ and λ_j, $(j = 1, \dots, k)$ such that:

(i) *Stationary conditions*:
For maximizing $f(\mathbf{x})$:

$$\frac{\partial f}{\partial x_i}(\mathbf{x}^*) = \sum_{i=1}^{m} \mu_i \frac{\partial g}{\partial x_i}(\mathbf{x}^*) + \sum_{j=1}^{n} k\lambda_j \frac{\partial h}{\partial x_i}(\mathbf{x}^*). \tag{4.3.2}$$

For minimizing $f(\mathbf{x})$:

$$-\frac{\partial f}{\partial x_i}(\mathbf{x}^*) = \sum_{i=1}^{m} \mu_i \frac{\partial g}{\partial x_i}(\mathbf{x}^*) + \sum_{j=1}^{n} k\lambda_j \frac{\partial h}{\partial x_i}(\mathbf{x}^*). \tag{4.3.3}$$

(ii) *Primal feasibility conditions*:

$$\begin{aligned} g_i(\mathbf{x}^*) &\le 0 \text{ for all } i = 1, \dots, m, \\ h_j(\mathbf{x}^*) &\le 0 \text{ for all } j = 1, \dots, k. \end{aligned} \tag{4.3.4}$$

(iii) *Dual feasibility condition*:

$$\mu_i \ge 0 \text{ for all } i = 1, \dots, m. \tag{4.3.5}$$

(iv) *Complementary slackness condition*:

$$\mu_i g_i(\mathbf{x}^*) = 0 \text{ for all } i = 1, \dots, m. \tag{4.3.6}$$

Note that in the absence of inequality constraints, i.e., when $m = 0$, the KKT conditions become the Lagrange conditions and the KKT multipliers become the Lagrange multipliers.

Further, if we add an additional multiplier μ_0, which may be zero, and rewrite the above KKT stationary conditions (i) as

$$\mu_0 \frac{\partial f}{\partial x_i}(\mathbf{x}^*) + \sum_{i=1}^{m} \mu_i \frac{\partial g_i}{\partial x_i}(\mathbf{x}^*) + \sum_{j=1}^{k} \lambda_j \frac{\partial h_j}{\partial x_j}(\mathbf{x}^*) = 0, \tag{4.3.7}$$

then the KKT conditions belong to a wider class of *first-order necessary conditions* (FONC) which allow non-smooth functions using subderivatives. Condition (4.3.7) is known as the *Fritz John condition*, to be discussed later in §5.3.

4.3.3 Regularity Conditions (or constraint qualifications). If a minimum point $\mathbf{x}^*$ were to satisfy the KKT conditions (4.3.2) through (4.3.6), the problem (4.3.1) must satisfy certain regularity conditions. The most used of such conditions are as follows:

1. *Linearity constrained qualifications* (LCQ): If g_i and h_j are affine functions, i.e., they are linear functions plus a constant, then no other condition is needed.

2. *Linear independence constraint qualification* (LICQ): The gradients of the active inequality constraints and those of the equality constraints are linearly independent at $\mathbf{x}^*$.

3. *Mangasarian-Fromovitz constraint qualification* (MFCQ): The gradients of the active inequality constraints and those of the equality constraints are positive-linearly independent at $\mathbf{x}^*$.

4. *Constant rank constraint qualification* (CRCQ): For each subset of the gradients of the active inequality constraints and those of the equality constraints, the rank at a vicinity of $\mathbf{x}^*$ is constant.

5. *Constant positive linear dependence constraint qualification* (CLPD): For each subset of the gradients of the active inequality constraints and those of the equality constraints, if it is positive-linearly dependent at $\mathbf{x}^*$ then it is positive-linearly dependent[1] at a vicinity of $\mathbf{x}^*$.

6. *Quasi-normality constraint qualification* (QNCQ): If the gradients of the active inequality constraints and those of the equality constraints are positive-linearly dependent at $\mathbf{x}^*$ with associated multipliers μ_i for inequalities and λ_j for equalities, then there exists no sequence $\{x_n\} \to \mathbf{x}^*$ such that $\mu_i \neq 0 \Rightarrow \mu_i g_i(x_n) > 0$ and $\lambda_j \neq 0 \Rightarrow \lambda_j h_j(x_n) > 0$.

7. *Slater conditions* (SC): For a convex problem, there exists a point x such that $h(\mathbf{x}) = 0$ and $g_i(\mathbf{x}) < 0$.

This is the most used condition in practice. Note that

$$\text{LICQ} \Rightarrow \text{MFCQ} \Rightarrow \text{CPLD} \Rightarrow \text{QNCQ},$$
$$\text{LICQ} \Rightarrow \text{CRCQ} \Rightarrow \text{CPLD} \Rightarrow \text{QNCQ},$$

but the converses are not true, and MFCQ $\not\Rightarrow$ CRCQ. In practice one must prefer weaker constraints since they provide stronger optimality conditions.

4.3.4 Sufficient Conditions. In general, the necessary conditions do not qualify as sufficient conditions for optimality. However, in some cases the necessary conditions also become sufficient conditions, with some additional information such as the *second-order sufficient condition* (SOSC). For example, for smooth functions, the SOSC involves the second derivative, hence the

[1] A set $(v_1, \dots, v_n)$ is positive-linearly dependent if there exists $a_1 \geq 0, \dots, a_n \geq 0$ not all zero such that $a_1 v_1 + \cdots + a_n v_n = 0$.

name. The *general rule* is: The necessary conditions are sufficient for optimality if the objective function f in an optimization (maximization) problem is a *concave* function, where the inequality constraints g_i are continuously differentiable convex functions and the equality constraints h_j are affine functions.

Example 4.5. (Nonlinear programming problem in $\mathbb{R}^2$) Consider the problem: maximize $f(x, y) = xy$ subject to the constraints $x + y^2 \leq 2$ for $x, y \geq 0$. Since the feasible region is bounded, a global maximum for this problem exists, because a continuous function on a closed and bounded (compact) set has a maximum there. We write the given constraints as $g_1(x, y) = x + y^2 \leq 2$, $g_2(x, y) = -x \leq 2$, $g_3(x, y) = -y \leq 0$. Then the KKT conditions can be written as

$$y - \lambda_1 + \lambda_2 = 0, \tag{4.3.10}$$

$$x - 2y\lambda_1 + \lambda_3 = 0, \tag{4.3.11}$$

$$\lambda_1(2 - x - y^2) = 0, \tag{4.3.12}$$

$$\lambda_2 x = 0, \tag{4.3.13}$$

$$\lambda_3 y = 0, \tag{4.3.14}$$

$$x + y^2 \leq 2, \tag{4.3.15}$$

$$x, y, \lambda_1, \lambda_2, \lambda_3 \geq 0. \tag{4.3.16}$$

Note that in the case of $\mathbb{R}^n$, in equations of the form $\lambda_i(b_i - g_i(x_1, \dots, , x_n))$ at least one of the two factors must be zero. If there are n such conditions there are at most 2^n possible cases to consider. In this example there are only $2^2 = 4$ cases, which are as follows:

CASE 1. Suppose $\lambda_1 = 0$. Then from (4.3.10) and (4.3.11) we get $y+\lambda+2 = 0$ and $x + \lambda_3 = 0$, respectively. Since each term is nonnegative, the only solution of these two equations is $x = 0 = y = \lambda_2 = \lambda_3$. Check that the KKT conditions are satisfied when $x = y = \lambda_1 = \lambda_2 = \lambda_3 = 0$. However, these values do not provide a local maximum since $f(0, 0) = 0$ and $f(x, y) > 0$ at points inside the feasible region.

CASE 2. Suppose $x + y^2 = 2$. Then at least one of $x = 2 - y^2$ and y must be positive.

CASE 2A. Suppose $x > 0$. Then $\lambda_2 = 0$, and in view of (4.3.10) we get $\lambda_1 = y$. Then (4.3.11) gives $x - 2y\lambda_1 + \lambda_3 = 2 - 3y^2 + \lambda_3$, or $3y^2 = 2 + \lambda_3 > 0$, and $\lambda_3 = 0$. Hence, $y = \sqrt{2/3}$, and $x = 2 - 2/3 = 4/3$. Note that these values satisfy all the KKT conditions.

CASE 2B. Suppose $x = 0$, which gives $y = \sqrt{2}$. Since $y > 0$ we get $\lambda_3 = 0$. Then (4.3.11) gives $\lambda_1 = 0$. But this is precisely Case 1.

Thus, we have only two local maxima at $(0, 0)$ and $(4/3, \sqrt{2/3})$, and one global maximum at $(4/3, \sqrt{2/3})$ ■

Example 4.6. (Utility maximization with one simple constraint) Consider the utility maximization problem with a budget constraint:

$$\text{Maximize } u(x,y) \text{ subject to } B = P_x x + P_y y, \text{ and } x^* \geq x. \tag{4.3.17}$$

This problem has two constraints. Using the Lagrange method we can set up this problem with two constraints so that the Lagrange problem becomes

$$\max_{x,y} u(x,y) + \lambda_1(B - P_x x - P_y y) + \lambda_2(x^* - x). \tag{4.3.18}$$

This type of problem is found in cases of marginal utility of budget-ration constraints, wartime rationing, and peak load pricing. We will provide examples in each of these situations.

However, in this utility maximization problem we know that the budget constraint may or may not be binding since it depends on the size of x^*.

Solution. We will vary slightly the first-order condition. Thus, using the KKT conditions which can be applied to cases where the constraints may sometimes be non-binding. The KKT conditions for problems (4.3.17)-(4.3.18) are

$$\begin{aligned} L_x &= u_x - \lambda_1 P_x - \lambda_2 = 0, \; x \geq 0, \\ L_y &= u_y - \lambda_1 P_y = 0, \; y \geq 0, \\ L_{\lambda_1} &= B - P_x x - P_y y \geq 0, \; \lambda_1 \geq 0, \\ L_{\lambda_2} &- x^* - x \geq 0, \; \lambda_2 \geq 0. \end{aligned} \tag{4.3.19}$$

Using the Lagrange problem (4.3.18) we must have $\lambda_1(B - P_x x - P_y y) = 0$, which yields either $\lambda_1 = 0$ or $B - P_x x - P_y y = 0$. Thus, if we regard λ_1 as the marginal utility of the budget (i.e., income), then in case the budget constraint is not satisfied, the marginal utility of the additional B is zero ($\lambda_1 = 0$).

Similarly for the ration constraint, either $x^* - x = 0$ or $\lambda_2 = 0$, where λ_2 can be regarded as the marginal utility of relaxing the ration constraint.

Solutions of such problems need the trial-and-error method. This method enumerates the points on the boundary. Since there is more than one possible outcome, we must try them all, but with the understanding that we are making an educated guess as to which constraint is more likely to be non-binding. Since in this problem we know that the budget constraint will always be binding, we will concentrate on the ration constraint, and go through the following steps:

Step 1. (Simply ignore the second constraint) We are assuming that $\lambda_2 = 0, \lambda_1 > 0$. Then the first-order KKT conditions (4.3.19) become

$$\begin{aligned} L_x &= u_x - \lambda_1 P_x - \lambda_2 = 0, \\ L_y &= u_y - \lambda_1 P_y = 0, \\ L_{\lambda_1} &= B - P_x x - P_y y = 0. \end{aligned} \tag{4.3.20}$$

We find a solution for x^* and y^*, and then check if the constraint that was ignored (i.e., λ_2) has been violated. If the answer is yes, then go to Step 2.

STEP 2. (Use both constraints, assuming that they are binding) We take $\lambda_1 > 0, \lambda_2 > 0$. Then the first-order KKT conditions (4.3.19) become

$$\begin{aligned} L_x &= u_x - \lambda_1 P_x - \lambda_2 = 0, \\ L_y &= u_y - \lambda_1 P_y = 0, \\ L_{\lambda_1} &= B - P_x x - P_y y = 0, \\ L_{\lambda_2} &= x^* - x = 0. \end{aligned} \tag{4.3.21}$$

Then the solution will be the point where the two constraints intersect.

STEP 3. (Use the second constraint, ignore the first one) We assume $\lambda_2 > 0, \lambda_1 = 0$, and the first-order KKT conditions (4.3.19) become

$$\begin{aligned} L_x &= u_x - \lambda_1 P_x - \lambda_2 = 0, \\ L_y &= u_y - \lambda_1 P_y = 0, \\ L_{\lambda_2} &= x^* - x = 0. \end{aligned} \tag{4.3.22}$$

We will explain these steps using a numerical example, which is as follows:

$$\text{Maximize } u(x,y) = xy \text{ subject to } x + y \leq 90, \text{and } x \leq 30. \tag{4.3.23}$$

From (4.3.18) the Lagrangian is

$$L(x,y) = xy + \lambda_1(90 - x - y) + \lambda_2(30 - x),$$

and the KKT conditions (4.3.19) are

$$\begin{aligned} L_x &= y - \lambda_1 - \lambda_2 = 0, \ x \geq 0, \\ L_y &= x - \lambda_1 = 0, \ y \geq 0, \\ L_{\lambda_1} &= 90 - x - y \geq 0, \ \lambda_1 \geq 0, \\ L_{\lambda_2} &= 30 - x \geq 0, \ \lambda_2 \geq 0. \end{aligned} \tag{4.3.24}$$

So we have four equations and four unknowns x, y, λ_1 and λ_2. To solve, we use the above steps, i.e., we first ask if any λ_i $(i = 1,2)$ could be zero. First try $\lambda_2 = 0$, because, in view of the form of the utility function, $\lambda_1 = 0$ does not make sense. This gives $x - \lambda_1 = y - \lambda_1$, or $x = y$, and from the constraint $90 - x - y$ we get $x^* = y^* = 0$, which could not be the case, as it violates the constraint that $x \leq 30$. Hence, $x^* = 30$ and $y^* = 60$, and thus, $\lambda_1^* = 30$ and $\lambda_2^* = 15$. ■

4.4 Inequality Constrained Optimization

So far we have studied optimization problems in which the constraints are strictly equalities. In certain economic problems there are weak constraints; for example, in problems in which the utility function is to be maximized subject to 'not more than x dollars or workweeks to minimize the costs.' Optimization problems of this kind are called *concave programming*, because the objective function and the constraint function are assumed to be concave. It is a form of nonlinear programming where the object function is optimized subject to inequality constraints.

As mentioned in Chapter 3, concave functions are also included since the convex functions are negatives of the concave functions. This implies that in maximization problems, concave programming can also be used to minimize a convex objective (or utility) function by maximizing the negative of that function.

CASE 4. ONLY INEQUALITY CONSTRAINTS. Consider the following optimization problem:

$$\text{Maximize } f(x_1, x_2) \ \text{ subject to } \ g(x_1, x_2) \geq 0, \quad x_1, x_2 \geq 0, \tag{4.4.1}$$

and the Lagrangian is

$$F(x_1, x_2) = f(x_1, x_2) + \lambda g(x_1, x_2). \tag{4.4.2}$$

The first-order KKT conditions, which are necessary and sufficient conditions for maximization, give the following six relations:

$$\begin{aligned}
&\text{1a.} \quad \frac{\partial F}{\partial x_i} = f(x_1^*, x_2^*) + \lambda g(x_1^*, x_2^*) \leq 0,\\
&\text{1b.} \quad x_i \geq 0,\\
&\text{1c.} \quad x_i^* \frac{\partial F}{\partial x_i} = 0, \quad i = 1, 2,\\
&\text{2a.} \quad \frac{\partial F}{\partial x_i} = g(x_1^*, x_2^*) \geq 0,\\
&\text{2b.} \quad \lambda^* \geq 0,\\
&\text{2c.} \quad \lambda^* \frac{\partial F}{\partial \lambda} = 0,
\end{aligned} \tag{4.4.3}$$

where x_1^*, x_2^* are the critical points of f. Conditions (1.c) and (2.c) are called the *complementary-slackness conditions*, which implies that both x_1^*, x_2^*, and $\dfrac{\partial f(x_1^*)}{\partial x_i}$ and $\dfrac{\partial f(x_2^*)}{\partial x_i}$ cannot both be zero.

For a linear objective function which is concave or convex, but not strictly concave or strictly convex, the concave programming that satisfies the KKT conditions will always satisfy the necessary and sufficient conditions for a maximum.

The significance of the KKT conditions (4.4.3) is as follows:

(i) Condition (1.a) requires that the Lagrangian function F be maximized with respect to x_1 and x_2, while condition (2.a) demands that the function F be minimized with respect to λ. This means that the concave programming is designed to seek out a saddle point in the function F in order to optimize the objective function f subject to the inequality constraints g.

(ii) In optimization problems with equality constraints which are set equal to zero, the quantity λg related to the constraint can be either subtracted or added to the objective function f to define the Lagrangian function as in Eqs (4.2.1) and (4.2.2). However, in concave programming with inequality constraints, the order of subtraction is very important, since the constraint in the KKT conditions is always expressed in the ≥ 0 form.

The KKT conditions with inequality constraints can be explained in the single variable case as follows. Suppose we want to find a local maximum for an objective function $f(x)$ in the first quadrant $x \geq 0$ (this is the inequality constraint). There are three cases: (i) The critical point is an interior point of the first quadrant (Figure 4.1(a)): $f'(x) = 0$ and $x > 0$; (ii) the critical point is on the boundary (at G) (Figure 4.1(b)): $f'(x) = 0$ and $x = 0$; and (iii) critical point is at H or J (Figure 4.1(c)): $f'(x) < 0$ and $x = 0$. Thus, all these three cases can be stated concisely in one statement:

$$f'(x) \leq 0, \quad x \geq 0 \ \text{ and } \ xf'(x) = 0,$$

which is precisely contained in the KKT conditions. Note that these conditions exclude a point like A in Figure 4.1(a) where it is not a maximum, because $f'(K) > 0$.

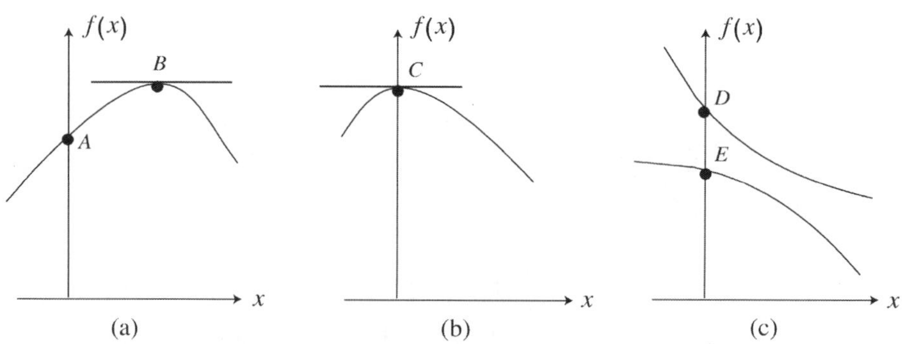

Figure 4.1 KKT Conditions in single variable.

Example 4.7. Consider the problem:

Maximize $f(x, y) \leq k$, where $x \geq 0, y \geq 0$, subject to the constraints:

$$
\begin{aligned}
&1.\ g(x^*, y^*) < k,\ x^* > 0, y^* > 0,\\
&2.\ g(x^*, y^*) = k,\ x^* > 0, y^* = 0,\\
&3.\ g(x^*, y^*) < k,\ x^* > 0, y^* = 0,\\
&4.\ g(x^*, y^*) < k,\ x^* = 0, y^* = 0,\\
&5.\ g(x^*, y^*) < k,\ x^* = 0, y^* > 0,\\
&6.\ g(x^*, y^*) = k,\ x^* = 0, y^* > 0,\\
&7.\ g(x^*, y^*) = k,\ x^* > 0, y^* > 0.
\end{aligned}
\tag{4.4.4}
$$

See Figure 4.2 for the feasible region and the sets where these constraints are applied.

Consider the non-negativity constraint $z \geq 0$. There are two basic types of solutions with this kind of constraint: (i) $h_1(z)$, and $h_2(z)$. We will consider $h_1(z)$ with a maximum at $z^* > 0$. In this case the maximum is at a point on a flat part of the function. Hence, $h_1'(z^*) = 0$.

Next consider $h_2(z)$ with a maximum at z^{**}. In this case the function could be flat at $z^{**} = 0$. But it could also be downward sloping, so $h_2'(z^{**}) \leq 0$ (see Figure 4.2b).

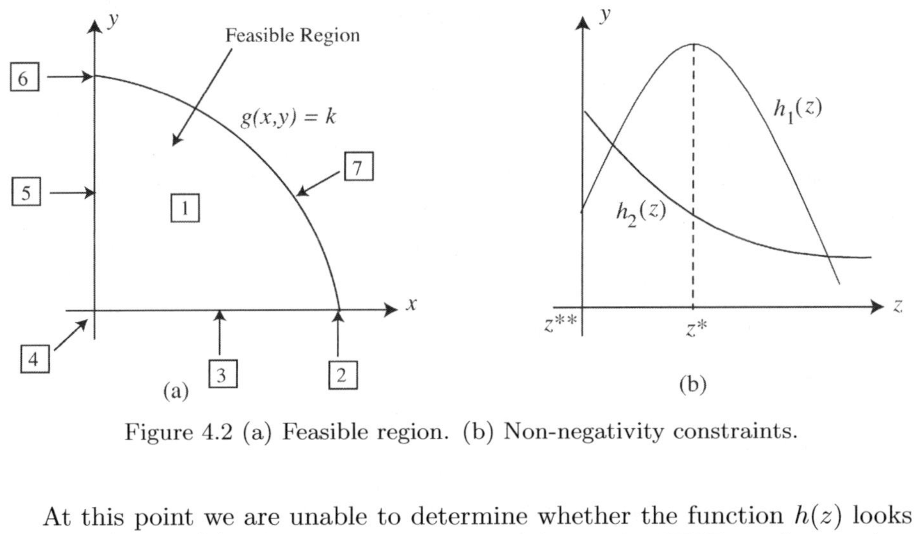

Figure 4.2 (a) Feasible region. (b) Non-negativity constraints.

At this point we are unable to determine whether the function $h(z)$ looks like $h_1(z)$ or $h_2(z)$. Thus, we need a set of first-order KKT conditions that allows for either case. These conditions can be written as

$$\begin{aligned}&\text{either } \; z^* = 0 \text{ and } h'(z^*0 \le 0,\\ &\text{or } \; z^* > 0 \text{ and } h'(z^*) = 0.\end{aligned}$$

These conditions can be written in short as

$$h'(z^*) \le 0 \text{ and } z^*(h'(z^*)) = 0.$$

Notice that the KKT conditions are similar to the complementary slackness conditions which are used in the Lagrangian formulation. Recall that $f(x, y)$ such that $g(x, y) < k$, without ignoring the non-negativity on x and y yet, the Lagrangian is

$$L = f(x, y) - \lambda(g(x, y) - k), \tag{4.4.5}$$

and the first-order conditions are

$$\begin{aligned}
L_x &= \frac{\partial f(x^*, y^*)}{\partial x} - \lambda g_x(x^*, y^*) = 0,\\
L_y &= \frac{\partial f(x^*, y^*)}{\partial y} - \lambda g_y(x^*, y^*) = 0,\\
&g(x^*, y^*) - k \le 0,\\
&\lambda^* g(x^*, y^*) - k = 0.
\end{aligned} \tag{4.4.6}$$

The last condition in (4.4.6) implies that either the constraints bind, or else the Lagrange multiplier λ is zero. This means that the solution could be in region $\boxed{7}$ or region $\boxed{1}$ (see Figure 4.2a).

The KKT conditions are relevant in problems where we add the non-negativity constraints $x \ge 0$ and $y \ge 0$ to the constrained maximization problem. This adds the restriction implied by these constraints directly into the first-order conditions, i.e., they capture the way the first-order conditions change when the solution is in the region $\boxed{2} - \boxed{6}$ in Figure 4.2(a).

Example 4.8. A company wants to maximize utility while spending no more than a predetermined budget. Suppose the concave programming problem is posed as:

$$\text{Maximize } \; u(x, y) \; \text{ subject to } B - p_x x - p_y y \ge 0, \;\; x, y \ge 0.$$

The Lagrangian is

$$L(x, y) = u(x, y) + \lambda(B - p_x x - p_y y).$$

First, L is maximized with respect to the variables x, y, subject to the KKT conditions:

$$\begin{aligned}
&\text{1a.} \quad \frac{\partial L}{\partial x} = u_x - \lambda^* p_x \le 0, \quad \frac{\partial L}{\partial y} = u_y - \lambda^* p_y \le 0,\\
&\text{1b.} \quad x^* \ge 0, \quad y^* \ge 0,\\
&\text{1c.} \quad x^*(u_x - \lambda^* p_x) = 0, \quad y^*(u_y - \lambda^* y_p) = 0.
\end{aligned}$$

Next, the Lagrangian L is minimized with respect to the variable λ and the related conditions:

$$\begin{aligned}
&\text{2a.} \quad \frac{\partial L}{\partial \lambda} = B - p_x x - p_y y,\\
&\text{2b.} \quad \lambda^* \ge 0,\\
&\text{2c.} \quad \lambda^*(B - p_x x - p_y y) = 0.
\end{aligned}$$

Thus, we have three cases of nontrivial solutions, which are as follows.

CASE 1. If $x^*, y^* > 0$, then from (1c) we have

$$u_x - \lambda^* p_x = 0, \quad u_y - \lambda^* p_y = 0,$$

which gives

$$\lambda^* = \frac{u_x}{p_x} = \frac{u_y}{p_y}. \tag{4.4.7}$$

Since $p_x, p_y > 0$ and assuming that the customer is unsatisfied (i.e., $u_x, u_y > 0$), we have $\lambda^* > 0$. But then from (2c) we will have $B - p_x x - p_y y = 0$. Thus, the budget constraint behaves exactly like an equality constraint (and not a weak inequality). Hence, the optimal points (x^*, y^*) will lie somewhere on the budget line and not below it. Further, from Eq (4.4.7) we also get $\dfrac{u_x}{u_y} = \dfrac{p_x}{p_y}$. Since $\dfrac{u_x}{u_y}$ is simply the slope of the indifference curve, and $\dfrac{p_x}{p_y}$ is the slope of the budget line, whenever both $x^*, y^* > 0$ (this case), the indifference curve will be tangent to the budget line at the point of optimization, and this provides an interior solution as in Figure 4.1a (see Figure 4.3a). This case, in fact, reduces the problem to the constrained optimization problem as discussed in previous section.

CASE 2. If $x^* = 0, y^* > 0$, then from (1.c) we have

$$u_x - \lambda^* p_x \le 0, \quad u_y - \lambda^* p_y = 0; \quad \text{thus} \quad \lambda^* \ge \frac{u_x}{p_x}, \quad \lambda^* = \frac{u_y}{p_y}. \tag{4.4.8}$$

Assuming that $u_x, u_y, p_x, p_y > 0$, we get $\lambda^* > 0$. Thus, from (2c) the budget constraint behaves exactly like an equality constraint, and not a weak inequality, even though only one variable is greater than zero and the other equals zero. Hence, as in case 1, the optimal point (x^*, y^*) will lie on the budget line and not below it (Figure 4.3a).

Now, substituting $\lambda^* = \dfrac{u_x}{u_y}$ on the right into (4.4.8) we get

$$\frac{u_x}{p_x} \le \frac{u_y}{p_y}, \quad \text{or} \quad \frac{u_x}{u_y} \le \frac{p_y}{p_x}.$$

This means that the indifference curves along the budget are everywhere flatter than the budget line, which leads to a corner solution in the upper left (Figure 4.3b), and at the corner solution the slope of the budget indifference curve that just touches the budget line may be flatter or equal to the slope of the budget line.

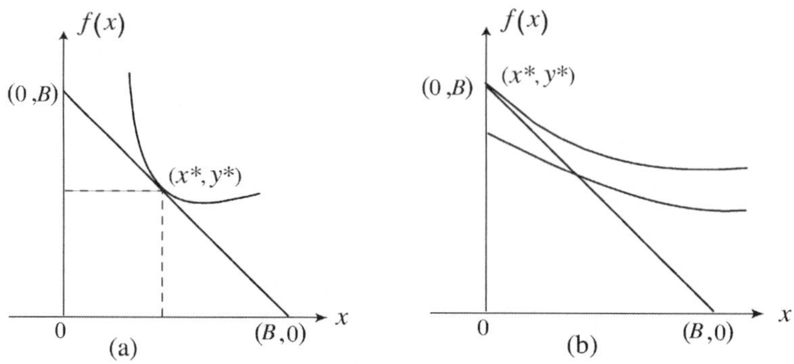

Figure 4.3 Two cases.

CASE 3. If $x^* = 0 = y^*$, then $L = u(x, y) + \lambda B$, and (1.a) gives $u_x \le 0, u_y \le 0$, and (1.b) and (1.c) are satisfied. Also, (2.a) becomes $L_\lambda = B > 0$, and then (2.c) gives $\lambda^* = 0$. Thus, the problem reduces to that of optimization with no constraints, which are discussed in §3.6.

This analysis provides an insight into the necessary and sufficient KKT conditions. However, their use in problems of practical applications is explained in the following examples, where x^*, y^*, λ^* still denote the critical (or optimal) values.

Example 4.9. Maximize the profit $\pi(x, y) = 30x - x^2 + 60y - 2y^2 - 11$ subject to the production constraint $x + y \le 36$.

First, the constraint is $36 - x - y \ge 0$. Then the Lagrangian is

$$\Pi(x, y) = 30x - x^2 + 60y - 2y^2 - 11 + \lambda(36 - x - y). \tag{4.4.9}$$

The KKT conditions are:

$$\begin{aligned}
&\text{1a.} \quad \Pi_x = 30 - 2x^* - \lambda^* \leq 0, \quad \Pi_y = 64 - 4y^* - \lambda^* \leq 0,\\
&\text{1b.} \quad x^* \geq 0, \quad y^* \geq 0,\\
&\text{1c.} \quad x^*(30 - 2x^* - \lambda^*) = 0, \quad y^*(64 - 4y^* - \lambda^*) = 0,\\
&\text{2a.} \quad \Pi_\lambda = 36 - x^* - y^*,\\
&\text{2b.} \quad \lambda^* \geq 0,\\
&\text{2c.} \quad \lambda^*(36 - x^* - y^*) = 0.
\end{aligned}$$

Next, we check these conditions as follows:

(i) Check if $\lambda^* = 0$ or $\lambda^* > 0$: If $\lambda^* = 0$, then (1a) gives $30 - 2x^* \leq 0$ and $64 - 4y^* \leq 0$, i.e., $x^* \geq 15, y^* \geq 16$, or $x^* + y^* = 31 > 36$, which violates the constraint. Thus, $\lambda^* \neq 0$, and we conclude that $\lambda^* > 0$.

(ii) If $\lambda^* > 0$, the constraint holds with equality, i.e.,

$$36 - x^* - y^* = 0.$$

(iii) If $\lambda^* > 0$, then we check if x^* or y^* can be zero. If $x = 0, y^* = 36$, then the second condition in (1c) is violated, because $36(64 - 4(36) - \lambda^*) \neq 0$ (since $\lambda^* > 0$). Similarly, if $y^* = 0, x^* = 36$, then the first condition in (1c) is violated, because $36(30 - 2(36) - \lambda^*) \neq 0$. Thus, neither x^* nor y^* can be zero, and from (1b) we conclude that both $x^* > 0, y^* > 0$.

(iv) Now, since $x^*, y^*, \lambda^* > 0$, we have the following equations from (1a) and (2a):

$$\begin{aligned}
30 - 2x^* - \lambda^* &= 0,\\
64 - 4y^* - \lambda^* &= 0,\\
36 - x^* - y^* &= 0,
\end{aligned}$$

to be solved simultaneously. This system is written in matrix form $\mathbf{Ax} = \mathbf{b}$ as

$$\begin{bmatrix} -2 & 0 & -1 \\ 0 & -4 & -1 \\ -1 & -1 & 0 \end{bmatrix} \begin{bmatrix} x^* \\ y^* \\ \lambda^* \end{bmatrix} = \begin{bmatrix} -30 \\ -64 \\ -36 \end{bmatrix}.$$

Using Cramer's rule, we get

$$|\mathbf{A}| = \begin{vmatrix} -2 & 0 & -1 \\ 0 & -4 & -1 \\ -1 & -1 & 0 \end{vmatrix} = 6, \qquad |\mathbf{A}|_1 = \begin{vmatrix} -30 & 0 & -1 \\ -64 & -4 & -1 \\ -36 & -1 & 0 \end{vmatrix} = 110,$$

$$|\mathbf{A}|_2 = \begin{vmatrix} -2 & -30 & -1 \\ 0 & -64 & -1 \\ -1 & -36 & 0 \end{vmatrix} = 36, \quad |\mathbf{A}|_3 = \begin{vmatrix} -2 & 0 & -30 \\ 0 & -4 & -64 \\ -1 & -1 & -36 \end{vmatrix} = 248.$$

Hence, the optimal values are $x^* = 110/6 = 18.33$, $y^* = 36/6 = 6$, $\lambda^* = 248/6 = 41.33$. Also, note that with $\lambda^* = 41.33$, a 1-unit increase in the constant of the constraint will increase the profit by approximately 43.33. ■

Example 4.10. In the foreign trade market, let the income be determined by

$$Y = C + I_0 + G_0 + X_0 + Z, \quad C = C_0 + bY, \quad Z = Z_0 + zY,$$

where X denotes the exports, Z the imports, the subscript zero indicates an exogenous fixed variable, b the marginal propensity to consume, and z the marginal propensity to import.

First, the Lagrangians for the above three functions are

$$\begin{aligned} F_1(Y, C, Z; C_0, I_0, G_0, X_0, Z_0, b, z) &= Y - C - I_0 - G_0 - X_0 - Z = 0, \\ F_2(Y, C, Z; C_0, I_0, G_0, X_0, Z_0, b, z) &= C - c_0 - bY, \\ F_3(Y, C, Z; C_0, I_0, G_0, X_0, Z_0, b, z) &= Z - Z_0 = zY. \end{aligned} \tag{4.4.10}$$

We will discuss three cases:

CASE 1. The partial derivatives of the unknown functions Y, C, Z with respect to X_0 are written in the matrix form as

$$\begin{bmatrix} \frac{\partial F_1}{\partial Y} & \frac{\partial F_1}{\partial C} & \frac{\partial F_1}{\partial Z} \\ \frac{\partial F_2}{\partial Y} & \frac{\partial F_2}{\partial C} & \frac{\partial F_2}{\partial Z} \\ \frac{\partial F_3}{\partial Y} & \frac{\partial F_3}{\partial C} & \frac{\partial F_3}{\partial Z} \end{bmatrix} \begin{bmatrix} \frac{\partial Y^*}{\partial x_0} \\ \frac{\partial C^*}{\partial X_0} \\ \frac{\partial Z^*}{\partial X_0} \end{bmatrix} = \begin{bmatrix} -\frac{\partial F_1}{\partial X_0} \\ -\frac{\partial F_2}{\partial X_0} \\ -\frac{\partial F_3}{\partial X_0} \end{bmatrix}, \tag{4.4.11}$$

where Y^*, C^*, Z^* denote the optimal values of the unknown functions. Substituting the values of the partial derivatives from (4.4.10) we obtain

$$\begin{bmatrix} 1 & -1 & 1 \\ -b & 1 & 0 \\ -z & 0 & 1 \end{bmatrix} \begin{bmatrix} \frac{\partial Y^*}{\partial x_0} \\ \frac{\partial C^*}{\partial X_0} \\ \frac{\partial Z^*}{\partial X_0} \end{bmatrix} = [1 \quad 0 \quad 0]. \tag{4.4.12}$$

Denoting this system as $\mathbf{Ax} = \mathbf{b}$, and using Cramer's rule, we find that

$$|\mathbf{A}| = 1 - b + z > 0, \ |\mathbf{A}_1| = \begin{vmatrix} 1 & -1 & 1 \\ 0 & 1 & 0 \\ 0 & 0 & 1 \end{vmatrix} = 1,$$

$$|\mathbf{A}_2| = \begin{vmatrix} 1 & 1 & 1 \\ -b & 0 & 0 \\ -z & 0 & 1 \end{vmatrix} = b, \quad |\mathbf{A}_3| = \begin{vmatrix} 1 & -1 & 1 \\ -b & 1 & 0 \\ -z & 0 & 0 \end{vmatrix} = z.$$

Thus,

$$\frac{\partial Y^*}{\partial x_0}=\frac{|\mathbf{A}_1|}{\mathbf{A}|}=\frac{1}{1-b+z}>0,$$
$$\frac{\partial C^*}{\partial x_0}=\frac{|\mathbf{A}_2|}{\mathbf{A}|}=\frac{b}{1-b+z}>0,$$
$$\frac{\partial Z^*}{\partial x_0}=\frac{|\mathbf{A}_3|}{\mathbf{A}|}=\frac{z}{1-b+z}>0.$$

CASE 2. The partial derivatives of the functions Y^*, C^*, Z^* with respect to b are obtained from (4.4.12) by replacing the matrix $\mathbf{x}$ by $\begin{bmatrix} \frac{\partial Y^*}{\partial b} & \frac{\partial C^*}{\partial b} & \frac{\partial Z^*}{\partial b} \end{bmatrix}^T$, thus giving

$$\begin{bmatrix} 1 & -1 & 1 \\ -b & 1 & 0 \\ -z & 0 & 1 \end{bmatrix} \begin{bmatrix} \frac{\partial Y^*}{\partial b} \\ \frac{\partial C^*}{\partial b} \\ \frac{\partial Z^*}{\partial b} \end{bmatrix} = [0 \quad Y^* \quad 0]. \tag{4.4.13}$$

Then using Cramer's rule we get

$$\frac{\partial Y^*}{\partial b}=\frac{Y^*}{1-b+z}>0, \quad \frac{\partial C^*}{\partial b}=\frac{(1+z)Y^*}{1-b+z}>0, \quad \frac{\partial Z^*}{\partial b}=\frac{zY^*}{1-b+z}>0.$$

CASE 3. The partial derivatives of the functions Y^*, C^*, Z^* with respect to z are obtained from (4.4.12) by replacing the matrix $\mathbf{x}$ by $\begin{bmatrix} \frac{\partial Y^*}{\partial z} & \frac{\partial C^*}{\partial z} & \frac{\partial Z^*}{\partial z} \end{bmatrix}^T$, thus giving

$$\begin{bmatrix} 1 & -1 & 1 \\ -b & 1 & 0 \\ -z & 0 & 1 \end{bmatrix} \begin{bmatrix} \frac{\partial Y^*}{\partial b} \\ \frac{\partial C^*}{\partial b} \\ \frac{\partial Z^*}{\partial b} \end{bmatrix} = [0 \quad 0 \quad Y^*]. \tag{4.4.14}$$

Then using Cramer's rule we get

$$\frac{\partial Y^*}{\partial z}=\frac{-Y^*}{1-b+z}<0, \quad \frac{\partial C^*}{\partial z}=\frac{(-bY^*}{1-b+z}<0, \quad \frac{\partial Z^*}{\partial z}=\frac{(1-b)Y^*}{1-b+z}>0. \blacksquare$$

4.5 Application to Mathematical Economics

The KKT method is often used in theoretical models in order to obtain qualitative results.

Example 4.11. (Value function) Consider the optimization problem with constant inequality constraints:

$$\text{Maximize } f(\mathbf{x}) \text{ subject to } g_i(\mathbf{x}) \le a_i,\ h_j(\mathbf{x}) = 0. \tag{4.5.1}$$

The value function is defined as

$$V(a_1, \dots, a_n) = \sup_x f(\mathbf{x}) \text{ subject to } g_i(x) \le a_i,\ h_j(x) = 0,$$
$$i = 1, \dots, m,\ j = 1, \dots, k. \tag{4.5.2}$$

Note that $\operatorname{dom} V = \{a_i \in \mathbb{R}^m\}$ for some $\mathbf{x} \in \mathbb{R}^n, g_i(\mathbf{x}) \le a_i,\ i = 1, \dots, m$.

SOLUTION. Each coefficient μ_i is the rate at which the value function V increases as a_i increases. Thus, if we interpret each a_i as a resource constraint, the coefficients μ_i determine the size of increase in the optimum value of the function V that will result from an increase in a resource. This sort of interpretation is important in economics and is used in utility maximization problems. ■

Example 4.12. Consider, for example, the case of a firm that wants to maximize its sales revenue subject to a minimum profit constraint. Let Q denote the quantity of output produced (to be chosen), $R(Q)$ the sales revenue with a positive first derivative and with zero value at zero output, $C(Q)$ the production costs with a positive first derivative and with a nonnegative value at zero output, and $G_{\min}$ the positive minimal acceptable level of profit. Then the problem is of the type (4.3.1) provided that the revenue function levels off so it eventually is less steep than the cost function. This problem can be formulated as follows:

$$\text{Maximize } R(Q) \text{ subject to } G_{\min} \le R(Q) - C(Q),\ Q > 0, \tag{4.5.3}$$

where the KKT conditions are

$$\begin{aligned}
&(1+\mu)\frac{dR}{dQ} - \mu\frac{dC}{dQ} \le 0,\ Q \ge 0\\
&(1+\mu)Q\frac{dR}{dQ} - \mu\frac{dC}{dQ} = 0,\\
&R(Q) - C(Q) - G_{\min} \ge 0,\\
&\mu \ge 0,\\
&\mu\big[R(Q) - C(Q) - G_{\min}\big] = 0.
\end{aligned} \tag{4.5.4}$$

SOLUTION. We cannot take $Q = 0$, because this choice violates the minimum profit constraint. Hence, $Q > 0$, and the third KKT condition implies that the first condition holds with equality. Solving this equality we get

$$\frac{dR}{dQ} = \frac{\mu}{1+\mu}\frac{dC}{dQ}. \tag{4.5.5}$$

Since, by assumption, both dR/dQ and dC/dQ are strictly positive, the inequality in the non-negativity condition on μ implies that $\mu > 0$. Thus, this firm operates at a level of output at which the marginal revenue dR/dQ is less than the marginal cost dC/dQ. This conclusion is interesting because it is in contradiction with the behavior of a profit maximizing firm which operates at a level at which they are equal.

4.5.1 Peak Load Pricing. (Supply-and-demand Pricing) Peak and off-peak pricing requires planning as a maximization problem for firms that invest in capacity in order to target a primary market. However, there is always a secondary market in which the firm can often sell its product. Once the capacity has been purchased to service the firm's primary market, the capital is freely available (up to the capacity) to be used in the secondary market. Typical examples include: schools and universities who build to meet day-time needs (peak) but may offer night-school classes (off-peak); theaters that offer shows in the evening (peak) and matinees (off-peak); or trucking companies who have dedicated routes but may choose to enter 'back-haul' markets. Since the capacity price is a factor in the profit maximizing decision for the peak market and is already paid, it is normally not a factor in calculating optimal price and quality for the smaller (off-peak) market. However, if the secondary market's demand is close to the same size as the primary market, capacity constraints may be an issue, especially given that it is common practice to price discriminate and charge lower prices in off-peak periods. Even though the secondary market is smaller than the primary, it is possible at the lower (profit maximizing) price that off-peak demand exceeds capacity. In such cases, choices must be made taking both markets into account, making the problem a classic application of KKT conditions.

Consider a profit maximizing company which faces two demand curves

$$\begin{aligned} P_1 &= D^1(Q_1) \quad \text{in the peak period,} \\ P_2 &= D^2(Q_2) \quad \text{in the off-peak period.} \end{aligned} \tag{4.5.6}$$

To operate, the firm must pay b per unit of output, whether it is peak (day) or off-peak (night). Moreover, the firm must purchase capacity at a cost of c per unit of output. Let K denote the total capacity measured in units of Q. The firm must pay for capacity, regardless of whether it operates in the off-peak period or not. Then the question is: Who should be charged for the capacity costs, peak or off-peak customers? Thus, the firm's maximization problem becomes

$$\begin{aligned} &\text{Maximize } P_1Q_1 + P_2Q_2 - b(Q_1 - Q_2) - cK \\ &\quad \text{subject to } K \geq Q_1, \ K \geq Q_2, \end{aligned} \tag{4.5.7}$$

where P_1 and P_2 are defined in (4.5.6). The Lagrangian of this problem is

$$L = D^1(Q_1)Q_1 + D^2(Q_2)Q_2 - b(Q_1 + Q_2) - cK + \lambda_1(K - Q_1)\lambda_2(K - Q_2), \tag{4.5.8}$$

and the KKT conditions are

$$\begin{aligned}
&L_1 = D^1 + Q_1 \frac{\partial D^1}{\partial Q_1} - b - \lambda_1 = 0, \quad (MR_1 - b - \lambda_1 = 0),\\
&L_2 = D^2 + Q_1 \frac{\partial D^2}{\partial Q_2} - b - \lambda_2 = 0, \quad (MR_2 - b - \lambda_2 = 0),\\
&L_K = -c + \lambda_1 + \lambda_2 = 0, \qquad (c = \lambda_1 + \lambda_2),\\
&L_{\lambda_1} = K - Q_1 \ge 0, \;\; \lambda_1 \ge 0,\\
&L_{\lambda_2} = K - Q_2 \ge 0, \;\; \lambda_2 \ge 0.
\end{aligned} \tag{4.5.9}$$

To find a solution, we follow the following steps:

STEP 1. Since $D^2(Q_2) < D^1(Q_1)$, we choose and try $\lambda_2 = 0$. Then the KKT conditions (4.5.9) give

$$MR_1 = b + c - \lambda_2 = b + c, \quad MR_2 = b + \lambda_2 = b,$$

which implies that $K + Q_1$. Then we check to see if $Q_2^* \le K$. If this inequality is true, then we have a valid solution. Otherwise, the second constraint is violated and the assumption $\lambda_2 = 0$ becomes false. Then we go to the next step.

STEP 2. If $Q_2^* > K$, then $Q_1^* = Q_2^* = K$, and $MR_1 = b + \lambda_1$, $MR_2 = b + \lambda_2$. Since $c = \lambda_1 + \lambda_2$, then λ_1 and λ_2 represent the share of c each group must pay.

We will illustrate this optimization problem by a numerical example, as follows.

Suppose the demand during the peak period is $P_1 = 22 - 10^{-5} Q_1$, and during the off-peak period is $P_2 = 18 - 10^{-5} Q_2$.

To produce a unit of output per half-day requires a unit of capacity costing 8 cents per day. The cost of a unit capacity is the same whether it is used during the peak period only or off-peak period also. In addition to the costs of capacity, it costs 6 cents in operating cost (labor and fuel) to produce 1 unit per half day (both during peak and off-peak periods).

If we assume that the capacity constraint is binding ($\lambda_2 - 0$), then the KKT conditions (4.5.9) become

$$\begin{aligned}
&\lambda_1 c = 8,\\
&\overbrace{22 - 2 \times 10^{-5} Q_1}^{\text{MR}} = \overbrace{b + c}^{\text{MC}} = 14,\\
&18 - 2 \times 10^{-5} Q_2 \; = \; b \quad = 6.
\end{aligned}$$

Solving this system we get $Q_1 = 40,000$, $Q_2 = 60,000$, which violates the assumption $Q_2 > Q_1 = K$ (i.e., the second constraint is non-binding). Hence, assuming that both constraints are binding, i.e., $Q_1 = Q_2 = Q$, the KKT conditions become

$$\begin{aligned} &\lambda_1 + \lambda_2 = 8, \\ &22 - 2 \times 10^{-5} Q = 6 + \lambda_1, \\ &18 - 2 \times 10^{-5} Q = 6 + \lambda_2, \end{aligned} \tag{4.5.10}$$

solving which we get $Q = K = 50,000$, $\lambda_1 = 5$, $\lambda_2 = 2$, $P_1 = 17$, $P_2 = 13$. Since the capacity constraint is binding in both markets, the peak market pays $\lambda_1 = 6$ and the off-peak market pays $\lambda_2 = 2$ of the capacity cost. ■

4.6 Comparative Statics

SUPPLY-DEMAND MODEL. Consider the case when there is one endogenous variable (like price p) and one exogenous variable (like consumers' income y). Let the supply Q_S and the demand Q_D of a commodity be defined by

$$\begin{aligned} &Q_S = a + bp, \ a, b > 0, \\ &Q_D = m - np + ky, \end{aligned} \tag{4.6.1}$$

subject to the equilibrium condition $Q_s = Q_d$. We will solve this system for the equilibrium price level p^* by writing the above equations as

$$\begin{aligned} &a + bp = m - np + ky, \\ &(b + a)p = m - a + ky, \end{aligned}$$

which gives $p \equiv p^* = \dfrac{m - a + ky}{b + n}$. To determine the equilibrium level of p^*, we find the change in p^* with respect to y, or any of the other five parameters a, b, m, n, k, we have $\dfrac{dp^*}{dy} = \dfrac{k}{b_n} > 0$. This means that an increase in consumers' income will result in an increase in the equilibrium price of the commodity.

This analysis can also be carried out by defining Eqs (4.6.1) explicitly as $Q_s - Q_D = 0$, i.e., the implicit function $F = a + bp - m + np - ky$. Then, assuming $F_p \neq 0$, we get

$$\frac{dp^*}{dy} = -\frac{F_y}{F_p}.$$

Since by differentiating we get $F_p = b + n$ and $F_y = k$, the ratio is given by

$$\frac{dp^*}{dy} = \frac{k}{b + n} > 0.$$

Next, consider the case when there is more than one endogenous variable. Thus, let

$$\begin{aligned} &F_1(y_1, y_2; x_1, x_2) = 0, \\ &F_2(y_1, y_2; x_1, x_2) = 0. \end{aligned} \tag{4.6.2}$$

To find the partial derivatives of this system with respect to one variable, say y_1, the total derivative of both functions with respect to x_1 is given by

$$\begin{aligned}\frac{\partial F_1}{\partial y_1}\frac{\partial y_1}{\partial x_1}+\frac{\partial F_1}{\partial y_2}\frac{\partial y_2}{\partial x_1}+\frac{\partial F_1}{\partial x_1}=0,\\ \frac{\partial F_2}{\partial y_1}\frac{\partial y_1}{\partial x_1}+\frac{\partial F_2}{\partial y_2}\frac{\partial y_2}{\partial x_1}+\frac{\partial F_2}{\partial x_1}=0.\end{aligned}\tag{4.6.3}$$

These equations can be written in matrix form as

$$\begin{bmatrix}\frac{\partial F_1}{\partial y_1} & \frac{\partial F_1}{\partial y_2}\\ \frac{\partial F_2}{\partial y_1} & \frac{\partial F_2}{\partial y_2}\end{bmatrix}\begin{bmatrix}\frac{\partial y_1^*}{\partial x_1}\\ \frac{\partial y_2^*}{\partial x_1}\end{bmatrix}=\begin{bmatrix}-\frac{\partial F_1}{\partial x_1}\\ -\frac{\partial F_2}{\partial x_1}\end{bmatrix},$$

or

$$\mathbf{JX}=\mathbf{B},\tag{4.6.4}$$

where y_1^*, y_2^* denote the incomes at the equilibrium point. Since

$$|\mathbf{J}|=\frac{\partial F_1}{\partial y_1}\frac{\partial F_2}{\partial y_2}-\frac{\partial F_2}{\partial y_1}\frac{\partial F_1}{\partial y_2}\neq 0,$$

the optimal values of the endogenous values y_1^*, y_2^* are determined as implicit functions of exogenous variables x_1:

$$\frac{\partial y_i}{\partial x_i}=\frac{|\mathbf{J}_i|}{|\mathbf{J}|}.\tag{4.6.5}$$

Using Cramer's rule, the first derivative $\frac{\partial y_1^*}{\partial x_1}$ is obtained by replacing the first column of $\mathbf{J}$ with the column vector $\mathbf{B}$, and the second derivative $\frac{\partial y_2^*}{\partial x_1}$ is obtained by replacing the second column of $\mathbf{J}$ with the column vector $\mathbf{B}$:

$$\frac{\partial y_1^*}{\partial x_1}=\frac{|\mathbf{J}_1|}{|\mathbf{J}|}=\frac{\begin{vmatrix}-\frac{\partial F_1}{\partial x_1} & \frac{\partial F_1}{\partial y_2}\\ -\frac{\partial F_2}{\partial x_1} & \frac{\partial F_2}{\partial y_2}\end{vmatrix}}{\begin{vmatrix}\frac{\partial F_1}{\partial y_1} & \frac{\partial F_1}{\partial y_2}\\ \frac{\partial F_2}{\partial y_1} & \frac{\partial F_2}{\partial y_2}\end{vmatrix}}=\frac{-\left(\frac{\partial F_1}{\partial x_1}\frac{\partial F_2}{\partial y_2}-\frac{\partial F_2}{\partial x_1}\frac{\partial F_1}{\partial y_2}\right)}{\frac{\partial F_1}{\partial y_1}\frac{\partial F_2}{\partial y_2}-\frac{\partial F_2}{\partial y_1}\frac{\partial F_1}{\partial y_2}},$$

and

$$\frac{\partial y_2^*}{\partial x_1} = \frac{|\mathbf{J}_2|}{|\mathbf{J}|} = \frac{\begin{vmatrix} -\dfrac{\partial F_1}{\partial y_1} & -\dfrac{\partial F_1}{\partial x_1} \\ -\dfrac{\partial F_2}{\partial y_1} & -\dfrac{\partial F_2}{\partial x_1} \end{vmatrix}}{\begin{vmatrix} \dfrac{\partial F_1}{\partial y_1} & \dfrac{\partial F_1}{\partial y_2} \\ \dfrac{\partial F_2}{\partial y_1} & \dfrac{\partial F_2}{\partial y_2} \end{vmatrix}} = \frac{-\left(\dfrac{\partial F_1}{\partial y_1}\dfrac{\partial F_2}{\partial x_1} - \dfrac{\partial F_2}{\partial y_1}\dfrac{\partial F_1}{\partial x_1}\right)}{\dfrac{\partial F_1}{\partial y_1}\dfrac{\partial F_2}{\partial y_2} - \dfrac{\partial F_2}{\partial y_1}\dfrac{\partial F_1}{\partial y_2}}.$$

The partial derivatives $\dfrac{\partial y_1^*}{\partial x_2}$ and $\dfrac{\partial y_2^*}{\partial x_2}$ are determined by the same method.

Example 4.13. Let the equilibrium in the goods and service market (IS curve) and the money market (LM curve) be given, respectively, by

$$F_1(Y, i; C_0, M_0, P_0) = Y - C_0 - C(Y, i) = 0, \quad 0 < C_Y < 1,\ C_i < 0, \tag{4.6.6}$$

$$F_2(Y, i; C_0, M_0, P_0) = L(Y, i) - M_0/P = 0, \quad L_Y > 0, L_i < 0, \tag{4.6.7}$$

where $L(Y, i)$ denotes the demand for money, M_0 the supply of money, C_0 the autonomous consumption, i the interest, and P the price level; thus M_0/P becomes the supply of real money. For the sake of simplicity, hold P as constant. Then the equilibrium level of P and i is affected by a change in C_0. Using the above method and Cramer's rule, we get from (4.6.6) and (4.6.7):

$$\begin{aligned} \frac{\partial Y}{\partial C_0} - 1 - C_Y\frac{\partial Y}{\partial C_0} - C_i\frac{\partial i}{\partial C_0} &= 0, \\ L_Y\frac{\partial Y}{\partial C_0} + L_i\frac{\partial i}{\partial C_0} &= 0, \end{aligned}$$

which in matrix form $\mathbf{JX} = \mathbf{B}$ is

$$\begin{bmatrix} 1 - C_Y & -C_i \\ L_Y & L_i \end{bmatrix} \begin{bmatrix} \dfrac{\partial Y^*}{\partial C_0} \\ \dfrac{\partial i^*}{\partial C_0} \end{bmatrix} = \begin{bmatrix} 1 \\ 0 \end{bmatrix}.$$

Then, using Cramer's rule, we get

$$\begin{aligned} \frac{\partial Y^*}{\partial C_0} &= \frac{L_i}{(1 - C_Y)L_i + C_iL_Y} > 0, \\ \frac{\partial i^*}{\partial C_0} &= \frac{-L_i}{(1 - C_Y)L_i + C_iL_Y} > 0. \end{aligned}$$

This means that an increase in C_0 will produce an increase in the equilibrium level of interest i. ■

Example 4.14. (Wartime Rationing) The civilian population is subjected to rationing of basic consumer goods during wartime. The rationing process and control is effected through the use of coupons issued by the government, which ensures that each consumer is allotted coupons every month. The consumers on their part redeem a certain number of coupons at the time of purchasing the food items they need. The consumer, however, pays two prices at the time of purchase, one for the purchase of rationed goods and the other for the price of the coupon. This practice requires that the consumer must have sufficient money and sufficient coupons at the time of purchasing a rationed item. As an example, we will analyze this rationing process by considering two rationed goods, say x and y.

Let a consumer's utility function be $u(x, y)$, and assume that the consumer has a fixed money budget of B dollars, and the prices of the two goods be P_x and P_y, respectively. Also assume that the consumer is allotted coupons C to be used to purchase both goods x and y at coupon prices of c_x and c_y. This leads to the following consumer's maximization problem:

$$\begin{aligned}&\text{Maximize } u(x, y)\\ &\qquad \text{subject to } B \geq P_x x + P_y y \text{ and } C \geq c_x x + c_y y,\ x \geq 0, y \geq 0.\end{aligned} \tag{4.6.8}$$

The Lagrangian for this problem is

$$L(x, y) = u(x, y) + \lambda_1(B - P_x x - P_y y) + \lambda_2(C - c_x x - x_y y), \tag{4.6.9}$$

where λ_1, λ_2 are the Lagrange multipliers on the budget and coupon constraints, respectively. The KKT conditions are

$$\begin{aligned}L_x &= u_x - \lambda_1 P_x - \lambda_2 c_y = 0,\\ L_y &= u_y - \lambda_1 P_y - \lambda_2 c_y = 0,\\ L_{\lambda_1} &= B - P_x x - P_y y \geq 0,\ \lambda_1 \geq 0,\\ L_{\lambda_2} &= C - c_x x - c_y y \geq 0,\ \lambda_2 \geq 0.\end{aligned} \tag{4.6.10}$$

Example 4.15. We will solve this optimization problem using the following data: Let the utility function be $u(x, y) = xy^2$, $B = 50$, $P_x = P_y = 1$, $C = 60$, and $c_x = 2$, $c_y = 1$. The Lagrangian (4.6.9) becomes

$$L = xy^2 + \lambda_1(50 - x - y) + \lambda_2(60 - 2x - y), \tag{4.6.11}$$

and the KKT conditions (4.6.10) become

$$\begin{aligned}L_x &= y^2 - \lambda_1 - 2\lambda_2 = 0,\ x \geq 0,\\ L_y &= 2xy - \lambda_1 - \lambda_2 = 0,\ y \geq 0,\\ L_{\lambda_1} &= 50 - x - y \geq 0,\ \lambda_1 \geq 0,\\ L_{\lambda_2} &= 60 - 2x - y \geq 0,\ \lambda_2 \geq 0.\end{aligned} \tag{4.6.12}$$

Using the trial-and-error method we proceed as follows: we choose one of the constraints to be non-binding and solve for x and y. After the solution, use these values to test if the constraint that was chosen to be non-binding is violated. If so, then re-do the process choosing the other constraint to be non-binding. If we find that this chosen non-binding constraint is again violated, then we can assume both constraints bind and the solution is determined by the constraints. This leads to the following steps:

STEP 1. (Ignoring the coupon constraint) Assume $\lambda_2 = 0, \lambda_1 > 0$. Then the first-order KKT conditions (4.6.12) become

$$\begin{aligned} L_x &= y^2 - 2\lambda_2 = 0, \\ L_y &= 2xy - \lambda_2 = 0, \\ L_{\lambda_2} &= 60 - 2x - y = 0. \end{aligned} \tag{4.6.13}$$

Solving for x and y we get $x^* = 10$, $y^* = 40$.

STEP 2. (Ignoring the budget constraint) Assume $\lambda_1 = 0, \lambda_2 > 0$. Then the first-order KKT conditions (4.6.12) become

$$\begin{aligned} L_x &= y^2 - \lambda_1 = 0, \\ L_y &= 2xy - \lambda_1 = 0, \\ L_{\lambda_1} &= 50 - x - y = 0. \end{aligned} \tag{4.6.14}$$

Solving for x and y we get $x^* = 16.67$, $y^* = 33.33$. But these values when substituted into the coupon constraint (last equation in (4.6.13)) we find that $2x^* + y^* = 2(10) + 40 = 60$. Thus, this solution does not violate the budget constraint. In fact, it just meets this constraint. However, this result is unusual in the sense that although the budget constraint is met, it is not binding due to the particular location of the coupon constraint. ■

4.7 Exercises

4.1. For the following function, find the critical points and determine if at these points the function is a relative maximum, relative minimum, inflection point, or saddle point.

(a) $f(x, y) = 48y - 3x^2 - 6xy - 2y^2 + 72x$.
(b) $f(x, y) = 5x^2 - 3y^2 - 30x + 7y + 4xy$.
(c) $f(x, y) = 3x^3 - 5y^2 - 225x + 70y + 20$.
(d) $f(x, y) = x^3 + 2y^3 - 3x^2 + 9y^2 - 45x - 60y$

ANS. (a) Critical point $(0, 12)$; inflection point at $(0, 12)$.

(b) Critical point $(2, 5/2)$; saddle point.

(c) Critical points $(5, 7), (-5, 7)$; relative maximum at $(-5, 7)$, saddle point at $(5, 7)$.

(d) Critical points $(-3, 2), (-3, -5), (5, 2), (5, -5)$; relative maximum at

$(-3,-5)$ and a relative minimum at $(5,2)$; saddle points may be at $(-3,2)$ and $5,-5)$.

4.2. Consider $f(x,y) = 52x + 36y - 4xy - 6x^2 - 3y^2 + 5$. The first-order partial derivatives, equated to zero, give $f_x = 52 - 4y - 12x = 0$, $f_y = 36 - 4x - 6y = 0$, so the critical point is $(3,4)$. The second-order partial derivatives are: $f_{xx} = -12, f_{xy} = -4, f_{yy} = -6$. Since both $f_{xx} < 0$, $F_{yy} < 0$, and $(f_{xy})^2 = (-4)^2 > 0$, and $f_{xx} \cdot f_{yy} > (f_{xy})^2$ at the point $(3,4)$, the function f has a global maximum at $(3,4)$.

4.3. Consider $f(x,y) = 48y - 3x^2 - 6xy - 2y^2 + 60x$. The first-order partial derivatives, equated to zero, give $f_x = -6x - 6y + 60 = 0$, $f_y = -6x - 4y + 48 = 0$, so the critical point is $(4,6)$, The second-order partial derivatives are: $f_{xx} = -6 < 0, f_{xy} = -6 = f_{yx} < 0, f_{yy} = -4 < 0$, and $f_{xx} \cdot f_{xy} = (-6)(-4) = 24 < (f_{xY})^2 = 36$. The function f has an inflection point at $(4,6)$.

4.4. Use the method of Lagrange multipliers to solve the problem: Given a budget constraint of \$110 when $P_K = 3$ and $P_L = 4$, maximize the generalized Cobb-Douglas production function $q = K^{0.4}L^{0.5}$.

HINT. The Lagrangian is $Q = K^{0.4}L^{0.5} + \lambda(162 - 3K - 4L)$. The critical values are $K^* = 24, L^* = 22.5$. Next, $|H_2| > 0, |\bar{\mathbf{H}}|$ is negative definite, so Q is maximized at the critical values.

4.5. Use the method of Lagrange multipliers to solve the problem: Maximize $f(x,y) = \frac{1}{2}x^2 = \frac{1}{2}y^2 - 2xy - y$ subject to the equality constraint $g(x,y) = x + y - 2$. SOLUTION. The Lagrangian for the problem is

$$L(x,y,\lambda) = \frac{1}{2}x^2 + \frac{1}{2}y^2 - 2xy - y + \lambda(x + y - 2).$$

Using the KKT conditions we get $L_x = x - 2y + \lambda = 0$, $L_y = y - 2x - 1 + \lambda = 0$, $L_\lambda = x + y - 2$, solving which we get $x^* = \frac{1}{2}, y^* = 1$, so $\lambda^* = 0$, and $f(x^*,y^*) = -1.375$. The Hessian for the problem is

$$|\mathbf{H}| = \begin{vmatrix} L_{xx} & L_{xy} \\ L_{yx} & L_{yy} \end{vmatrix} = \begin{vmatrix} 1 & -2 \\ -2 & 1 \end{vmatrix} = -3 < 0.$$

Hence $f(x,y)$ has a local maximum at $(\frac{1}{2}, 1)$.

4.6. Maximize the utility function $u(x,y) = x^{0.3}y^{0.4}$ subject to the budget constraint $g(x,y) = 2x + 8y = 172$.

HINT. Let $u(x,y,\lambda) = x^{0.3}y^{0.4} + \lambda(172 - 2x - 8y)$. The critical values are $x^* = 6, y^* = 20, \lambda^* = 0.14$, and the bordered Hessian $|\bar{\mathbf{H}}|$ is negative definite and the utility is maximized at the critical values.

4.7. If the equilibrium in the goods and service market (IS curve) and the money market (LM curve) are defined as in Example 4.13, what effect will y^* and i^* have on a change in M_0.

ANS. Take p as a constant. Then

$$\frac{\partial y}{\partial M_0} - \left(C_y \frac{\partial y}{\partial M_0}\right) - \left(C_i \frac{\partial i}{\partial M_0}\right) = 0,$$
$$\left(L_y \frac{\partial y}{\partial M_0}\right)\left(L_i \frac{\partial i}{\partial M_0}\right) - \frac{1}{P} = 0.$$

or

$$\begin{bmatrix} 1 - C_y & -C_i \\ L_y & L_i \end{bmatrix} \begin{bmatrix} \dfrac{\partial y^*}{\partial M_0} \\ \dfrac{\partial i^*}{\partial M_0} \end{bmatrix} = \begin{bmatrix} 0 \\ 1/p \end{bmatrix}.$$

Then using Cramer's rule we get

$$\frac{\partial y^*}{\partial M_0} = \frac{C_i}{P(1 - C_y)L_i + C_i L_y} > 0,$$
$$\frac{\partial i^*}{\partial M_0} = \frac{1 - C_i}{P(1 - C_y)L_i + C_i L_y} < 0.$$

This means that an increase in the money supply M_0 will increase the equilibrium level of income, but decrease the equilibrium interest rate.

4.8. Use Lagrange multipliers to optimize $f(x, y, z) = xyz^2$ subject to the constraint $x + y + z = 20$. The Lagrangian is

$$F(x, y, z) = xyz^2 + \lambda(20 - x - y - z).$$

Then $F_x = yz^2 - \lambda = 0$, $F_y = xz^2 - \lambda = 0$, $F_z = 2xyz - \lambda = 0$, $F_\lambda = 20 - x - y - z = 0$. To solve these equations simultaneously, we equate λ from the first two equations, and from the first and the third equation, giving:

$$yz^2 = xz^2, \quad yz^2 = 2xyz,$$

or $y = x$ and $z = 2x$. Substituting these in the fourth equation we get: $20 - x - x - 2x = 0$, or $x^* = 5$, which gives $y^* = 5, z^* = 10, \lambda^* = 500$ as critical values. Thus, $F(5, 5, 10) = 2500$.

The second-order derivatives are: $F_{xx} = 0, F_{yy} = 0, F_{zz} = 2xy, F_{xy} = z^2, F_{yz} = 2xz, F_{xz} = 2yz$. Also, from $g(x, y, z) = x + y + z - 20$, we get

$g_x = 1 = g_y = g_z$. Then the bordered Hessian from Eq (A.24), using the second form, is

$$|\mathbf{H}| = \begin{bmatrix} 0 & 1 & 1 & 1 \\ 1 & 0 & z^2 & 2yz \\ 1 & z^2 & 0 & 2xz \\ 1 & 2yz & 2xz & 2xy \end{bmatrix}.$$

The second principal minor is $|\bar{H}_2| = 0 - 1(-z^2) + 1(z^2) = 2z^2$. Thus, $|\bar{H}_2|_{10} > 0$. The third principal minor is

$$\begin{aligned} |\bar{H}_3| = |\mathbf{H}| = 0 - 1 \begin{vmatrix} 1 & z^2 & 2yz \\ 1 & 0 & 2xz \\ 1 & 2xz & 2xy \end{vmatrix} + 1 \begin{vmatrix} 1 & 0 & 2yz \\ 1 & z^2 & 2xz \\ 1 & 2yz & 2xy \end{vmatrix} - 1 \begin{vmatrix} 1 & 0 & z^2 \\ 1 & z^2 & 0 \\ 1 & 2yz & 2xz \end{vmatrix} \\ = -\big[1(0 - 2xz \cdot 2xz) - z^2(2xy - 2xz) + 2yz(2yz - 0)\big] \\ + \big[1(z^2 \cdot 2xy - 2yz \cdot 2xz) - 0 + 2yz(2yz - z^2)\big] \\ - \big[1(z^2 \cdot 2xz - 0) - 0 + z^2(2yz - z^2)\big] \\ = z^4 - 4xz^3 - 4yz^3 - 4xyz^2 + 4x^2z^2 + 4y^2z^2. \end{aligned}$$

Thus, $|\bar{H}_3|_{5,5,10} = -20000 < 0$. Hence, $|\bar{H}_2| > 0$ and $|\bar{H}_3| < 0$ imply that $|\mathbf{H}|$ is negative definite, and the function f is maximized at the critical values.

4.9. Maximize the total utility defined by $u(x, y) = 10x^2 + 15xy - 3y^2$ when the firm meets the quota $g(x, y)$ equal to $2x + y = 23$.

ANS. Critical values: $x^* = 9$, $y^* = 5$, $\lambda^* = 105$; $|\mathbf{H}|$ is negative definite and u is maximized at $(x^*, y^*) = (9, 5)$.

4.10. Maximize the utility function $u = Q_1Q_2$ when $P_1 = 1, P_2 = 3$, and the firm's budget is $B = 60$. Also estimate the effect of a 1-unit increase in the budget. The budget constraint is $Q_1 + Q_2 = 60$, and the constraint is $Q_1 + 3Q_2 = 60$. We consider the Lagrangian

$$L = Q_1Q_2 + \lambda(60 - Q_1 - 3Q_2).$$

The first-order partial derivatives equated to zero give: $L_{Q_1} = Q_2 - \lambda = 0$, $L_{Q_2} = Q_1 - 3\lambda = 0$, $L_\lambda = 60 - Q_1 - 4Q_2 = 0$. Solving these equations simultaneously we obtain the critical values: $Q_1^* = 30, Q_2^* = 10 = \lambda^*$. The second-order partial derivatives are: $L_{Q_1Q_1} = 0 = L_{Q_2Q_2}, L_{Q_1Q_2} = 1 = L_{Q_2Q_1}$, giving the Hessian

$$|\mathbf{H}| = \begin{vmatrix} 0 & 1 \\ 1 & 0 \end{vmatrix} = -1 < 0.$$

Hence, L is maximized at the critical values.

With $\lambda^* = 10$, a \$1 increase in the budget will change the constant of the constraint to 61, so that the new Lagrangian is

$$L = Q_1Q_2 + \lambda(61 - Q_1 - 4Q_2),$$

which yields: $L_{Q_1} = Q_2 - \lambda = 0$, $L_{Q_2} = Q_1 - 3\lambda = 0$, $L_\lambda = 121 - Q_1 - 3Q_2 = 0$. Solving these equations simultaneously we obtain the critical values: $Q_1^* = 30.5, Q_2^* = 10.167 = \lambda^*$. Thus, the utility function increases from $u = (30)(10) = 300$ to $u = (30.5)(10.167) = 310.083$, i.e., there is an increase in the utility function of about 10.

4.11. Consider the model of Example 4.13 and assume that

$$\begin{aligned} Y - C_0 - C(Y,i) = 0, &\quad 0 < C_Y < 1,\ C_i < 0, \\ L(Y,i) = M_0/P0, &\quad L_Y > 0,\ L_i < 0, \end{aligned}$$

where P is constant. As in Example 4.13, we will use the comparative statics method to determine the effect on the equilibrium levels of Y and i by a change in the money supply M_0. The first-order partial derivatives of Y and i with respect to M_0 are

$$\begin{aligned} \frac{\partial Y}{\partial M_0} - C_Y\,\frac{\partial Y}{\partial M_0} - C_i\,\frac{\partial i}{\partial M_0} &= 0, \\ L_Y\,\frac{\partial Y}{\partial M_0} + L_i\,\frac{\partial i}{\partial M_0} - \frac{1}{P} &= 0, \end{aligned}$$

which in the matrix form $\mathbf{JX} = \mathbf{B}$ is

$$\begin{bmatrix} 1 - C_Y & -C_i \\ L_Y & L_i \end{bmatrix} \begin{bmatrix} \dfrac{\partial Y}{\partial M_0} \\ \dfrac{\partial i}{\partial M_0} \end{bmatrix} = \begin{bmatrix} 0 \\ 1/P \end{bmatrix},$$

where $\mathbf{J}$ is the same as in Example 4.13. Using Cramer's rule, as in Example 4.13, we find that

$$\frac{\partial y_1}{\partial x_2} = \frac{|\mathbf{J}_1|}{|\mathbf{J}|} = -\frac{\dfrac{\partial F_1}{\partial x_2}\dfrac{\partial F_2}{\partial x_2} - \dfrac{\partial F_2}{\partial x_2}\dfrac{\partial F_1}{\partial x_2}}{\dfrac{\partial F_1}{\partial y_1}\dfrac{\partial F_2}{\partial y_2} - \dfrac{\partial F_2}{\partial y_1}\dfrac{\partial F_1}{\partial y_2}},$$

$$\frac{\partial y_1}{\partial x_2} = \frac{|\mathbf{J}_2|}{|\mathbf{J}|} = -\frac{\dfrac{\partial F_1}{\partial y_1}\dfrac{\partial F_2}{\partial x_2} - \dfrac{\partial F_2}{\partial y_1}\dfrac{\partial F_1}{\partial x_2}}{\dfrac{\partial F_1}{\partial y_1}\dfrac{\partial F_2}{\partial y_2} - \dfrac{\partial F_2}{\partial y_1}\dfrac{\partial F_1}{\partial y_2}}.$$

4.12. We will discuss a combination of foreign trade and our national market by combining the (i) goods market, (ii) foreign trade market, and (ii) money market. Let these three markets be defined by

$$\begin{aligned} &(i) && I = I(i), \quad I_i < 0, \\ & && S = S(Y,i), \quad 0 < S_Y < 1, \; S_i > 0, \\ &(ii) && Z = Z(Y,i), \quad 0 < Z_Y < 1, \; Z_i < 0, \\ & && X = X_0, \\ &(iii) && M_D = L(Y,i), \quad L_Y > 0, \; L_i < 0, \\ & && M_S = M_0, \end{aligned} \tag{4.8.4}$$

where Z denotes the imports, S the savings, X_0 the autonomous exports, M_D the demand for money, M_S the money supply, and the other symbols are defined in the above example.

The combined goods market remains in equilibrium (i.e., stable) when the injections are equal to the leakages, i.e., when

$$I(i) + X_0 = S(Y,i) + Z(Y,i),$$

and the money market is in equilibrium when the demand for money is equal tp the money supply, i.e., when

$$L(Y,i) = M_0.$$

Combining these two market situations, the Lagrangian is defined by

$$\begin{aligned} F_1(Y,i;M_0,X_0) &= I(i) + X_0 - S(Y,i) - Z(Y,i), \\ F_2(Y,i;M_0,X_0) &= L(Y,i) - M_0. \end{aligned} \tag{4.8.5}$$

We will consider the following two cases:

CASE 1. The partial derivatives of the functions Y^*, i^* with respect to M_0, expressed in the matrix form $\mathbf{Ax} = \mathbf{b}$, are given by

$$\begin{bmatrix} \dfrac{\partial F_1}{\partial Y} & \dfrac{\partial F_1}{\partial i} \\ \dfrac{\partial F_2}{\partial Y} & \dfrac{\partial F_2}{\partial i} \end{bmatrix} \begin{bmatrix} \dfrac{\partial Y^*}{\partial M_0} \\ \dfrac{\partial i^*}{\partial M_0} \end{bmatrix} = \begin{bmatrix} -\dfrac{\partial F_1}{\partial M_0} \\ -\dfrac{\partial F_2}{\partial M_0} \end{bmatrix},$$

or

$$\begin{bmatrix} -S_Y - Z_Y & I_i - S_i - Z_i \\ L_Y & L_i \end{bmatrix} \begin{bmatrix} \dfrac{\partial Y^*}{\partial M_0} \\ \dfrac{\partial i^*}{\partial M_0} \end{bmatrix} = \begin{bmatrix} 0 \\ 1 \end{bmatrix}.$$

Using Cramer's rule, as in Example 4.10, we obtain

$$\frac{\partial Y^*}{\partial M_0} = \frac{I_i - S_i - Z_i}{L_i(S_Y - Z_y) + L_Y(I_i - S_i - Z_i)} > 0,$$
$$\frac{\partial i^*}{\partial M_0} = \frac{S_Y + Z_Y}{L_i(S_Y - Z_y) + L_Y(I_i - S_i - Z_i)} < 0.$$

CASE 2. The partial derivatives of the functions Y^*, i^* with respect to X_0, expressed in the matrix form $\mathbf{Ax} = \mathbf{b}$, are given by

$$\begin{bmatrix} \frac{\partial F_1}{\partial Y} & \frac{\partial F_1}{\partial i} \\ \frac{\partial F_2}{\partial Y} & \frac{\partial F_2}{\partial i} \end{bmatrix} \begin{bmatrix} \frac{\partial Y^*}{\partial X_0} \\ \frac{\partial i^*}{\partial X_0} \end{bmatrix} = \begin{bmatrix} -\frac{\partial F_1}{\partial X_0} \\ -\frac{\partial F_2}{\partial X_0} \end{bmatrix},$$

or

$$\begin{bmatrix} -S_Y - Z_Y & I_i - S_i - Z_i \\ L_Y & L_i \end{bmatrix} \begin{bmatrix} \frac{\partial Y^*}{\partial X_0} \\ \frac{\partial i^*}{\partial X_0} \end{bmatrix} = \begin{bmatrix} -1 \\ 0 \end{bmatrix}.$$

Using Cramer's rule, as in Example 4.10, we obtain

$$\frac{\partial Y^*}{\partial X_0} = \frac{L_i}{L_i(S_Y - Z_y) + L_Y(I_i - S_i - Z_i)} > 0,$$
$$\frac{\partial i^*}{\partial X_0} = \frac{L_Y}{L_i(S_Y - Z_y) + L_Y(I_i - S_i - Z_i)} < 0. \blacksquare$$

4.13. A constant elasticity of substitution (CES) function is normally defined as $q(K,L) = A\big[\alpha K^{-\beta} + (1-\alpha)L^{-\beta}\big]^{-1/\beta}$, where $A > 0$ is the efficiency parameter, α, $0 < \alpha < 1$ the distribution parameter denoting relative factor shares, and $\beta > -1$ is the substitution parameter that determines the value of the elasticity of substitution. A CES production function is defined by

$$\begin{aligned} q(K,L) &= 60\big[0.4K^{-0.5} + (1-0.4)L^{-0.5}\big]^{-1/0.5} \\ &= 60\big[0.4K^{-0.5} + 0.6L^{-0.5}\big]{-2}, \end{aligned}$$

subject to the equality constraint $2K + 5L = 80$. The Lagrangian is

$$Q(K,L,\lambda) = 60\big[0.4K^{-0.5} + 0.6L^{-0.5}\big]{-2} + \lambda(80 - 2K - 5L).$$

The first-order partial derivatives of Q equated to zero are

$$\begin{aligned} Q_K &= -120\big[0.4K^{-0.5} + 0.6L^{-0.5}\big]^{-3}(-0.2K^{-1.5}) - 2\lambda \\ &= 24K^{-1.5}\big[0.4K^{-0.5} + 0.6L^{-0.5}\big]^{-3} - 2\lambda = 0; \\ Q_L &= -120\big[0.4K^{-0.5} + 0.6L^{-0.5}\big]^{-3}(-0.3L^{-1.5}) - 5\lambda \\ &= 36L^{-1.5}\big[0.4K^{-0.5} + 0.6L^{-0.5}\big]^{-3} - 5\lambda; \\ Q_\lambda &= 80 - 2K - 5L = 0. \end{aligned}$$

The first two equations yield

$$\frac{24K^{-1.5}\big[0.4K^{-0.5}+0.6L^{-0.5}\big]^{-3}}{36L^{-1.5}\big[0.4K^{-0.5}+0.6L^{-0.5}\big]^{-3}}=\frac{2\lambda}{5\lambda},$$

or $\frac{2}{3}\frac{K^{-1.5}}{L^{-1.5}}=\frac{2}{5}$, which simplifies to $L^{1.5}=0.4(1.5)K^{1.5}$, or $L\approx 0.7K$. Then using the last equation $80-2K-5L=0$, we get the critical values as $K^*\approx 14.45$ and $L^*\approx 10.18$.

Next, the second-order partial derivatives of Q are:

$$\begin{aligned}
Q_{KK}&=-36K^{-2.5}\big[0.4K^{-0.5}+0.6L^{-0.5}\big]^{-3}-14.4K^{-3}\big[0.4K^{-0.5}+0.6L^{-0.5}\big]^{-4},\\
Q_{LL}&=-54L^{-2.5}\big[0.4K^{-0.5}+0.6L^{-0.5}\big]^{-3}+32.4L^{-3}\big[0.4K^{-0.5}+0.6L^{-0.5}\big]^{-4},\\
Q_{KL}&=21.6K^{-1.5}L^{-1.5}\big[0.4K^{-0.5}+0.6L^{-0.5}\big]^{-4}=Q_{LK}.
\end{aligned}$$

Using the values of K^* for K and L^* for L, and carrying out some computations, we find that $Q_{KK}\approx -1.09$, $Q_{LL}\approx -2.24$, and $Q_{KL}\approx 1.5$. Thus, since $Q_{KK}<0, Q_{LL}<0$, and $Q_{KK}Q_{LL}<(Q_{KL})^2$, we conclude that q is maximized at the point (K^*,L^*). Alternatively, from the Hessian is $|\mathbf{H}|=\begin{vmatrix}-1.09 & 1.5\\ 1.5 & -2.24\end{vmatrix}$, we find that $|H_1|=-1.09<0$, and $|H_2|=|\mathbf{H}|=-3.85<0$, so the Hessian is negative definite (ND), and q is maximized at (K^*,L^*). ■

5

Convex Programming

As we have seen, optimization problems deal with finding the maximum or minimum of a function, called the *objective function*, subject to certain prescribed constraints. As opposed to concave programming, we minimize a given objective function with or without constraints in convex programming. Thus, given functions $f, g_1, \dots, g_m$ and $h_1, \dots, h_k$ defined on some domain $D \subset \mathbb{R}^n$, the minimization problem is stated as follows: Determine $\min_{\mathbf{x} \in D} f(\mathbf{x})$ subject to the constraints $g_i(\mathbf{x}) \le 0$ for all $i = 1, \dots, m$ and all $h_j(\mathbf{x}) = 0$ for all $j = 1, \dots, k$.

5.1 Minimization Problems

Some minimization problems without any constraints have already been presented in previous chapters. However, we will again discuss the unconstrained case, and then consider the cases of necessary and sufficient conditions for (local) optimality with only equality constraints, only inequality constraints, and combined equality and inequality constraints.

5.1.1 Unconstrained Minimization. Assume that the function $f : D \mapsto \mathbb{R}^n$ is a continuously differentiable function. Then for unconstrained minimization, the necessary and sufficient conditions for a local minimum $\mathbf{x}^*$ of $f(\mathbf{x})$ are: (i) the first-order partial derivatives with respect to each x_i, $i = 1, 2, \dots, n$, must be zero, i.e.,

$$\frac{\partial f}{\partial x_i}(\mathbf{x}) = 0, \tag{5.1.1}$$

where $\mathbf{x}^*$ is obtained by solving equations (5.1.1) simultaneously; and (ii) the Hessian $|\mathbf{H}|$ of f at $\mathbf{x}^*$ is positive semidefinite (PSD), i.e.,

$$|\mathbf{H}|(f) \ge 0 \quad \text{for all } \mathbf{x}^*, \tag{5.1.2}$$

where the Hessian is defined in §1.6.2, and definite and semidefinite matrices are discussed in §1.5.

Example 5.1. Given the function $f(x,y) = 3x^2 - xy + 2y^2 - 4x - 7y + 10$, we have $f_x = 6x - y - 4 = 0$, $f_y = -x + 4y - 7 = 0$, solving which we get the critical point $(x^*, y^*) = (1,2)$. Next, we take the second-order direct partial derivatives and evaluate them at the above critical point. Thus, $f_{xx} = 6, f_{yy} = 4, f_{xy} = -1 = f_{yx}$, and we get

$$f_{xx}(1,2) = 6 > 0, \quad f_{yy}(1,2) = 4 > 0.$$

Also, it is easy to check that $f_{xx} \cdot f_{yy} > (f_{xy})^2$. Hence, f has a global minimum at (1,2). ■

5.1.2 Equality Constraints. The method of Lagrange multipliers is used in such cases. Thus, given the problem to minimize a function $f(x,y)$ subject to a constraint $g(x,y) = k$ (constant), the Lagrangian can be defined by either of the following two equivalent forms:

$$L(x,y,\lambda) = f(x,y) + \lambda(g(x,y) - k), \tag{5.1.3}$$

or

$$L(x,y,\lambda) = f(x,y) - \lambda(k - g(x,y)), \tag{5.1.4}$$

where $\lambda > 0$ is known as the *Lagrange multiplier*, $f(x,y)$ as the *original function* or the *objective function*, and $g(x,y)$ as the *constraint*. Since the constraint is always set equal to zero, the product $\lambda(g(x,y) - k)$, or $\lambda(k - g(x,y))$ is zero, and therefore, the addition of this term does not change the value of the objective function $f(x,y)$. The critical values at which the objective function is optimized are denoted by x^*, y^*, λ^*, and are determined by taking the first-order partial derivatives with respect to x, y, and λ, equating them to zero, and solving them simultaneously:

$$L_x(x,y,\lambda) = 0, \quad L_y(x,y,\lambda) = 0, \quad L_\lambda(x,y,\lambda) = 0. \tag{5.1.5}$$

The second-order conditions, which are obviously different from those of the unconstrained optimization, are similar to those discussed in the previous chapter.

The first-order conditions (5.1.5) are similar to the KKT conditions, which are the necessary conditions, discussed in detail in §4.3; they are applicable in equality and inequality constraints to the Lagrangian $L(x,y,\lambda) = f(x,y) + \lambda(g(x,y) - k)$.

Example 5.2. Minimize $0.5x^2 + y^2 - xy - y$ such that $x + y = 5$. The Lagrangian is

$$L(x,y,\lambda) = 0.5x^2 + y^2 - xy - y + \lambda(x + y - 5),$$

so we have

$$\frac{\partial L}{\partial y} = 2y - x - 5 + \lambda = 0, \quad \frac{\partial L}{\partial x} = x - y + \lambda = 0, \text{ and } \frac{\partial L}{\partial \lambda} = x + y - 5 = 0.$$

Solving these equations simultaneously, we get the optimal values: $x^* = 1.5, y^* = 3.5$, and $\lambda^* = 2$. Then $f(x^*, y^*.\lambda^*) = 4.625$. Notice that the factor $g(x, y) - k$ with these optimal values is zero, as expected. To check the sufficient conditions: the Hessian $|\mathbf{H}| = \begin{vmatrix} 1 & -1 \\ -1 & 1 \end{vmatrix} = 2 > 0$, and the first-order principal $|H_1| = 1 > 0$, while the second-order principal $|H_2| = |\mathbf{H}| > 0$. Thus, the Hessian positive definite, and the conditions for a minimum are satisfied. ■

5.1.3 Equality Constraints: General Case. Consider the problem:

Minimize $f(\mathbf{x})$ such that $g_j(\mathbf{x}) = 0$, $j = 1, 2, \dots, m$, $x \in \mathbb{R}^n$.

CASE 1 $m = 1$ (single constraint). This case corresponds to the case in §5.1.2, and the method of Lagrange multiplier is used. In this case since the (equality) constraint is $g(\mathbf{x}) = 0$, the point $\mathbf{x}$ lies on the graph of the nonlinear equation $g(\mathbf{x}) = 0$ (see Figure 5.1). This necessary condition reduces to

$$\frac{\partial f}{\partial \mathbf{x}}(\mathbf{x}^*) + \lambda g(\mathbf{x}^*) = 0, \tag{5.1.6}$$

and the point $\mathbf{x}^*$ where the minimum occurs is called a *minimizer* for the problem. Notice that condition (5.1.6) can also be expressed as $\nabla f(\mathbf{x}^*) + \lambda g(\mathbf{x}^*) = 0$.

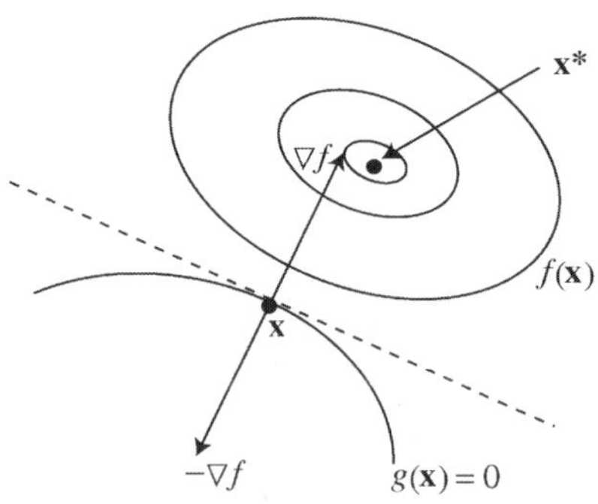

Figure 5.1 Equality constraint in $\mathbb{R}$.

CASE 2 (general case). In $\mathbb{R}^n$ the necessary condition (5.1.6) holds for each constraint $g_j(\mathbf{x}) = 0$. The Lagrangian for this problem is

$$L(f, \boldsymbol{\lambda}) = f(\mathbf{x}) + \sum_{j=1}^{m} \lambda_j g_j(\mathbf{x}). \tag{5.1.7}$$

The KKT (necessary) conditions are: If the point $\mathbf{x}^* \in \mathbb{R}^n$ is a minimizer for the problem, then for some $\boldsymbol{\lambda}^* \in \mathbb{R}^n$:

$$\frac{\partial L}{\partial x_i}(\mathbf{x}^*) + \sum_{j=1}^{m} \lambda_j \frac{\partial h_j}{\partial x_j}(\mathbf{x}^*) = 0, \quad \text{and} \quad g_j(\mathbf{x}^*) = 0 \text{ for all } i, j. \tag{5.1.8}$$

In addition to conditions (5.1.8), suppose that $\mathbf{z} \neq \mathbf{0}$ for an arbitrary point $\mathbf{z} \in \mathbb{R}^n$. Then we have

$$\mathbf{z}^T \frac{\partial g_j}{\partial x_j}(\mathbf{x}^*) = 0, j = 1, \dots, m \implies \mathbf{z}^T |\mathbf{H}|\, \mathbf{z} > 0, \tag{5.1.9}$$

$$\mathbf{z}^T \nabla g_j(\mathbf{x}^*) = 0, j = 1, \dots, m \implies \mathbf{z}^T \nabla^2_{\mathbf{x}} L(\mathbf{x}^*, \lambda^*)\mathbf{z} > \mathbf{0}, \tag{5.1.9}$$

where $|\mathbf{H}|$ is the Hessian for the Lagrangian $L(\mathbf{x}^*, \lambda^*)$, $\mathbf{z}^T = [\, z_1, z_2, \dots, z_m\,]$, and f has a strict local minimum at $\mathbf{x}^*$ such that $g_j(\mathbf{x}^*) = \mathbf{0}$ for $j = 1, \dots, m$. Note that the sufficient condition (5.1.9) is useful when the Hessian $|\mathbf{H}| = 0$.

!!! *Note that condition (5.1.8) can also be expressed as*

$$\nabla f(\mathbf{x}^*) + \sum_{j=1}^{m} \lambda_j \nabla h_j(\mathbf{x}^*) = 0, \quad \textit{and} \quad g_j(\mathbf{x}^*) = 0 \textit{ for all } i, j.$$

Example 5.3. Minimize $f(x, y) = x^2 + y^2 - xy + 4y$ subject to the equality constraint $x + y = 2$, i.e., $g(x, y) = x + y - 2$. The Lagrangian is

$$L(x, y, \lambda) = x^2 + y^2 - xy + 4y + \lambda(x + y - 2).$$

The KKT conditions give

$$\frac{\partial L}{\partial x} = 2x - y + \lambda = 0, \quad \frac{\partial L}{\partial y} = 2y - x + 4 + \lambda = 0, \quad \frac{\partial L}{\partial \lambda} = x + y - 2 = 0,$$

solving which we get $x^* = 2, y^* = 0, \lambda^* = 4$, giving $f(x^*, y^*) = 4$. Thus, the point (x^*, y^*) satisfies the necessary conditions to be a minimum.

The Hessian for the problem is

$$|\mathbf{H}| = \begin{vmatrix} L_{xx} & L_{xy} \\ L_{yx} & L_{yy} \end{vmatrix} = \begin{vmatrix} 2 & -1 \\ -1 & 2 \end{vmatrix} = 3 > 0.$$

Hence $f(x, y)$ has a local minimum at $(2, 0)$. ■

Example 5.4. Minimize $f(x, y) = \frac{1}{2}x^2 + \frac{1}{2}y^2 - xy - 4y$ subject to the equality constraint $g(x, y) = x + y - 3$. The Lagrangian for the problem is

$$L(x, y, \lambda) = \frac{1}{2}x^2 + \frac{1}{2}y^2 - xy + 4y + \lambda(x + y - 2).$$

Using the KKT conditions we get $L_x = x - y + \lambda = 0$, $L_y = y - x - 4 + \lambda = 0$, $L_\lambda = x + y - 3$, solving which we get $x^* = \frac{5}{2}, y^* = \frac{1}{2}$, so $\lambda^* = 2$, and $f(x^*, y^*) = 5$. The Hessian for the problem is

$$|\mathbf{H}| = \begin{vmatrix} L_{xx} & L_{xy} \\ L_{yx} & L_{yy} \end{vmatrix} = \begin{vmatrix} 1 & -1 \\ -1 & 1 \end{vmatrix} = 0.$$

Since the Hessian test fails, we use condition (5.1.9). Since $\dfrac{\partial g}{\partial x_i}(x, y) = \begin{bmatrix} 1 \\ 1 \end{bmatrix}$, we have $\mathbf{z}^T \dfrac{\partial g_j}{\partial x_i}(\mathbf{x}) = z_1 + z_2 = 0$, i.e., $z_2 = -z_1$. Next, consider $\mathbf{z} = \begin{bmatrix} z_1 \\ z_2 \end{bmatrix} \neq \begin{bmatrix} 0 \\ 0 \end{bmatrix}$. Then

$$\mathbf{z}^T |\mathbf{H}| \mathbf{z} = [\, z_1 \quad z_2 \,] \begin{bmatrix} 1 & -1 \\ -1 & 1 \end{bmatrix} \begin{bmatrix} z_1 \\ z_2 \end{bmatrix} = (z_1 - z_2)^2 = (2z_1)^2 > 0.$$

Hence, $f(x, y)$ has a strict local minimum at $(\frac{5}{2}, \frac{1}{2})$. ■

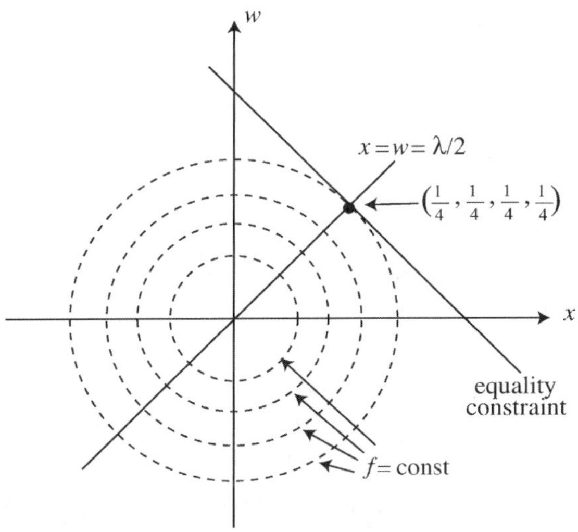

Figure 5.2 For Example 5.5.

Example 5.5. (Only one constraint) Minimize $f = x^2 + y^2 + z^2 + w^2$ subject to $x + y + z + w = 1$. The Lagrangian is $F = x^2 + y^2 + z^2 + w^2 + \lambda(1 - x - y - z - w)$. From $F_{\mathbf{x}} = 0$, where $\mathbf{x} = \{x, y, z, w\}$, we get $2x - \lambda = 0$, $2y - \lambda = 0$, $2z - \lambda = 0$, and $2w - \lambda = 0$. Thus, $x = y = z = w = \lambda/2$, and so $x + y + z + w = 2\lambda$, and using the constraint we get $\lambda = \frac{1}{2}$. Hence, $x = y = z = w = \frac{1}{4}$, and $f = \frac{1}{4}$. The function f is represented, for convenience, in two dimensions (x, w) in Figure 5.2, where function f is a circle, and the constraint is the straight line $x + 1 + w = 1$. The four dimension case can be easily visualized from Figure 5.2, where the circles can be replaced by spheres, and we are seeking the smallest sphere that intersects with the equality constraint which will be a three-dimensional plane in a four dimensional space. ■

5.1.4 Inequality Constraints. Consider the problem: Minimize $f(\mathbf{x})$ such that $g_j(\mathbf{x}) \leq 0$, $\mathbf{x} \in \mathbb{R}^n$, and $j = 1, \dots, n$. The inequality constraint is replaced by $g(\mathbf{x}) + S^2 = 0$, where S is called the *slackness condition*. Then the necessary KKT conditions (5.1.5) is applied.

Example 5.6. Minimize $f(x, y) = x^2 + \frac{1}{2}y^2 - 12x - 4y - 60$ subject to the constraint $30x + 2y \leq 120$. SOLUTION. The Lagrangian with the slackness variable S is

$$L(x, y, \lambda) = x^2 + \frac{1}{2}y^2 - 12x - 4y - 60 + \lambda(30x + 20y + S^2 - 120).$$

Then

$$\begin{aligned} \frac{\partial L}{\partial x} &= 2x - 12 + 30\lambda = 0, \\ \frac{\partial L}{\partial y} &= y - 4 + 20\lambda = 0, \\ \frac{\partial L}{\partial \lambda} &= 30x + 20y + S^2 - 120. \end{aligned}$$

This is a nonlinear system of equations, which defines only the necessary KKT conditions. We now consider two cases:

CASE 1. If $\lambda = 0$, then solving the first two equations, we get $x = 6$ and $y = 4$. Then the third equation gives $S^2 = -140$, which is infeasible.

CASE 2. If $S = 0$, then the first two equations give $x = \frac{60}{17} = 3.529$ and $y = \frac{12}{17} = 0.706$. Then from the third equation we get $\lambda = \frac{28}{17} = 1.647$. Thus, the only feasible solution is the point$(\frac{60}{17}, \frac{28}{17})$, yielding $f(x^*, y^*) = -92.47$. Next, to check the sufficient conditions, we have the Hessian $|\mathbf{H}| = \begin{vmatrix} 2 & 0 \\ 0 & 1 \end{vmatrix} = 8 > 0$, and the first-order and second-order principals are $|H_1| = 2 > 0$, and $|H_2| = |\mathbf{H}| > 0$. Thus, the Hessian positive definite, and the conditions for a minimum are satisfied. ■

Example 5.7. Minimize $0.5x^2 + 0.5y^2 - 8x - 2y - 80$ subject to the constraint $g(x, y) = 20x + 10y \leq 130$. Using the slackness variable $S \geq 0$, the constraint becomes $g(x, y) = 20x + 10y + S - 130 = 0$. The Laplacian is

$$L(x, y, S, \lambda) = 0.5x^2 + 0.5y^2 - 8x - 2y - 80 + \lambda(20x + 10y + S - 130).$$

Then the KKT conditions give

$$\frac{\partial L}{\partial x} = x - 8 + \lambda = 0, \quad \frac{\partial L}{\partial y} = y - 2 + 10\lambda,$$
$$\frac{\partial L}{\partial \lambda} = 20x + 10y + S - 130, \quad \frac{\partial L}{\partial S} = \lambda.$$

We have a system of nonlinear equations. To solve it, we consider the following two cases:

CASE 1. Let $\lambda = 0$. Then we get $x = 8$ and $y = 2$, which give $S = -50$ (infeasible).

CASE 2. Let $S = 0$. Then $x - 8 + 20\lambda = 0$, $y - 2 + 10\lambda = 0$. Solving these two equations by eliminating λ we get $x - 2y = 4$, which when used with the third equation gives $x = 6, y = 1$, and $\lambda = 0.05$. Hence, the only solution is the point $\mathbf{x} = (6, 1)$, and if the given problem has a solution, then this must be the required solution. ■

5.1.5 General Linear Case. Consider the problem: Minimize $f(\mathbf{x})$ subject to the constraint $g(\mathbf{x}) = g_j(\mathbf{x}) \leq 0$, $j = 1, 2, \dots, m$, $\mathbf{x} \in \mathbb{R}^n$. The geometric interpretation of this problem is as follows: As in the case of the equality constraint, in this case we also have the necessary conditions defined by Eq (5.1.6) for some value of λ which in the equality case is simply a positive or negative real number. However, in the case of inequality constraints the sign of λ is known in advance depending on the direction of $-\nabla f$, as shown in Figure 5.3, where ∇g represents the direction of increase of $g(\mathbf{x})$.

Thus, for minimization, we have:

$$g(\mathbf{x}) \leq \mathbf{0} \Longrightarrow \boldsymbol{\lambda} \geq \mathbf{0}, \quad \text{and} \quad g(\mathbf{x}) \geq \mathbf{0} \Longrightarrow \boldsymbol{\lambda} \leq \mathbf{0}.$$

Then the Lagrangian is $L(\mathbf{x}, \lambda) = f(\mathbf{x}) + \boldsymbol{\lambda} g(\mathbf{x})$, and the KKT conditions are

$$\begin{aligned}
&(1)\ \frac{\partial L}{\partial x_i}(\mathbf{x}^*, \boldsymbol{\lambda}^*) = \frac{\partial f}{\partial x_i}(\mathbf{x}^*) + \boldsymbol{\lambda}^* \frac{\partial g}{\partial x_i}(\mathbf{x}^*) = 0, \text{ for all } i = 1, \dots, n, \\
&(2)\ g(\mathbf{x}^*) \leq \mathbf{0}, \\
&(3)\ \boldsymbol{\lambda}^* g(\mathbf{x}^*) = \mathbf{0}, \quad \boldsymbol{\lambda}^* \geq 0.
\end{aligned} \tag{5.1.10}$$

Note that condition (2) yields the given inequality constraint, and condition (3) is complementary slackness, i.e.,

$$g(\mathbf{x}^*) < \mathbf{0} \Longrightarrow \boldsymbol{\lambda}^* = \mathbf{0}; \ \boldsymbol{\lambda}^* > \mathbf{0} \Longrightarrow g(\mathbf{x}^*) = \mathbf{0}.$$

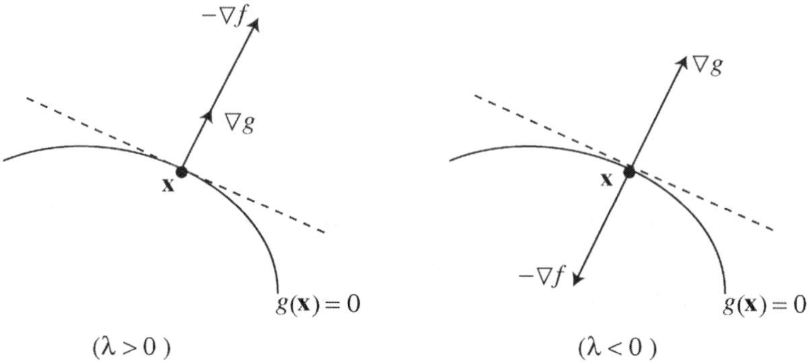

Figure 5.3 Inequality constraint in $\mathbb{R}$ and sign of λ.

In the general case of m inequality constraints, if $\mathbf{x}^*$ is a minimizer, then for all $\lambda_j^* \geq 0$ for $j = 1, \dots, m$, such that for $L(\mathbf{x}, \boldsymbol{\lambda}) = f(\mathbf{x}) + \sum\limits_{j=1}^{m} \lambda_j^* g_j(\mathbf{x})$, the KKT necessary conditions are

$$
\begin{aligned}
&(1)\ \frac{\partial L}{\partial \mathbf{x}}(\mathbf{x}^*, \boldsymbol{\lambda}^*) = \frac{\partial f}{\partial \mathbf{x}}(\mathbf{x}^*) + \sum_{j=1}^{m} \lambda_j^* \frac{\partial g_j}{\partial \mathbf{x}}(\mathbf{x}^*) = 0 \quad (n \text{ equations}),\\
&(2)\ g_j(\mathbf{x}^*) \leq 0 \quad \text{for } j = 1, 2, \dots, m,\\
&(3)\ \{g(\mathbf{x}^*)\}\, \boldsymbol{\lambda}_j^* = \mathbf{0},\ j = 1, 2, \dots, m.
\end{aligned}
\tag{5.1.11}
$$

Note that if the inequality constraint are of the form $g(\mathbf{x}) \geq 0$, then we must have $\lambda_j^* \leq \mathbf{0}$; or, alternatively, use $-g(\mathbf{x})$ and retain $\boldsymbol{\lambda}^* \geq \mathbf{0}$.

!!! *Note that condition (1) in (5.1.10) and (5.1.11) can also be expressed respectively as* $\nabla_{\mathbf{x}} L(\mathbf{x}^*, \boldsymbol{\lambda}^*) = \nabla f(\mathbf{x}^*) + \boldsymbol{\lambda}^* \nabla g(\mathbf{x}^*) = \mathbf{0}$, *and* $\nabla L_{\mathbf{x}}(\mathbf{x}^*, \boldsymbol{\lambda}^*) = \nabla f(\mathbf{x}^*) + \sum\limits_{j=1}^{m} \lambda_j^* \nabla g_j(\mathbf{x}) = 0$ *(n equations).*

5.2 Nonlinear Programming

Consider the nonlinear program: Minimize $f(\mathbf{x})$ subject to the nonlinearity constraints

$$
\begin{aligned}
g_j(\mathbf{x}) &\leq 0, \quad j = 1, 2, \dots, m,\\
g_j(\mathbf{x}) &\geq 0, \quad j = m+1, \dots p,\\
h_j(\mathbf{x}) &= 0, \quad j = p+1, \dots, q, \quad \mathbf{x} \in \mathbb{R}^n,
\end{aligned}
$$

where f is a convex function on $\mathbb{R}^n$; g_j, $j = 1, \dots, m$, are convex functions on $\mathbb{R}^n$; g_j, $j = m+1, \dots, p$, are concave functions on $\mathbb{R}^n$; and h_j, $j = 1, \dots, q$, are linear functions of the form $h_j(\mathbf{x}) = \sum_{k=1}^{n} a_{jk}x_k - b_j$. Note that the domain of each one of these functions is nonempty. Thus, a convex minimization program has a convex objective, and the set of feasible solutions is a convex set.

The Lagrangian is

$$L(\mathbf{x}, \boldsymbol{\lambda}) = f(\mathbf{x}) + \sum_{j=1}^{m} \lambda_j g_j(\mathbf{x}) - \sum_{j=m+1}^{p} \lambda_j g_j(\mathbf{x}) + \sum_{j=p+1}^{q} \mu_j g_j(\mathbf{x}). \tag{5.2.1}$$

If $\mathbf{x}^*$ minimizes $f(\mathbf{x})$ while the above conditions (5.1.11) are satisfied, then provided certain *regularity conditions*, or *constraint qualifications* to be discussed in the sequel, are met, there exist vectors $\boldsymbol{\lambda}^*$ and $\boldsymbol{\mu}^*$ such that the following KKT necessary conditions are satisfied:

$$\begin{aligned}
&(1)\ \frac{\partial f}{\partial \mathbf{x}}(\mathbf{x}^*) + \sum_{j=1}^{m} \lambda_j \frac{\partial g_j}{\partial \mathbf{x}}(\mathbf{x}^*) - \sum_{j=m+1}^{p} \lambda_j \frac{\partial g_j}{\partial \mathbf{x}}(\mathbf{x}^*) + \sum_{j=p+1}^{q} \mu_j \frac{\partial g_j}{\partial \mathbf{x}}(\mathbf{x}^*),\\
&(2)\ \text{all constraints given in (5.2.1) are satisfied,}\\
&(3)\ \lambda_j^* \geq 0, \quad j = 1, 2, \dots, p,\\
&(4)\ \lambda_j^* \geq 0, \quad j = 1, 2, \dots, p,\\
&(5)\ \mu_j^* \text{ are unrestricted in sign for } j = p+1, \dots, q.
\end{aligned} \tag{5.2.2}$$

!!! *Condition (1) in (5.2.2) can also be expressed as*

$$\nabla f(\mathbf{x}^*) + \sum_{j=1}^{m} \lambda_j \nabla g_j(\mathbf{x}^*) - \sum_{j=m+1}^{p} \lambda_j \nabla g_j(\mathbf{x}^*) + \sum_{j=p+1}^{q} \mu_j \nabla g_j(\mathbf{x}^*).$$

Example 5.8. (One equality constraint and one inequality constraint) Minimize $f = x^2 + y^2 + z^2 + w^2$, subject to $x + y + z + w = 1$ and $w \leq C$, where C is a scalar which will also be determined. There are two possible scenarios for this problem, which are presented in Figure 5.4, depending on the value of C, whether small or large; the figures represent the two-dimensional case.

Note that the shaded region is outside the feasible region because $w > C$ there. The Lagrangian is defined as

$$F = x^2 + y^2 + z^2 + w^2 + \lambda(1 - x - y - z - w) + \mu(w - C),$$

where λ and μ are the Lagrange multipliers. Then the KKT conditions are: (a) $\dfrac{\partial F}{\partial x} = 0$; (b) $x + y + z + w = 1$; (c) $w \leq C$; (d) $\mu \geq 0$; and (e) $\mu(w - C) = 0$.

Condition (a) yields $2x - \lambda = 0, 2y - \lambda = 0, 2z - \lambda = 0, 2w - \lambda = 0$, which gives $x = y = z = \frac{\lambda}{2}$, $w = \frac{\lambda - \mu}{2}$.

Using condition (b) we get $x + y + z + w = 4 \cdot \frac{1}{2}\lambda - \frac{1}{2}\mu = 1$, which gives $4\lambda - \mu = 2$, or $\lambda = \frac{1}{2}(2 + \mu)$. Hence,

$$x = y = z = \frac{2+\mu}{8} = \frac{1}{4} + \frac{\mu}{8}, \; w = \frac{2+\mu}{2} - \frac{\mu}{2} = \frac{1}{4} - \frac{3\mu}{8}. \tag{5.2.3}$$

Next, from condition (c), we get $\frac{1}{4} - \frac{3\mu}{8} \le C$, or

$$\frac{3\mu}{8} \ge \frac{1}{4} - C. \tag{5.2.4}$$

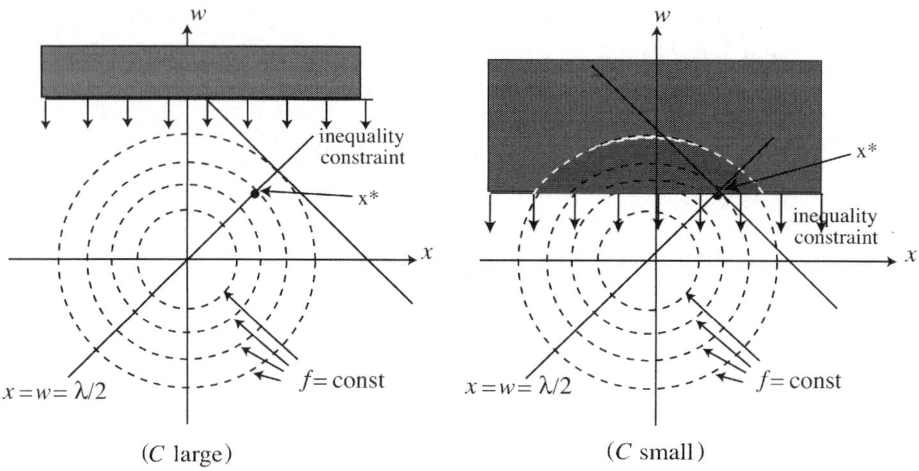

Figure 5.4 Example 5.8: (a) C small; (b) C large.

CASE 1. If $C > \frac{1}{4}$, we have the interior case (see Figure 5.4). Since $\frac{1}{4} - C \le 0$, we find from (5.2.4) that condition (d) is satisfied. Thus, from (5.2.3) we get

$$x = y = z \ge \frac{1}{4}; \quad w = 1 - (x + y + z) \le \frac{1}{4}.$$

But by condition (e), we have $\mu = 0$, and hence, $x = y + z = w = \frac{1}{4}$. This is the optimal case, even if we require $w < C$ and $C > \frac{1}{4}$.

CASE 2. If $C = \frac{1}{4}$, this is similar to Case 1, and the unconstrained optimum lies on the boundary.

CASE 3. If $C < \frac{1}{4}$, and if $w < C$, then condition (e) would require that $\mu = 0$. But then (5.2.3) would give $x = \frac{1}{4}$, which would violate condition (c). Hence, $w = C$ and $x = y = z = \frac{1}{3}(1 - C)$. Further,

$$f = 3\left[\frac{1}{9}(1-C)^2\right] + c^2 = \frac{1}{3}(1-C)^2 + C^2 = \frac{1}{3}(1 - 2C + 4C^2),$$

and thus, $f \geq \frac{1}{4}$; and $f = \frac{1}{4}$ when $C = \frac{1}{4}$. The graph of f is presented in Figure 5.5. ■

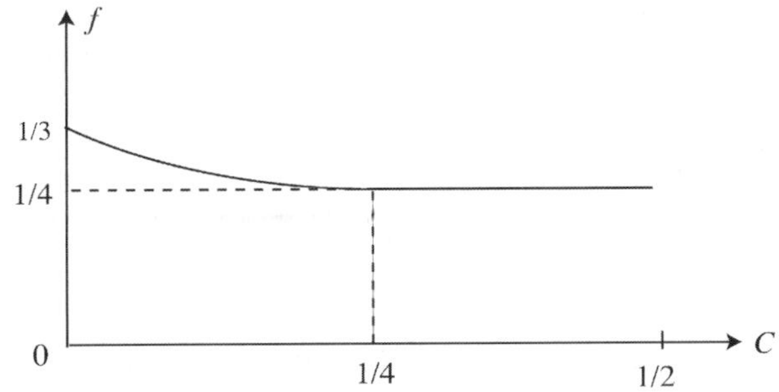

Figure 5.5 Graph of f.

5.2.1 Two Inequality Constraints and One Equality Constraint: Minimize $f(\mathbf{x})$ subject to the constraints

$$\begin{aligned} &g_j(\mathbf{x}) \leq 0, \quad j = 1, 2, \dots, m, \\ &g_j(\mathbf{x}) \geq 0, \quad j = m+1, \dots p, \\ &g_j(\mathbf{x}) = 0, \quad j = p+1, \dots, q, \quad \mathbf{x} \in \mathbb{R}^n. \end{aligned}$$

The Lagrangian is

$$L(\mathbf{x}, \boldsymbol{\lambda}) = f(\mathbf{x}) + \sum_{j=1}^{m} \lambda_j g_j(\mathbf{x}) - \sum_{j=m+1}^{p} \lambda_j g_j(\mathbf{x}) + \sum_{j=p+1}^{q} \mu_j g_j(\mathbf{x}). \tag{5.2.5}$$

If $\mathbf{x}^*$ minimizes $f(\mathbf{x})$ while the above conditions (5.1.11) are satisfied, then provided certain *regularity conditions*, or *constraint qualifications* to be discussed in the sequel, are met, there exist vectors $\boldsymbol{\lambda}^*$ and $\boldsymbol{\mu}^*$ such that the following KKT necessary conditions are satisfied:

$$\begin{aligned} &(1)\ \frac{\partial f}{\partial \mathbf{x}}(\mathbf{x}^*) + \sum_{j=1}^{m} \lambda_j \frac{\partial g_j}{\partial \mathbf{x}}(\mathbf{x}^*) - \sum_{j=m+1}^{p} \lambda_j \frac{\partial g_j}{\partial \mathbf{x}}(\mathbf{x}^*) + \sum_{j=p+1}^{q} \mu_j \frac{\partial g_j}{\partial \mathbf{x}}((\mathbf{x}^*), \\ &(2)\ \text{all constraints given in (5.2.1) are satisfied,} \\ &(3)\ \lambda_j^* \geq 0, \quad j = 1, 2, \dots, p, \\ &(4)\ \lambda_j^* \geq 0, \quad j = 1, 2, \dots, p, \\ &(5)\ \mu_j^* \text{ are unrestricted in sign for } j = p+1, \dots, q. \end{aligned} \tag{5.2.6}$$

!!! *Condition (1) in (5.2.6) can also be expressed as*

$$\nabla f(\mathbf{x}^*) + \sum_{j=1}^{m} \lambda_j \nabla g_j(\mathbf{x}^*) - \sum_{j=m+1}^{p} \lambda_j \nabla g_j(\mathbf{x}^*) + \sum_{j=p+1}^{q} \mu_j \nabla g_j(\mathbf{x}^*).$$

5.2.2 Two Inequality Constraints. The general case is: Minimize $f(\mathbf{x})$ subject to $g_j(\mathbf{x}) \leq 0$ for $j = 1, 2, \dots, m$, and $q_i(\mathbf{x}) = \mathbf{x}_i \geq 0$ for $i = 1, 2, \dots, n$. Let the Lagrange multipliers be $\mu_1, \mu_2, \dots, \mu_n$ associated with each of the non-negativity constraints. Then, using the slackness in the KKT conditions we will have $\mu_i^* x_i^* = 0$ for $i = 1, 2, \dots, n$, and condition (1) in (5.2.6) becomes

$$\text{(1a)} \qquad \frac{\partial f}{\partial \mathbf{x}}(\mathbf{x}^*) + \sum_{j=1}^{m} \lambda_j \frac{\partial g_j}{\partial \mathbf{x}}(\mathbf{x}^*) - \sum_{i=1}^{n} \mu_i^* \frac{\partial q_i j}{\partial \mathbf{x}}(\mathbf{x}^*) = 0,$$

which implies that

$$\frac{\partial f}{\partial \mathbf{x}}(\mathbf{x}^*) + \sum_{j=1}^{m} \lambda_j \frac{\partial g_j}{\partial \mathbf{x}}(\mathbf{x}^*) - \mu_1^* \begin{bmatrix} 1 \\ 0 \\ 0 \\ \vdots \\ 0 \\ 0 \end{bmatrix} - \mu_2^* \begin{bmatrix} 0 \\ 1 \\ 0 \\ \vdots \\ 0 \\ 0 \end{bmatrix} - \cdots - \mu_{n-1}^* \begin{bmatrix} 0 \\ 0 \\ 0 \\ \vdots \\ 1 \\ 0 \end{bmatrix} - \mu_n^* \begin{bmatrix} 0 \\ 0 \\ 0 \\ \vdots \\ 0 \\ 1 \end{bmatrix} = 0, \tag{5.2.7}$$

or

$$\frac{\partial f}{\partial \mathbf{x}}(\mathbf{x}^*) + \sum_{j=1}^{m} \lambda_j \frac{\partial g_j}{\partial \mathbf{x}}(\mathbf{x}^* \boldsymbol{\mu}^*),$$

where $\boldsymbol{\mu}^* \in \mathbb{R}^n = [\,\mu_1^* \quad \mu_2^* \quad \cdots \quad \mu_n^*\,]^T$.

Note that we must have $\boldsymbol{\mu}^* \geq 0$. Thus, the above KKT necessary conditions (5.2.6) become

$$\begin{aligned} &(1)\ \frac{\partial f}{\partial \mathbf{x}}(\mathbf{x}^*) + \sum_{j=1}^{m} \lambda_j \frac{\partial g_j}{\partial \mathbf{x}}(\mathbf{x}^*) \geq 0, \\ &(2)\ \text{all constraints given in (5.2.1) are satisfied}, \\ &(3)\ \lambda_j^* \{g_j(\mathbf{x}^*)\} = 0, \quad j = 1, 2, \dots, m, \\ &(4)\ \lambda_j^* \geq 0, \quad j = 1, 2, \dots, m, \\ &(5)\ \mu_i^* \mathbf{x}^* = 0 \text{ for all } i, \end{aligned} \tag{5.2.8}$$

where condition (5) means that

$$\sum_{i=1}^{n} \mathbf{x}^* \mu^* = [\mathbf{x}^*]^T \left[\frac{\partial f}{\partial \mathbf{x}}(\mathbf{x}^*) + \sum_{j=1}^{m} \lambda_j^* \frac{\partial g_j}{\partial \mathbf{x}}(\mathbf{x}^*) \right] = 0. \tag{5.2.9}$$

$$\sum_{i=1}^{n} \mathbf{x}^* \mu^* = [\mathbf{x}^*]^T \left[\nabla f(\mathbf{x}^*) + \sum_{j=1}^{m} \lambda_j^* \nabla g_j(\mathbf{x}^*) \right] = 0. \tag{5.2.10}$$

Note that no explicit Lagrange multipliers are used for non-negativity constraints. Also, although the KKT conditions are generally used to check optimization, they are, however, not valid under all situations. There is another set of necessary conditions, known as *Fritz John conditions*, discussed in the next section, which are valid at all times, but in many cases they do not provide the same information as the KKT conditions.

!!! *In the above discussion, in view of Theorem 2.18, the expression (1a) can be expressed as*

$$\nabla f(\mathbf{x}^*) + \sum_{j=1}^{m} \lambda_j \nabla g_j(\mathbf{x}^*) - \sum_{i=1}^{n} \mu_i^* \nabla q_i(\mathbf{x}^*) = 0;$$

the equation (5.2.7) can be expressed as

$$\nabla f(\mathbf{x}^*) + \sum_{j=1}^{m} \lambda_j \nabla g_j(\mathbf{x}^*) - \mu_1^* \begin{bmatrix} 1 \\ 0 \\ 0 \\ \vdots \\ 0 \\ 0 \end{bmatrix} - \mu_2^* \begin{bmatrix} 0 \\ 1 \\ 0 \\ \vdots \\ 0 \\ 0 \end{bmatrix} - \cdots - \mu_{n-1}^* \begin{bmatrix} 0 \\ 0 \\ 0 \\ \vdots \\ 1 \\ 0 \end{bmatrix} - \mu_n^* \begin{bmatrix} 0 \\ 0 \\ 0 \\ \vdots \\ 0 \\ 1 \end{bmatrix} = 0,$$

or

$$\nabla f(\mathbf{x}^*) + \sum_{j=1}^{m} \lambda_j \nabla g_j(\mathbf{x}^* \mu^*);$$

and condition (1) in (5.2.8) can be expressed as

$$\nabla f(\mathbf{x}^*) + \sum_{j=1}^{m} \lambda_j \nabla g_j(\mathbf{x}^*) \geq 0.$$

Example 5.9. Consider the inequality constrained optimization problem

$$\min_{\mathbf{x} \in \mathbb{R}^2} f(\mathbf{x}) \text{ subject to } g(\mathbf{x}) \leq 0,$$

where $f(\mathbf{x}) = x^2 + y^2$ and $g(\mathbf{x}) = x^2 + y^2 - 1$.

How can we determine whether $\mathbf{x}^*$ is at a local minimum? Since the unconstrained minimum of $f(\mathbf{x})$ lies within the feasible region, the necessary and sufficient conditions for a constrained local minimum are the same as those for an unconstrained local minimum, i.e., $\frac{\partial f}{\partial \mathbf{x}}(\mathbf{x}^*) = 0$ (or equivalently, $\nabla_{\mathbf{x}} f(\mathbf{x}^*) = 0$) , and the Hessian is positive definite. Note that for this optimization problem the constraint is not active at the local minimum since

$g(\mathbf{x}^*) < 0$. Hence, the local minimum is identified by the same conditions as in case 2, Eq(5.2.2). ■

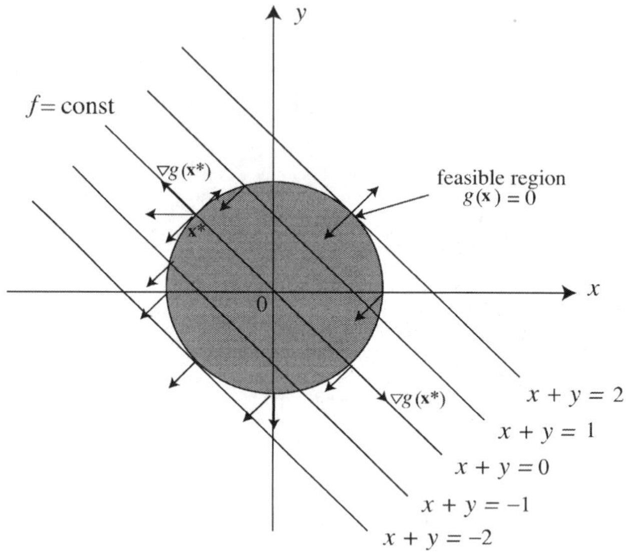

Figure 5.6 (a) Isoclines of $f(\mathbf{x})$. (b) Domain $h(\mathbf{x}) = 0$.

Example 5.10. Consider the constrained optimization problem of Example 5.9 except that $f(\mathbf{x})$ is now defined by

$$f(x, y) = (x - 1.2)^2 + (y - 1.2)^2,$$

with the same $g(x, y) = x^2 + y^2 - 1$ (see Figure 5.6). First, we determine whether $\mathbf{x}^*$ is a local minimizer. Since the unconstrained local minimum of $f(\mathbf{x})$ lies outside the feasible region, this is definitely an optimization problem with an equality constraint $g(\mathbf{x}) = 0$. Hence, a local optimum occurs when $\frac{\partial f}{\partial \mathbf{x}}(\mathbf{x})$ and $\frac{\partial g}{\partial \mathbf{x}}(\mathbf{x})$ are parallel, so that we have

$$-\frac{\partial f}{\partial \mathbf{x}}(\mathbf{x}) = \lambda \frac{\partial g}{\partial \mathbf{x}}(\mathbf{x}). \tag{5.2.11}$$

Also, we determine if a constrained local minimum as $-\nabla_{\mathbf{x}}$ points away from the feasible region. Thus, the constrained local minimum occurs when $\frac{\partial f}{\partial \mathbf{x}}(\mathbf{x})$ and $\frac{\partial g}{\partial \mathbf{x}}(\mathbf{x})$ are in the same direction), which gives

$$-\frac{\partial f}{\partial \mathbf{x}}(\mathbf{x}) = \lambda \frac{\partial g}{\partial \mathbf{x}}(\mathbf{x}),\ \lambda > 0. \blacksquare \tag{5.2.12}$$

!!! *Eqs (5.2.11) and (5.2.12) can be expressed, respectively, as* $-\nabla_{\mathbf{x}} f(\mathbf{x}) = \lambda \nabla_{\mathbf{x}} g(\mathbf{x})$, *and* $-\nabla_{\mathbf{x}} f(\mathbf{x}) = \lambda \nabla_{\mathbf{x}} g(\mathbf{x}),\ \lambda > 0$.

5.3 Fritz John Conditions

For the problem: Minimize $f(\mathbf{x})$ subject to the inequality constraints $g_j(\mathbf{x}) \leq 0$, $j = 1, 2, \dots, m$, the Fritz John conditions are weaker necessary conditions, which are based on the *weak* Lagrangian

$$L(\mathbf{x}, \boldsymbol{\lambda}) = \lambda_0^* f(\mathbf{x}^*) + \sum_{j=1}^{m} \lambda_j^* g_j(\mathbf{x}^*)). \tag{5.3.1}$$

If $\mathbf{x}^*$ is the minimizer, then there exists a $\boldsymbol{\lambda}^* \in \mathbb{R}^{m+1}$, and the Fritz John conditions are

$$\begin{aligned}
&(1)\ \frac{\partial \tilde{L}}{\partial \mathbf{x}}(\mathbf{x}^*, \boldsymbol{\lambda}^*) = \lambda_0^* \frac{\partial f}{\partial \mathbf{x}}(\mathbf{x}^*) + \sum_{j=1}^{m} \lambda_j^* \frac{\partial g_j}{\partial \mathbf{x}}(\mathbf{x}^*) = 0, \\
&(2)\ g_j(\mathbf{x}^*) \leq 0, \quad j = 1, 2, \dots, m, \\
&(3)\ \lambda_j^* \{g_j(\mathbf{x}^*)\} = 0, \quad j = 1, 2, \dots, m, \\
&(4)\ \boldsymbol{\lambda}^* \geq 0 \quad \text{and} \quad \boldsymbol{\lambda}^* \neq 0.
\end{aligned} \tag{5.3.2}$$

!!! *The first Fritz John condition in (5.3.2) can also be written as*

$$\nabla_{\mathbf{x}} \tilde{L}(\mathbf{x}^*, \boldsymbol{\lambda}^*) = \lambda_0^* \nabla f(\mathbf{x}^*) + \sum_{j=1}^{m} \lambda_j^* \nabla g_j(\mathbf{x}^*) = 0.$$

The Fritz John (FJ) conditions are *always* necessary for $\mathbf{x}^*$ to be a solution. However, the KKT conditions are necessary *provided* certain conditions known as *constraint qualifications* (CQ) are satisfied. This can be represented as

$$\boxed{\text{Local optimum}} \Longrightarrow \boxed{\text{Fritz John}} \overset{\text{CQ}}{\Longrightarrow} \boxed{\text{KKT}}.$$

One of the examples of a constraint qualification (CQ) is as follows: The set $S = \left\{ \frac{\partial g_k}{\partial \mathbf{x}}(\mathbf{x}^*) \,|\, k \in K \right\}$ (or equivalently, $S = \{\nabla g_k(\mathbf{x})^* \,|\, k \in K\}$) is linearly independent, where $K = \{k \,|\, g_k(\mathbf{x}^*)\} = 0$. That is, the gradient vectors of all the constraints that are satisfied as strict equalities at $\mathbf{x}^*$ must be linearly independent.

In Fritz John conditions (5.3.2), suppose that $\lambda_0^* = 0$. Then these conditions are satisfied for *any* function f at the point $\mathbf{x}^*$, regardless of whether or not the function f has a minimum at $\mathbf{x}^*$. This is the main weakness of the

Fritz John conditions, because if $\lambda_0^* = 0$, then these conditions do not use the objective and they are of no practical use in locating the optimal point $\mathbf{x}^*$.

Remember that the CQs are essentially the constraints that ensure the $\lambda_0^* > 0$. Thus, if we redefine λ_j^* as λ_j^*/λ_0^*, $j = 0, 1, \dots, m$, then the Fritz John conditions reduce to the KKT conditions, so that $\lambda_0^* = 1$ can be ignored.

Example 5.11. Minimize $f(x, y) = -x$ subject to $g_1(x, y) = y-(1-x)^5 \leq 0$, and $g_2(x, y) = -y \leq 0$, where $\mathbf{x} = (x, y)$. The graphs are presented in Figure 5.7, where the feasible region and the optimal point $\mathbf{x}^*$ are identified for this problem.

In matrix notation, $\mathbf{x} = \begin{bmatrix} x \\ y \end{bmatrix}$. The optimal solution is $\mathbf{x}^* = (1, 0)$, which in matrix form is written as $\mathbf{x}^* = \begin{bmatrix} 1 \\ 0 \end{bmatrix}$. Now, in view of $\mathbf{x}^*$, we have $g(x^*, y^*) = y$, and $g_2(x^*, y^*) = -y$, and so we get

$$\frac{\partial f}{\partial \mathbf{x}}(\mathbf{x}^*) = \begin{bmatrix} -1 \\ 0 \end{bmatrix}, \quad \frac{\partial g_1}{\partial \mathbf{x}}(\mathbf{x}^*) = \begin{bmatrix} 0 \\ 1 \end{bmatrix}, \quad \frac{\partial g_2}{\partial \mathbf{x}}(\mathbf{x}^*) = \begin{bmatrix} 0 \\ -1 \end{bmatrix}.$$

Note that the CQs are not met, since the first partials of both constraints that are satisfied as strict equalities at $\mathbf{x}^*$ are not linearly independent. Next, the FJ conditions are

$$\lambda_0 \frac{\partial f}{\partial \mathbf{x}}(\mathbf{x}^*) + \lambda_1 \frac{\partial g_1}{\partial \mathbf{x}}(\mathbf{x}^*) + \lambda_2 \frac{\partial g_1}{\partial \mathbf{x}}(\mathbf{x}^*) = 0,$$

i.e.,

$$\lambda_0 \begin{bmatrix} -1 \\ 0 \end{bmatrix} + \lambda_1 \begin{bmatrix} 0 \\ 1 \end{bmatrix} + \lambda_2 \begin{bmatrix} 0 \\ -1 \end{bmatrix} = \begin{bmatrix} 0 \\ 0 \end{bmatrix},$$

which are satisfied if $\lambda_0 = 0$ and $\lambda_1 = \lambda_2$. On the other hand, the KKT conditions are

$$-\frac{\partial f}{\partial \mathbf{x}}(\mathbf{x}^*) = \begin{bmatrix} 1 \\ 0 \end{bmatrix} = \lambda_1 \begin{bmatrix} 0 \\ 1 \end{bmatrix} + \lambda_2 \begin{bmatrix} 0 \\ -1 \end{bmatrix},$$

i.e., $\lambda_1(0) + \lambda_2(0) = 1$ and $\lambda_1(1) + \lambda_2(-1) = 0$, which are inconsistent, that is, these equations have no solution for λ_1 and λ_2. ■

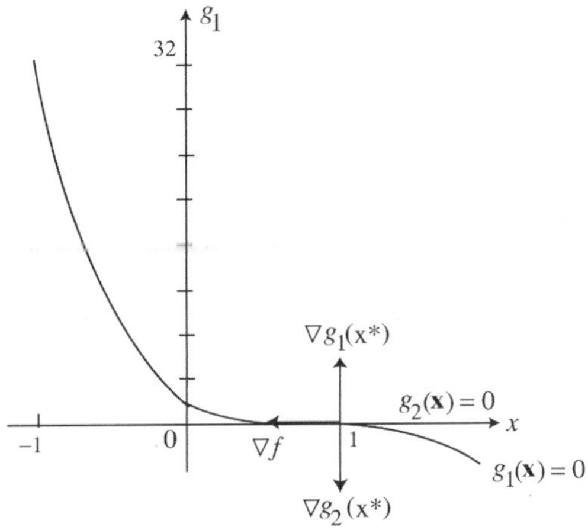

Figure 5.7 Example 5.11.

Example 5.12. Minimize $f(x,y) = -y$ such that $g_1(\mathbf{x}) = x-(1-y)^3 \le 0$, and $g_2(\mathbf{x}) = -x \le 0$, where $\mathbf{x} = (x, y)$. The graph is similar to Figure 5.7, where the feasible region and the optimal point $\mathbf{x}^*$ are identified for this problem.

In matrix notation, $\mathbf{x} = \begin{bmatrix} x \\ y \end{bmatrix}$. The optimal solution is at $\mathbf{x}^* = \begin{bmatrix} 0 \\ 1 \end{bmatrix}$. Also,

$$\frac{\partial f}{\partial \mathbf{x}}(\mathbf{x}^*) = \begin{bmatrix} 0 \\ -1 \end{bmatrix}, \quad \frac{\partial g_1}{\partial \mathbf{x}}(\mathbf{x}^*) = \begin{bmatrix} 1 \\ 0 \end{bmatrix}, \quad \frac{\partial g_2}{\partial \mathbf{x}}(\mathbf{x}^*) = \begin{bmatrix} -1 \\ 0 \end{bmatrix}.$$

Note that the CQs are not met as they are not the necessary conditions. However, the FJ conditions identify an optimal problem. The gradient vectors of both constraints that are satisfied as strict equalities at $\mathbf{x}^*$ are not linearly independent. Next, the FJ conditions that provide an optimal solution are

$$\lambda_0 \frac{\partial f}{\partial \mathbf{x}}(\mathbf{x}^*) + \lambda_1 \frac{\partial g_1}{\partial \mathbf{x}}(\mathbf{x}^*) + \lambda_2 \frac{\partial g_1}{\partial \mathbf{x}}(\mathbf{x}^*) = 0,$$

i.e.,

$$\lambda_0 \begin{bmatrix} 0 \\ -1 \end{bmatrix} + \lambda_1 \begin{bmatrix} 1 \\ 0 \end{bmatrix} + \lambda_2 \begin{bmatrix} -1 \\ 0 \end{bmatrix} = \begin{bmatrix} 0 \\ 0 \end{bmatrix},$$

which are satisfied if $\lambda_0 = 0$ and $\lambda_1 = \lambda_2$. On the other hand, the KKT conditions are

$$-\frac{\partial f}{\partial \mathbf{x}}(\mathbf{x}^*) = \begin{bmatrix} 0 \\ 1 \end{bmatrix} = \lambda_1 \begin{bmatrix} 1 \\ 0 \end{bmatrix} + \lambda_2 \begin{bmatrix} -1 \\ 0 \end{bmatrix},$$

i.e., $\lambda_1(1) + \lambda_2(-1) = 0$ and $\lambda_1(0) + \lambda_2(0) = 1$, which are inconsistent, that is, they cannot be solved for λ_1 and λ_2. ■

5.3.1 Feasibility. The following four cases for the feasibility problem are considered.

CASE 1. A convex optimization problem with equality and inequality constraints is to find

$$\min_{\mathbf{x}} f(\mathbf{x}) \text{ subject to } g_i(\mathbf{x}) = 0,\ i = 1, \dots, m;\ h_j(\mathbf{x}) \le 0,\ j = 1, \dots, k, \tag{5.3.3}$$

where $f : \mathbb{R}^n \mapsto \mathbb{R}$ is the objective or cost function, $\mathbf{x} \in \mathbb{R}^n$ is the optimization variable, $g_i(\mathbf{x})$ are the equality constraints, and $h_j(\mathbf{x})$ are the inequality constraints. The optimal value of $\mathbf{x}$, denoted by $\mathbf{x}^*$, is

$$f(\mathbf{x}^*) = \inf\{f(x) \,|\, g_i(\mathbf{x}) = 0, h_j(\mathbf{x}) \le 0.\} \tag{5.3.4}$$

Note that

$$\mathbf{x}^* = \begin{cases} \infty & \text{if problem is infeasible,} \\ -\infty & \text{if problem is unbounded below,} \end{cases}$$

where the infeasibility of the problem means that no $\mathbf{x}$ satisfies the constraints.

Then (i) $\mathbf{x}$ is feasible if $\mathbf{x} \in \text{dom}(f)$ and it satisfies the constraints.

(ii) A feasible $\mathbf{x}$ is optimal if $f(\mathbf{x}) = f(\mathbf{x}^*)$.

CASE 2. Find

$$\min_{\mathbf{x}} f(\mathbf{x}) \text{ subject to } g_i(\mathbf{x}) = 0, i = 1, \dots, m; h_j(\mathbf{x}) \le 0, j = 1, \dots, k. \tag{5.3.5}$$

This problem can be regarded as a special case of the above general problem (5.3.3) with $f(\mathbf{x}) = 0$. For this problem

$$f(\mathbf{x}^*) = \begin{cases} 0 & \text{if constraints are feasible,} \\ \infty & \text{if constraints are infeasible,} \end{cases}$$

where feasibility of constraints implies that any feasible $\mathbf{x}$ is optimal.

(iii) $\mathbf{x}^*$ is locally optimal if there is an $A > 0$ such that x is optimal for the following problem: Find

$$\min_{\mathbf{x}} f(\mathbf{x}) \text{ subject to } g_i(\mathbf{x}) = 0,\ i = 1, \dots, m;\ h_j(\mathbf{x}) \le 0,\ j = 1, \dots, k \tag{5.3.6}$$

where $\|\mathbf{z} - \mathbf{x}\|_2 \le A$.

Example 5.13. For $n = 1, k = m = 0$, consider

(a) $f(x) = 1/x : \text{dom}(f) = \mathbb{R}_+$: it has no optimal point.

(b) $f(x) = -\log x : \text{dom}(f) = \mathbb{R}_+$: we find that $f(x^*) = -\infty$.

(c) $f(x) = x \log x : \text{dom}(f) = \mathbb{R}_{++}$: we have $f(x^*) = -1/e$, and $x = 1/e$ is optimal.

(d) $f(x) = x^3 - 3x : f(x^*) = -\infty$: local optima are at $x = \pm 1$, local maximum at $x = -1$, and local minimum at $x = 1$. ■

CASE 3. An optimization problem has an *implicit constraint* if

$$\mathbf{x} = \bigcap_{i=0}^{m} \text{dom}(g_i) \cap \bigcap_{j=1}^{k} \text{dom}(h_j), \tag{5.3.7}$$

where $g_i(\mathbf{x}) = 0$ and $h_j(\mathbf{x}) \le 0$ are explicit constraints. A problem is unconstrained if it has no explicit constraints ($m = k = 0$).

Example 5.14. The problem

$$\min f(\mathbf{x}) = -\sum_{i=1}^{k} \log(b_i - a_i^T \mathbf{x}) \tag{5.3.8}$$

is an unconstrained problem with implicit constraints $a_i^T \mathbf{x} < b_i$. ■

CASE 4. An optimization problem with affine inequality constraints is:

$$\text{Minimize } f(\mathbf{x}) \text{ subject to } g_i(\mathbf{x}) \le 0, i = 1, \dots, m; a_i^T \mathbf{x} = b_j, j = 1, \dots, k, \tag{5.3.9}$$

where f and g_i ($i = 1, \dots, m$) are convex, and equality constraints are affine. The problem (5.3.9) is *quasi-convex* if the function f is quasi-convex and the functions $g_i(\mathbf{x})$ are convex. The problem (5.3.9) is often written as

$$\text{Minimize } f(\mathbf{x}) \text{ subject to } g_i(\mathbf{x}) \le 0, i = 1, \dots, m;\ a_i^T \mathbf{x} = b_i. \tag{5.3.10}$$

Note that the feasible set of a convex optimization problem is a convex set.

Example 5.15. Find

$$\min_x \{f(x) = x^2 + y^2\} \text{ subject to } g_1(x) = (x+y)^2;\ h_1(x) = \frac{x}{1+y^2} \le 0.$$

Note that f is convex, and the feasible set $\{(x, y) \mid x = -y \le 0\}$ is convex. But h_1 is not convex, and g_1 is not affine. Hence, it is not a convex optimization problem. ■

An equivalent, but not identical, problem to Example 5.15 is

$$\text{minimize } \{x^2+y^2\} \text{ subject to } x+y=0,\ x \le 0.$$

5.3.2 Slater's Condition. For convex programs, if there exists an $\mathbf{x}' \in \mathbb{R}^n$ such that

$$g_j(\mathbf{x}') < 0 \text{ for } j=1,2,\dots,m, \quad \text{and} \quad g_j(\mathbf{x}') > 0 \text{ for } j=m+1,\dots,p,$$

then the CQ holds at $\mathbf{x}$ since the relative interior is nonempty. Moreover, for a convex program where the CQ holds, the KKT necessary conditions are also sufficient.

5.4 Lagrangian Duality

Like linear programming, the nonlinear programming has a duality theory. For the problem

Minimize $f(\mathbf{x})$ subject to the constraints

$$g_j(\mathbf{x}) \le 0, j=1,2,\dots,m, \quad \text{and} \quad h_j(\mathbf{x})=0, j=1,2,\dots,p,$$

the Lagrangian is given by

$$L(\mathbf{x},\boldsymbol{\lambda},\boldsymbol{\mu}) = f(\mathbf{x}) = \sum_j \lambda_j g_j(\mathbf{x}) + \sum_j \mu_j h_j(\mathbf{x}).$$

Then the *Lagrangian duals* are as follows:

Primal Problem (minmax):

Minimize $\bar{L}(\mathbf{x})$, where

$$\begin{aligned}\bar{L}(\mathbf{x}) &= \max_{\boldsymbol{\lambda},\boldsymbol{\mu}} L(\mathbf{x},\boldsymbol{\lambda},\boldsymbol{\mu}) \\ &= \begin{cases} f(\mathbf{x}) \text{ if } g_j(\mathbf{x}) \le 0 \text{ for all } j \text{ and } h_j(\mathbf{x})=0 \text{ for all } j, \\ +\infty \text{ if } g_j(\mathbf{x}) > 0 \text{ or } h_j(\mathbf{x}) \ne 0 \text{ for some } j, \end{cases}\end{aligned}$$

where $\bar{L}$ is the primal Lagrangian. Then, the original problem is

$$\begin{aligned}\min_{x\in D} \bar{L}(\mathbf{x}) &= \min_{x\in D} \Big\{ \max_{\boldsymbol{\lambda},\boldsymbol{\mu}} L(\mathbf{x},\boldsymbol{\lambda},\boldsymbol{\mu}) \Big\} \\ &= \min_{x\in D} \Big\{ f(\mathbf{x}) \,|\, g_j(\mathbf{x}) \le 0, h_j(\mathbf{x})=0 \text{ for all } j. \Big\}\end{aligned}$$

Dual Problem (maxmin):

Maximize $\hat{L}(\boldsymbol{\lambda},\boldsymbol{\mu})$ subject to $\lambda \ge 0$, where $\hat{L}(\boldsymbol{\lambda},\boldsymbol{\mu}) = \min_{\mathbf{x}\in D} L(\mathbf{x},\boldsymbol{\lambda},\boldsymbol{\mu})$.

In particular, under conditions of convexity and differentiability, this reduces to

$$\max L(\mathbf{x}, \boldsymbol{\lambda}, \boldsymbol{\mu}), \text{ subject to } \frac{\partial L}{\partial \mathbf{x}}(\mathbf{x}, \boldsymbol{\lambda}, \boldsymbol{\mu}) = \mathbf{0},\ \boldsymbol{\lambda}, \boldsymbol{\mu} \geq \mathbf{0}, \mathbf{x} \in D,$$

where $\hat{L}$ is the dual Lagrangian.

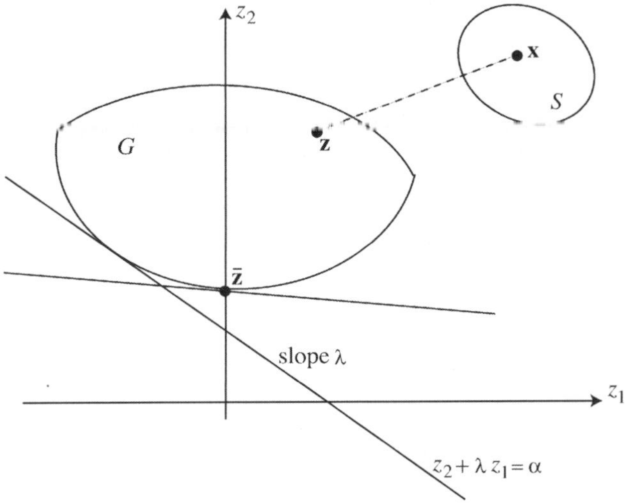

Figure 5.8 Duality.

Theorem 5.1. (Weak duality theorem) *The dual Lagrangian* $\hat{L}(\boldsymbol{\lambda}, \boldsymbol{\mu})$ *and the primal Lagrangian* $\bar{L}(\mathbf{x}, \boldsymbol{\lambda}, \boldsymbol{\mu})$ *are related by the inequalities*

$$\begin{aligned} &\hat{L}(\boldsymbol{\lambda}, \boldsymbol{\mu}) \leq L(\mathbf{x}, \boldsymbol{\lambda}, \boldsymbol{\mu}) \leq \bar{L}(\mathbf{x}, \boldsymbol{\lambda}, \boldsymbol{\mu}) \quad \text{for all } \boldsymbol{\lambda} \geq \mathbf{0} \text{ and } \mathbf{x} \in D, \\ &\max_{\boldsymbol{\lambda}, \boldsymbol{\mu}} \big\{ \min_{\mathbf{x}} L(\mathbf{x}, \boldsymbol{\lambda}, \boldsymbol{\mu}) \big\} \leq \min_{\mathbf{x}} \big\{ \max_{\boldsymbol{\lambda}, \boldsymbol{\mu}} L(\mathbf{x}, \boldsymbol{\lambda}, \boldsymbol{\mu}) \big\}. \end{aligned} \tag{5.4.1}$$

The quantity $\big[\bar{L}(\mathbf{x}) - \hat{L}(\boldsymbol{\lambda}, \boldsymbol{\mu})\big]$ is called the *duality gap*.

5.4.1 Geometrical Interpretation of Duality. Consider the problem: Minimize $f(\mathbf{x})$ subject to $g(\mathbf{x}) \leq 0$, $\mathbf{x} \in D$. Let $z_1 = g(\mathbf{x})$ and $z_2 = f(\mathbf{x})$. Then, as in Figure 5.8, G is the image of the set D under the map $z_1 = g(\mathbf{x}), z_2 = f(\mathbf{x})$. The original problem is equivalent to

$$\text{Minimize } z_2, \text{ subject to } z_1 \leq 0,\ \mathbf{z} = (z_1, z_2) \in G,$$

which yields the optimum solution $\bar{\mathbf{z}} = (z_1, z_2)$ shown in Figure 5.8.
The dual problem is: For $\boldsymbol{\lambda} \geq 0$, we have

$$\hat{L}(\boldsymbol{\lambda}) = \min_{\mathbf{x} \in D} L(\mathbf{x}, \boldsymbol{\lambda}) = \min_{\mathbf{x} \in D} \big\{ f(\mathbf{x}) + \boldsymbol{\lambda} g(\mathbf{x}) \big\} = \min_{\mathbf{z} \in G} z_2 + \lambda z_1 = \alpha.$$

Then, the dual problem is to find the slope λ of the tangential line (or plane) for which the intercept with the z_2-axis is maximized.

Example 5.16. (Min-max problem) Minimize $5x^2 + 2xy + y^2$ subject to $3x + y \geq k$, $x, y \geq 0$, where $k > 0$ is an integer. Let $\mathbf{x} = (x, y)$, and $D = \{\mathbf{x} \,|\, x \geq 0, y \geq 0\}$. The Lagrangian is

$$L(\mathbf{x}, \boldsymbol{\lambda}) = 5x^2 + 2xy + y^2 + \lambda(k - 3x - y).$$

Then the primal problem is:

$$\min_{\mathbf{x}\in D} \big\{ \max_{\boldsymbol{\lambda}\geq 0} L(\mathbf{x}, \boldsymbol{\lambda}) \big\} = \min_{\mathbf{x}\in D} \big\{ \bar{L}(\mathbf{x}) \big\},$$

that is,

$$\text{Minimize } 5x^2 + 2xy + y^2 \quad \text{subject to} \quad 3x + y \geq k,\ x, y \geq 0.$$

The associated dual problem is

$$\max_{\boldsymbol{\lambda}\geq 0} \big\{ \min_{\mathbf{x}\in D} L(\mathbf{x}, \boldsymbol{\lambda}) \big\} = \max_{\boldsymbol{\lambda}\geq \mathbf{0}} \big\{ \hat{L}(\boldsymbol{\lambda}) \big\}.$$

Consider the dual objective

$$\hat{L}(\boldsymbol{\lambda}) = \min_{\mathbf{x}\in D} \big\{ L(\mathbf{x}, \lambda) \big\} = 5x^2 + 2xy + y^2 + \lambda(k - 3x - y).$$

Verify that this objective function $L(\mathbf{x}, \lambda)$ is convex, thus, the minimum is obtained by using the necessary and sufficient conditions, which are

$$\frac{\partial L}{\partial x} = 10x + 2y - 3\lambda = 0, \qquad \frac{\partial L}{\partial y} = 2x + 2y - \lambda = 0,$$

which give the optimal point $x^* = \frac{\lambda}{4} = y^*$, which lie in D provided $\lambda \geq 0$. Hence, $\hat{L}(\boldsymbol{\lambda}) = L(\lambda/4, \lambda/4) = k\lambda - \frac{1}{2}\lambda^2$. Then the dual problem reduces to

$$\min_{\lambda\geq 0} \big\{ k\lambda - \frac{1}{2}\lambda^2 \big\}.$$

Verify that the objective function for this minimum problem is concave. Hence,

$$\frac{\partial \hat{L}}{\partial \lambda} = k - \lambda = 0, \quad \text{which yields} \quad \lambda = k > 0.$$

Since $\lambda > 0$, we get the dual solution:

$$\lambda^* = k, \qquad \text{which gives} \qquad \hat{L}(\lambda^*) = k^2/2,$$

and the primal solution:

$$x^* = y^* = \lambda^*/4 = k/4 > 0, \text{ and } 3x^* + y^* = k, \text{ for any } k > 0,$$

which is feasible. The complementary slackness is satisfied since $\lambda^*(k-3x^*-y^*)=0$. Hence, $(x^*,y^*)=\left(\frac{k}{4},\frac{k}{4}\right)\in D$ is the optimal point for the primal problem. Moreover, since $f(\mathbf{x}^*)=5x^{*2}+2x^*y^*+y^{*2}=k^2/2=L(\lambda^*)$, there is no duality gap. ■

Example 5.17. Minimize x^2+xy+y^2 subject to $x+y\geq 6$, $x,y\geq 0$. Let $\mathbf{x}=(x,y)$, and $D=\{\mathbf{x}\,|\,x\geq 0, y\geq 0\}$. The Lagrangian is

$$L(\mathbf{x},\boldsymbol{\lambda})=x^2+xy+y^2+\lambda(6-x-y).$$

Then the primal problem is:

$$\min_{\mathbf{x}\in D}\big\{\max_{\boldsymbol{\lambda}\geq 0}L(\mathbf{x},\boldsymbol{\lambda})\big\}=\min_{\mathbf{x}\in D}\big\{\bar{L}(\mathbf{x})\big\},$$

i.e.,

$$\text{Minimize } x^2+xy+y^2 \quad \text{subject to} \quad x+y\geq 6,\ x,y\geq 0.$$

The associated dual problem is

$$\max_{\boldsymbol{\lambda}\geq 0}\big\{\min_{\mathbf{x}\in D}L(\mathbf{x},\boldsymbol{\lambda})\big\}=\max_{\boldsymbol{\lambda}\geq 0}\big\{\hat{L}(\boldsymbol{\lambda})\big\}.$$

Consider the dual objective

$$\hat{L}(\boldsymbol{\lambda})=\min_{\mathbf{x}\in D}\big\{L(\mathbf{x},\boldsymbol{\lambda})=x^2+xy+y^2+\lambda(6-x-y).$$

Verify that this objective function $L(\mathbf{x},\boldsymbol{\lambda})$ is convex, thus, the minimum is obtained by using the necessary and sufficient conditions which are

$$\frac{\partial L}{\partial x}=2x+y-\lambda=0,\qquad \frac{\partial L}{\partial y}=x+2y-\lambda=0,$$

which give the optimal point $x^*=\frac{\lambda}{3}=y^*$, which lie in D provided $\lambda\geq 0$. Hence, $\hat{L}(\boldsymbol{\lambda})=L(\lambda/3,\lambda/3)=6\lambda-\frac{1}{3}\lambda^2$. Then the dual problem reduces to

$$\min_{\lambda\geq 0}\big\{6\lambda-\tfrac{1}{3}\lambda^2\big\}.$$

Verify that the objective function for this min problem is concave. Hence,

$$\frac{\partial\hat{L}}{\partial\lambda}=6-\frac{2}{3}\lambda=0,\quad \text{which yields}\quad \lambda=9>0.$$

Since $\lambda>0$, we get the dual solution:

$$\lambda^*=9,\qquad \text{which gives}\qquad \hat{L}(\lambda^*)=27>0,$$

and the primal solution:

$$x^* = y^* = \frac{\lambda^*}{3} = 3 > 0, \text{ and } x^* + y^* = 6,$$

which is feasible. The complementary slackness is satisfied since $\lambda^*(6 - x^* - y^*) = 0$. Hence, $(x^*, y^*) = (3, 3) \in D$ is the optimal point for the primal problem. Moreover, since $f(\mathbf{x}^*) = x^{*2} + x^*y^* + y^{*2} = (3)^2 + (3)(3) + (3)^2 = 27 \neq \lambda^*$, there is a no duality gap. ■

Example 5.18. Minimize $(-2x^2 - 3x^3)$ subject to $x^2 \leq 1$ (see Figure 5.9).

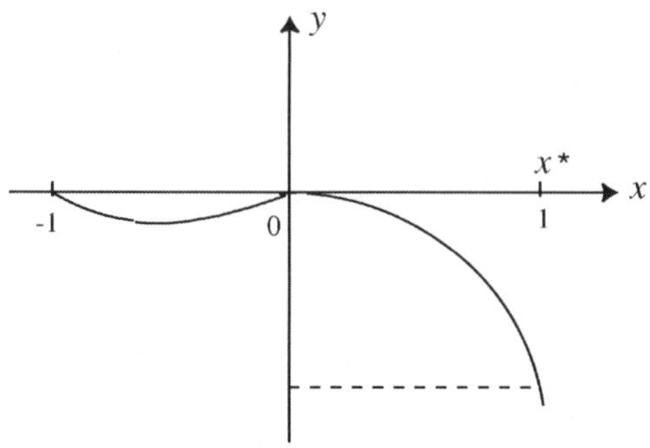

Figure 5.9 $f(x) = -2x^2 - 3x^3$.

The Lagrangian is $L(x, \lambda) = -2x^2 - 3x^3 + \lambda(x^2 - 1)$. For the primal problem the optimality condition is

$$\frac{\partial L}{\partial x} = 0, \quad \text{which yields} \quad -4x - 9x^2 + 2\lambda x = 0, \ \lambda(x^2 - 1) = 0, \ \lambda \geq 0, \quad x^2 \leq 1.$$

Case 1. $\lambda = 0$. Then $x = -4/9$, or $x = 0$.

Case 2. $\lambda > 0$. Then $x^2 - 1 = 0$, or $x = \pm 1$, where for $x = 1$ we get $\lambda = 13/2$, and for $x = -1$ we get $\lambda = -5/2$ which is not feasible, and therefore, rejected. Thus, the solutions are:

$$(x, \lambda) = (-4/9, 0) \quad \text{or} \quad (0, 0) \quad \text{or} \quad (1, 13/2),$$
$$f(x) = 16/243 \quad \text{or} \quad 0 \quad \text{or} \quad -5.$$

Hence, the primal solution is $x^* = 1, f(x^*) = -5$.

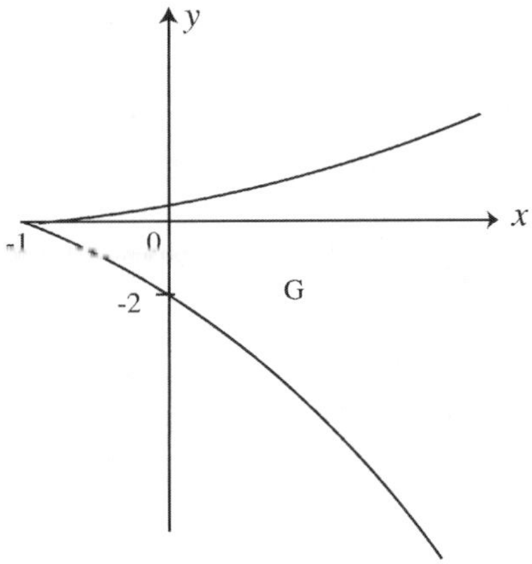

Figure 5.10 Duality gap.

The dual problem is $\max_{\lambda \geq 0} \{\hat{L}(\lambda)\}$, that is,

$$\hat{L}(\lambda) = \min_x L(x,\lambda) = \min_x \{-2x^2 - 3x^3 + \lambda(x^2-1)\} = -\infty \text{ for all values of } \lambda.$$

Thus, the primal objective $= -5$, and the dual objective $= -\infty$. Hence, there is a duality gap. Geometrically, let $z_1 = g(x) = x^2 - 1$ and $z_2 = f(x) = -2x^2 - 3x^3$. Then the supporting plane (line) runs through $(-1, 0)$ and intersects the z_2-axis at $-\infty$ (see Figure 5.10). ■

5.4.2 Saddle Point Sufficiency. The weak duality is defined by (5.4.1). The definition of a saddle point of $L(\mathbf{x}, \boldsymbol{\lambda})$ is as follows: The point $(\bar{\mathbf{x}}, \bar{\boldsymbol{\lambda}})$ is called a *saddle point* of $L(\mathbf{x}, \boldsymbol{\lambda})$ if

$$L(\bar{\mathbf{x}}, \bar{\boldsymbol{\lambda}}) \leq L(\mathbf{x}, \bar{\boldsymbol{\lambda}}) \text{ for all } \mathbf{x} \in D, \text{ i.e., } L(\mathbf{x}, \bar{\boldsymbol{\lambda}}) = \min_{\mathbf{x} \in D} L(\mathbf{x}, \bar{\boldsymbol{\lambda}}) = \hat{L}(\bar{\boldsymbol{\lambda}}), \tag{5.4.2a}$$

$$L(\bar{\mathbf{x}}, \bar{\boldsymbol{\lambda}}) \geq L(\mathbf{x}, \bar{\boldsymbol{\lambda}}) \text{ for all } \boldsymbol{\lambda} \geq 0, \text{ i.e., } L(\bar{\mathbf{x}}, \bar{\boldsymbol{\lambda}}) = \max_{\boldsymbol{\lambda} \geq \mathbf{0}} L(\bar{\mathbf{x}}, \boldsymbol{\lambda}) = \bar{L}(\bar{\mathbf{x}}). \tag{5.4.2b}$$

It means that if $(\bar{\mathbf{x}}, \bar{\boldsymbol{\lambda}})$ is a *saddle point* of $L(\mathbf{x}, \boldsymbol{\lambda})$, then

$$\bar{L}(\bar{\mathbf{x}}) = L(\bar{\mathbf{x}}, \bar{\boldsymbol{\lambda}}) = \hat{L}(\bar{\boldsymbol{\lambda}}). \tag{5.4.3}$$

This shows that the primal objective is equal to the dual objective and the duality gap is zero.

Consider the primal problem

$$\min \mathbf{x} \in D\{f(\mathbf{x})\} \quad \text{subject to} \quad g_j(\mathbf{x}) \le 0,\ j = 1, 2, \dots, m, \tag{5.4.4}$$

where $f(\mathbf{x})$ and $g_j(\mathbf{x})$ are all convex functions, and D is a convex set.

The *saddle point sufficiency condition* is: If $\mathbf{x} \in D$ and $\boldsymbol{\lambda} \ge \mathbf{0}$, then $(\bar{\mathbf{x}}, \bar{\boldsymbol{\lambda}})$ is a saddle point of $L(\mathbf{x}, \boldsymbol{\lambda})$ iff

(i) $\bar{\mathbf{x}}$ minimizes $L(\mathbf{x}, \boldsymbol{\lambda}) = f(\mathbf{x}) + \boldsymbol{\lambda}^T g_j(\mathbf{x})$ over D;

(ii) $g_j(\mathbf{x}) \le 0$ for each $j = 1, 2, \dots, m$;

(iii) $\bar{\boldsymbol{\lambda}} g_j(\bar{\mathbf{x}}) = 0$, which implies that $f(\bar{\mathbf{x}}) = L(\bar{\mathbf{x}}, \bar{\boldsymbol{\lambda}})$.

If $(\bar{\mathbf{x}}, \bar{\boldsymbol{\lambda}})$ is a saddle point of $L(\mathbf{x}, \boldsymbol{\lambda})$, then $\bar{\mathbf{x}}$ solves the primal problem (5.4.4) and $\bar{\boldsymbol{\lambda}}$ solves the dual problem which is

$$\max_{\boldsymbol{\lambda} \ge \mathbf{0}} \hat{L}(\boldsymbol{\lambda}) = \min_{\mathbf{x} \in D} L(\mathbf{x}, \boldsymbol{\lambda}). \tag{5.4.5}$$

5.4.3 Strong Duality. Consider the primal problem: Find

$$\begin{aligned} \Phi = \inf f(\mathbf{x}) \text{ subject to } g_j(\mathbf{x}) \le 0,\ j = 1, 2, \dots, m_1, \\ \text{and } h_j(\mathbf{x}) = 0, j = 1, 2, \dots, m_2,\ \mathbf{x} \in D, \end{aligned} \tag{5.4.6}$$

where $D \subseteq \mathbb{R}^n$ is a nonempty convex set, $f(\mathbf{x})$ and $g_j(\mathbf{x})$ are convex and $h_j(\mathbf{x})$ are linear functions.

The dual problem is defined as follows: Find

$$\Psi = \sup \hat{L}(\boldsymbol{\lambda}, \boldsymbol{\mu}) \text{ subject to } \boldsymbol{\lambda} \ge \mathbf{0}, \tag{5.4.7}$$

where $\hat{L}(\boldsymbol{\lambda}, \boldsymbol{\mu}) = \inf\limits_{\mathbf{x} \in D} \{f(\mathbf{x}) + \boldsymbol{\lambda}^T g(\mathbf{x}) = \boldsymbol{\mu}^T h(\mathbf{x})\}$. In (5.4.6) and (5.4.7), inf may be replaced by min, and sup by max.

Theorem 5.2. (Strong duality theorem) *Assuming that the following CQ holds: There exists an* $\hat{\mathbf{x}}$ *in* D *such that* $g_j(\hat{\mathbf{x}}) < 0$ *for* $j = 1, 2, \dots, m_1$, *and* $h_j(\hat{\mathbf{x}}) = 0$ *for* $j = 1, 2, \dots, m_2$, *and* $0 \in int\{h(D)\}$, *where* $h(D) = h(\mathbf{x})$, $\mathbf{x} \in D$, *then*

$$\Phi = \Psi, \tag{5.4.8}$$

i.e., there is no duality gap. Moreover, if $\Phi > -\infty$, *then*

(a) $\Psi = L(\boldsymbol{\lambda}^*, \boldsymbol{\mu}^*)$ *for some* $\boldsymbol{\lambda}^* \ge \mathbf{0}$; *and*

(b) if $\mathbf{x}^*$ *solves the primal, then it satisfies the complementary slackness:* $\lambda_j^* g_j(\mathbf{x}^*) = 0$ *for all* $j = 1, 2, \dots, m_1$.

5.5 Exercises

5.1. Let the production function be a Cobb-Douglas function with decreasing returns to scale, so that the firm's profit function is defined by

$$\pi = PAK^\alpha L^\beta - rK - wL. \tag{5.5.1}$$

Using the first-order partial derivatives π_K and π_L, the first-order conditions are

$$\begin{aligned} F_1(K, L; r, w, P, A, \alpha, \beta) &= \alpha PAK^{\alpha-1}L^\beta - r = 0,\\ F_2(K, L; r, w, P, A, \alpha, \beta) &= \beta PAK^\alpha L^{\beta-1} - w = 0. \end{aligned}$$

The first-order conditions can be expressed by partial derivatives with respect to w (wages) in the matrix form $\mathbf{Ax} = \mathbf{b}$ as

$$\begin{bmatrix} \alpha(\alpha-1)PAK^{\alpha-2}L^\beta & \alpha\beta PAK^{\alpha-1}L^{\beta-1}\\ \alpha\beta PAK^{\alpha-1}L^{\beta-1} & \beta(\beta-1)PAK^\alpha L^{\beta-2} \end{bmatrix} \begin{bmatrix} \dfrac{\partial K^*}{\partial w}\\ \dfrac{\partial L^*}{\partial w} \end{bmatrix} = \begin{bmatrix} 0\\ 1 \end{bmatrix},$$

solving which we get

$$\begin{aligned} |\mathbf{A}| &= \alpha(\alpha-1)PAK^{\alpha-2}L^\beta\,\beta(\beta-1)K^\alpha L^{\beta-2} - (\alpha\beta PAK^{\alpha-1}L^{\beta-1})^2\\ &= \alpha\beta(1-\alpha-\beta)P^2A^2K^{2\alpha-2}L^{2\beta-2} > 0 \quad \text{for } \alpha+\beta < 1. \end{aligned}$$

Since this is an unconstrained optimization problem, we have $|\mathbf{A}| = |\mathbf{H}|$, and $|\mathbf{H}| = |H_2| > 0$, which implies that the profit is minimized, and this profit minimizing firm operates under decreasing returns to scale. We will consider two cases:

CASE 1. For computing change in the demand for capital and for labor due to an increase in wages, i.e., $\dfrac{\partial K^*}{\partial w}$ and $\dfrac{\partial L^*}{\partial w}$, we have

$$\begin{aligned} |\mathbf{A}_1| &= \begin{vmatrix} 0 & \alpha\beta PAK^{\alpha-1}L^{\beta-1}\\ 1 & \beta(\beta-1)PAK^\alpha L^{\beta-2} \end{vmatrix} = -\alpha\beta PAK^{\alpha-1}L^{\beta-1} < 0,\\ |\mathbf{A}_2| &= \begin{vmatrix} \alpha(\alpha-1)PAK^{\alpha-2}L^\beta & 0\\ \alpha\beta PAK^{\alpha-1}L^{\beta-1} & 1 \end{vmatrix} < 1, \end{aligned}$$

Thus,

$$\frac{\partial K^*}{\partial w} = \frac{|\mathbf{A}_1|}{|\mathbf{A}|} = \frac{-\alpha\beta PAK^{\alpha-1}L^{\beta-1}}{\alpha\beta(1-\alpha-\beta)P^2A^2K^{2\alpha-2}L^{2\beta-2}} = \frac{-KL}{(1-\alpha-\beta)\,\mathrm{tr}(\mathbf{A})} < 0,$$

where tr($\mathbf{A}$) is the trace of the matrix $\mathbf{A}$ (i.e., sum of the diagonal elements of $\mathbf{A}$). This means that an increase in wages will decrease the demand for capital. Similarly,

$$\frac{\partial L^*}{\partial w} = \frac{|\mathbf{A}_2|}{|\mathbf{A}|} = \frac{\alpha(\alpha-1)PAK^{\alpha-2}L^{\beta}}{\alpha\beta(1-\alpha-\beta)P^2A^2K^{2\alpha-2}L^{2\beta-2}} = \frac{-(1-\alpha)L^2}{(1-\alpha-\beta)\,\mathrm{tr}(\mathbf{A})} < 0.$$

This shows that an increase in wages will reduce the optimal level of labor used.

CASE 2. For computing change in the demand for capital and for labor due to an increase in output price, i.e., $\frac{\partial K^*}{\partial P}$ and $\frac{\partial L^*}{\partial P}$, the first-order conditions can be expressed by partial derivatives with respect to w (wages) in the matrix form $\mathbf{Ax} = \mathbf{b}$ as

$$\begin{bmatrix} \alpha(\alpha-1)PAK^{\alpha-2}L^{\beta} & \alpha\beta PAK^{\alpha-1}L^{\beta-1} \\ \alpha\beta PAK^{\alpha-1}L^{\beta-1} & \beta(\beta-1)PAK^{\alpha}L^{\beta-2} \end{bmatrix} \begin{bmatrix} \frac{\partial K^*}{\partial P} \\ \frac{\partial L^*}{\partial P} \end{bmatrix} = \begin{bmatrix} -\alpha AK^{\alpha-1}L^{\beta} \\ -\beta AK^{\alpha}L^{\beta-1} \end{bmatrix}.$$

Note that the matrix $\mathbf{A}$ is the same as in Case 1, while in this case we have

$$|\mathbf{A}_1| = \begin{bmatrix} -\alpha AK^{\alpha-1}L^{\beta} & \alpha\beta PAK^{\alpha-1}L^{\beta-1} \\ -\beta AK^{\alpha}L^{\beta-1} & \beta(\beta-1)PAK^{\alpha}L^{\beta-2} \end{bmatrix} = \alpha\beta PA^2K^{2\alpha-1}L^{2\beta-2} > 0,$$

$$|\mathbf{A}_2| = \begin{bmatrix} \alpha(\alpha-1)PAK^{\alpha-1}L^{\beta} & -\alpha AK^{\alpha-1}L^{\beta} \\ \alpha\beta PAK^{\alpha-1}L^{\beta-1} & -\beta AK^{\alpha}L^{\beta-1} \end{bmatrix} = \alpha\beta PA^2K^{2\alpha-2}L^{2\beta-1} > 0.$$

This yields

$$\frac{\partial K^*}{\partial P} = \frac{|\mathbf{A}_2|}{|\mathbf{A}|} = \frac{\alpha\beta PA^2K^{2\alpha-2}L^{2\beta-1}}{\alpha\beta(1-\alpha-\beta)^2P^2A^2K^{\alpha-2}L^{\beta-2}} = \frac{K}{(1-\alpha-\beta)P} > 0,$$

which shows that an increase in the output will increase the demand for capital Similarly,

$$\frac{\partial L^*}{\partial P} = \frac{|\mathbf{A}_2|}{|\mathbf{A}|} = \frac{\alpha\beta PA^2K^{2\alpha-2}L^{2\beta-1}}{\alpha\beta(1-\alpha-\beta)^2P^2A^2K^{\alpha-2}L^{\beta-2}} = \frac{L}{(1-\alpha-\beta)P} > 0,$$

which shows that an increase in the output will increase the optimal level of labor used. ■

5.2. Consider $f(x,y) = 3x^2 - xy + 2y^2 - 4x - 7y + 8$. The first-order partial derivatives, equated to zero, give: $f_x = 6x - y - 4 = 0$, $f_y = -x + 4y - 7 = 0$, solving which we get $x = 1, y = 2$. Thus, the critical number is $(1,2)$. The second-order partial derivatives are: $f_{xx} = 6, f_{xy} = f_{yx} = -1, f_{yy} = 4$.

Checking the condition $f_{xx} \cdot f_{yy} > (f_{xy})^2$, we have $6 \cdot 4 > (-1)^2$. Since both f_{xx} and f_{yy} are positive, we have a global minimum at $(1, 2)$. ■

5.3. Optimize $f(x, y) = 4x^2 + 3xy + 6y^2$, subject to the constraint $x + y = 28$. The Lagrangian in the form (5.1.2) is

$$F(x, y, \lambda) = 4x^2 + 3xy + 6y^2 + \lambda(28 - x - y). \tag{5.5.2}$$

The first-order partial derivatives are: $F_x = 8x + 3y - \lambda = 0$, $F_y = 3x + 12y - \lambda$ $F_\lambda = 28 - x - y$, which when solved simultaneously give the critical values: $x^* = 18, y^* = 10, \lambda^* = 174$. Substituting these values in (5.5.2) we get $F(18, 10) = 2436 = f(18, 10)$. Notice that both functions $f(x, y)$ and $F(x, y)$ are equal at the critical values, since the constraint is equal to zero at those values.

The second-order derivatives are $F_{xx} = 8, F_{yy} = 12, F_{xy} = F_{yx} = 3$. Also, from the constraint $g(x, y) = x + y - 28$, we have $g_x = 1, g_y = 1$. Then the bordered Hessian (§1.6.3) is

$$|\bar{\mathbf{H}}| = \begin{vmatrix} 8 & 3 & 1 \\ 3 & 12 & 1 \\ 1 & 1 & 0 \end{vmatrix},$$

and its second principal minor is $|\bar{H}_2| = |\bar{\mathbf{H}}| = 8(-1) - 3(-1) + 1(3 - 12) = -14 < 0$. Since $|\bar{H}_2| < 0$, $|\mathbf{H}|$ is positive definite, $F(x, y)$ is at a local minimum. ■

5.4. Optimize the following Cobb-Douglas production functions subject to the given constraints by (i) using the Lagrange function and finding the critical points, and (b) by using the Hessian.

(a) $Q = K^{0.4}L^{0.5}$ subject to $6K + 2L = 270$. We get

$$\begin{aligned} Q_K &= 0.4K^{-0.6}L^{0.5} - 6\lambda = 0, \\ Q_L &= 0.5K^{0.4}L^{-0.5} - 2\lambda = 0, \\ Q_\lambda &= 270 - 6K - 2L = 0. \end{aligned}$$

From the first two equations we find that $\dfrac{0.4K^{-0.6}L^{0.5}}{0.5K^{0.4}L^{-0.5}} = \dfrac{6\lambda}{2\lambda}$, or

$$\frac{L}{K} = \frac{(3)(0.5)}{0.4} = 3.75, \text{ which gives } L = 3.75K.$$

Substituting $L = 3.75K$ into the third equation above, we get

$$270 - 6K - 7.5K = 0, \text{ which gives the critical points: } K = 20, L = 75.$$

(b) Using formula (16.6), the Hessian is

$$\begin{aligned}|\bar{\mathbf{H}}| &= \begin{vmatrix} -0.24K^{-1.6}L^{0.5} & 0.2K^{-0.6}L^{-0.5} & 6 \\ 0.2K^{-0.6}L^{-0.5} & -0.25K^{0.4}L^{-1.5} & 2 \\ 6 & 2 & 0 \end{vmatrix} \\ &= 4.8K^{-0.6}L^{-0.5} + 9K^{0.4}L^{-1.5} + 0.96K^{-1.6}L^{0.5} > 0.\end{aligned}$$

Since $\bar{H}_2| > 0$, $|\bar{\mathbf{H}}|$ is negative definite, and Q is maximized at the point $(20, 75)$.

5.5. Maximize the utility function $u = x^{0.6}y^{0.3}$ subject to the budget constraints $8x - 5y - 300$.

ANS. Since $U(x,y) = x^{0.6}y^{0.3} + \lambda(600 - 8x - 5y)$, we have $U_x = 0.6x^{-0.4}y^{0.3} - 8\lambda$, $U_y = 0.3x^{0.6}y^{0.7} - 5\lambda$, $U_\lambda = 300 - 8x - 5y$. Then from the first two equations we get $y = \frac{4}{5}x$, which after substituting in the last equation gives the critical points $(25, 20)$.

5.6. Minimize the total costs defined by $c = 15x^2 + 30xy + 30y^2$ when the firm meets the quota $g(x,y)$ equal to $2x + 3y = 20$. Define

$$C(x,y) = 15x^2 + 30xy + 30y^2 + \lambda(20 - 2x - 3y).$$

Then $C_x = 30x + 30y - 2\lambda = 0$, $C_y = 30x + 60y - 3\lambda = 0$, $C_\lambda = 20 - 2x - 3y$. Solving these three equations simultaneously, we get the critical values: $x^* = 4$, $y^* = 4$, $\lambda^* = 120$.

The second-order partial derivatives are: $C_{xx} = 30$, $C_{yy} = 60$, $C_{xy} = 30 = C_{yx}$, and $g_x = 2, g_y = 3$. Thus, the bordered Hessian (§1.6.3) is

$$|\bar{\mathbf{H}}| = \begin{vmatrix} 30 & 30 & 2 \\ 30 & 60 & 3 \\ 2 & 3 & 0 \end{vmatrix}.$$

The second principal minor is $|\bar{H}_2| = -150 < 0$. Thus, $|\bar{\mathbf{H}}|$ is positive definite and C is minimized when $x = y = 12$. ■

5.7. Minimize the utility $u = x^{1/}2y^{3/5}$ subject to the budget constraint $3x + 9y = 66$. Define

$$U(x,y) = x^{1/2}y^{3/5} + \lambda(66 - 3x - 9y).$$

Then $U_x = \frac{1}{2}x^{-1/2}y^{3/5} - 3\lambda = 0$, $U_y = \frac{3}{5}x^{1/2}y^{-2/5} - 9\lambda = 0, U_\lambda = 66 - 3x - 9y = 0$. Solving these equations simultaneously, we get the critical values: $x^* = 10, y^* = 20, \lambda^* = 0.3$. The second-order partial derivatives are:

$U_{xx} = -\frac{1}{4}x^{-3/2}y^{3/5}$, $U_{xy} = \frac{3}{10}x^{-1/2}y^{-2/5} = U_{yx}$, $U_{yy} = -\frac{6}{25}x^{1/2}y^{-7/5}$. Then the bordered Hessian is

$$|\bar{\mathbf{H}}| = \begin{vmatrix} -\frac{1}{4}x^{-3/2}y^{3/5} & \frac{3}{10}x^{-1/2}y^{-2/5} & 3 \\ \frac{3}{10}x^{-1/2}y^{-2/5} & -\frac{6}{25}x^{1/2}y^{-7/5} & 9 \\ 3 & 9 & 0 \end{vmatrix}.$$

The second principal minor is

$$|\bar{H}_2| = |\bar{\mathbf{H}}| = \frac{81}{4}x^{-3/2}y^{3/5} + \frac{81}{5}x^{-1/2}y^{-2/5} + \frac{54}{25}x^{1/2}y^{-7/3} > 0,$$

since all terms are positive. Hence, $|\bar{\mathbf{H}}|$ is positive definite, and U is minimized at the critical values. ■

5.8. Minimize $x^2 + 2y^2$ subject to $x + y \geq 3$ and $y - x^2 \geq 1$. SOLUTION. The Lagrangian is

$$L(x, y, \lambda_1, \lambda_2) = x^2 + 2y^2 + \lambda_1(3 - x - y) + \lambda_2(1 - y + x^2), \quad \lambda_1, \lambda_2 \geq 0.$$

Then

$$\begin{aligned} L_x(x, y, \lambda_1, \lambda_2) &= 2x - \lambda_1 + 2\lambda_2 = 0, \\ L_y(x, y, \lambda_1, \lambda_2) &= 4y - \lambda_1 - \lambda_2 = 0, \\ L_{\lambda_1}(x, y, \lambda_1, \lambda_2) &= 3 - x - y = 0, \\ L_{\lambda_2}(x, y, \lambda_1, \lambda_2) &= 1 - y + x^2 = 0. \end{aligned} \tag{5.5.3}$$

Solving the last two equations in (5.5.3), we get $(x, y) = (-2, 5)$ or $(x, y) = (1, 2)$. Using $(x, y) = (-2, 5)$ in the first two equations in (5.5.3) we find that $\lambda_1 = 28, \lambda_2 = -8$, which is not feasible. Next, using $(x, y) = (1, 2)$ in the first two equations in (5.5.3) we get $\lambda_1 = 6, \lambda_2 = 2$, which is feasible and the point $(1, 2)$ is the global minimizer.

5.9. Minimize $x^2 + y^2 - 4x - 4y$ subject to $x^2 \leq y$, $x + y \leq 2$. SOLUTION. The Lagrangian is

$$L(x, y, \lambda_1, \lambda_2) = x^2 + y^2 - 4x - 4y + \lambda_1(x^2 - y) + \lambda_2(x + y - 2), \quad \lambda_1, \lambda_2 \geq 0.$$

Then

$$\begin{aligned} L_x(x, y, \lambda_1, \lambda_2) &= 2x - 2\lambda_1 x + \lambda_2 = 0, \\ L_y(x, y, \lambda_1, \lambda_2) &= 2y - 4 - \lambda_1 + \lambda_2 = 0, \\ L_{\lambda_1}(x, y, \lambda_1, \lambda_2) &= x^2 - y = 0, \\ L_{\lambda_2}(x, y, \lambda_1, \lambda_2) &= x + y - 2 = 0. \end{aligned} \tag{5.5.4}$$

Solving the last two equations in (5.5.4), we get $(x, y) = (-2, 4)$ or $(x, y) = (1, 1)$. Using $(x, y) = (-2, 4)$ in the first two equations in (5.5.3) we find that $\lambda_1 = -4, \lambda_2 = -0$, which is not feasible. Next, using $(x, y) = (1, 1)$ in the first two equations in (5.5.3) we get $\lambda_1 = 0, \lambda_2 = 2$, which is feasible and the point $(1, 1)$ is the global minimizer.

5.10. The Constant Elasticity of Substitution (CES) production function is defined by

$$q = A\big[\alpha K^{-\beta} + (1-\alpha)L^{-\beta}\big]^{-1/\beta}, \tag{5.5.1}$$

where $A > 0$ is the coefficient parameter, α $(0 < \alpha < 1)$ the distribution parameter denoting relative factor shares, and $\beta > -1$ the substitution parameter that determines the value of elasticity of substitution (see Exercise 2.35). Consider $q = 100\big[0.4K^{-0.5} + 0.6L^{-0.5}\big]^{-2}$, and determine the relative minimum. Using the Lagrangian, the first-order partial derivatives of $Q \equiv q$ are:

$$\begin{aligned} Q_K &= 40K^{-1.5}\big[0.4K^{-0.5} + 0.6L^{-0.5}\big]^{-3} = 0 \\ Q_L &= 60L^{-1.5}\big[0.4K^{-0.5} + 0.6L^{-0.5}\big]^{-3} = 0. \end{aligned}$$

Solving these two equations we get $L^{1.5} = 1.5K^{1.5}$, or $L \approx 1.3K$. The second-order partial derivatives of Q are:

$$\begin{aligned} Q_{KK} &= -60K^{-2.5}\big[0.4K^{-0.5} + 0.6L^{-0.5}\big]^{-3} + 24K^{-3}\big[0.4K^{-0.5} + 0.6L^{-0.5}\big]^{-4}, \\ Q_{LL} &= -90L^{-2.5}\big[0.4K^{-0.5} + 0.6L^{-0.5}\big]^{-3} + 54L^{-3}\big[0.4K^{-0.5} + 0.6L^{-0.5}\big]^{-4}, \\ Q_{KL} &= 36K^{-1.5}L^{-1.5}\big[0.4K^{-0.5} + 0.6L^{-0.5}\big]^{-4} = Q_{LK}. \end{aligned}$$

We will use some numerical computation: We take $K = 1$, then $L = 1.3$. Then $\big[0.4K^{-0.5}+0.6L^{-0.5}\big]^{-3} \approx 1.26$, $\big[0.4K^{-0.5}+0.6L^{-0.5}\big]^{-4} = 1.36$, and $Q_{KK} \approx -42.96 < 0$, $Q_{LL} \approx -25.55$, and $Q_{KL} \approx 30.38 > 0$. Thus, $Q_{KK}Q_{LL} \approx 1098 > 0$ and $(Q_{KL})^2 = 923 > 0$. Since $Q_{kk}Q_{LL} > 0$ and $Q_{KK}Q_{LL} > (Q_{KL})^2$, we have a relative minimum at $(1, 1.73)$.

6

Quasi-Concave Functions

6.1 Quasi-Concave Functions

Most of the objective functions used in optimization problems are generally quasi-concave (or quasi-convex). In many problems, both quasi-concave and quasi-convex functions characterize a constraint set, which is a convex set. As mentioned before, there are quasi-concave (or quasi-convex) functions which are not concave (or convex) functions, although the converse need not be true. However, both quasi-concavity and quasi-convexity are defined in terms of convex sets, and they hold a symmetric relationship:

$f : \mathbb{R}^n \mapsto \mathbb{R}$ is (strictly) quasi-concave iff $(-f)$ is (strictly) quasi-convex.

Let $f : \mathbb{R}^n \mapsto \mathbb{R}$ be a real-valued function at $\mathbf{x} = (x_1, \dots, x_n) \in \mathbb{R}^n$. Then the *upper-level set* of f, denoted by U_α for all α, is defined by

$$U_\alpha = \{\mathbf{x} \in \text{dom}(f) \mid f(\mathbf{x}) \geq \alpha\}. \tag{6.1.1}$$

Upper-level sets, also known as upper contour sets, are convex sets for quasi-concave functions, and they are used in problems involving consumer's utility maximization and a company's cost minimization. For example, let an input requirement in the case of a production function correspond to the upper-level set (6.1.1), where α denotes an output level, $\mathbf{x}$ an input vector, and f a single-output production function. Then, in the case of utility maximization, where u denotes a utility function, the set of all consumption bundles $\{\mathbf{x}\}$ that are preferable to a given consumption bundle $\{\mathbf{x}^*\}$ is also an upper-level set $U_\alpha = \{\mathbf{x}^* \mid u(\mathbf{x}^*) \geq \alpha\}$.

A function f is said to be *quasi-concave* iff the upper-level set U_α is a convex set for every $\mathbf{y} \in \mathcal{R}(f)$.

A real-valued function $f : \mathbb{R}^n \mapsto \mathbb{R}$ is *strictly quasi-concave* iff

$$f(t\mathbf{x} + (1-t)\mathbf{x}') > \min\{f(\mathbf{x}), f(\mathbf{x}')\} \tag{6.1.2}$$

for all $\mathbf{x}, \mathbf{x}' \in \mathbb{R}^n$ and for all $t \in [0, 1]$. This definition differs from the above definition of quasi-concavity in that only strict inequality is used.

Let $x, x' \in \mathbb{R}$ be two distinct points on the x-axis (a convex set) such that the interval $[x, x']$ supports an arc AB on the curve, and B is higher than A (see Figure 6.1(a)). Since all the points between A and B on the arc are strictly higher than A, it satisfies the condition of quasi-concavity. The curves are strictly quasi-concave if all possible $[x, x']$ intervals have arcs that satisfy this same condition. Notice that this function also satisfies the condition of non-strict quasi-concavity, but does not satisfy the condition of quasi-convexity, because some points on the arc AB are higher than A, and this is not acceptable for a quasi-convex function. Figure 6.1(b) presents the case where a horizontal line segment $A'B'$ exists on which all points have the same height. This curve meets the condition of quasi-concavity, but does not satisfy that of strict quasi-concavity.

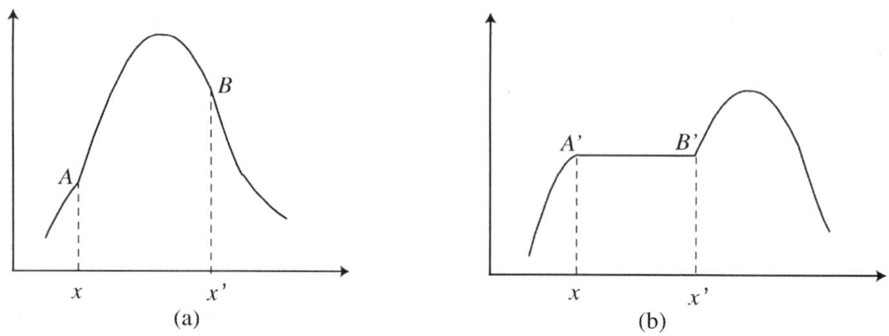

Figure 6.1 Quasi-concavity.

Note that in general a quasi-concave function that is also concave has its graph approximately shaped like a bell, or part thereof, and a quasi-convex function has its graph shaped like an inverted bell, or a part of it. Thus, quasi-concavity (or quasi-convexity) is a weaker condition than concavity (or convexity).

The above geometrical characterization leads to the following algebraic definition: A function f is $\left\{\begin{matrix}\text{quasi-concave}\\ \text{quasi-convex}\end{matrix}\right\}$ iff, for any pair of distinct points x and x' in the (convex-set) domain of f, and for $0 < t < 1$,

$$f(x') \ge f(x) \implies f(tx + (1-t)x') \left\{\begin{matrix} \ge f(x) \\ \le f(x') \end{matrix}\right\}. \tag{6.1.3}$$

A linear function $f(x)$ is both quasi-concave and quasi-convex. To prove, note that multiplying an inequality by -1 reverses the sign of inequality. If f is quasi-concave, with $f(x) \ge f(x')$, then $f(tx + (1-t)x') \ge f(x)$. Now, for the function $-f$, we will have $-f(x) \geq -f(x')$ and $f(tx+(1-t)x') \leq -f(x)$. Thus, $-f$ satisfies the condition of quasi-convexity.

Concavity implies quasi-concavity. To prove, let f be concave. Then $f(tx + (1-t)x') \geq tf(x) + (1-t)f(x')$. Now, assume that $f(x') \geq f(x)$. Then any weighted average of $f(x)$ and $f(x')$ cannot possibly be less than $f(x)$, i.e., $tf(x) + (1-t)f(x') \geq f(x)$. Combining these two results we find that $f(tx + (1-t)x') \geq f(x)$ for $f(x') \geq f(x)$, which satisfies the definition of quasi-concavity.

The condition of quasi-concavity does not guarantee concavity.

In the case of concave (and convex) functions, there is a very useful result: the sum of concave (convex) functions is also concave (convex). However, this result cannot be generalized to quasi-concave and quasi-convex functions.

Sometimes quasi-concavity and quasi-convexity can be checked by using the following definition:

A function $f(\mathbf{x})$, where $\mathbf{x} = (x_1, \dots, x_n) \in \mathbb{R}^n$, is $\left\{ \begin{matrix} \text{quasi-concave} \\ \text{quasi-convex} \end{matrix} \right\}$ iff, for any constant k, the set

$$\left\{ \begin{matrix} S^{\geq} \equiv \{x \,|\, f(x) \geq k\} \\ S^{\leq} \equiv \{x \,|\, f(x) \leq k\} \end{matrix} \right\} \quad \text{is a convex set.} \tag{6.1.4}$$

The sets $S^{\geq}$ and $S^{\leq}$ are presented in Figure 6.2.

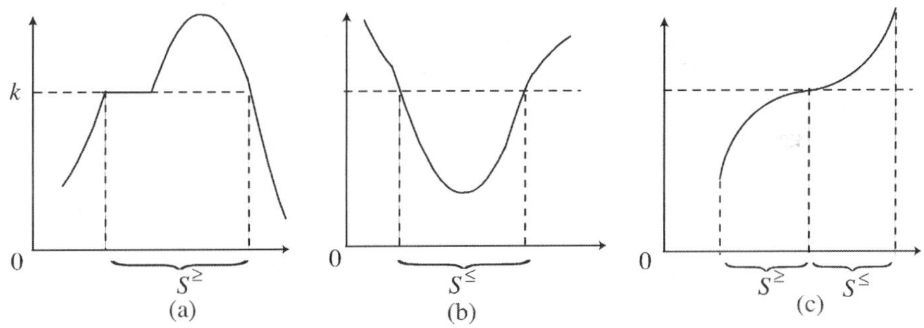

Figure 6.2 Sets $S^{\geq}$ and $S^{\leq}$.

The three functions in Figure 6.2 all contain concave as well as convex segments and therefore they are neither concave nor convex. However, the function in Figure 6.2(a) is quasi-concave because for any value of k, although the figure shows only one value of k, and the set $S^{\geq}$ is convex. The function in Figure 6.2(b) is, however, quasi-convex since the set $S^{\leq}$ is convex. The function in Figure 6.2(c) is a monotone function and it differs from the other two functions in that both $S^{\geq}$ and $S^{\leq}$ are convex sets. Hence, the function is both quasi-concave and quasi-convex. Note that formula (6.1.4) can be used to check quasi-concavity and quasi-convexity, but it cannot verify whether they are strict or nonstrict.

Example 6.1. Check $f(x) = x^2$, $x \geq 0$, for quasi-concavity and quasi-convexity. The graph of the function shows that it is a convex and a strictly convex function. It is also quasi-concave because its graph is a U-shaped curve, starting at the origin and increasing; it is similar to Figure 6.2(c) generating a convex $S^{\geq}$ as well as a convex $S^{\leq}$ set. Instead we use formula (6.1.3). If x and x' are two distinct nonnegative values of x, then $f(x) = x^2$, $f(x') = x'^2$, and $f(tx + (1-t)x') = (tx + (1-t)x')^2$. Now, suppose $f(x') \geq f(x)$, i.e., $x'^2 \geq x^2$; then, $x' \geq x$, or specifically $x' > x$, since x and x' are distinct points. Thus, the weighted average $tx + (1-t)x'$ must lie between x and x', and we have for $0 < t < 1$,

$$x'^2 > (tx + (1-t)x')^2 > x^2 \quad \text{or} \quad f(x') > f(tx + (1-t)x') > f(x).$$

But in view of (6.1.3), this result implies that f is both strictly quasi-concave and strictly quasi-convex. ■

Example 6.2. Show that $f(x, y) = xy$, $x, y \geq 0$ is quasi-concave. Use the criterion in (6.1.4) and show that the set $S^{\geq} = \{(x, y) \,|\, xy \geq k\}$ is a convex set for any k. Note that the curve $xy = k$ with $k \geq 0$ is a different curve for each k. If $k > 0$, this curve is a rectangular hyperbola in the first quadrant of the xy-plane, and the set consisting of all points on and above this hyperbola is a convex set. But if $k = 0$, the curve is defined by $xy = 0$, which constitutes the nonnegative parts of the x and y axes, and it is again a convex set. Hence, the function $f(x, y) = xy$, $x, y \geq 0$, is quasi-concave. Be careful not to confuse the given curve with $z = xy$ which is a surface in the (x, y, z)-space. In this example we are examining the characteristics of the surface which is quasi-concave in $\mathbb{R}^3$. ■

Example 6.3. Show that the function $f(x, y) = (x-a)^2 + (y-b)^2$ is convex and so it is quasi-convex. Use the criterion (6.1.4), and set $(x-a)^2 + (y-b)^2 = k$, $k \geq 0$. For each k, the curve is a circle in the xy-plane with center at (a, b) and radius $\sqrt{k}$. Since the set $\{(x, y) \,|\, (x-a)^2 + (y-b)^2 = k\}$ is the set of all points on and inside this circle, it is a convex set, even when $k = 0$ in which case the circle degenerates into a single point (a, b), and a set with a single point is a convex set. Hence, the given function is quasi-convex. ■

6.2 Differentiable Functions

A differentiable function $f(x)$ in $\mathbb{R}$ is $\left\{\begin{matrix}\text{quasi-concave} \\ \text{quasi-convex}\end{matrix}\right\}$ iff, for any pair of distinct points x and x' in dom(f),

$$f(x') \geq f(x) \implies \left\{\begin{matrix} f'(x)(x'-x) \\ f'(x')(x'-x) \end{matrix}\right\} \geq 0. \tag{6.2.1}$$

Quasi-concavity and quasi-convexity are strict if the inequality on the right side in (6.2.1) is changed to strict inequality > 0.

A differentiable function $f(\mathbf{x})$, where $\mathbf{x}=(x_1, \dots ,_n) \in \mathbb{R}^n$, is $\left\{\begin{array}{c}\text{quasi-concave}\\ \text{quasi-convex}\end{array}\right\}$ iff, for any pair of distinct points $\mathbf{x}$ and $\mathbf{x}'$ in dom(f),

$$f(\mathbf{x}') \geq f(\mathbf{x}) \Longrightarrow \left\{\begin{array}{c}\sum_{j=1}^{n} \dfrac{\partial f(\mathbf{x})}{\partial x_j}(x'_j - x_j)\\ \sum_{j=1}^{n} \dfrac{\partial f(\mathbf{x}')}{\partial x_j}(x'_j - x_j)\end{array}\right\} \geq 0. \tag{6.2.2}$$

For strict quasi-concavity and quasi-convexity, the right side of (6.2.2) must be changed to strict inequality > 0.

If a function $f(\mathbf{x})$, $\mathbf{x} \in \mathbb{R}^n$, is twice continuously differentiable, we can check quasi-concavity and quasi-convexity by using the bordered Hessian $|\mathbf{B}|$ (single function, §1.6.4) defined by

$$|\mathbf{B}| = \begin{vmatrix} 0 & f_1 & f_2 & \cdots & f_n \\ f_1 & f_{11} & f_{12} & \cdots & f_{1n} \\ f_2 & f_{21} & f_{22} & \cdots & f_{2n} \\ \cdots & \cdots & \cdots & \cdots & \cdots \\ f_n & f_{n1} & f_{n2} & \cdots & f_{nn} \end{vmatrix}, \tag{6.2.3}$$

where $f_i = \dfrac{\partial f}{\partial x_i}$ and $f_{ij} = \dfrac{\partial^2 f}{\partial x_i \partial x_j}$, $i,j = 1, \dots . n$. Note that, unlike the bordered Hessian $|\bar{\mathbf{H}}|$ described in §1.6.3 and used for optimization problems involving an extraneous constraint g, the above-defined bordered Hessian $|\mathbf{B}|$ is composed of the first derivatives of the function f only, without any extraneous constraint g. The leading principal minors of $|\mathbf{B}|$ are

$$|B_1| = \begin{vmatrix} 0 & f_1 \\ f_1 & f_{11} \end{vmatrix}, \; |B_2| = \begin{vmatrix} 0 & f_1 & f_2 \\ f_1 & f_{11} & f_{12} \\ f_2 & f_{21} & f_{22} \end{vmatrix}, \; |B_n| = |\mathbf{B}|. \tag{6.2.4}$$

We will state two conditions, one of which is necessary and the other is sufficient, and both relate to quasi-concavity on a domain consisting only of the nonnegative orthant (the n-dimensional analogue of the nonnegative quadrant) which is defined by $x_1, \dots , x_n \geq 0$. These conditions are as follows:

The necessary condition for a function $z = f(\mathbf{x})$ to be quasi-concave on the nonnegative orthant is

$$|B_1| \leq 0, \quad |B_2| \geq 0, \quad \text{and } |B_n| \left\{\begin{array}{c}\leq\\ \geq\end{array}\right. 0 \text{ if } n \text{ is } \left\{\begin{array}{l}\text{odd}\\ \text{even,}\end{array}\right. \tag{6.2.5}$$

where the partial derivatives are evaluated in the nonnegative orthant. Recall that the first condition in (6.2.5) is automatically satisfied since $|B_1| = -f_1^2 = -\left(\dfrac{\partial f}{\partial x_1}\right)^2$.

The sufficient condition for f to be strictly quasi-concave on the nonnegative orthant is that

$$|B_1| < 0, \quad |B_2| > 0, \quad |B_n| \begin{Bmatrix} \leq \\ \geq \end{Bmatrix} 0 \ \text{ if } n \ \text{ is } \begin{cases} \text{odd} \\ \text{even,} \end{cases} \tag{6.2.6}$$

where the partial derivatives are evaluated in the nonnegative orthant. The details of these conditions are available in Arrow and Enthoven [1961:797], and Takayama [1993:65].

Example 6.4. The function $f(x_1, x_2) = x_1 x_2$, $x_1, x_2 \geq 0$ is quasi-concave (compare Example 6.2). We will check it using (6.2.2). Let $\mathbf{u} = (u_1, u_2)$ and $\mathbf{v} = (v_1, v_2)$ be two points in dom(f). Then $f(\mathbf{u}) = u_1 u_2$ and $f(\mathbf{v}) = v_1 v2$. Assume that

$$f(\mathbf{v}) \geq f(\mathbf{u}), \quad \text{or} \quad v_1 v_2 \geq u_1 u_2, \tag{6.2.7}$$

where $u_1, u_2, v_1, v_2 \geq 0$. Since the partial derivatives of f are $f_1 = x_2$ and $f_2 = x_1$, condition (6.2.2) implies that $f_1(\mathbf{u})(v_1 - u_1) + f_2(\mathbf{u})(v_2 - u_2) = u_2(v_1 - u_1) + u_1(v_2 - u_2) \geq 0$, which after rearranging the terms is

$$u_2(v_1 - u_1) \geq u_1(u_2 - v_2). \tag{6.2.8}$$

Now there are four cases to consider depending on the values of u_1 and u_2:

(1) If $u_1 = u_2 = 0$, then (6.2.8) is trivially satisfied.

(2) If $u_1 = 0$ and $u_2 > 0$, then (6.2.8) reduces to $u_2 v_1 \geq 0$, which is again satisfied since u_2 and v_1 are both nonnegative.

(3) If $u_1 > 0$ and $u_2 = 0$, then (6.2.8) reduces to $0 \geq -u_1 v_2$, which is satisfied.

(4) Suppose $u_1, u_2 > 0$, so that $v_1, v_2 > 0$ also. Subtracting $v_2 u_1$ from both sides of (6.2.7), we obtain

$$v_2(v_1 - u_1) \geq u_1(u_2 - v_2). \tag{6.2.9}$$

This leads to the following three possibilities: (a) If $u_2 = v_2$, then $v_1 \geq u_1$. Since (u_1, u_2) and (v_1, v_2) are distinct points, and $u_2 = v_2$ and $v_1 > u_1$, so condition (6.2.8) is satisfied.

(b) If $u_2 > v_2$, then we have $v_1 > u_1$ by (6.2.9). Multiplying both sides of (6.2.9) by u_2/v_2, we get

$$u_2(v_1 - u_1) \geq \frac{u_2}{v_2} u_1(u_2 - v_2) > u_1(u_2 - v_2), \quad \text{since } u_2 > v_2. \tag{6.2.10}$$

Hence, (6.2.8) is satisfied.

(c) If $u_2 < v_2$, i.e., u_2/v_2 is a positive fraction, the first inequality of (6.2.10) still holds in this case. The second inequality also holds because a fraction u_2/v_2 of a negative number $(u_2 - v_2)$ is greater than the number itself. Hence the given function is quasi-concave.

As for the sufficient condition, note that the partial derivatives are $f_1 = x_2, f_2 = x_1, f_{11} = f_{22} = 0, f_{12} = f_{21} = 1$, giving

$$|B_1| = \begin{vmatrix} 0 & x_2 \\ x_2 & 0 \end{vmatrix} = -x_2^2 \le 0, \quad |B_2| = \begin{vmatrix} 0 & x_2 & x_1 \\ x_2 & 0 & 1 \\ x_1 & 1 & 0 \end{vmatrix} = 2x_1x_2 \ge 0.$$

Thus, (6.2.6) is satisfied in the positive orthant. ■

Example 6.5. Show that the function $f(x, y) = x^a y^b$, $(x, y > 0; a > 0, b > 1)$ is strictly quasi-concave. The partial derivatives are: $f_x = ax^{a-1}y^b$, $f_y = bx^a y^{b-1}$, $f_{xx} = a(a-1)x^{a-2}y^b$, $f_{yy} = b(b-1)x^a y^{b-2}$, $f_{xy} = f_{yx} = abx^{a-1}y^{b-1}$. Then the minors of the bordered Hessian $|\mathbf{B}|$ are

$$|B_1| = \begin{vmatrix} 0 & f_x \\ f_x & f_{xx} \end{vmatrix} = -(ax^{a-1}y^b)^2 < 0,$$

$$|B_2| = \begin{vmatrix} 0 & f_x & f_y \\ f_x & f_{xx} & f_{xy} \\ f_y & f_{yx} & f_{yy} \end{vmatrix} = [2a^2b^2 - a(a-1)b^2 - a^2b(b-1)]\, x^{3a-2}y^{3b-2} > 0,$$

thus satisfying the condition for strict quasi-concavity in (6.2.6). ■

Note that it is improper to define strict quasi-concavity by saying that the upper-level sets are strictly convex, since an upper-level set can be strictly convex even when the function has flat parts. Moreover, a function f is strictly quasi-concave iff $-f$ is strictly quasi-convex.

Quasi-concavity is a weaker assumption than concavity in the sense that, although every concave function is quasi-concave, the converse is not true. However, economists sometimes demand something more than quasi-concavity. For example, a quasi-concave function does not imply risk aversion, while a concave utility function does.

6.3 Theorems on Quasi-Concavity

Let S denote the convex set on which a concave function is defined. Thus, $(tx + (1-t)x', ty + (1-t)y') \in S$. Then quasi-concavity of f means that

$$tf(x) + (1-t)f(x') = ty + (1-t)y' \le f(tx + (1-t)x'). \tag{6.3.1}$$

A real-valued function f on a metric space X is called *upper semicontinuous* if for each real number α the set $\{x : f(x) < \alpha\}$ is open. If f is continuous, both f and $-f$ are upper semicontinuous. This definition leads to the proposition: Let f be an upper semicontinuous and real-valued function on a countably compact space, then f is bounded from above and assumes its maximum (Royden [1968: 161]).

Theorem 6.1. *Let f be an upper semicontinuous function on a convex set $S \in \mathbb{R}^2$. If f is strictly quasi-concave on S, then f is quasi-concave on S.*

PROOF. Let (x, x') be in S, where $y = f(x)$ and $y' = f(x')$. If S is a convex set, then the convex combination $(tx + (1-t)x', ty + (1-t)y')$ is also in S for any $t \in [0, 1]$. Thus, the inequality (6.3.1) holds for all $x, x' \in S$. Conversely, assume that $tf(x)+(1-t)f(x') \leq f(tx+(1-t)x')$. Choose y and y' such that $y \leq f(x)$ and $y' \leq f(x')$. Obviously, (x, y) and (x', y') are both in S. Thus, $ty \leq tf(x)$, and $(1-t)y' \leq (1-t)f(x')$ for any $0 \leq t \leq 1$. But this implies that $ty + (1-t)y' \leq tf(x) + (1-t)f(x')$. Since the right-hand side of this inequality is assumed to be not greater than $tf(x) + (1-t)f(x')$, we get from inequality (6.3.1)

$$ty + (1-t)y' \leq f(tx + (1-t)x').$$

Hence, $(tx + (1-t)x', ty + (1-t)y') \in \text{hyp}(f)$, i.e., S is an upper-level set, and the inequality (6.3.1) implies (6.1.2), which completes the proof. ■

This means that strict quasi-concavity and upper semicontinuity imply quasi-concavity, but the converse is not true. Another useful result is

Theorem 6.2. *Let f be a strictly quasi-concave function on a convex set $S \in \mathbb{R}^2$. If x^* is a local maximizer of f on S, then x^* is a global maximizer of f on S.*

Note that a sum of two concave functions is a concave function. However, the sum of two quasi-concave functions is not necessarily a quasi-concave function. Also, the sum of a concave function and a quasi-concave function is not necessarily either a concave function or a quasi-concave function.

Theorem 6.3. *Let $f : \mathbb{R}^2 \mapsto \mathbb{R}$ be a quasi-concave function, and let $g : \mathbb{R} \mapsto \mathbb{R}$ be a nondecreasing function whose domain contains $\mathcal{R}(f)$. Then the composite function $F(x) := g \circ f : \mathbb{R}^2 \mapsto \mathbb{R}$ is a quasi-concave function.*

PROOF. This follows from the fact that strictly monotone functions have strictly monotone inverse functions. Thus, for any α we get

$$\{x | F(x) \geq \alpha\} = \{x \,|\, g(f(x)) \geq \alpha\} = \{x \,|\, f(x) \geq g^{-1}(\alpha)\},$$

which is a convex set since f is quasi-concave. ■

Theorem 6.4. *Let $\mathcal{B}_\alpha^U := X \cap \{x \,|\, f(x) \geq \alpha\} = X \cap U_\alpha$. Then $f(x)$ is quasi-concave iff $\mathcal{B}_\alpha^U$ is a convex set for all α.*

PROOF. (Necessity ⇒) Let $x, y \in \mathcal{B}_\alpha^U$, and let $t \in [0, 1]$. Then by definition of quasi-concavity we have $x \in X$ and both $f(x) \geq \alpha$ and $f(y) \geq \alpha$. Using the quasi-concavity of f, we have $f(tx + (1-t)y) \geq \min\{f(x), f(y)\} \geq \alpha$. Thus, $tx + (1-t)y \in \mathcal{B}_\alpha^U$, which implies that $\mathcal{B}_\alpha^U$ is a convex set.

(Sufficiency $\Leftarrow$) Let $\alpha = \min\{f(x), f(y)\}$. Then both $x \in \mathcal{B}_\alpha^U$ and $y \in \mathcal{B}_\alpha^U$ hold. Since $\mathcal{B}_\alpha^U$ is a convex set, we get $tx + (1-t)y \in \mathcal{B}_\alpha^U$, which implies that $f(tx + (1-t)y) \geq \alpha = \min\{f(x), f(y)\}$. This proves quasi-concavity of f. ■

This proposition states that a function f is quasi-concave iff for any α the upper-level set of f is a convex set.

To check whether a given function f is quasi-concave, just check the property in Theorem 6.4, which states that a function f is quasi-concave iff all of its upper-level sets are convex sets; i.e., for any α, the set of points, where $f(x) \geq \alpha$ is true, is a convex set.

We will consider functions of a single variable $x \in \mathbb{R}$. Recall that monotone functions in $\mathbb{R}$ are both quasi-concave and quasi-convex, since both upper- and lower-level sets are convex sets; in fact, they are intervals. Hence, monotone functions are *quasi-linear*. Some examples follow.

Example 6.6. Consider the simplest case: $f(x) = x$, which is a strictly increasing linear function (see Figure 6.3). ■

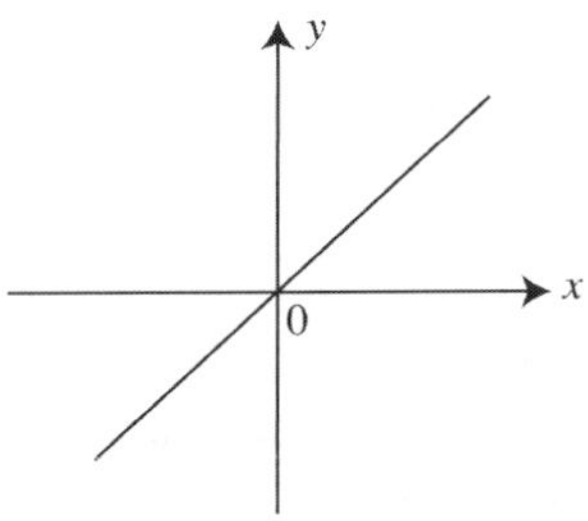

Figure 6.3 Strictly increasing function.

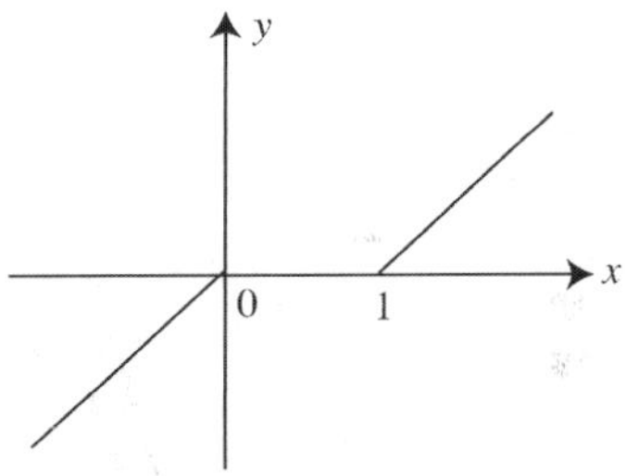

Figure 6.4 Quasi-concave function.

Example 6.7. Consider the function $f(x) = \begin{cases} x & \text{if } x \leq 0, \\ 0 & \text{if } 0 \leq x \leq 1, \\ x-1 & \text{if } x \geq 1. \end{cases}$

This function is not linear, but monotone increasing; thus, it is quasi-concave (see Figure 6.4, where the horizontal part of the graph corresponds to the interval $[0, 1]$. ■

Example 6.8. Consider the function $f(x) = x^3$, which is strictly increasing (see Figure 6.5). ■

Example 6.9. Consider the function $f(x) = \begin{cases} -x^2 & \text{if } x \leq 0, \\ 0 & \text{if } 0 \leq x \leq 1, \\ -(x-2)^2+1 & \text{if } x \geq 1. \end{cases}$
This function is neither increasing nor decreasing and has a horizontal part on its graph (see Figure 6.6). This function is obviously quasi-concave since all its upper-level sets are convex sets (intervals); it is not quasi-convex because the lower-level sets are not convex sets. The plot of this function shows the general shape of quasi-concave functions of a single variable. ■

Note that each continuous quasi-concave function of a single variable belongs to one of the following classes of functions (Martos [1975]): (i) Either f is a monotone increasing function on X, or a monotone decreasing function on X; and (ii) there exist $a, b \in X$, $a \leq b$, such that f is monotonically increasing for $x < a$, constant for $x \in [a, b]$, and monotonically decreasing for $x > b$.

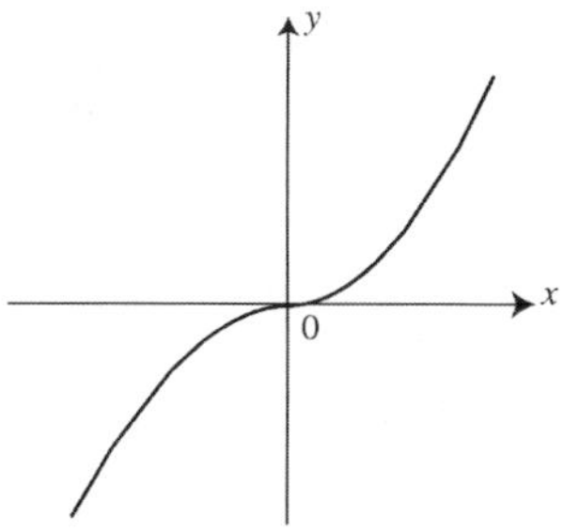

Figure 6.5 Strictly increasing function.

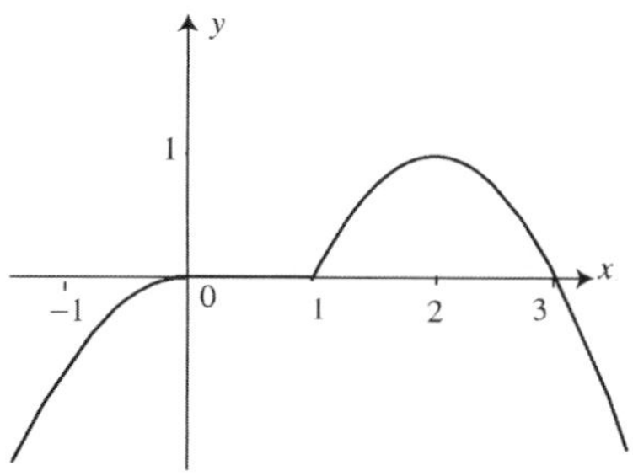

Figure 6.6 Quasi-concave function.

6.4 Three-Dimensional Case

We will consider an analogy with a mountain, and try to explain the concept of concavity in reference to a high mountain by asking if the surface of the mountain is *concave*. The answer is 'yes' if every straight line connecting any two points on the surface lies everywhere on or under the surface. The ideal situation is if the mountain was a perfect dome (semi-sphere); in this case the condition is satisfied and the function defined by its surface is concave. The condition is also satisfied if the mountain is a perfect cone, since in this case every straight line connecting two points on its surface lies exactly on the surface, and the function defined by the cone is concave.

The function in the case of a perfect dome or a cone has a common property: each level line is a circle. Similarly, on a topological map of the above-mentioned mountain, the set of points inside each contour (i.e., the set of points at which the height of the mountain exceeds a given number) is a convex set. The spacing of the contour lines differs, but the set of points inside every contour has the same shape for each mountain: it is a disk, and each such set is convex. However, not every mountain has this property, as one can see from the topological map of any actual mountain. In fact, the contours of mountains do not generally enclose convex sets.

In reality, mountains generally come in different forms. For example, a mountain may be a deformation of a cone that gets progressively steeper at higher altitudes, becoming harder to climb. In this case a straight line from the top of the mountain to any other point on its surface will not lie on or under the surface, but rather pass through the air. The function defined by the surface of such a mountain is *not* concave.

Let us consider the surface of a mountain as function $f(x, y)$, where x denotes longitude and y the latitude. Then a contour is a *level curve* of f. A function with the property that, for every value of a, the set of points (x, y) such that $f(x, y) \geq a$ is a convex set is said to be *quasi-concave*. The set of points (x, y) in this definition lie inside every contour on a topographical map.

Example 6.10. Let $f(x, y) = -x^2 - y^2$. The upper-level set of f for α is the set of points (x, y) such that $-x^2 - y^2 \geq \alpha$, or $x^2 + y^2 \leq -\alpha$. Thus, if $\alpha > 0$, the upper-level set U_α is empty, whereas if $\alpha < 0$, it is a disk of radius $\alpha^{1/2}$. ■

6.5 Multivariate Case

We can now generalize to $\mathbb{R}^n$. Let f be a multivariate function defined on the set S. Then f is said to be *quasi-concave* if, for any number a, the set of the points for which $f(\mathbf{x}) \geq a$ is a convex set.

Let f be a multivariate function defined on a set S. For any real number a, the set U_α, defined by (6.1.1), is the *upper-level set* of $f \in S$ for all α. In the case of a mountain, U_α is the set of all points at which the altitude is at least α.

Note that f is quasi-convex iff $-f$ is quasi-concave. The notion of quasi-concavity is weaker than that of convexity, in the following sense.

Theorem 6.5. *Every concave function is quasi-concave, and every convex function is quasi-convex.*

PROOF. Let the function be f, and the convex set on which it is defined be S. Let a be a real number and let x and y be two pints in the upper-level set U_α, where $x \in U_\alpha$ and $y \in U_\alpha$. We must show that U_α is convex, i.e., we must show that for every $t \in [0, 1]$ we have $tx + (1 - t)y \in U_\alpha$.

First, note that the set S on which f is defined is convex; thus, we have $tx + (1 - t)y \in S$, and so f is defined at the point $tx + (1 - t)y$. Next, the concavity of f means that

$$f(tx + (1 - t)y) \geq tf(x) + (1 - t)f(y). \tag{6.5.1}$$

Moreover, since $x \in U_\alpha$ means $f(x) \geq a$, and $y \in U_\alpha$ means $f(y) \geq a$, we have

$$tf(x) + (1-t)f(y) \geq ta + (1-t)a = a. \tag{6.5.2}$$

Then combining the inequalities (6.5.1) and (6.5.2) we get

$$f(tx + (1-t)y) \geq a,$$

which means that $tx + (1-t)y \in U_\alpha$, thereby proving that every upper-level set is convex and hence f is quasi-concave. The other part of the theorem follows simply by using the fact that $-f$ is convex. ■

Example 6.11. Consider the function $f(x, y) = -x^2 - y^2$. This is a concave function, and also quasi-concave. Take $\psi(y) = e^y$. Then, using Theorem 6.3, the function $g(x, y) = e^{f(x,y)}$ will be quasi-concave (see Figure 6.7). Notice that although the function $g(x, y)$ is obtained from a concave function using a strictly monotonically increasing transformation, g is not at all a concave function. ■

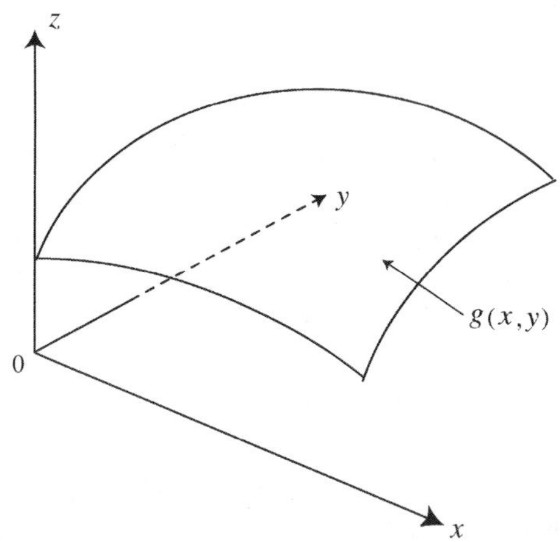

Figure 6.7 Function $g(x, y)$.

6.6 Sums of Quasi-Concave Functions

The following property holds for the sum of concave functions: If $f_1, \dots, f_m$ are concave functions and λ_i, $i = 1, \dots, m$, then $F(x) = \lambda_1 f_1(x) + \cdots + \lambda_m f_m(x)$ is a concave function. But this property does not hold for quasi-concave functions.

Example 6.12. The function $g_1(x) = e^x$, $x \geq 0$, is a strictly monotonically increasing function and thus, it is quasi-concave. Further, the function $g_2(x) = e^{-x}$, $x \leq 0$, is a strictly monotonically decreasing function, and thus it is also quasi-concave. But their sum $g(x) = g_1(x) + g_2(x)$ is strictly convex and not quasi-concave (see Figure 6.8). ■

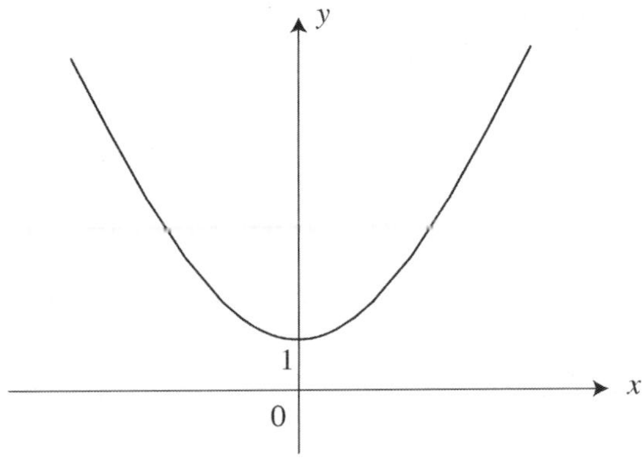

Figure 6.8 Function $g_1(x)$ and $g_2(x)$.

Example 6.13. Consider the bell-shaped curves $g(x, y) = e^{-x^2-y^2}$ and $h(x, y) = e^{-(x-1)^2-(y-1)^2}$. Both functions are quasi-concave. Let their sum be $G(x, y) = g(x, y) + h(x, y)$. The function $G(x, y)$ is not quasi-concave as it consists of two disjoint bell-shaped surfaces, representing the functions $g(x, y)$ and $h(x, y)$, respectively. The indifference curves and the upper-level set (shaded region) of $G(x, y)$ corresponding to $\alpha = 0.7$ are shown in Figure 6.9. It is obvious that the upper-level set is not convex because it consists of two parts. This example shows that, in general, the sum of quasi-concave functions is not necessarily a quasi-concave function.

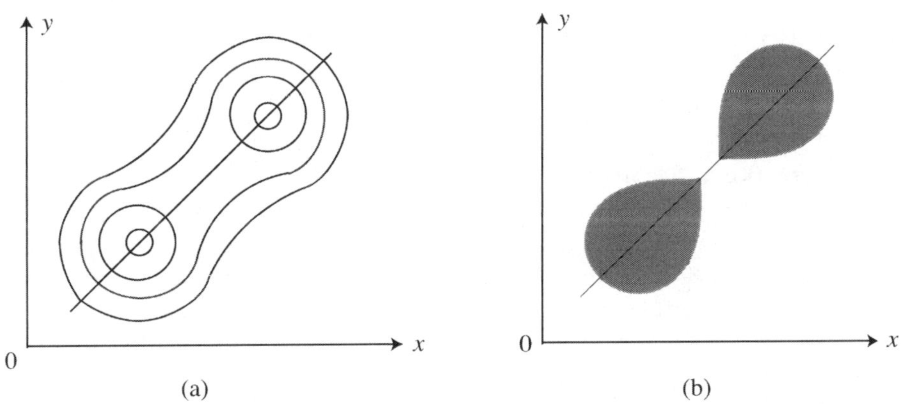

Figure 6.9 (a) Indifference curves, and (b) upper-level set of the function $G(x, y)$.

6.7 Strictly Quasi-Concave Functions

Using the definition (3.8.8) of strict concavity of a function f, which states that f is a strictly concave function over X if for any y, $y \in X, x \neq y$, and for any t, $0 < t < 1$, we have the inequality

$$f(tx + (1-t)y) > tf(x) + (1-t)f(y). \tag{6.7.1}$$

This implies that

$$f(tx + (1-t)y) > tf(x) + (1-t)f(y) \geq \min\{f(x), f(y)\}, \tag{6.7.2}$$

which leads to the generalization of concavity: $f(x)$ *is strictly quasi-concave on* X*, if for all* $x, y \in X, x \neq y$, $0 < t < 1$,

$$\min\{f(x), f(y)\} < f(tx + (1-t)y\}. \tag{6.7.3}$$

$f(x)$ is strictly quasi-convex on $X \iff -f$ is strictly quasi-concave on X.

Hence,

$f(x)$ is strictly concave on $X \implies f$ is strictly quasi-concave on X.

$f(x)$ is strictly quasi-concave on $X \implies f$ is quasi-concave on X.

A set $C \subset \mathbb{R}^n$ is said to be *strictly convex* if for any $x, y \in C$, $0 < t < 1$ we have $tx + (1-t)y \in \text{Int}(C)$, where $\text{Int}(C)$ is the set of all interior points of C.

Consider the upper-level sets of strictly quasi-concave functions. Generally one would think that strict quasi-concavity is equivalent to strict convexity of the upper-level sets. This is partially true; however, we have

Theorem 6.6. *If f is a continuous strictly quasi-concave function, then its upper-level sets are either empty or they are strictly convex sets.*

PROOF. As defined earlier, let $L_\alpha^U := \{x \mid f(x) \geq \alpha\}$. Since f is continuous, any point y with $f(y) > 0$ must belong to $\text{Int}(L_a^U)$. Since f is quasi-concave, L_α^U is a convex set. If this set is not strictly convex, then there exist $x, y \in L_\alpha^U$ and $0 < t < 1$ such that $x_t \equiv tx + (1-t)y \notin \text{Int}(L_\alpha^U)$. Moreover, because of continuity of f, we must have $f(x_t) = \alpha$. Again, since f is quasi-concave, we have $\min\{f(x), f(y)\} \leq f(x_t) = \alpha$. Hence, $\min\{f(x), f(y)\} = f(x_t)$, which contradicts the strict-concavity of f. ■

Note that the converse of this theorem is not true, because there exist functions with all their nonempty upper-level sets which are strictly convex but not strictly quasi-concave.

Example 6.14. The function

$$\hat{f}(x) = \begin{cases} -x^2 & \text{if } x \le 0, \\ -(x-1)^2 + 1 & \text{if } x \ge 0 \end{cases}$$

is strictly quasi-concave (see Figure 6.10). ■

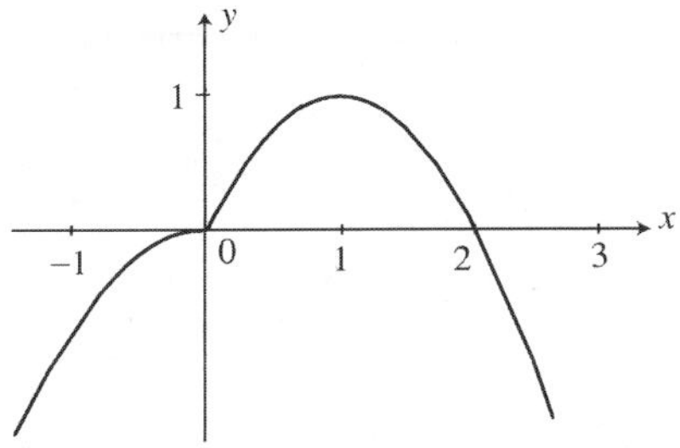

Figure 6.10 Function $f(x)$.

Theorem 6.7. *Let f be a strictly quasi-concave function and let Ψ be a strictly increasing function defined on $\mathcal{R}(f)$. Then $F(x) := \Psi(f(x))$ is a strictly quasi-concave function.*

PROOF. For any $x_1, x_2 \in X$, $t \in (0,1)$ and $x_t = tx_1 + (1-t)x_2$, we have

$$\begin{aligned} F(x_t) &= \Psi(f(x_t)) = \Psi(f(tx_1 + (1-t)x_2)) \\ &> \min\{\Psi(f(x_1), f(x_2)\}) = \min\{\Psi(f(x_1)), \Psi(f(x_2))\}) \\ &= \min\{F(x_1), F(x_2)\}, \end{aligned}$$

which proves the strict quasi-concavity of F. ■

It seems from these situations that the graphs of strictly quasi-concave functions must be 'curved.' But, in general this is not true, as shown, for example, by the function f_1 (Figure 6.3) which is a strictly quasi-concave linear function. Thus, strictly concave functions are also strictly quasi-concave: the function f_1 shows that there are strictly quasi-concave functions which are concave but not strictly concave.

6.7.1 Sums of Strictly Quasi-Concave Functions. If $g_1, \dots, g_m$ are strictly quasi-concave functions and if $k_i \ge 0$, $i = 1, \dots, m$, then there exist $k_i > 0$ for all i such that $F(x) := k_1 g_1(x) + \cdots + k_m g_m(x)$ is a strictly concave function. However, it does not hold for strictly quasi-concave functions. In fact, in general, we have the following example.

Example 6.15. Consider $g(x) = g_1(x) + g_2(x) = e^x + e^{-x}$, which is the sum of two strictly monotone (and therefore two strictly quasi-concave) functions; but $g(x)$ is not strictly quasi-concave; in fact, it is not even quasi-concave, although it is strictly convex (see Figure 6.8). ■

6.8 Quasi-Concave Programming

We will consider the optimization problem:

$$\max\{f(x)\},\ x \in X. \tag{6.8.1}$$

First, assume that f is a quasi-concave function. Then the problem (6.8.1) may have local maxima, which are not global. For example, consider the function f_4 (Figure 6.6): for this function, any $x \in (0,1)$ is a local maximum whereas its global maximum is at $x^* = 2$. The difficulty is due to the horizontal parts on its graph, which can be eliminated by requiring strict quasi-concavity of the function. The following result is well known.

Theorem 6.8. *Let X be a nonempty convex set and assume that f is strictly quasi-concave. Then any local maximum is a global solution of the problem (6.8.1).*

PROOF. Assume that $\bar{x} \in X$ is a local maximum, i.e., there exists an $\varepsilon > 0$ such that $f(x) \le f(\bar{x})$ holds for any $x \in X$ with $\|x - \bar{x}\| \le \varepsilon$. Assume that $\bar{x}$ is not a global maximum. Then there exists a $y \in X$ such that $f(y) > f(\bar{x})$. Let $x_t := ty + (1-t)\bar{x}$, $0 \le t \le 1$. Then the convexity of X implies that $x_t \in X$ for all $t \in [0,1]$. On the other hand, if t is small enough, we clearly have $\|x_t - \bar{x}\| \le \varepsilon$. Using strict quasi-concavity of f, we get

$$f(x_t) = f(ty + (1-t)\bar{x}) > \min\{f(\bar{x}), f(y)\} = f(\bar{x}),$$

which holds for $t > 0$ and t small enough. But this contradicts the assumption that $\bar{x}$ is a local maximum. ■

Note that the strict quasi-concavity of f also implies that optimal solutions of the problem (6.8.1) are unique provided they exist. This fact has important implications in economics: for example, if strictly quasi-concave utility functions are used, the solution of consumer's optimization problem will be unique.

Theorem 6.9. *Let X be a nonempty convex set and assume that f is strictly quasi-concave. In case an optimal solution of the problem (6.8.1) exists, it is unique.*

PROOF. Assume that $x^* \in X$ is an optimal solution of the problem (6.8.1), and let $y \in X$, $y \ne x^*$ be another maximum. Then we have $f(x^*) = f(y)$ and the strict quasi-concavity of f implies that

$$f(tx^* + (1-t)y) > \min\{f(x^*), f(y)\} = f(x^*)$$

holds for any $t \in (0,1)$. But this contradicts the optimality of x^* since, in view of convexity of X, we get $tx^* + (1-t)y \in X$ for $t \in [0,1]$. ■

Next, we assume that

$$X = \{x \in \mathbb{R}^n \mid g_1(x) \geq 0, \dots, g_m(x) \geq 0\},$$

where each g_i, $i = 1, \dots, m$, is quasi-concave. Then X is a convex set. In fact, $\{x \mid g_i(x) \geq 0\}$ is an upper-level set of g_i that corresponds to $\alpha = 0$, and so it is convex for all i. We have

$$X = \bigcap_{i=1}^{m} \{x \mid g_i(x) \geq 0\},$$

and thus, X is a convex set.

Now we formulate the general quasi-concave optimization problem:

$$\max\{f(x) \mid g_i(x) \geq 0, \, i = 1, \dots, m\}, \tag{6.8.2}$$

where $f, g_1, \dots, g_m$ are quasi-concave functions. In view of the above discussion, the set of feasible (admissible) solutions of the problem (6.8.2) is a convex set. For the objective function we have assumed only quasi-concavity, and hence, the problem (6.8.2) may have local solutions which are not global.

If we assume that f is strictly quasi-concave, then the problem (6.8.2) has a unique optimal solution and no local maximum exists. Assuming continuous differentiability, the Karush-Kuhn-Tucker (KKT) conditions for the problem (6.8.2) are

$$\begin{aligned} &\frac{\partial f(x)}{\partial x_i} + \sum_{i=1}^{m} \frac{\partial g(x)}{\partial x_i} = 0, \\ &g_i(x) \geq 0, \, i = 1, \dots, m, \\ &t_i g_i(x) \geq 0, \, i = 1, \dots, m, \\ &t_i \geq 0, \, i = 1, \dots, m. \end{aligned} \tag{6.8.3}$$

These are necessary conditions of optimality under the usual regularity conditions for general nonlinear programming problems. More details about the KKT conditions are described in §4.3. Besides these conditions, we have the following results especially for quasi-concave optimization problems.

Theorem 6.10. (Arrow and Enthoven [1961]) *Assume that $g_1, \dots, g_m$ are quasi-concave functions and that the following regularity conditions hold:*

(a) there exists an $\bar{x} \in \mathbb{R}^n$ such that $g_i(\bar{x}) > 0$ for all $i = 1, \dots, m$ (Slater condition), and

(b) for each i, either g_i is concave or otherwise $\dfrac{\partial g_i}{\partial x} \neq 0$, $i = 1, \dots, m$, for each feasible (admissible) solution of the problem (6.8.2).

Then, if x^* is a locally optimal solution of the problem (6.8.2), there exists a t^* such that with (x^*, t^*) the KKT conditions (6.8.3) hold.

The KKT conditions are also sufficient optimality conditions under appropriate assumptions (see §4.3). The following theorem is useful.

Theorem 6.11. (Arrow and Enthoven [1961]) *Assume that $f, g_1, \dots, g_m$ are quasi-concave functions and that the Kuhn-Tucker conditions (6.8.3) hold for (x^*, t^*). If f is twice continuously differentiable on the feasible (admissible) set and if $\nabla f(x^*) \neq 0$, then x^* is an optimal solution of the problem (6.8.2).*

Example 6.16. Suppose there is a 1-unit change (decrease) in the constant of the constraint. We want to determine what change it will make in L and λ^* in the above example. Then with the new constraint $g(x, y) = x + y - 27$, we have

$$L(x, y, \lambda) = 4x^2 + 3xy + 6y^2 + \lambda(27 - x - y),$$

which gives $L_x = 8x + 3y - \lambda = 0, L_y = 3x + 12y - \lambda = 0, L_\lambda = 27 - x - y = 0$. Solving these equations simultaneously we get $x^* = 17.36$, $y^* = 9.64$, $\lambda^* = 167.78$; thus, $L(17.36, 9.64) = 2265.10 = f(17.36, 9, 64)$, which is approximately 171 smaller than the previous value of $L = 2436$ and close to the previous value of $\lambda^* = 174$. ■

This is the reason why Lagrange multipliers are called *shadow prices*. Also, in utility optimization subject to a budget constraint, the value of λ will estimate the marginal utility of an extra dollar of income.

Example 6.17. Minimize the utility function $u(x, y) = x^{1/2}y^{3/5}$ subject to the budget constraint $3x + 9y = 66$. Define

$$L(x, y) = x^{1/2}y^{3/5} + \lambda(66 - 3x - 9y).$$

Then $L_x = \frac{1}{2}x^{-1/2}y^{3/5} - 3\lambda = 0$, $L_y = \frac{3}{5}x^{1/2}y^{-2/5} - 9\lambda = 0, L_\lambda = 66 - 3x - 9y = 0$. Solving these equations simultaneously, we get the critical values: $x^* = 10, y^* = 20, \lambda^* = 0.3$. The second-order partial derivatives are: $L_{xx} = -\frac{1}{4}x^{-3/2}y^{3/5}$, $L_{xy} = \frac{3}{10}x^{-1/2}y^{-2/5} = L_{yx}$, $L_{yy} = -\frac{6}{25}x^{1/2}y^{-7/5}$. Then the bordered Hessian is

$$|\mathbf{H}| = \begin{vmatrix} -\frac{1}{4}x^{-3/2}y^{3/5} & \frac{3}{10}x^{-1/2}y^{-2/5} & 3 \\ \frac{3}{10}x^{-1/2}y^{-2/5} & -\frac{6}{25}x^{1/2}y^{-7/5} & 9 \\ 3 & 9 & 0 \end{vmatrix}.$$

The second principal minor is

$$|H_2| = |\mathbf{H}| = \tfrac{81}{4}x^{-3/2}y^{3/5} + \tfrac{81}{5}x^{-1/2}y^{-2/5} + \tfrac{54}{25}x^{1/2}y^{-7/3} > 0,$$

since all terms are positive. Hence, $|\mathbf{H}|$ is positive definite, and L is minimized at the critical values. ■

6.9 Summary

Recall that a concave function f over X satisfies the inequality (3.2.2), i.e.,

$$f(tx + (1-t)y) \geq tf(x) + (1-t)f(y)$$

for any $x, y \in X$ and for $t \in [0,1]$. From this inequality we obtain

$$f(tx + (1-t)y) \geq tf(x) + (1-t)f(y) \geq \min\{f(x), f(y)\}.$$

This leads to the following generalization of concavity: A function f is called *quasi-concave* on X, if for all $x, y \in X$ and for all $t \in [0,1]$,

$$\min\{f(x), f(y)\} \leq f(tx + (1-t)y),$$

or equivalently

$$f(x) \leq f(y) \quad \text{implies that} \quad f(x) \leq f(tx + (1-t)y).$$

The definition of quasi-convexity is analogous: If f is a convex function over X then for all $x, y \in X$ and for all $t \in [0,1]$,

$$f(x) \leq f(tx + (1-t)y) \leq tf(x) + (1-t)f(y) \leq \max\{f(x), f(y)\},$$

which leads to the following generalization: A function f is called *quasi-convex* on X, for all $x, y \in X$ and for all $t \in [0,1]$,

$$\max\{f(x), f(y)\} \geq f(tx + (1-t)y),$$

or equivalently $f(x) \geq f(y)$ implies $f(x) \geq f(tx + (1-t)y)$. This definition of quasi-convexity is a consequence of the fact that if f is concave then $-f$ is convex. Moreover, a function f is *quasi-linear* if it is both quasi-concave and quasi-convex. The following relations hold:

f is concave over $X \implies f$ is quasi-concave over X.

f is convex over $X \implies f$ is quasi-convex over X.

f is linear over $X \implies f$ is quasi-linear over X.

f is quasi-concave over $X \implies -f$ is quasi-convex over X.

f is quasi-convex over $X \implies -f$ is quasi-concave over X.

$-\min\{f(x), f(y)\} = \max\{-f(x), -f(y)\}$.

$-\max\{f(x), f(y)\} = \min\{-f(x), -f(y)\}$.

The empty set is considered as convex, by convention.

6.10 Exercises

6.1. Is every quasi-concave function concave? If so, prove it; otherwise provide a counterexample.

ANS. Not every quasi-concave function is concave. COUNTEREXAMPLE: Define the function $f(x) = x^2$ with $\mathrm{dom}(f) = \mathbb{R}_+$ (the set of positive real numbers). We must show that (i) this function f is quasi-concave, and (ii) f is not concave. To see (i), note that f is a strictly increasing function on $\mathbb{R}_+$. Thus, if $f(x') \geq f(x)$ for $x, x' \in \mathrm{dom}(f)$, then $x' \geq x$, and therefore, for any $t \in [0,1]$, we have $tx' + (1-t)x \geq x$. Hence, f is quasi-concave. To see (ii), note that $f(0) = 0, f(2) = 4$, but $f(\frac{1}{2} \cdot 0 + \frac{1}{2} \cdot 2) = f(1) = 1 \leq \frac{1}{2}f(0) + \frac{1}{2}f(2) = 2$, which cannot be true if f is a concave function. Note that $f(x) = x^2$ defined on the entire real line would not be quasi-concave (why?).

6.2. Let f and g be real-valued concave functions with the same domain D. Define a function h such that for all $x \in D$, $h(x) = f(x) + g(x)$. Is h a concave function?

SOLUTION. Since f and g are both concave functions with domain D, we have for $x, x' \in D$ and for all $t \in [0,1]$,

$$f(tx+(1-t)x') \geq tf(x)+(1-t)f(x'), \quad g(tx+(1-t)x') \geq tg(x)+(1-t)g(x').$$

Thus, from these two inequalities we get

$$\begin{aligned} h(tx + (1-t)x') &= f(tx + (1-t)x') + f(tx + (1-t)x') \geq tg(x) + (1-t)g(x') \\ &\geq tf(x) + (1-t)f(x') + tg(x) + (1-t)g(x') \\ &= t(f(x) + g(x)) + (1-t)(f(x') + g(x')) \\ &= th(x) = (1-t)h(x'), \end{aligned}$$

which holds for all $t \in [0,1]$, i.e., h is a concave function.

6.3. Let f and g be real-valued concave functions defined on the same domain D. Define a function h so that $h(x) = f(x)g(x)$ for all $x \in D$. (a) Is h a concave function? If so, prove it; otherwise provide a counterexample. (b) Is h a quasi-concave function? If so, prove it; otherwise provide a counterexample.

SOLUTION. (i) h is not necessarily concave. For example, let D be the real line, and let $f(x) = x = g(x)$. Both f and g are concave functions. But since $h(x)x^2$, with $h''(x) = 2 > 0$, the function h is not a concave function. (For h to be concave, $h''(x)$ must be negative or zero everywhere on D).

(ii) h is not in general quasi-concave, except in the case when both $f(x) > 0$ and $g(x) > 0$ for all $x \in D$. If this is the case, then $\log h(x) = \log f(x) +$

$\log g(x)$, where both $\log f(x)$ and $\log g(x)$ are concave functions, by virtue of the property that a concave function of a concave function is concave. Thus, $\log h(x)$, being the sum of two concave functions, is concave. Hence, $h(x)$ is a monotone transformation of a concave function, and therefore, it is quasi-concave. On the other hand, if one of the functions f and g is negative valued, then $\log h(x)$ is not well-defined, and so the above argument fails. In fact, $h(x)$ is, in general, not quasi-concave if both f and g are concave. For example, let D be the set of all real numbers $\mathbb{R}$, and $f(x) = -1$ and $g(x) = -x^{-2}$ for all $x \in \mathbb{R}$, whence $h(x) = -g(x) = x^{-2}$. Note that although both of the functions f and g are concave, since $f''(x) < 0$ and $g''(x) < 0$, the function $h(x)$ is not quasi-concave because $h(1) = 1 = H(-1)$, but $h(\frac{1}{2} \cdot 1 + \frac{1}{2} \cdot (-1)) = h(0) = 0 < h(1)$.

6.4 A consumer utility function is defined as

$$u(x, y) = \min\{v_1(x, y), v_2(x, y)\},$$

where v_1 and v_2 are both quasi-concave functions. Is u quasi-concave? If it is, prove it; otherwise provide a counterexample.

SOLUTION. By assumption v_i, $i = 1, 2$, are concave functions, which means that for $t \in [0, 1]$,

$$v_i(tx + (1-t)y) \geq \min\{v_i(x), v_i(y)\}, \quad i = 1, 2.$$

Now,

$$\begin{aligned} u(tx + (1-t)y) &= \min\{v_1(tx + (1-t)y), v_2((tx + (1-t)y)\} \\ &\geq \min\{\min\{v_1(x), v_1(y)\}, \min\{v_2(x), v_2(y)\}\} \\ &= \min\{\min\{v_1(x), v_2(x)\}, \min\{v_1(y), v_2(y)\}\} \\ &= \min\{u(x), u(y)\}, \end{aligned}$$

which implies that u is quasi-concave.

6.5. Use Lagrange multipliers to optimize $f(x, y, z) = xyz^2$ subject to the constraint $x + y + z = 20$. Define

$$L(x, y, z) = xyz^2 + \lambda(20 - x - y - z).$$

Then $L_x = yz^2 - \lambda = 0$, $L_y = xz^2 - \lambda = 0$, $L_z = 2xyz - \lambda = 0$, $L_\lambda = 20 - x - y - z = 0$. To solve these equations simultaneously, we equate λ from the first two equations, and from the first and the third equation, giving:

$$yz^2 = xz^2, \quad yz^2 = 2xyz,$$

or $y = x$ and $z = 2x$. Substituting these in the fourth equation we get: $20 - x - x - 2x = 0$, or $x^* = 5$, which gives $y^* = 5, z^* = 10, \lambda^* = 500$ as critical values. Thus, $L(5, 5, 10) = 2500$.

The second-order derivatives are: $L_{xx} = 0, L_{yy} = 0, L_{zz} = 2xy, L_{xy} = z^2, L_{yz} = 2xz, L_{xz} = 2yz$. Also, from $g(x, y, z) = x + y + z - 20$, we get $g_x = 1 = g_y = g_z$. Then the bordered Hessian from Eq (1.6.7), using the second form, is

$$|\mathbf{H}| = \begin{bmatrix} 0 & 1 & 1 & 1 \\ 1 & 0 & z^2 & 2yz \\ 1 & z^2 & 0 & 2xz \\ 1 & 2yz & 2xz & 2xy \end{vmatrix}.$$

The second principal minor is $|\bar{H}_2| = 0 - 1(-z^2) + 1(z^2) = 2z^2$. Thus, $|\bar{H}_2|_{10} > 0$. The third principal minor is

$$|\bar{H}_3| = |\mathbf{H}| = 0 - 1 \begin{vmatrix} 1 & z^2 & 2yz \\ 1 & 0 & 2xz \\ 1 & 2xz & 2xy \end{vmatrix} + 1 \begin{vmatrix} 1 & 0 & 2yz \\ 1 & z^2 & 2xz \\ 1 & 2yz & 2xy \end{vmatrix} - 1 \begin{vmatrix} 1 & 0 & z^2 \\ 1 & z^2 & 0 \\ 1 & 2yz & 2xz \end{vmatrix}$$
$$= -\big[1(0 - 2xz \cdot 2xz) - z^2(2xy - 2xz) + 2yz(2yz - 0)\big]$$
$$+ \big[1(z^2 \cdot 2xy - 2yz \cdot 2xz) - 0 + 2yz(2yz - z^2)\big]$$
$$- \big[1(z^2 \cdot 2xz - 0) - 0 + z^2(2yz - z^2)\big]$$
$$= z^4 - 4xz^3 - 4yz^3 - 4xyz^2 + 4x^2z^2 + 4y^2z^2.$$

Thus, $|\bar{H}_3|_{5,5,10} = -20000 < 0$. Hence, $|\bar{H}_2| > 0$ and $|\bar{H}_3| < 0$ imply that $|\mathbf{H}|$ is negative definite, and the function f is maximized at the critical values. ■

6.6. Minimize the total costs defined by $c = 15x^2 + 30xy + 30y^2$ when the firm meets the quota $g(x, y)$ equal to $2x + 3y = 20$. Define

$$L(x, y) = 15x^2 + 30xy + 30y^2 + \lambda(20 - 2x - 3y).$$

Then $L_x = 30x + 30y - 2\lambda = 0$, $L_y = 30x + 60y - 3\lambda = 0$, $L_\lambda = 20 - 2x - 3y$. Solving these three equations simultaneously, we get the critical values: $x^* = 4$, $y^* = 4$, $\lambda^* = 120$.

The second-order partial derivatives are: $L_{xx} = 30$, $L_{yy} = 60$, $L_{xy} = 30 = L_{yx}$, and $g_x = 2, g_y = 3$. Thus, the bordered Hessian (1.6.7) is

$$|\mathbf{H}| = \begin{vmatrix} 30 & 30 & 2 \\ 30 & 60 & 3 \\ 2 & 3 & 0 \end{vmatrix}.$$

The second principal minor is $|H_2| = -150 < 0$ Thus, $|\mathbf{H}|$ is positive definite and L is minimized when $x = y = 12$. ■

6.7. Maximize the utility function $u = Q_1Q_2$ when $P_1 = 1, P_2 = 3$, and the firm's budget is $B = 60$. Also estimate the effect of a 1-unit increase in the budget. The budget constraint is $Q_1 + Q_2 = 60$, and the constraint is $Q_1 + 3Q_2 = 60$. We consider the Lagrangian

$$L = Q_1Q_2 + \lambda(60 - Q_1 - 3Q_2).$$

The first-order partial derivatives equated to zero give: $L_{Q_1} = Q_2 - \lambda = 0$, $L_{Q_2} = Q_1 - 3\lambda = 0$, $L_\lambda = 60 - Q_1 - 4Q_2 = 0$. Solving these equations simultaneously we obtain the critical values: $Q_1^* = 30, Q_2^* = 10 = \lambda^*$. The second-order partial derivatives are: $L_{Q_1Q_1} = 0 = L_{Q_2Q_2}, L_{Q_1Q_2} = 1 = L_{Q_2Q_1}$, giving the Hessian

$$|\mathbf{H}| = \begin{vmatrix} 0 & 1 \\ 1 & 0 \end{vmatrix} = -1 < 0.$$

Hence, L is maximized at the critical values.

With $\lambda^* = 10$, a \$1 increase in the budget will change the constant of the constraint to 61, so that the new Lagrangian is

$$L = Q_1Q_2 + \lambda(61 - Q_1 - 4Q_2),$$

which yields $L_{Q_1} = Q_2 - \lambda = 0$, $L_{Q_2} = Q_1 - 3\lambda = 0$, $L_\lambda = 121 - Q_1 - 3Q_2 = 0$. Solving these equations simultaneously we obtain the critical values $Q_1^* = 30.5, Q_2^* = 10.167 = \lambda^*$. Thus, the utility function increases from $u = (30)(10) = 300$ to $u = (30.5)(10.167) = 310.083$, i.e., there is an increase in the utility function of about 10. ■

6.8. Let f be a function of n variables with continuous partial derivatives of all orders in an open convex set S, and let $|B|_k$ be the determinant of its kth order bordered Hessian $\mathbf{B}$. Verify that the criteria to check quasi-concavity and quasi-convexity, described in §6.2, can be summarized as follows.

(i) If f is quasi-concave, then

$$\begin{cases} |B|_1(\mathbf{x}) \le 0, |B|_2(\mathbf{x}) \ge 0, \dots, |B|_n(\mathbf{x}) \le 0 & \text{if } n \text{ is odd}, \\ |B|_1(\mathbf{x}) \le 0, |B|_2(\mathbf{x}) \ge 0, \dots, |B|_n(\mathbf{x}) \ge 0 & \text{if } n \text{ is even}, \end{cases}$$

for all $\mathbf{x} \in S$;

(ii) if f is quasi-convex, then $|B|_k(\mathbf{x}) \le 0$ for all $\mathbf{x} \in S$, $k = 1, 2, \dots, n$;

(iii) if for all $\mathbf{x} \in S$,

$$\begin{cases} |B|_1(\mathbf{x}) < 0, |B|_2(\mathbf{x}) > 0, \dots, |B|_n(\mathbf{x}) < 0 & \text{if } n \text{ is odd}, \\ |B|_1(\mathbf{x}) < 0, |B|_2(\mathbf{x}) > 0, \dots, |B|_n(\mathbf{x}) > 0 & \text{if } n \text{ is even}, \end{cases}$$

then f is quasi-concave; and

(iv) if $|B|_k(\mathbf{x}) < 0$ for all $\mathbf{x} \in S$, $k = 1, 2, \dots, n$, then f is quasi-convex.

This theorem can also be stated as follows:

$$\begin{cases} |B|_1(\mathbf{x}) \le 0, |B|_2(\mathbf{x}) \ge 0, \dots, |B|_n(\mathbf{x}) \le 0 & \text{if } n \text{ is odd,} \\ |B|_1(\mathbf{x}) \le 0, |B|_2(\mathbf{x}) \ge 0, \dots, |B|_n(\mathbf{x}) \ge 0 & \text{if } n \text{ is even,} \end{cases}$$

for all $\mathbf{x} \in S$, is a necessary condition for quasi-concavity, and

$$\begin{cases} |B|_1(\mathbf{x}) < 0, |B|_2(\mathbf{x}) > 0, \dots, |B|_n(\mathbf{x}) < 0 & \text{if } n \text{ is odd,} \\ |B|_1(\mathbf{x}) < 0, |B|_2(\mathbf{x}) > 0, \dots, |B|_n(\mathbf{x}) > 0 & \text{if } n \text{ is even,} \end{cases}$$

for all $\mathbf{x} \in S$, is a sufficient condition for quasi-concavity.

Note that unlike an analogous result for concave functions, the above conditions do not cover all cases of quasi-concave functions. For example, in the case when $|B|_k(\mathbf{x}) \le 0$ for all $k \le n$ but $|B|_j(\mathbf{x}) = 0$ for some j and some $\mathbf{x} \in S$, the above result may include the possibility that the function f is quasi-convex, but it fails to tell if it is.

6.9. Prove that a function f of a single variable defined on an interval I is quasi-concave iff there exists a number x' such that f is nondecreasing on $\{x \in I : x < x'\}$ and nonincreasing on $\{x \in I : x > x'\}$. HINT. Suppose f satisfies the condition and suppose that for some number a the points x_1 and x_2 belong to the upper-level set U_α. Then $f(x_1) \ge a$ and $f(x_2) \ge a$, so that $f(x) \ge a$ for every point x between x_1 and x_2. Thus, $x \in U_\alpha$, so that U_α is convex and therefore, f is quasi-concave. If f does not satisfy the condition, we can find x_1, x_2 and x_3 in I such that $x_1 < x_2 < x_3$ and $f(x_2) < \min\{f(x_1), f(x_3)\}$. Then the upper-level set U_α for $a = \min\{f(x_1), f(x_3)\}$ includes x_1 and x_3, but not x_2, and therefore, U_α is not convex, and f is not quasi-concave.

6.10. Prove that a differentiable function f of n variables defined on a convex set S is quasi-concave on S iff

$$x, x' \in S, \text{ and } f(x) \ge f(x') \implies \sum_{j=1}^{n} f'_j(x')(x_j - x'_j) \ge 0.$$

HINT. Use (6.2.1).

6.11. Prove that all extrema (critical points) of a concave function are global maxima. HINT. The definition of concavity $f(tx + (1-t)f(y)) \ge tf(x) + (1-t)f(y)$ can be written as $f(y + t(x-y)) \ge f(y) + t[f(x) - f(y)]$, or $\dfrac{f(y+t(x-y))-f(y)}{t} \geq f(x)-f(y)$. Taking the limits as $t \to 0$, and noticing that the right-hand side does not depend on t, we get

$$\lim_{t\to 0} \frac{f(y+t(x-y))-f(y)}{t} = f'(y) \geq f(x)-f(y). \tag{6.10.2}$$

Suppose y is a critical point so that $f'(y)=0$. Then Eq (6.10.2) gives $f(x)-f(y) \leq 0$, which implies that $f(x) \leq f(y)$ for all x. Hence, the function f is lower at any point other than the critical point y, i.e., y is a global maximum of f.

6.12. Prove that a monotone composite transformation $(g \circ f) = g(f)(x)$ of a quasi-concave function f is itself quasi-concave.

PROOF. Let $\alpha \in \mathbb{R}$. Since g is monotonic, there exists an $\alpha' \in \mathbb{R}$ such that $\alpha = g(\alpha')$. Then the upper-level set of f

$$U_\alpha(g(f))) = \{x, g(f(x)) \geq \alpha\} = \{x, g(f) \geq g(\alpha')\} = \{x, f(x) \geq \alpha'\} = U_{\alpha'}(f)$$

is a convex set. Note that a composite transformation of a concave function is not necessarily concave; for example, $f(x) = x$ is concave, $g(x) = x^3$ is monotonically increasing, but $g(f(x)) = x^3$ is not concave in $\mathbb{R}$.

6.13. Prove that every Cobb-Douglas utility function $u(x,y) = Ax^2y^2$, $a, b > 0$, is quasi-concave.

HINT. We know that the DRS (decreasing return to scale) Cobb-Douglas function, such as $f(x,y) = x^{1/3}y^{1/3}$, is concave. Also, an IRS (increasing return to scale) Cobb-Douglas function, such as $g(x,y) = x^{2/3}y^{2/3}$, is quasi-concave. This means that the IRS Cobb-Douglas function is the monotonic composite transformation of the DRS Cobb-Douglas function:

$$f(x,y) = x^{2/3}y^{2/3} = (x^{1/3}y^{1/3})^2,$$

so $x^{2/3}y^{2/3} = g((f(x,y))$, where $f(x,y) = x^{1/3}y^{1/3}$ and $g(z) = z^2$, where $g(z)$ is a monotonic transformation.

6.13. Any CES utility function $u(x,y) = (ax^r + by^r)^{1/r}$, $0 < r < 1$, is quasi-concave, since $u(x,y) = g(h(x,y))$, where $h(x,y) = ax^r + by^r$ is a concave function because it is a positive linear combination of concave functions, and $h(z) = z^{1/r}$ is a monotonic transformation.

6.14. Show, by an example, that quasi-concave functions do not have the same implications for continuity and differentiability as concave functions.

ANS. Let $f : \mathbb{R}_+ \mapsto \mathbb{R}$ be defined by

$$f(x) = \begin{cases} x^3 & \text{if } 0 \le x \le 1, \\ 1 & \text{if } 1 \le x \le 2, \\ x^3 & \text{if } x > 2. \end{cases}$$

Since f is non-decreasing, it is both quasi-concave and quasi-convex on $\mathbb{R}$. But f is discontinuous at $x = 2$, and not differentiable there. Moreover, f is constant on $(1, 2)$, and thus, every point in this open interval is a relative maximum as well as a relative minimum. However, no point in the interval $(1, 2)$ is neither a global maximum nor a global minimum. Finally, $f'(0) = 0$, but 0 is neither a relative maximum nor a relative minimum, and f is not differentiable at $x = 1$.

6.15. Show that the function $f : \mathbb{R}^2_+ \mapsto \mathbb{R}$ given by $f(x, y) = 50 \ln x \ln y$ is stricyly concave on $[e, \infty) \subset \mathbb{R}^2_+$, but only quasi-concave on the larger domain $[1, \infty)$. HINT: Use 3-D plot. Show that the second principal minor $|H_2| = f_{xx} f_{yy} - (f_{xy})^2 = 0$ at the point (e, e).

7

Quasi-Convex Functions

Since a function f is quasi-convex in a domain $D \in \mathbb{R}^n$, iff $(-f)$ is quasi-concave in that domain, we can repeat most of the results of the previous chapter by replacing the upper-level set by the lower-level set, the $\geq$ sign by $\leq$, and the operation 'min' by 'max'. However, there are some useful and interesting results in quasi-convex function theory, and we will risk some repetition and provide all relevant information on the topic.

7.1 Quasi-Convex Functions

A real-valued function $f : \mathbb{R}^n \mapsto \mathbb{R}$ is *quasi-convex* if $\text{dom}(f)$ is convex and the lower-level sets $L_\alpha = \{\mathbf{x} \in \text{dom}(f) | f(\mathbf{x}) \leq \alpha\}$ are convex for all α such that for all $\mathbf{x}, \mathbf{x}' \in \text{dom}(f)$ and all $t \in [0, 1]$,

$$f(t\mathbf{x} + (1-t)\mathbf{x}') \leq \max\{f(\mathbf{x}), f(\mathbf{x}')\}. \tag{7.1.1}$$

A real-valued function $f : \mathbb{R}^n \mapsto \mathbb{R}$ is *strictly quasi-convex* iff

$$f(t\mathbf{x} + (1-t)\mathbf{x}') < \max\{f(\mathbf{x}), f(\mathbf{x}')\}, \tag{7.1.2}$$

for all $\mathbf{x}, \mathbf{x}' \in \text{dom}(f)$ and all $t \in [0, 1]$. The inequality (7.1.2) is the defining property of a quasi-convex function, with an additional property that the negative of a (quasi-) convex function is a (quasi-) concave function. Since 'quasi' means 'as if', we expect quasi-convex functions to have some special properties similar to those for convex functions (and similarly in the case of quasi-concave functions). Moreover, since every convex function is quasi-convex, we expect the convex functions to be more highly structured. Although De Finetti [1949] was the first person to recognize some of these characteristics of functions having convex level sets, it was Fenchel [1983] who was the pioneer in formalizing, naming, and developing the class of quasi-convex functions. Later, Slater [1950] generalized the KKT saddle-point equivalence theorem, and Arrow and Enthoven [1961] laid the foundation of quasi-convex programming with applications to consumer demand.

A strictly quasi-convex function need not be quasi-convex. An example is: $f(x) = \begin{cases} 1 \text{ if } x = 0, \\ 0 \text{ if } x \neq 0. \end{cases}$. The lower-level set $L_\alpha = \{x : f(x) \leq 0\}$ for $\alpha = 0$ is not convex, but f is strictly quasi-convex.

Note that it is not proper to define strict quasi-convexity by requiring that the lower contour sets should be strictly convex, because a lower contour set can be strictly convex even when f has some flat portions. Further, a function is strictly quasi-convex iff $-f$ is strictly quasi-concave; and a strictly quasi-convex function is quasi-convex. The lower-level set L_α and quasi-convex functions are presented in Figure 7.1. Also, f is said to be *quasi-linear* if it is quasi-convex and quasi-concave.

The inequality (7.1.1) became the defining property of a convex and quasi-convex function, with an additional property that the negative of a (quasi-) convex function is a (quasi-) concave function. Thus, we will find that quasi-convex functions have some special properties similar to those for convex functions (and similarly in the case of quasi-concave functions).

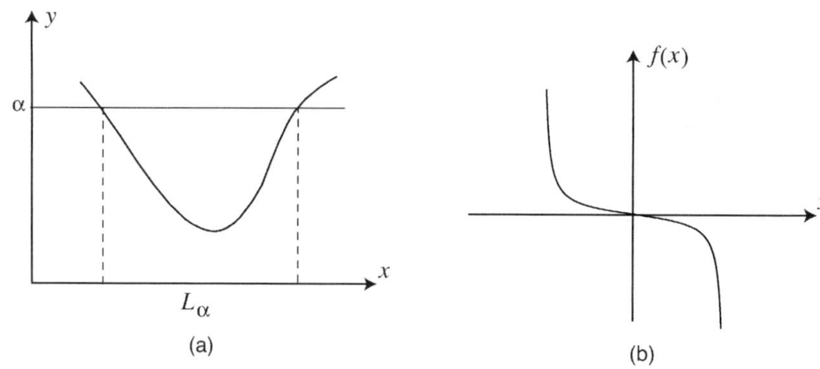

Figure 7.1 Quasi-convex function.

Thus, we can prove that a convex function is quasi-convex. The proof is as follows: Let the function f have the domain S (a convex set). Let a be a real number and x and y be points in the lower-level set L_α with $x, y \in L_\alpha$. First, we show that the set L_α is convex. For this, we need to show that for every $t \in [0, 1]$ we have $tx + (1-t)y \in L_\alpha$. Since S on which f is defined is convex, we have $tx+(1-t)y \in S$, and thus f is defined at the point $tx+(1-t)y$. Now, convexity of f implies that $f(tx + (1-t)y) \leq tf(x) + (1-t)f(y)$. Moreover, the fact that $x \in L_\alpha$ means that $f(x) \leq a$, and similarly, $y \in L_\alpha$ means that $f(y) \leq a$. Hence, $tf(x) + (1-t)f(y) \leq ta + (1-t)a = a$. Combining these two inequalities we get $f(tx + (1-t)y) \leq a$, so that $tx + (1-t)y \in L_\alpha$. Thus, every upper-level set is convex and hence, f is quasi-convex. ■

Note that quasi-concavity is weaker than concavity, in the sense that every convex function is quasi-convex. This is similar to the case that every concave function is quasi-concave. The following result relates to both quasi-concave and quasi-convex functions.

Theorem 7.1. Let F be a function defined on $\mathbb{R}^n$, and g be a function defined on $\mathbb{R}$. If F is quasi-concave and g is decreasing, then the function $f(x) = g(F(x)$ is quasi-convex for all x.

Theorem 7.2. A function f defined on a convex set $S \in \mathbb{R}^n$ is quasi-convex iff for all $\mathbf{x}, \mathbf{x}' \in S$ such that $f(\mathbf{x}) \geq f(\mathbf{x}')$, we have for $t \in [0,1]$,

$$f(t\mathbf{x} + (1-t)\mathbf{x}') \leq f(\mathbf{x}'). \tag{7.1.3}$$

PROOF. First, suppose that f is quasi-convex. Let $\mathbf{x}, \mathbf{x}' \in S$, and $f(\mathbf{x}) \leq f(\mathbf{x}')$. Then for $\alpha = f(\mathbf{x}')$ we have $\mathbf{x} \in L_\alpha$ and $\mathbf{x}' \in L_\alpha$. Since f is quasi-convex, L_α is convex, and thus, $t\mathbf{x} + (1-t)\mathbf{x}' \in L_\alpha$ for all $t \in [0,1]$. Hence, $f(t\mathbf{x} + (1-t)\mathbf{x}') \leq a = f(x')$ for all $t \in [0,1]$.

Next, suppose that for all $\mathbf{x}, \mathbf{x}' \in S$ with $f(\mathbf{x}) \leq f(\mathbf{x}')$ and for all $t \in [0,1]$, we have $f(t\mathbf{x} + (1-t)\mathbf{x}') \leq f(\mathbf{x}')$. For any α, the set L_α is either empty, in which case it is convex, or consists of a single point, in which case it is convex, or contains two or more points, in which case we choose $\mathbf{x}, x' \in L_\alpha$ with $f(\mathbf{x}) \leq f(\mathbf{x}')$. Then $f(\mathbf{t}x + (1-t)\mathbf{x}')) \leq f(\mathbf{x}') \leq \alpha$ for all $t \in [0,1]$, because $\mathbf{x}' \in L_\alpha$. Hence, $t\mathbf{x} + (1-t)\mathbf{x}' \in L_\alpha$, so that L_α is convex, and f is quasi-convex. ■

Note that a similar theorem holds for quasi-concave functions (Exercise 6.2).

Example 7.1. Consider the function $f(x,y) = \sqrt{x^2+y^2}$, which is convex and quasi-convex (see Figure 7.2).

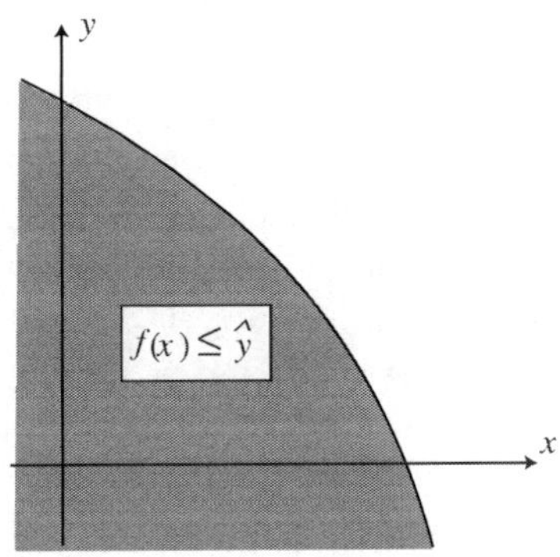

Figure 7.2 $f(x,y) = \sqrt{x^2+y^2}$.

Recall that the sum of two convex functions is a convex function. However, the sum of two quasi-convex functions is not necessarily a quasi-convex function, i.e., if f and g are quasi-convex, then $(f+g)(x) = f(x) = g(x)$ need not be quasi-convex. Also, the sum of a convex function and a quasi-convex function is not necessarily either a convex function or a quasi-convex function.

A concave function can be quasi-convex. For example $x \mapsto \log x$ is concave and quasi-convex.

Any monotone function is both quasi-convex and quasi-concave. More generally, a function that decreases up to a point and increases thereafter is quasi-convex (compare unimodality, i.e., there is a single highest value).

Let S be a convex subset of $\mathbb{R}^n$, and let $f : S \mapsto \mathbb{R}$ be a function. Then the following statements are equivalent:

1. f is quasi-convex on S.
2. For all $\mathbf{x}, \mathbf{y} \in S$ and all $t \in (0,1)$,

$$f(\mathbf{x}) \leq f(\mathbf{y}) \implies f(t\mathbf{x} + (1-t)\mathbf{y}) \leq f(\mathbf{y}). \tag{7.1.4}$$

3. For all $\mathbf{x}, \mathbf{y} \in S$ and all $t \in (0,1)$,

$$f(t\mathbf{x} + (1-t)\mathbf{y}) \leq \max\{f(\mathbf{x}), f(\mathbf{y})\}. \tag{7.1.5}$$

Let $f : \mathbb{R}^n \mapsto \mathbb{R}$ be a quasi-convex function, and let $g : \mathbb{R} \mapsto \mathbb{R}$ be a nondecreasing function whose domain contains $\mathcal{R}(f)$. Then the composite function $g \circ f : \mathbb{R}^n \mapsto \mathbb{R}$ is a quasi-convex function.

Let $f : Y \mapsto \mathbb{R}$ be a real-valued function defined on the open set $Y \in \mathbb{R}^n$, and let $X \subset Y$ be a convex set. We will assume that f is continuous or differentiable on Y. The assumption that $Y \subset \mathbb{R}^n$, but not $Y = \mathbb{R}^n$, is justified in view of the following two examples.

Example 7.2. Consider the Cobb-Douglas utility function $u(\mathbf{x}, \mathbf{y}) = \mathbf{x}^a \mathbf{y}^b$, $a, b > 0$, $a + b = 1$, with $\text{dom}(u) = \mathbb{R}^n_+ = \{\mathbf{x} \,|\, x_i \geq 0, i = 1, \dots, n\}$ with $n = 2$. Take $a = b = \frac{1}{2}$, then $u(-1, 1) = i \cdot 1 = i$, where $i = \sqrt{-1}$, is not defined on $\mathbb{R}$, but is defined on $\mathbb{R}^n_+$, although the derivatives do not exist on the boundary. Hence, the function $u(\mathbf{x}, \mathbf{y})$ is differentiable on $Y = \mathbb{R}^n_{++} = \{\mathbf{x} | x_i > 0, i = 1, \dots, n\}$ with $n = 2$. ■

Example 7.3. The additively separable logarithmic utility function $u(x, y) = \log x + \log y$, with $\text{dom}(u) = \mathbb{R}^2_{++}$, is differentiable on $Y = \mathbb{R}^n_{++} = \{\mathbf{x} | x_i > 0, i = 1, \dots, n\}$ with $n = 2$. ■

Quasi-convexity is a generalization of convexity. The set of all quasi-convex (quasi-concave) functions contains the set of all convex (concave) functions.

Example 7.4. Let $f : \mathbb{R} \mapsto \mathbb{R}$ be an increasing function. Then f is both quasi-concave and quasi-convex. To prove this, consider $x, y \in \mathbb{R}$, and assume without loss of generality that $x > y$. Then for any $t \in (0,1)$, we have

$$x > tx + (1-t)y > y. \tag{7.1.6}$$

Since f is increasing, we have

$$f(x) \geq f(tx + (1-t)y) \geq f(y). \tag{7.1.7}$$

Since $f(x) = \max\{f(x), fy\}$, the inequality (7.1.6) shows that f is quasi-convex.

It is always possible to choose a nondecreasing function that is neither concave nor convex on $\mathbb{R}$, like $f(x) = x^3$. However, we have shown that not every quasi-convex) function is convex.

Example 7.5. Consider the *indirect utility function*

$$v(p, m) \equiv \max\{u(x) \,|\, px \leq m\}, \tag{7.1.8}$$

where $v(p, m)$ denotes the decrease in the north-east direction; p is the commodity price and m the income (money) of a consumer; and u is a utility function. Figure 7.3 shows that in the case when u is quasi-concave and monotone in consumption bundles, then v is monotone increasing in m and decreasing in p, and quasi-convex in (p, m). Its gradient is in the south-west direction, and its lower contour set is the shaded region in Figure 7.3.

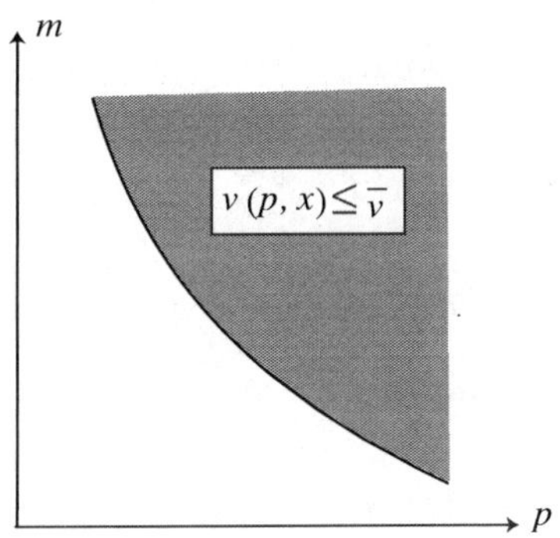

Figure 7.3 $v(p, m)$.

Note that, contrary to quasi-concavity, the quasi-convexity provides a condition that allows minimization of $v(p, x)$ on the lower contour set $\{p : v(p, m) \leq \bar{v}\}$, such that the indirect utility function plays the duality role in the theory of consumption and demand. ■

7.2 Properties of Quasi-Convex Functions

1. f is quasi-convex iff it is quasi-convex on lines, i.e., $f(x_0+th)$ is quasi-convex in t for all x_0 and h.

2. The *modified Jensen's inequality* is defined as follows: f is quasi-convex iff for all $x, y \in \text{dom}(f)$ and $t \in [0, 1]$

$$f(tx + (1-t)y) \leq \max\{f(x), f(y)\}. \tag{7.2.1}$$

This is presented in Figure 7.4.

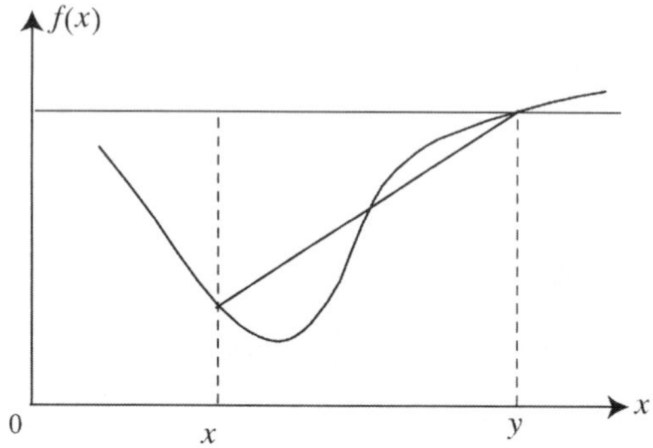

Figure 7.4 Modified Jensen's inequality.

3. Positive multiples: If f is quasi-convex and $\alpha \geq 0$, then αf is quasi-convex.

4. Positive maximum: If f_1, f_2 are quasi-convex, then $\max\{f_1, f_2\}$ is quasi-convex. This property extends to a supremum over an arbitrary set.

5. Affine transformation of the domain: If f is quasi-convex, then $f(\mathbf{Ax}+\mathbf{b})$ is quasi-convex.

6. Composition with a monotone increasing function: If f is quasi-convex and g a monotone increasing function, then $(g \circ f) = g(f(\mathbf{x}))$ is quasi-convex.

7. If f is quasi-convex, then $g(\mathbf{x}) = \inf_{\mathbf{y}} f(\mathbf{x}, \mathbf{y})$ is quasi-convex in $\mathbf{x}$.

8. In general, the sums of quasi-convex functions are not necessarily quasi-convex.

9. For a quasi-convex function f there holds the contraction property: $f(tx) \leq tf(x)$ for $t \in [0,1]$ if $f(0) \geq 0$. It follows directly from $f(tx) = f[tx + (1-t)0]$ $\leq tf(x) + (1-t)0 \leq tf(x)$. This implies that the inequality for quasi-convex functions is a weaker upper bound, since $f(x) > tf(x)$ for $t < 1$ and $f(x) > 0$. Thus, $f(tx) \leq f(x)$, but contradiction does not hold in that $f(tx)$ may be equal to $f(x)$ for $t \in (0,1)$.

10. A differentiable function f on a convex domain is quasi-convex iff the following first-order condition is satisfied:

$$f(y) \leq f(x) \Longrightarrow \frac{\partial f(x)}{\partial x} \cdot (y - x) \leq 0. \tag{7.2.2}$$

This is known as the *first-order condition* for quasi-convexity.

Theorem 7.3. A function f defined on a convex set $S \in \mathbb{R}^n$ is quasi-convex iff for all $x, x' \in S$ such that $f(x) \geq f(x')$, we have for $t \in [0,1]$,

$$f(tx + (1-t)x') \leq f(x'). \tag{7.2.3}$$

PROOF. First, suppose that f is quasi-convex. Let $x, x' \in S$, and $f(x) \leq f(x')$. Then for $a = f(x')$ we have $x \in L_\alpha$ and $x' \in L_\alpha$. Since f is quasi-convex, L_α is convex, and thus, $tx + (1-t)x' \in L_\alpha$ for all $t \in [0,1]$. Hence, $f(tx + (1-t)x') \leq a = f(x')$ for all $t \in [0,1]$.

Next, suppose that for all $x, x' \in S$ with $f(x) \leq f(x')$ we have for all $t \in [0,1]$,

$$f(tx + (1-t)x') \leq f(x'). \tag{7.2.4}$$

Then for any α, the set L_α is either empty, in which case it is convex, or consists of a single point, in which case it is convex, or contains two or more points, in which case we choose $x, x' \in L_\alpha$ with $f(x) \leq f(x')$. Then $f(tx + (1-t)x')) \leq f(x') \leq a$ for all $t \in [0,1]$, because $x' \in L_\alpha$. Hence, $tx + (1-t)x' \in L_\alpha$, so that L_α is convex, and f is quasi-concave. ■

Example 7.6. Consider the function $f(x) = x^2$, $x > 0$. We have $|B_1| = \begin{vmatrix} 0 & f_1 \\ f_1 & f_{11} \end{vmatrix} = \begin{vmatrix} 0 & 2x \\ 2x & 2 \end{vmatrix} = -4x^2 < 0$ for all $x > 0$. Hence, this function is both quasi-convex and quasi-concave on the set $\{x : x > 0\}$. ■

Example 7.7. Consider the function $f(x) = x^2, x \geq 0$. Then $|B_1(0)| = 0$, and this function does not satisfy the sufficient condition for either quasi-concavity or quasi-convexity. ■

Let S be a convex subset of $\mathbb{R}^n$, and let $f : S \mapsto \mathbb{R}$ be a function. Then the following statements are equivalent:

1. f is quasi-convex on S.
2. For all $x, y \in S$ and all $t \in (0,1)$,

$$f(x) \leq f(y) \implies f(tx + (1-t)y) \leq f(y). \tag{7.2.5}$$

3. For all $x, y \in S$ and all $t \in (0,1)$,

$$f(tx + (1-t)y) \leq \max\{f(x), f(y)\}. \tag{7.2.6}$$

Let S be a convex subset of $\mathbb{R}$, and let $f : S \mapsto \mathbb{R}$ be a function. Then f is said to be *strictly quasi-convex* if for all $x, y \in S$ with $x \neq y$, and for all $t \in (0,1)$,

$$f(tx + (1-t)y) < \max\{f(x), f(y)\}. \tag{7.2.7}$$

It is always possible to choose a nondecreasing function that is neither concave nor convex on $\mathbb{R}$, like $f(x) = x^3$. However, we have shown that not every quasi-convex (quasi-concave) function is convex (concave).

7.3 Bordered Hessian Test

We can use the bordered Hessian **B**, presented in §1.6.4, to check if a given function is quasi-convex. Unlike the bordered Hessian $\bar{\mathbf{H}}$ that is used in optimization problems subject to extraneous constraints, the bordered Hessian **B**, defined for the function f alone, is useful to test if the function f is quasi-concave or quasi-convex. Thus, for a function f with domain in $\mathbb{R}^n$, the determinant of the bordered Hessian **B** is defined by

$$|\mathbf{B}| = \begin{vmatrix} 0 & f_1 & f_2 & \cdots & f_n \\ f_1 & f_{11} & f_{12} & \cdots & f_{1n} \\ f_2 & f_{21} & f_{22} & \cdots & f_{2n} \\ \cdots & \cdots & \cdots & \cdots & \cdots \\ f_n & f_{n1} & f_{n2} & \cdots & f_{nn} \end{vmatrix}, \tag{7.3.1}$$

where f_i are the first-order derivatives of f, and f_{ij} are the second-order derivatives of f. The leading principal minors of **B** are

$$|B_1| = \begin{vmatrix} 0 & f_1 \\ f_1 & f_{11} \end{vmatrix}, \quad |B_2| = \begin{vmatrix} 0 & f_1 & f_2 \\ f_1 & f_{11} & f_{12} \\ f_2 & f_{21} & f_{22} \end{vmatrix}, \quad |B_n| = |\mathbf{B}|. \tag{7.3.2}$$

7.3.1 Properties of the Bordered Hessian. There are two properties which can be described as follows.

1. If f is $\begin{Bmatrix} \text{quasi-concave} \\ \text{quasi-convex} \end{Bmatrix}$ on a open set $D \in \mathbb{R}$, then $\begin{Bmatrix} (-1)^{k-1}|B_k| \ge 0 \\ |B_k| \le 0 \end{Bmatrix}$ for $k = 2, \dots, n+1$.

2. If f is $\begin{Bmatrix} (-1)^{k-1}|B_k| > 0 \\ |B_k| < 0 \end{Bmatrix}$ on a open set $D \in \mathbb{R}$ for $k = 2, \dots, n+1.$, then f is $\begin{Bmatrix} \text{quasi-concave} \\ \text{quasi-convex} \end{Bmatrix}$.

The top choice in property 2 requires that the signs of the leading principal minors alternate, starting with the negative sign for the 2×2 matrix $|B_2|$.

Example 7.8. Consider the function $f(x) = x^2$, $x > 0$. We have $|B_1| = \begin{vmatrix} 0 & f_1 \\ f_1 & f_{11} \end{vmatrix} = \begin{vmatrix} 0 & 2x \\ 2x & 2 \end{vmatrix} = -4x^2 < 0$ for all $x > 0$. Hence, this function is both quasi-concave and quasi-convex on the set $\{x : x > 0\}$. ■

Example 7.9. Consider the function $f(x) = x^2, x \ge 0$. Then $|B_1(0)| = 0$, and this function does not satisfy the sufficient condition for either quasi-concavity or quasi-convexity. ■

Example 7.10. Let $f : \mathbb{R} \mapsto \mathbb{R}$ be an increasing function. Then f is both quasi-concave and quasi-convex. To prove this, consider $x, y \in \mathbb{R}$, and assume without loss of generality that $x > y$. Then for any $t \in (0, 1)$, we have

$$x > tx + (1 - t)y > y. \tag{7.3.3}$$

Since f is increasing, we have

$$f(x) \geq f(tx + (1 - t)y) \geq f(y). \tag{7.3.4}$$

Since $f(x) = \max\{f(x), fy\}$, the inequality (7.3.3) shows that f is quasi-convex. Similarly, since $f(x) = \min\{f(x), f(y)\}$, the inequality (7.3.4) shows that f is quasi-concave.

7.4 Quasi-Convex Optimization

7.4.1 No Constraints. Some examples are as follows.

Example 7.11. Minimize $f(x, y, z) = 2x^2 - 5x - xy + 3y^2 - 4y + 3yz + 4z^2 + 3z - 2xz$. Equating the first-order partial derivatives, we get

$$f_x = 4x - 5 - y - 2z = 0, \; f_y = -x + 6y - 4 + 3z = 0, \; f_z = 3y + 8z + 3 - 2x = 0,$$

which in the matrix form $\mathbf{Ax} = \mathbf{b}$ is written as

$$\begin{bmatrix} 4 & -1 & -2 \\ -1 & 6 & 3 \\ -2 & 3 & 8 \end{bmatrix} \begin{bmatrix} x \\ y \\ z \end{bmatrix} = \begin{bmatrix} 5 \\ 4 \\ -3 \end{bmatrix}$$

which, by using Cramer's rule gives $|\mathbf{A}| = 136$, $|\mathbf{A}_1| = 176$, $|\mathbf{A}_2| = 216$, $|\mathbf{A}_3| = -76$, giving the critical point (x^*, y^*, z^*) approximately as $(1.29, 1.59, -0.56)$. Next, the second-order partial derivatives are:

$$\begin{aligned} &f_{xx} = 4, \; f_{xy} = -1, \; f_{xz} = -2; \\ &f_{yx} = -1, \; f_{yy} = 6, \; f_{yz} = 3; \\ &f_{zx} = -2, \; f_{zy} = 3, \; f_{zz} = 8. \end{aligned}$$

Then the Hessian is

$$|\mathbf{H}| = \begin{bmatrix} 4 & -1 & -2 \\ -1 & 6 & 3 \\ -2 & 3 & 8 \end{bmatrix},$$

and the minors are: $|H_1| = 4 > 0$, $|H_2| = \begin{bmatrix} 4 & -1 \\ -1 & 6 \end{bmatrix} = 25 > 0$, and $|H_3| = |\mathbf{H}| = |\mathbf{A}| = 176 > 0$. Thus, the Hessian is positive definite (PD), and the function f is minimized at the critical point (or, the critical point is the minimizer of f). ■

Example 7.12. Consider $f(x,y) = 5x^3 - 15xy + 5y^3$. Equating the first-order derivatives to zero, we get $f_1 5x^2 - 15y = 0$, $f_y = -15x + 15y^2 = 0$. Solving these two equations, we get the critical points as $(0,0)$ and $(1,1)$. Next, the second-order derivatives are $f_{xx} = 30x$, $f_{xy} = -15 = f_{yx}$, $f_{yy} = 30y$. At the point $(0,0)$, we find that $f_{xx} 0 = f_{yy}$, $f_{xy} = f_{yx} = -15$. Notice that f_{xx} and f_{yy} have the same sign, and $f_{xx} f_{yy} = 0 < (f_{xy})^2 = (-15)^2$; hence, the function has a point of inflection at $(0,0)$. Next, at the point $(1,1)$, we have $f_{xx} = 30 > 0$, $f_{xy} = -15 = f_{yx}$, $f_{yy} = 30 > 0$, and $f_{xx} f_{yy} = 900 > (f_{xy})^2 = (-15)^2 = 225$; hence, the function has a relative minimum at $(1,1)$. ■

7.4.2 Equality Constraints. Some examples follow.

Example 7.13. The total costs of a firm are given by $c(x,y) = 25x^2 + 50xy + 50y^2$ when the firm is required to meet the production quota $g(x,y) = 2x + 3y = 40$ (i.e., $c(x,y)$ is subject to the equality constraint g). The Laplacian is given by

$$C(x,y) = 25x^2 + 50xy + 50y^2 + \lambda(40 - 2x - 3y).$$

To find the critical values, we equate the first-order partial derivatives of C to zero, thus yielding $C_x = 50x + 50y - 2\lambda = 0$, $C_y = 50x + 100y - 3\lambda = 0$, $C_\lambda = 4 - 2x - 3y = 0$. Solving these three equations, using Cramer's rule, we get the critical values as $x^* = 8 = y^*$, $\lambda^* = 400$. Next, the second-order partial derivatives of C are: $C_{xx} = 50 = C_{xy} = C_{yx}$, $C_{yy} = 100$, and the first-order partial derivatives of g are: $g_x = 2$, $g_y = 3$. Thus, the bordered Hessian is

$$|\bar{\mathbf{H}}| = \begin{vmatrix} 50 & 50 & 2 \\ 50 & 100 & 3 \\ 2 & 3 & 0 \end{vmatrix}.$$

Then $|\bar{H}_2| = |\bar{\mathbf{H}}| = -250 < 0$; thus, the bordered Hessian is positive definite (PD), and the costs $c(x,y)$ are minimized at the critical values. ■

Example 7.14. Minimize the cost of production of 400 units of a good when $Q = 5K^{0.6}L^{0.2}$, and $P_K = 20$, $P_L = 10$, i.e., minimize $c(K,L) = 20K + 10L$ subject to the constraints $5K^{0.6}L^{0.2} = 400$. The Laplacian is

$$C(K,L) = 20K + 10L + \lambda(400 - 5K^{0.6}L^{0.2}).$$

To find the critical values, we equate the first-order partial derivatives of C to zero, thus yielding $C_K = 20 - 5\lambda(0.6)K^{-0.4}L^{0.2} = 0$, $C_L = 10 -$

$5\lambda K^{0.6}(0.2)L^{-0.8}$, $C_\lambda = 400 - 5K^{0.6}L^{0.2}$. The first two equations yield $\dfrac{20}{10} = \dfrac{3\lambda K^{-0.4}L^{0.2}}{\lambda K^{0.6}L^{-0.8}}$, or $K = 2L/3$, which when substituted in the third equation gives $K \approx 379, L \approx 569$. Next, the second-order partial derivatives of C are: $C_{KK} = 1.2\lambda K^{-1.4}L^{0.2}$, $C_{LL} = \lambda K^{0.6}L^{-1.8}$, $C_{KL} = -0.6\lambda K^{-0.4}L^{-0.8} = C_{LK}$. Thus, the bordered Hessian is

$$|\mathbf{H}| = \begin{vmatrix} 1.2\lambda K^{-1.4}L^{0.2} & -0.6\lambda K^{-0.4}L^{-0.8} & 3K^{-0.4}L^{0.2} \\ -0.6\lambda K^{-0.4}L^{-0.8} & \lambda K^{0.6}L^{-1.8} & K^{0.6}L^{-0.8} \\ 3K^{-0.4}L^{0.2} & K^{0.6}L^{-0.8} & 0 \end{vmatrix}.$$

Then

$$\begin{aligned} |\bar{H}_2| &= 3K^{-0.4}L^{0.2}[-0.6\lambda K^{0.2}L^{-1.6} - 3\lambda K^{0.2}L^{-1.6}] \\ &\quad - K^{0.6}L^{-0.2}[1.2\lambda K^{-0.8}L^{0.4} + 1.8\lambda K^{-0.8}L^{0.4}] \\ &= -10.8\lambda K^{-0.2}L^{-1.4} - 3\lambda K^{-0.2}L^{0.2} < 0, \end{aligned}$$

since $K, L, \lambda > 0$. Hence, the bordered Hessian is positive definite (PD), and the cost $c(x, y)$ is minimized at the critical values. ■

7.4.3 Inequality Constraints. Minimize a quasi-convex function $f(\mathbf{x})$ subject to $g_i(\mathbf{x}) \le 0$, $i = 1, \dots, m$, and $\mathbf{Ax} = \mathbf{b}$, where the inequality constraints g_i, are convex. If the objective function f is differentiable, the first-order condition for quasi-convexity implies that $\mathbf{x}$ is optimal if $\nabla f(\mathbf{x})^T(\mathbf{y} - \mathbf{x}) > 0$ for all $\mathbf{x} \ne \mathbf{y}$. This condition is only sufficient for optimality, and it requires that $\nabla f \ne 0$. Note that $\nabla f = 0$ holds in the convex case for $\mathbf{x}$ to be optimal. Figure 7.5 shows that the simple optimality condition $f'(x) = 0$, valid for convex functions, does not hold for quasi-convex function.

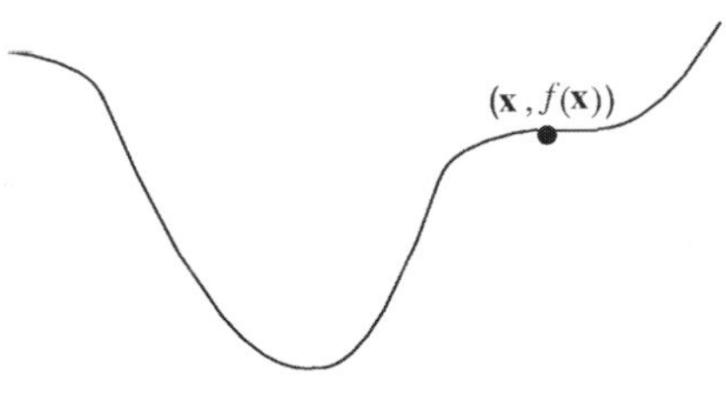

Figure 7.5 Quasi-convex optimality.

7.4.4 Convex Feasibility Method. A general method to solve quasi-convex optimization problem uses the representation of the sublevel sets of a quasi-convex function via a family of convex inequalities. Let $\phi_t : \mathbb{R}^n \mapsto \mathbb{R}$, $t \in \mathbb{R}$, be a family to convex functions such that $f(\mathbf{x}) \le t \iff \phi_t(\mathbf{x}) \le 0$, and for each $\mathbf{x}$, the functions $\phi_t(\mathbf{x})$ are nonincreasing, i.e., $\phi_s(\mathbf{x}) \le \phi_t(\mathbf{x})$ whenever

$s \geq t$. Let $\mathbf{x}^*$ denote the optimal point for the quasi-convex problem. If the feasibility problem is:

Find $\mathbf{x}$ subject to $\phi_t(\mathbf{x}) \leq 0, g_i(\mathbf{x}) \leq 0, \mathbf{Ax} = \mathbf{b}, i = 1, \dots, m,$ (7.4.1)

is feasible, then we have $\mathbf{x}^* \leq t$. Conversely, if this problem is infeasible, then we cannot conclude that $\mathbf{x}^* \geq t$. The problem (7.4.1) is known as the *convex feasibility problem*, since the inequality constraints are all convex functions and the equality constraints are linear.

This leads to a simple algorithm for solving quasi-convex optimization problem by using *bisection* that solves the convex feasibility problem at each step. Assuming that the problem is feasible, start with the interval $[l, r]$ that contains the optimal value $\mathbf{x}^*$. Then solve the convex feasibility problem at the mid-point $t = (l + r)/2$, by determining the half interval that contains the optimal value, and continue halving the half-interval each time until the width of the interval that contains the optimal value is small enough, say $\varepsilon > 0$. This is known as the *bisection method*. Note that the length of the interval after k iterations is $2^{-k}(r - l)$, which means that exactly $\ln((r - l)/\varepsilon)$ iterations will be required before the algorithm terminates.

Example 7.15. Minimize $f(x, y) = y^2 - \log(1+x) - y$ subject to $x - 2y \leq 3$, $x \geq 0$, $y \geq 0$. Take $F(x, y) = y^2 - \log(1+x) + \lambda_1(3 - x + 2y) + \lambda_2(-x) + \lambda_3(-y)$. Then the first-order KKT conditions give

$$F_x = -1/(1+x) - \lambda_1 - \lambda_2 = 0,\ F_y = 2y + 2\lambda_1 - \lambda_3 = 0,$$
$$F_{\lambda_1} = 3 - x + 2y = 0,\ F_{\lambda_2} = -x = 0,\ F_{\lambda_3} = -y = 0.$$

Since the last equation gives $y = 0$, we get $x = 3$ from $F_{\lambda_1} = 0$. Thus, the critical point is (3.0). The second-order conditions are: $F_{xx} = 1/(1+x)^2$, $F_{xy} = 0 = F_{yx}$, $F_{yy} = 2$, which at the point (3.0) give the Hessian

$$|\mathbf{H}| = \begin{vmatrix} \frac{1}{16} & 0 \\ 0 & 2 \end{vmatrix} = \frac{1}{8} > 0.$$

Also, the first minor is $|H_1| = \frac{1}{16} > 0$, and the second minor $|H_2| = |\mathbf{H}| = F_{xx}F_{yy} - (F_{xy})^2 = \frac{1}{8} > 0$. Hence, the function $f(x, y)$ has a minimum at $(3, 0)$. ■

7.4.5 Equality and Inequality Constraints. Minimize $f(x)$ subject to equality constraint $g_i(x) = 0$ and inequality constraint $h_j(x) \leq 0$. The domain D of the optimization problem is

$$D = \bigcap_{i=1}^{n} \text{dom}(g_i) \cap \bigcap_{j=1}^{m} h_j(x). \tag{7.4.2}$$

A point $x \in D$ is called a *critical point* x^* if it satisfies the above equality and inequality constraints, and the problem is said to be feasible if there exists at least one critical point. The critical value x^* is said to be the *optimal value* of the above problem and is defined by

$$x^* \in \{f(x) \mid g_i(x) = 0, i = 1, \dots, n; h_j(x) \leq 0, j = 1, \dots, m\}, \tag{7.4.3}$$

where x^* can take the values $\pm\infty$. The set of all optimal points is denoted by

$$X^* = \{x \mid g_i(x) = 0, i = 1, \dots, n; h_j(x) \leq 0, j = 1, \dots, m; f(x) = x^*\}. \tag{7.4.4}$$

If $X^* = \emptyset$, the optimization problem is not solvable. A point x^* is *locally optimal* if there exists an $M > 0$ such that

$$\begin{aligned} f(x) = \inf\{f(x) \mid g_i(z) = 0, i = 1, \dots, n; \\ h_j(z) = 0, j = 1, \dots, m; \|z - x\|_2 \leq M\}, \end{aligned} \tag{7.4.5}$$

i.e., x^* solves the optimization problem.

Example 7.16. Let $\text{dom}(f) = \mathbb{R}_{++}$. Then (a) $f(x) = 1/x$: $x^* = 0$ but the optimal value is not achieved; (b) $f(x) = -\log x$: $x^* = -\infty$, but this problem is unbounded from below; (c) $f(x) = x \log x$: $x^* = -1/e$ which is achieved. ∎

7.4.6 Minmax Theorem. In microeconomics, quasi-concave utility functions imply that consumers have convex preferences. Quasi-convex functions are important in game theory, industrial organization, and general equilibrium theory, particularly for applications of Sion's minmax theorem, which is a generalization of von Neumann's minmax theorem (zero-sum games).

Theorem 7.4. (Minmax Theorem; von Neumann [1928]) Let $X \subset \mathbb{R}^n$ and $Y \subset \mathbb{R}^m$ be compact sets. If $f : X \times Y \mapsto \mathbb{R}$ is a continuous function that is convex-concave, i.e., $f(\cdot, y) : X \mapsto \mathbb{R}$ is convex for fixed y, and $f(x, \cdot) : Y \mapsto \mathbb{R}$ is concave for fixed x, then we have

$$\min_{x \in X} \max_{y \in Y} \{f(x, y)\} = \max_{y \in Y} \min_{x \in X} \{f(x, y)\}. \tag{7.4.6}$$

Example 7.17. Martos [1969, 1971] has shown that there is a class of quasi-convex functions which need not be convex, namely, the quadratic function $f(x) = \frac{1}{2}xQx + cx$, where Q is a symmetric matrix. If $X = \mathbb{R}^n$, then $f(x)$ is convex iff $f(x)$ is quasi-convex; thus, Q is positive and semidefinite. However, if $X = \mathbb{R}^n_+$, the function $f(x)$ may be quasi-convex, yet not be convex. For example, $f(x_1, x_2) = -x_1 x_2$ is quasi-convex over $\mathbb{R}^2_+$, but it is not convex there.

Example 7.18. One can combine convex, concave and linear functions to form quasi-convex functions. For example, let f and g be defined on a convex set X such that $f(x) \neq 0$ for all $x \in X$. Then g/f is quasi-convex on X if any of following *six* conditions holds:

I. g is convex and $f(x) > 0$ for all $x \in X$, OR g is concave and $f(x) < 0$ for all $x \in X$; AND

II. f is linear on X, OR f is convex on X and $g(x) \leq 0$ for all $x \in X$, OR f is concave on X and $g(x) \geq 0$ for all $x \in X$.

7.5 Summary

Let S be a nonempty convex set in $\mathbb{R}^n$, and let $f : S \mapsto \mathbb{R}$. Then there are the following types of convexity and quasi-convexity at a point.

1. CONVEXITY AT A POINT. The function f is convex at $\mathbf{x} \in S$ if $f(t\mathbf{x} + (1-t)\mathbf{x}') \leq tf(\mathbf{x}) + (1-t)f(\mathbf{x}')$ for each $t \in (0,1)$ and each $\mathbf{x}' \in S$.

2. STRICT CONVEXITY AT A POINT. The function f is strictly convex at $\mathbf{x} \in S$ if $f(t\mathbf{x} + (1-t)\mathbf{x}') < tf(\mathbf{x}) + (1-t)f(\mathbf{x}')$ for each $t \in (0,1)$ and each $\mathbf{x}' \in S$, $\mathbf{x} \neq \mathbf{x}^*$.

3. QUASI-CONVEXITY AT A POINT. The function f is quasi-convex at $\mathbf{x} \in S$ if $f(t\mathbf{x} + (1-t)\mathbf{x}') \leq \max\{f(\mathbf{x}), f(\mathbf{x}')\}$ for each $t \in (0,1)$ and each $\mathbf{x}' \in S$.

4. STRICT QUASI-CONVEXITY AT A POINT. The function f is strictly quasi-convex at $\mathbf{x} \in S$ if $f(t\mathbf{x} + (1-t)\mathbf{x}') < \max\{f(\mathbf{x}), f(\mathbf{x}')\}$ for each $t \in (0,1)$ and each $\mathbf{x}' \in S$, $f(\mathbf{x}') \neq f(\mathbf{x})$.

7.6 Exercises

7.1. Let $f : \mathbb{R} \mapsto \mathbb{R}$ be convex, and $a, b \in \mathrm{dom}(f)$, $a < b$. (a) Show that

$$f(x) \leq \frac{b-x}{b-a} f(a) + \frac{x-a}{b-a} f(b) \quad \text{for all } x \in [a,b].$$

(b) Show that

$$\frac{f(x) = f(a)}{x-a} \leq \frac{f(b) - f(a)}{b-a} \leq \frac{f(b) - f(x)}{b-x} \quad \text{for all } x \in (a,b).$$

(c) Suppose f is differentiable. Use the inequality in (b) to show that

$$f'(a) \leq \frac{f(b) - f(a)}{b-a} \leq f'(b).$$

(d) Suppose f is twice differentiable. Use the result in (c) to show that $f''(a) \geq 0$ and $f''(b) \geq 0$.

HINT. The first three inequalities follow from the definition of a convex function: Suppose f is differentiable. Then f is convex iff $\mathrm{dom}(f)$ is a convex set and $f(y) \geq f(x)+f'(x)(y-x)$ for all $x, y \in \text{dom}(f)$, which is the first-order Taylor's series approximation at x. Part (d) is obvious.

7.2. Suppose $f : \mathbb{R}^n \mapsto \mathbb{R}$ is convex with $\text{dom}(f) = \mathbb{R}^n$, and bounded above on $\mathbb{R}^n$. Show that f is constant.

7.3. Prove that a convex function is quasi-convex. PROOF. Let the function f have the domain S (a convex set). Let a be a real number and xy be points in the lower-level set L_a with $x, y \in L_a$. First, we show that the set L_a is convex. For this, we need to show that for every $t \in [0, 1]$ we have $tx + (1-t)y \in L_a$. Since S on which f is defined is convex, we have $tx + (1-t)y \in S$, and thus f is defined at the point $tx + (1-t)y$. Now, convexity of f implies that $f(tx + (1-t)y) \leq tf(x) + (1-t)f(y)$. Moreover, the fact that $x \in L_a$ means that $f(x) \leq a$, and similarly, $y \in L_a$ means that $f(y) \leq a$. Hence, $tf(x) + (1-t)f(y) \leq ta + (1-t)a = a$. Combining these two inequalities we get $f(tx+(1-t)y) \leq a$, so that $tx+(1-t)y \in L_a$. Thus, every upper-level set is convex and hence, f is quasi-convex.

Note that quasi-convexity is weaker than convexity, in the sense that every convex function is quasi-convex.

7.4. Prove that $g(t) = f(tx+(1-t)y)$ is quasi-convex for $t \in [0, 1]$ and for any $x, y \in \mathbb{R}^n$ iff f is quasi-convex. HINT. Use definition (7.1.1).

7.5. Prove that

$$\sup_x\{f(x) + g(x)\} \leq \sup_x\{f(x)\} = \sup_x\{g(x)\}.$$

HINT. This inequality follows from the defining inequality of quasi-convex functions and the triangle inequality.

7.6. Prove that the floor function $x \mapsto \lfloor x \rfloor$ is a quasi-convex function that is neither convex nor continuous.

7.7. Prove that if $x \mapsto f(x)$ and $y \mapsto g(y)$ are positive convex decreasing functions, then $(x, y) \mapsto f(x)g(y)$ is quasi-convex.

7.8. Prove that if f and g are both convex, both nondecreasing (or nonincreasing), and positive functions on an interval, then fg is convex.

7.9. Let f be a function of n variables with continuous partial derivatives of all order in an open convex set S, and let $\bar{\mathbf{H}}_k$ be the determinant of its kth-order bordered Hessian. Then

(i) if f is quasi-convex, then $|\bar{\mathbf{H}}|_k(\mathbf{x}) \le 0$ for all $\mathbf{x} \in S$, $k = 1, 2, \dots, n$; and

(ii) if $|\bar{\mathbf{H}}|_k(\mathbf{x}) < 0$ for all $\mathbf{x} \in S$, $k = 1, 2, \dots, n$, then f is quasi-convex.

This result can also be stated as follows:

$$\begin{cases} |\bar{\mathbf{H}}|_1(\mathbf{x}) \ge 0, |\bar{\mathbf{H}}|_2(\mathbf{x}) \le 0, \dots, |\bar{\mathbf{H}}|_n(\mathbf{x}) \ge 0 & \text{if } n \text{ is odd,} \\ |\bar{\mathbf{H}}|_1(\mathbf{x}) \ge 0, |\bar{\mathbf{H}}|_2(\mathbf{x}) \le 0, \dots, |\bar{\mathbf{H}}|_n(\mathbf{x}) \le 0 & \text{if } n \text{ is even,} \end{cases}$$

for all $\mathbf{x} \in S$, is a necessary condition for quasi-convexity, and

$$\begin{cases} |\bar{\mathbf{H}}|_1(\mathbf{x}) > 0, |\bar{\mathbf{H}}|_2(\mathbf{x}) < 0, \dots, |\bar{\mathbf{H}}|_n(\mathbf{x}) > 0 & \text{if } n \text{ is odd,} \\ |\bar{\mathbf{H}}|_1(\mathbf{x}) > 0, |\bar{\mathbf{H}}|_2(\mathbf{x}) < 0, \dots, |\bar{\mathbf{H}}|_n(\mathbf{x}) < 0 & \text{if } n \text{ is even,} \end{cases}$$

for all $\mathbf{x} \in S$, is a sufficient condition for quasi-convexity.

7.10. If f is convex, nondecreasing, and positive, and g is concave, nonincreasing and positive functions on an interval, then f/g is convex.

7.11. Minimize $C(x, y) = (x-3)^2 + (y-4)^2$ subject to $x + y \ge 4$ and $x, y \ge 0$.

7.12. Minimize $C(x, y) = 2x + y$ subject to $x^2 - 4x + y \ge 0$ and $x, y \ge 0$.

7.13. Plot the graphs of the following functions and check whether they are quasi-concave, quasi-convex, both, or neither: (a) $f(x) = x^3 - 2x$; (b) $f(x, y) = 6x - 9y$; (c) $f(x, y) = y - \ln x$.

7.14. Verify that the cubic function $f(x) = ax^3 + bx^2 + cx + d$ is in general neither quasi-concave nor quasi-convex.

7.15. Use definitions (6.1.3), (6.1.4), and (7.1.3) to check whether $f(x) = x^2, x \ge 0$ is quasi-concave or quasi-convex. (See Example 6.1).

7.16. Show that $f(x, y) = xy, x, y \ge 0$ is not quasi-convex.

7.17. Consider the parabolic cylinder $x = y^2$, $x \ge 0$. Determine if this function is convex, quasi-convex, concave, or quasi-concave. HINT. The graphs of $y = \pm\sqrt{x}, x \ge 0$ and the 3-D plot of the given functions are presented in Figures 7.6 (a), (b), and (c); the graphs do not exist for $x > 0$.

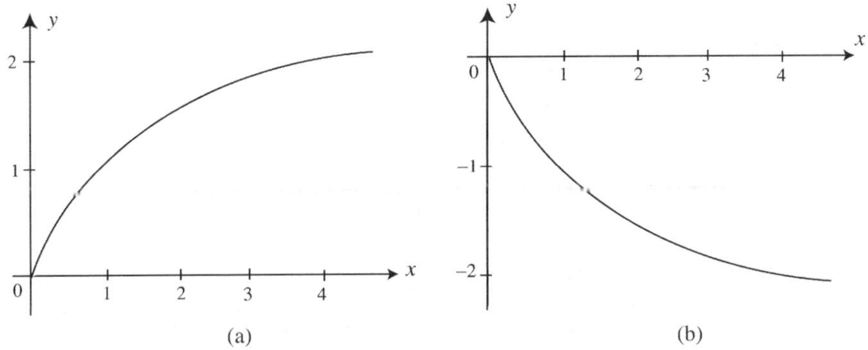

Figure 7.6 (a) $y = \sqrt{x}$, (b) $y = -\sqrt{x}$.

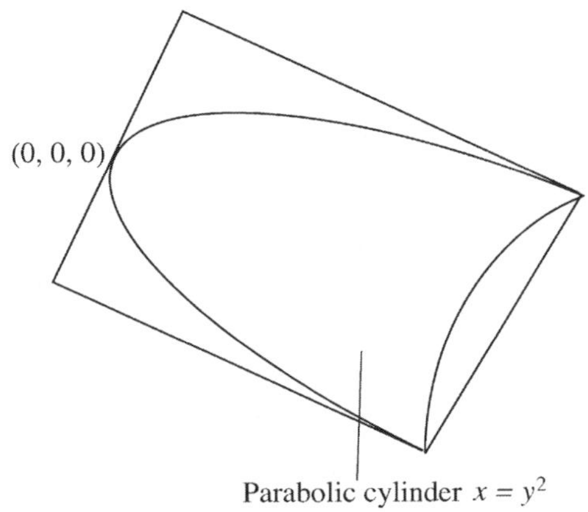

Figure 7.6 (c) 3-D Plot of $x = y^2$.

7.18. Use the bordered Hessian to check whether the following functions are quasi-concave or quasi-convex: (a) $f(x,y) = -x^2 - y^2$, $x, y > 0$; (b) $f(,y) = -(x+1)^2 - (y+2)^2$, $x, y > 0$.

7.19. Minimize $C(x,y) = (x-3)^2 + (y-4)^2$ subject to $x + y \geq 4$ and $x, y \geq 0$.

7.20. Minimize $C(x,y) = 2x + y$ subject to $x^2 - 4x + y \geq 0$ and $x, y \geq 0$.

7.21. Let f and g be real-valued convex functions with the same domain D. Define a function h so that $h(x) = f(x)g(x)$ for all $x \in D$. Show that

h is a quasi-convex function. ANS. h is necessarily convex; for example, let $f(x) = x$ and $g(x) = x$ for $x \in \mathbb{R}$ (the real line). Both f and g are convex, and so is $h(x) = x^2$, and also quasi-convex.

7.22. Show that $\mathbf{x}^* = (-\frac{1}{2}, \frac{1}{3}, 1)$ is the optimal point for the optimization problem: Minimize $\frac{1}{2}\mathbf{x}^T\mathbf{A}\mathbf{x} + \mathbf{b}^T\mathbf{x} + \mathbf{c}$, where

$$\mathbf{A} = \begin{bmatrix} 10 & 12 & 2 \\ 6 & 17 & 13 \\ 8 & 13 & 6 \end{bmatrix}, \mathbf{b} = \begin{bmatrix} 12 \\ -14 \\ -21 \end{bmatrix}, \mathbf{x} = \begin{bmatrix} x \\ y \\ z \end{bmatrix}, \mathbf{c} = [7].$$

HINT. The function $f(\mathbf{x}) = f(x,y,z) = \frac{1}{2}\big[24x^2 + 42y^2 + 21z^2\big] + 12x - 14y - 21z + 7$. Equating the first-order partial derivatives of f with respect to x, y, z to zero yields the critical point $\mathbf{x}^* = (-\frac{1}{2}, \frac{1}{3}, 1)$. Use Hessian $\mathbf{H}$, as in Example 7.11, to show that the function f is minimized at $\mathbf{x}^*$.

7.23. Show that the function $f(x,y) = -x^a y^b$, $x, y > 0$, $0 < a, b < 1$, is quasi-convex. HINT. Use the bordered Hessian $|\mathbf{B}|$ (§7.3.1) and show that $|B_1| < 0$ and $|B_2| < 0$.

8

Log-Concave Functions

Log-concavity is an important part of information economics. Since the logarithm of the cumulative distribution function (c.d.f.) of a random variable is a concave function, it turns out that the ratio of the probability density function (p.d.f.) to c.d.f. is a monotone decreasing function. Some readers may prefer to read §8.5 on log-concave distribution prior to starting this chapter.

8.1 Definitions

A nonnegative function $f : \mathbb{R}^n \mapsto \mathbb{R}$ is said to be *log-concave* (or *logarithmically concave*) if its domain is a convex set, and f satisfies the inequality

$$f(t\mathbf{x} + (1-t)\mathbf{x}') \geq f(\mathbf{x})^t f(\mathbf{x}')^{1-t} \tag{8.1.1}$$

for all $\mathbf{x}, \mathbf{x}' \in \text{dom}(f)$ and $0 < t < 1$. If f is strictly positive, then the logarithm of the function f, i.e., $\log f$, is concave:

$$\log(f(t\mathbf{x} + (1-t)\mathbf{x}') \geq t \log f(\mathbf{x}) + (1-t) \log f(\mathbf{x}') \tag{8.1.2}$$

for all $\mathbf{x}, \mathbf{x}' \in \text{dom}(f)$ and $0 < t < 1$.

Compare the above definition with the following definition of a log-convex function, which uses the property that f is convex if $-f$ is concave. Thus, a function f is *log-convex* if it satisfies the inequality

$$f(t\mathbf{x} + (1-t)\mathbf{x}') \leq f(\mathbf{x})^t f(\mathbf{x}')^{1-t} \tag{8.1.3}$$

for all $\mathbf{x}, \mathbf{x}' \in \text{dom}(f)$ and $0 < t < 1$. Hence, f is *log-convex* if $\log f$ is convex.

Example 8.1. (i) Exponential function $f(x) = e^{ax}$; (ii) $f(x) = x^a$, $a \geq 0$; and (iii) $f(x) = \dfrac{e^x}{1+e^x}$ (known as the *inverse logit function*),[1] are log-concave functions. ■

[1] Note that the logit function is used in statistics to determine the log-odds or logarithm of the order $p/(1-p)$, where p denotes the function power.

Some properties of log-concave functions are as follows:

(i) A positive log-concave function is also quasi-concave.

(ii) Every concave function that is nonnegative on its domain is log-concave. However, the converse is not necessarily true. An example is the Gaussian function $f(x) = e^{-x^2/2}$ which is log-concave since $\log f(x) = -x^2/2$ is a concave function of x. But f is not concave since its second derivative $f''(x) = e^{-x^2/2}(x^2 - 1) > 0$ for $|x| > 1$.

(iii) A nonnegative C^2-function f defined on a convex domain is log-concave iff for all $\mathbf{x} \in \mathbb{R}^n$ for which $f(\mathbf{x}) > 0$,

$$f(\mathbf{x})[\mathbf{H}] \preceq \begin{bmatrix} \dfrac{\partial f}{\partial x_1} \\ \vdots \\ \dfrac{\partial f}{\partial x_n} \end{bmatrix} \begin{bmatrix} \dfrac{\partial f}{\partial x_1} & \cdots & \dfrac{\partial f}{\partial x_n} \end{bmatrix}, \tag{8.1.4}$$

where $[\mathbf{H}]$ is the Hessian matrix of f (§1.6.2). In other words, condition (8.1.4) states that the difference between the left-hand side and the right-hand side expressions is negative semidefinite (NSD). In the case of functions f in $\mathbb{R}$ condition (8.1.4) simplifies to

$$f(x)f''(x) \le (f'(x))^2. \tag{8.1.5}$$

Since $f(\mathbf{x}) > 0$, condition (8.1.4) can also be written using the Schur complement (see Gill et al. [1990]). However, condition (8.1.4) is often written incorrectly as

$$f(x)\nabla^2 f(x) \preceq \nabla f(x)\nabla f(x)^T,$$

implying that $f(\mathbf{x})\nabla^2 f(\mathbf{x}) - \nabla f(\mathbf{x})\nabla f(\mathbf{x})^T$ is negative semidefinite (NSD).

Other properties of log-concave functions are as follows: (i) (sum:) sum of log-concave functions are not always log-concave; (ii) (product): if f and g are log-concave, then fg is log-concave; (iii) (integral): if $f(x, y)$ is log-concave in x, y, then $\int f(x, y)\, dy$ is log-concave; and (iv) (convolution:) if f and g are log-concave, then $\int f(x - y)g(y)\, dy$ is log-concave.

8.1.1 Log-Concavity Preserving Operations

(i) PRODUCTS. The product of log-concave functions is also log-concave. In fact, if f and g are log-concave, then $\log f$ and $\log g$ are concave, by definition. Hence,

$$\log f(x) + \log g(x) = \log(f(x)g(x)) \tag{8.1.6}$$

is concave, and so fg is also log-concave.

(ii) MARGINALS. If $f(x,y) : \mathbb{R}^{n+m} \mapsto \mathbb{R}$ is log-concave, then

$$g(x) = \int f(x,y)\,dy \tag{8.1.7}$$

is log-concave; see the Prékopa-Leindler inequality below (Theorem 8.10).

(iii) The above property implies that convolution preserves log-concavity, since $h(x,y) = f(x-y)g(y)$ is log-concave if f and g are log-concave, and hence the convolution $f \star g$ of the functions f and g, defined by

$$(f \star g)(x) = \int f(x-y)g(y)\,dy = \int h(x,y)\,dy, \tag{8.1.8}$$

is log-concave.

(iv) The Laplace transform of a nonnegative convex function is log-concave.

(v) If the random variable has a log-concave p.d.f., then the c.d.f. is a log-concave function.

(vi) If two independent random variables have log-concave p.d.f.s, then their sum has a log-concave p.d.f.

8.2 Theorems

Consider a real-valued function f which is log-concave on some interval I if $f(x) \ge 0$ and $x \mapsto \log f(x)$ is an extended real-valued concave function, where we have $\log 0 = -\infty$.

Theorem 8.1. (Marshall and Olkin [1979; 16B.3.a]) *A sufficient condition for a nonnegative function f to be log-concave on I is given by*

$$f(x+h)f(x') \le f(x)f(x'+h), \quad x < x';\ x+h, x'+h \in I, h > 0. \tag{8.2.1}$$

Log-concavity is a more difficult concept than log-convexity. It is known that the sum of two log-convex functions is again a log-convex function, but it need not be true for log-concavity. Before we prove an important theorem on convolution of log-concave functions, we will introduce a definition of a totally positive function of order k: A function F defined on $A \times B$, where A and B are subsets of $\mathbb{R}$, is said to be *totally positive of order* k, denoted by $(\text{TP})_k$, if for all $m, 1 \le m \le k$ and all $x_1, x_2 < \cdots < x_m$, $y_1 < y_2 < \cdots < y_m$, where $x_i \in A$ and $y_i \in B$ we have

$$\begin{vmatrix} F(x_1,y_1) & \cdots & F(x_1,y_m) \\ \vdots & \cdots & \vdots \\ F(x_m,y_1) & \cdots & F(x_m,y_m) \end{vmatrix} \ge 0. \tag{8.2.2}$$

This definition implies that if u and v are nonnegative functions and F is $(\text{TP})_k$, then $u(x)v(y)F(x,y)$ is also $(\text{TP})_k$.

In the particular case when $F(x,y) = f(y-x)$, first note that F is $(\mathrm{TP})_2$ on $A \times B$ iff f is nonnegative and

$$f(y_1 - x_1)f(y_2 - x_2) - f(y_2 - x_1)f(y_1 - x_2) \geq 0 \tag{8.2.3}$$

for all $x_1 < x_2$ in A and $y_1 < y_2$ in B.

Theorem 8.2. *Condition (8.2.3) is equivalent to log-concavity of* $x \mapsto f(x)$.

PROOF. Assume that $x_1 < x_2 < y_1 < y_2$. Let $y_1 - x_2 = r$, and $y_1 - x_1 = s, y_2 - y_1 = h$. Then $r < s$ and $h > 0$, and (8.2.3) becomes $f(r+h)f(s) \geq f(r)f(s+h)$, which is simply condition (8.2.1). ■

Schoenberg [1951] has shown that $F(x,y) = f(y-x)$ is $(\mathrm{TP})_2$ on $\mathbb{R} \times \mathbb{R}$ iff f is log-concave on $\mathbb{R}$.

Theorem 8.3. (Marshal and Olkin [1979:2]) *If K is $(TP)_m$ and L is $(TP)_n$ and σ is a sigma-finite measure,*[2] *then the convolution*

$$M(x,y) = \int K(x,z)L(z,y)\,d\sigma(z). \tag{8.2.4}$$

For proof, see Marshal and Okin [1979].

8.2.1 General Results on Log-Concavity.

Let g be a twice-differentiable real-valued function defined on an interval I of the extended real line. A function g is said to be *log-concave* on the interval (a,b) if the function $\log g$ is a concave function on (a,b).

Remark 1. Log-concavity of g on (a,b) is equivalent to each of the following two conditions: (i) $g'(x)/g(x)$ is monotone decreasing on (a,b); and (ii) $(\log g(x))'' < 0$.

Theorem 8.4. *Let g be strictly monotone (increasing or decreasing) defined on the interval (a,b). Suppose that either $g(a) = 0$ or $g(b) = 0$. If g' is a log-concave function on (a,b), then $g(x)$ is also a log-concave function on (a,b).*

PROOF. Following Bagnoli and Bergstrom [1989], suppose $g(a) = 0$. Using the generalized mean value theorem (or Cauchy mean value theorem) there exists $\xi \in (a,b)$ such that

$$\frac{g'(x) - g(a)}{g(x) - g(a)} = \frac{g''(\xi)}{g'(\xi)}. \tag{8.2.5}$$

[2] Let Σ be a σ-algebra over X. A function $\mu : \Sigma \mapsto \mathbb{R}$ is called a *measure* if (i) $\mu(E) = 0$ for all $E \in \Sigma$, and (ii) $\mu(\emptyset) = 0$. The Lebesgue measure on R is a complete translation-invariant measure of a σ-algebra containing the intervals in $\mathbb{R}$ such that $\mu[0,1] = 1$. A σ-finite measure is a finite number, and real numbers with standard Lebesgue measure are σ-finite but not finite.

Suppose g''/g' is monotone decreasing. Since $x > \xi$, we have $\dfrac{g''(\xi)}{g'(\xi)} > \dfrac{g''(x)}{g'(x)}$. Then it follows from (8.2.5) that

$$\frac{g'(x) - g'(a)}{g(x)} > \frac{g''(x)}{g'(x)}. \tag{8.2.6}$$

Since g is strictly monotone and $g(a) = 0$, then $g(x)$ is of the same sign as $g'(x)$ for all $x \in (a, b)$. Therefore multiplying both sides of (8.2.6) by $g(x)g'(x)$ preserves the direction of inequality, and yields $g'(x)^2 - g'(x)g'(a) > g(x)g''(x)$, and hence $g(x)g''(x) - g'(x)^2 < -g'(x)g'(a) < 0$. Thus,

$$\frac{g(x)g''(x) - g'(x)^2}{g'(x)^2} < 0, \quad \text{or} \quad \left(\frac{g'(x)}{g(x)}\right)' < 0. \tag{8.2.7}$$

Then using Remark 1 we conclude that g is log-concave.

Next, suppose that $g(b) = 0$. Using the generalized mean value theorem, there exists an $x \in (a, b)$ such that

$$\frac{g'(x) - g(b)}{g(x) - g(b)} = \frac{g''(\xi)}{g'(\xi)}. \tag{8.2.8}$$

If $g''(x)/g'(x)$ is monotone decreasing, then since $g(b) = 0$ and $x < \xi$, it follows from (8.2.8) that

$$\frac{g'(x) - g'(b)}{g(x)} < \frac{g''(x)}{g'(x)}. \tag{8.2.9}$$

Since $g(x)$ is monotone and $g(b) = 0$, we must have $g'(x)g(x) < 0$ for $x < b$. Multiplying both sides of (8.2.8) by $g(x)g'(x)$, we get $g'(x)^2 - g'(x)f'(x) > g(x)g''(x)$. As before, this inequality implies inequality (8.2.7), which in view of Remark 1 establishes the log-concavity of g. ■

According to Schoenberg [1951], any log-concave function f on $\mathbb{R}$ is either monotone or, if it is non-monotone, then $f(x) \to 0$ as $x \to \pm\infty$ at least exponentially. Thus, if f and g are non-monotone and log-concave functions on $\mathbb{R}$, then their convolution is well defined on $\mathbb{R}$.

An important theorem is as follows:

Theorem 8.5. (Lekkerkerker [1953]) *Convolution of log-concave functions defined on the interval* $[0, \infty)$ *under some additional requirements, mentioned in the next theorem, is log-concave.*

Lekkerkerker's proof is very long. A simpler proof of this result as published by Merkle [1998a,b] is as follows.

Theorem 8.6. (Merkle [1998a]) *Let f and g be log-concave functions on $\mathbb{R}$ such that the convolution*

$$h(x) = \int_{-\infty}^{+\infty} f(x-z)g(z)\,dz \tag{8.2.10}$$

is defined for all $x \in \mathbb{R}$. Then the function h is log-concave on $\mathbb{R}$.

PROOF. Let $F(x,z) = f(x-z)$ and $L(z,y) = g(z-y)$. Since f and g are log-concave functions, we find, by Theorem 8.3, that F and L are $(\mathrm{TP})_2$ on $\mathbb{R}\times\mathbb{R}$, and $M(x,y) = \int_{-\infty}^{+\infty} f(x-z)g(z-y)\,dz$ is $(\mathrm{TP})_2$ on $\mathbb{R}\times\mathbb{R}$. Now, setting $z-y=t$ we get $M(x,y) = \int_{-\infty}^{+\infty} f(x-y-t)g(t)\,dt = h(x-y)$. Since M is $(\mathrm{TP})_2$, we find that h is log-concave on $\mathbb{R}$. ■

Note that since we have set $t = z-y$, the differential dz in this theorem cannot be replaced by an arbitrary sigma-finite measure $d\sigma$. Also, if the function f is log-concave on an interval I, then the function f^* defined by

$$f^*(x) = \begin{cases} f(x) & \text{if } x \in I, \\ 0 & \text{otherwise} \end{cases}$$

is log-concave on $\mathbb{R}$. Hence, Theorem 8.6 can be applied to convolutions of functions defined on intervals of $\mathbb{R}$. The statement of Theorem 8.6 holds if the 'log-concave' is replaced by 'log-convex', and the proof is direct since the function $f \mapsto f(x-z)g(z)$ is log-convex in z for each x.

8.2.2 Log-Concavity of Density and Left-Side Integral. Let X be a real-valued random variable whose support is an interval (a,b) on the extended real line. Let X have a cumulative distribution function (c.d.f.) F and a density function f, where $f(x) = F'(x)$. For all $x \in (a,b)$, denote the left-side integral of the c.d.f. by $G(x) = \int_a^x F(t)\,dt$. We will show that log-concavity of f implies log-concavity of F, which in turn implies log-concavity of G. Note that we speak of 'left-side integral' here because $F(x)$ and $G(x)$ measure areas lying on the left of x on a histogram.

Theorem 8.7. *(i) If the density function f is log-concave on (a,b), then the c.d.f. F is also log-concave on (a,b); and (ii) if the c.d.f. F is log-concave on (a,b), then the left-side integral G of the c.d.f. is also a log-concave function on (a,b).*

PROOF. (i) Applying Theorem 8.4 to the function F, we find that $F(a)=0$ and F is strictly increasing on $[a,b]$, since F is a c.d.f. with support $[a,b]$, and that if F''/F' is monotone decreasing, then so is F'/F. But $F' = f$ and $F'' = f'$. Thus, if f'/f is monotone decreasing, then F'/F is monotone decreasing. Then part (i) follows from this fact and Remark 1 (§8.2.1). To prove part (ii), apply Theorem 8.4 to the function $G(x)$. Clearly, $G(a) = 0$ and G is a strictly increasing function on $[a,b]$. Thus, if G''/G' is monotone decreasing, then so is G'/G. But $G'' = f$ and $G' = F$. Thus, if f/F is monotone decreasing, then so is G'/G. Part (ii) then follows from this fact and Remark 1. ■

Corollary 8.1. *If the density function f is monotone decreasing, then its c.d.f. F and its left-side integral G are both log-concave.*

PROOF. Since F is a c.d.f., then F must be monotone increasing. Then if f is monotone decreasing, $f(x)/F(x)$ must be monotone decreasing. But $(f(x)/F(x))' = (\log F(x))''$. Thus, if f is monotone decreasing, F must be log-concave. Then G must also be log-concave since, by Theorem 8.7, log-concavity of F implies log-concavity of G. ■

8.2.3 Reliability Theory and Right-Side Integral. Reliability theory deals with the time patterns of survival probability of an existing machine or organism. The length of remaining life for an object is modeled as a random variable X with c.d.f. $F(x)$, and with support (a, b). This theory takes into account the properties of the 'right tail' of the distribution of X. The *reliability function* $\bar{F}$ on X is defined by $\bar{F}(x) = 1 - F(x)$, and the *failure rate* (also known as the *hazard function*) $r(x)$ is defined as $r(x) = f(x)/\bar{F}(x)$. Also, the right-side integral of the reliability function is defined by $R(x) = \int_x^h \bar{F}(t)\,dt$.

Theorem 8.8. *(i) If the density function f is log-concave on (a, b), then the reliability function $\bar{F}(x)$ is also log-concave on (a, b), and (ii) if the reliability function $\bar{F}$ is log-concave on (a, b), then the right-side integral $R(x)$ is a log-concave function on (a, b).*

PROOF. Note that this theorem is a dual of Theorem 8.7. To prove part (i), apply Theorem 8.4 to the function $\bar{F}(x) = 1 - F(x)$. Since F is a c.d.f., we must have $\bar{F}(b) = 0$ and $\bar{F}$ is a monotone decreasing function. Then, if $\bar{F}''(x)/\bar{F}'(x)$ is a decreasing function, so is $\bar{F}'(x)/\bar{F}(x)$. But $\bar{F}'(x) = -f(x)$ and $\bar{F}''(x) = -f'(x)$ for all $x \in [a, b]$. Therefore, $\bar{F}''(x)/\bar{F}'(x) = f'(x)/f(x)$ which is monotone decreasing, and so is $\bar{F}/(x)/\bar{F}(x)$. Hence, if f is log-concave, then $\bar{F}$ must be log-concave.

To prove part (ii), apply Theorem 8.4 to the function $R(x)$. Clearly, $R(b) = 0$ and $R(x)$ is monotone decreasing in x. Using Theorem 8.4, we find that if $R''(x)/R'(x)$ is monotone decreasing, then so is $R'(x)/R(x)$. In this case then $R'(x) = -\bar{F}(x)$ and $R''(x) = f(x)$. Thus, if $R''(x)/R'(x) = -f(x)/F(x)$ is monotone decreasing, then $R'(x)/R(x) = -\bar{F}(x)/R(x)$ must also be monotone decreasing. Then, if $\bar{F}$ is log-concave, so is R log-concave. ■

Since the failure rate is defined by $r(x) = f(x)/\bar{F}(x) = -\bar{F}'(x)/\bar{F}(x)$, we find that the reliability function $\bar{F}$ is log-concave iff the failure rate is monotone increasing in x. This fact and Theorem 8.8 lead to the following results.

Corollary 8.2. *If the density function f is log-concave on (a, b), then the failure rate is monotone increasing on (a, b). If the failure rate is monotone increasing on (a, b), then $R'(x)/R(x)$ is monotone increasing.*

Corollary 8.3. *If the density function f is monotone increasing, then the reliability function $\bar{F}$ is log-concave.*

PROOF. Since $\bar{F}$ is a reliability function, then it must be monotone decreasing. Thus, if f is monotone increasing, the failure rate $f/\bar{F}$ must be monotone increasing. But increasing failure rate is equivalent to a log-concave reliability function. ■

8.2.4 Mean Residual Lifetime. The *mean residual lifetime function* $\text{mrl}(x)$ evaluated at x is the expected length of remaining life for a machine of age x; it is defined as $\text{mrl}(x) = \int_x^b t f(t)\,dt - x$. If this function is monotone increasing, then a machine will age with the passage of time, in the sense that its expected remaining life time will diminish as it gets older.

Theorem 8.9. (Muth [1977]) *Let the random variable X represent the length of life. The sufficient condition for mean residual lifetime* $\text{mrl}(x)$ *to be a monotone decreasing function is either the p.d.f. $f(x)$ is log-concave, or the failure rate $r(x)$ is a monotone increasing function.*

PROOF. Integrating $\text{mrl}(x)$ by parts we get

$$\text{mrl}(x) = \int_x^b \bar{F}(t)\,dt/\bar{F}(x).$$

Since $R(x) = \int_x^h \bar{F}(t)\,dt$, we have $\text{mrl}(x) = R(x)/R'(x)$, so $\text{mrl}(x)$ is a decreasing function iff $R(x)$ is log-convex. By Theorem 8.8(ii), $R(x)$ will be log-convex if $r(x)$ is an increasing function, thereby proving the sufficiency of condition (ii). By Theorem 8.8(i), log-concavity of $f(x)$ implies that $r(x)$ is monotone increasing, which implies that $\text{mrl}(x)$ is monotone decreasing, which proves the sufficiency of condition (i). ■

8.3 Asplund Sum

The Asplund sum is used in adding log-concave functions, and in conjugate functions.

8.3.1 Derivatives of Integrals of Log-Concave Functions. Consider the function $f : \mathbb{R}^n \mapsto \mathbb{R}_+$ of the form $f = e^{-u}$, where $u : \mathbb{R}^n \mapsto (-\infty, \infty]$ is convex, with $u \neq \infty$. We assume that

$$\lim_{|x|\to\infty} u(x) = \infty \iff \lim_{|x|\to\infty} f(x) = 0.$$

Example 8.2. Let $f(x) = e^{-|x|^2/2}$, $u(x) = \dfrac{|x|^2}{2}$. Then

$$\begin{aligned} f(x) = \chi_K(x) &= \begin{cases} 1 & \text{if } x \in K, \\ 0 & \text{if } x \notin K, \end{cases} \\ u(x) = I_K(x) &= \begin{cases} 0 & \text{if } x \in K, \\ \infty & \text{if } x \notin K, \end{cases} \end{aligned} \tag{8.3.1}$$

where $K \subset \mathbb{R}^n$ is a convex set. ■

8.3.2 Adding Log-Concave Functions (Asplund sum). Let f and g be log-concave and $s > 0$. Set

$$\begin{aligned} (f \oplus g)(z) &= \sup_{z=x+y} \{f(x)g(y)\}, \\ (s \odot f)(x) &= sf\Big(\frac{x}{s}\Big). \end{aligned} \tag{8.3.2}$$

These operations preserve log-concavity since both $f \oplus g$ and $s \odot f$ are log-concave. If K and L are convex sets and $s, t > 0$, then

$$(s \odot \chi_K) \oplus (t \odot \chi_L) = \chi_{sK+tL}, \tag{8.3.3}$$

where $sK + tL = \{sx + ty : x \in K, y \in L\}$.

The result (8.3.3) follows from $(s \odot \chi_K)(x) = s\chi_K\big(\frac{x}{s}\big) = s \odot \frac{1}{s} = 1$ if $\frac{x}{s} \in K$, and similarly $(t \odot \chi_L)(x) = 1$ if $\frac{x}{t} \in L$. This may also say that the direct sum of $s \odot \chi_K = s$ if $x \in K$ and $t \odot \chi_L = t$ if $x \in L$ is equal to $\chi_{sK+tL} = s + t$ if x is both K and L (i.e., $x \in K \cap L$).

!!! *The $\oplus$ and $\odot$ notations are borrowed from physics, where, e.g., $\oplus$ means 'direct sum of two vector spaces.' For their definition, see https://www.physicsforums.com/threads/o-plus-symbol.362404/; also /o-dot-symbol. The physical definitions may not be adapted here rigorously, but since they are properly defined, they seem to be useful.*

8.3.3 Asplund Sum and Conjugate Functions. Let $u : \mathbb{R}^n \mapsto (-\infty, \infty]$ be a convex function. The conjugate function u^* of u is defined by

$$u^*(y) = \sup_{x \in \mathbb{R}^n} (x, y) = u(x). \tag{8.3.4}$$

The conjugate function $u^* : \mathbb{R}^n \mapsto (-\infty, \infty]$ is also a convex function.

Note that the $\oplus$ and $\odot$ operations can be defined equivalently as follows: For log-concave functions $f = e^{-u}$, $g = e^{-v}$, and $\alpha, \beta > 0$,

$$(\alpha \odot f) \oplus (\beta \odot g) = (\alpha \odot e^{-u}) \oplus (\beta \odot e^{-v}) = e^{-w}, \tag{8.3.5}$$

where $w = \alpha u + \beta v$. Here $\oplus$ and $\odot$ are linear, in the usual sense, with respect to the conjugates of the exponents (with reverse sign).

8.3.4 Integral Functional. For a log-concave function f that verifies the decay condition at infinity, let

$$I(f) = \int_{\mathbb{R}^n} f(x)\, dx \in [0, \infty). \tag{8.3.6}$$

Note that if $f = \chi_K$ (K a convex set), then $I(f) = V(K)$, where $V(K)$ is the volume of K.

We will now study the limit

$$\delta I(f, g) = \lim_{\varepsilon \to 0+} \frac{I((f \oplus \varepsilon) \odot g) - I(f)}{\varepsilon}, \tag{8.3.7}$$

in particular with respect to existence, and representation formulas in terms of f.

It is known that when f is the Gaussian function, i.e., $f(x) = e^{-|x|/2}$, the limit (8.3.7) is regarded as the *mean width of* g, and denoted by $M(g)$. The question is why the study of the limit (8.3.7) is important. If $f = \chi_K$ and $g = \chi_L$, this limit can be written as

$$\lim_{\varepsilon \to 0+} \frac{V(K + \varepsilon L) - \varepsilon(L)}{\varepsilon} = \int_{S^{n-1}} h_L\, d\sigma_K, \tag{8.3.8}$$

where h_L is the support function of L, and σ_K is the area measure of K. In particular, this limit identifies the area measure of K.

Theorem 8.10. (Prékopa-Leindler inequality) *For every f and g log-concave and $t \in (0, 1)$,*

$$I((t \odot f) \oplus ((1 - t) \odot g)) \geq I(f)^t I(g)^{1-t}. \tag{8.3.9}$$

Equality holds iff there exists x_0 such that $g(x) = f(x - x_0)$ for every $x \in \mathbb{R}^n$.

This means that the functional $\log(I)$ is concave in the class of log-concave functions, equipped with the linear structure given by the operations $\oplus$ and $\odot$ defined by (8.3.2). Note that for $f = \chi_K$ and $g = \chi_L$ (K and L convex sets), we get the Brunn-Minkowski inequality in its multiplicative form:

$$V(tK + (1 - t)L) \geq V(K)^t V(L)^{1-t}, \tag{8.3.10}$$

which is equivalent to the classical inequality.

Theorem 8.11. *Let f and g be log-concave, and assume that $I(f) > 0$. Then the limit (8.3.7) denoted by $\delta I(f, g)$ exists and belongs to $(-\infty, \infty]$.*

PROOF is based on the Prékopka-Leindler inequality (Theorem 8.10):

$$\varepsilon \to \log(I(f \oplus \varepsilon) \odot g)) \text{ is concave. } \blacksquare$$

Note that (i) the assumption $I(f) > 0$ can be removed in dimension $n = 1$.

(ii) Choose $f(x) = e^{-|x|^2/2}$ and $g = e^{-|x|}$ in dimension $n = 1$; then $\delta I(f, g) = +\infty$.

(iii) For a suitable choice of f and g, $\delta I(f, g) < 0$, in contrast with the idea that $\delta I(f, g)$ is a *mixed integral* of f and g (mixed volumes are always nonnegative).

(iv) Choose $g = f$; then we have the formula $\delta I(f, f) = nI(f) - E(f)$ (no homogeneity here!), where

$$E(f) = -\int_{\mathbb{R}^n} f \log(f)\, dx \tag{8.3.11}$$

is the entropy of f.

8.3.5 Area Measure of Log-Concave Functions. Comparing the formulas

$$\lim_{\varepsilon \to 0+} \frac{V(K + \varepsilon L) - V(L)}{\varepsilon} = \int_{\mathbb{S}^{n-1}} h_L \, d\sigma_K, \tag{8.3.12}$$

and

$$\lim_{\varepsilon \to 0+} \frac{I((e^{-u} \oplus \varepsilon) \odot e^{-v}) - I(e^{-v})}{\varepsilon} = \int_{\mathbb{R}^{n-1}} v^* \, d\mu_f, \tag{8.3.13}$$

seems to suggest that $d\mu_f$ is a sort of *area measure* of f. Note that μ_f determines f uniquely, i.e.,

$$\mu_f = \mu_g \implies \text{there exists an } x_0 \in \mathbb{R}^n \text{ such that } g(x) = f(x - x_0) \text{ for all } x \in \mathbb{R}^n.$$

The proof uses the characterization of equality in the Prékopa-Leindler inequality.

8.4 Log-Concavity of Nonnegative Sequences

Hoggar's theorem states that the sum of two independent discrete-valued log-concave random variables is itself log-concave. We will determine conditions under which this result still holds for dependent variables. Log-concavity of the Stirling numbers of the second kind and of the Eulerian numbers is established.

The property of log-concavity for nonnegative sequences is defined as follows: A sequence $\{u(i), i \geq 0\}$ is log-concave if, for all $i \geq 1$,

$$u(i)^2 \geq u(i-1)u(i+1). \tag{8.4.1}$$

Eq (8.4.1) is also known as the *quadratic Newton inequality*. Log-concavity is used in combinatorics, algebra, geometry, computer science, and econometrics. In probability and statistics it is related to the notion of negative association of random variables.

Definition 8.2. A random variable V taking values in $\mathbb{Z}_+$ is log-concave if the probability mass function $P_V(i) = \mathfrak{P}(V = i)$ forms a log-concave sequence, that is, V is log-concave iff for all i

$$P_V(i)^2 \geq P_V(i-1)P_V(i+1). \tag{8.4.2}$$

Example 8.3. (a) Let Geom(p) denote the geometric distribution with the probability mass function $P_X(i) = (1-p)p^i$ for $i \in \mathbb{Z}_+$. For each p, these random variables represent the 'edge case' of the log-concavity property, whenever Eq (8.4.2) is an identity for all i.

(b) The Poisson distribution $P(\lambda)$ is log-concave for any $\lambda \geq 0$.

(c) Any binomial distribution is log-concave. ■

Definition 8.2 can be generalized as follows: Given sequences $P_V(v)$ and $P_{V+W}(x)$, there exists a two-dimensional array of coefficients $P_{V|W}(x)$ such that

$$P_{V+W}(x) = \sum P_V(v)\, P_{V+W}(x-v|v). \tag{8.4.3}$$

In fact, the sequence $P_{V|W}$ acts like conditional probability without requiring the sequence to sum to 1.

Example 8.4. For some $p \in (0,1)$ and $\alpha \in (0,1)$, define the joint distribution of V and W by

$$\mathfrak{P}(V = i, W = j) = \binom{i+j}{i}(1-p)p^{i+j}\alpha^i(1-\alpha)^j, \quad \text{for } i,j \geq 0. \tag{8.4.4}$$

Using the identities

$$\sum_{i=0}^{k}\binom{k}{i}\alpha^i(1-\alpha)^{k-i} = 1, \quad \sum_{j=0}^{\infty}\binom{i+j}{i}t^j = (1-t)-i-1, \quad \text{for } 0 \leq t < 1,$$

we find that $V + W$ is Geom(p), V is Geom($\alpha p/(\alpha p + (1-p)$)), and W is *Geom*($p/((1-\alpha)p + (1-p))$). The conditional probabilities are negative binomials with

$$P_{W|V}(j\,|\,i) = \mathfrak{P}(W = j\,|\,V = i) = \binom{i+j}{i}(1-p+\alpha p)^{i+1}(p(1-\alpha))^j. \blacksquare \tag{8.4.5}$$

Definition 8.3. Given coefficients $P_{W|V}$ and fixed i, define

$$a_{r,s}^{(i)} = P_{W|V}(i-r|r)P_{W|V}(i-s|s) - P_{W|V}(i-r-1|r)P_{W|V}(i-s+1|s). \tag{8.4.6}$$

Then we have

Condition B. For the quantities $a_{r,s}^{(i)}$ defined by (8.4.6), the following two conditions must hold for all $0 < t \le m \le i$:

$$\text{(a)} \sum_{k=-t}^{t} a_{m+k,m-k}^{(i)} \ge 0, \quad \text{and (b)} \sum_{k=-t-1}^{t} a_{m+k+1,m-k}^{(i)} \ge 0. \tag{8.4.7}$$

Example 8.5. (Continued) Note that Condition B holds in Example 8.4, since for any given i, the quantity $a_{r,s}^{(i)}$ is proportional to $\binom{i}{r}\binom{i}{s} - \binom{i-1}{r}\binom{i+1}{s}$. Thus, in (8.4.7a), as t increases, the term $a_{m+k,m-k}^{(i)}$ is proportional to

$$2\binom{i}{m+t}\binom{i}{m-t} - \binom{i-1}{m+t}\binom{i+1}{m-t} - \binom{i+1}{m+t}\binom{i-1}{m-t} \begin{cases} > 0 \text{ for } t \le T, \\ < 0 \text{ for } t > T, \end{cases}$$

for some value of T. Thus, the partial sums $\sum\limits_{k=-t}^{t} a_{m+k,m-k}^{(i)}$ form a sequence which increases for $t \le T$ and decreases thereafter $(t < T)$. Using the identity $\sum_j \binom{a}{j}\binom{b}{r-j} = \binom{a+b}{r}$, we find that the sum $\sum\limits_{k=-m}^{m} a_{m+k,m-k}^{(i)} = 0$. Hence the sequence of partial sums is nonnegative for any t. Using this method, a similar result follows for the sums in (8.4.7b). ■

Lemma 8.1. *Fix $k \ge m$ and suppose that $\{c_j\}$ is a sequence such that $C_n := \sum\limits_{j=0}^{n} c_j \ge 0$ for all $0 \le n \le m$. Then for any log-concave sequence p, and for any $0 \le i \le m$,*

$$\sum_{j=0}^{i} p(k+j)p(m-j)c_j \ge 0. \tag{8.4.8}$$

Proof. Apply Abel's summation formula (summation by parts) which yields

$$\begin{aligned} \sum_{j=0}^{i} p(k+j)p(m-j)c_j &= \sum_{j=0}^{i} p(k+j)p(m-j)(C_j - C_{j-1}) \\ &= \sum_{j=0}^{i} [p(k+j)p(m-j) - p(k+j+1)p(m-j-1)]C_j \\ &\qquad + C_i p(k+i+1)p(m-i-1), \end{aligned} \tag{8.4.9}$$

where $C_{-1} = 0$. The log-concavity of p implies that $p(k+j)p(m-j) \ge p(k+j+1)p(m-j-1)$ for $j \ge 0$. Thus, since each $C_j \ge 0$, the result follows. ■

Theorem 8.12 (Hoggar [1974]) *If V and W are independent log-concave random variables, then their sum $V+W$ is also log-concave. Equivalently, the convolution of any two log-concave sequences is log-concave.*

The proof of Hoggar's theorem is a special case of the following theorem.

Theorem 8.13. (Johnson and Goldschmidt [2005]) *If V is a log-concave random variable, and random variables V and W satisfy Condition B, then the sum $V+W$ is also log-concave.*

PROOF. For any i, the sequence P_{V+W} defined by (8.4.3) satisfies the log-concavity definition, i.e.,

$$\begin{aligned} P_{V+W}(i)^2 &- P_{V+W}(i-1)P_{V+W}(i+1) \\ &= \sum_{j=0}^{i}\sum_{k=0}^{i+1} P_V(j)P_V(k)\big[P_{W|V}(i-j|j)P_{W|V}(i-k|k) \\ &\qquad\qquad -P_{W|V}(i-j-1|j)P_{W|V}(i-k+1|k)\big] \\ &= \sum_{j=0}^{i}\sum_{k=0}^{i+1} P_V(j)P_V(k)\, a_{j,k}^{(i)} \\ &\equiv \sum_{j=0}^{i}(S_1+S_2+S_3), \end{aligned} \tag{8.4.10}$$

where the sums S_1, S_2, S_3 correspond to the following three regions: S_1 to the region $\{j \ge k\}$; S_2 to the region $\{j = k-1\}$, and S_3 to the $\{j \le k-2\}$. Also, in region S_3 we use new coordinates $(k-1, j+1)$ for $j \le k-2$. Then the right-hand side of Eq (8.4.10) can be rewritten as

$$\begin{aligned} \sum_{j=0}^{i}\sum_{k=0}^{i+1} P_V(j)P_V(k)\, a_{j,k}^{(i)} &= \sum_{j=0}^{i} P_V(j)P_V(j+1)a_{j,j+1}^{(i)} \\ &+ \sum_{i\ge j\ge k}\Big(P_V(j)P_V(k)a_{j,k}^{(i)} + P_V(j+1)P_V(k-1)a_{k-1,j+1}^{(i)}\Big) \end{aligned} \tag{8.4.11}$$

where

$$\begin{aligned} S_1 = \sum_{k=0}^{m} \big[&P_V(m+k)P_V(m-k)a_{m+k,m-k}^{(i)} \\ &+ P_V(m+k+1)P_V(m-k)a_{m-k,m+k+1}^{(i)}\big], \end{aligned} \tag{8.4.12}$$

$$S_2 = \sum_{k=0}^{m} P_V(m)P_V(m+1)a_{m,m+1}^{(i)}, \tag{8.4.13}$$

$$S_3 = \sum_{i\ge j\ge k}\Big(P_V(j)P_V(k)a_{j,k}^{(i)} + P_V(j+1)P_V(k-1)a_{k-1,j+1}^{(i)}\Big). \tag{8.4.14}$$

The sum S_1 in (8.4.12) is further split according to whether $r = j + k$ is even or odd: If r is even, then $m = r/2$; if r is odd, then $m = (r-1)/2$. We will consider the case when r is even. In this case

$$\begin{aligned}
S_1 &= \sum_{k=0}^{m} \big[P_V(m+k)P_V(m-k)a^{(i)}_{m+k,m-k} \\
&\qquad + P_V(m+k+1P_V(m-k-1)a^{(i)}_{m-k-1,m+k+1}\big] \\
&= \sum_{k=0}^{m} \big[P_V(m+k)P_V(m-k)a^{(i)}_{m+k,m-k} \\
&\qquad + \sum_{k=0}^{m-1} P_V(m+k+1P_V(m-k-1)a^{(i)}_{m-k-1,m+k+1} \\
&= P_V(m)^2 a^{(i)}_{m,m} + \sum_{k=1}^{m} P_V(m+k)P_V(m-k)\{a^{(i)}_{m+k,m-k} + a^{(i)}_{m-k,m+k}\} \\
&= \sum_{k=0}^{m} P_V(m+k)P_V(m-k)c_k,
\end{aligned} \tag{8.4.15}$$

where $c_0 = a^{(i)}_{m,m}$ and $c_k = a^{(i)}_{m+k,m-k}$ for $1 \le k \le m$. Then condition B(a) tells us that $\sum_{k=0}^{t} c_k \ge 0$ for all $0 \le t \le m$, and so by Lemma 8.1 with $k = m, i = m$, Eq (8.4.12) is positive. In the same way we can show that the sum of the second and third terms in (8.4.11) equals

$$\sum_{k=0}^{m} P_V(m+k+1)P_V(m-k)\, d_k, \tag{8.4.16}$$

where $d_k = a^{(i)}_{m+k+1,m-k} + a^{(i)}_{m-k,m+k+1}$ for $0 \le k \le m$. Then condition B(b) tells us that $\sum_{k=0}^{t} d_k \ge 0$ for all $0 \le t \le m$, and so, by Lemma 8.1 with $k = m+1, i = m$, Eq (8.4.13) is positive. Hence, $P_{V+W}(i)^2 - P_{V+W}(i-1)P_{V+W}(i+1) \ge 0$ for all i. Other cases are similarly resolved. ∎

Thus, we have established that the sum of any two independent and identically distributed geometric random variables (both on the edge case) is a negative binomial distribution (still, log-concave, but no longer the edge case).

Next, the quantities $a^{(i)}_{j,k}$ have the following properties for independent random variables: V and W, (i) $a^{(i)}_{j,j+1} \equiv 0$ for all j; (ii) $a^{(i)}_{k-1,j+1} = -a^{(i)}_{j,k}$; and (iii) if W is log-concave, then $a^{(i)}_{j,k} \ge 0$ for $j \ge k$.

We fix i, define c_j as in Lemma 8.1, and define d_j as a sequence such that $\sum_{j=0}^{t} d_j \ge 0$ for all $0 \le t \le m$, and if V and W are independent and log-concave, then

$$
\begin{aligned}
& a^{(i)}_{m,m} + \sum_{j=1}^{t} a^{(i)}_{m+j,m-j} - \sum_{j=1}^{t} a^{(i)}_{m+j-1,m-j+1} \\
& = a^{(i)}_{m,m} + \sum_{j=1}^{t} a^{(i)}_{m+j,m-j} - \sum_{j=0}^{t-1} a^{(i)}_{m+j,m-j} = a^{(i)}_{m+t,m-t} \geq 0,
\end{aligned} \tag{8.4.17}
$$

and

$$
\begin{aligned}
& a^{(i)}_{m+j+1,m-j} - \sum_{j=0}^{t} a^{(i)}_{m+j,m-j+1} \\
& = a^{(i)}_{m,m} + \sum_{j=1}^{t} a^{(i)}_{m+j,m-j} - \sum_{j=1}^{t-1} a^{(i)}_{m+j,m-j} = a^{(i)}_{m+t+1,m-t} - a^{(i)}_{m,m-1} \\
& = a^{(i)}_{m+t+1,m-t} \geq 0.
\end{aligned} \tag{8.4.18}
$$

Hence, Condition B holds for independent and log-concave V and W.

8.5 Log-Concave Distributions

The following probability distributions are log-concave: (i) Normal distribution and multivariate normal distributions; (ii) exponential distribution; (iii) uniform distribution over any convex set; (iv) logistic distribution; (v) extreme value distribution; (vi) Laplace distribution; (vii) chi-distribution; (viii) Wishart distribution, where $n \geq p+1$ (Prékopa [1971]); (ix) Dirichlet distribution, where all parameters are ≥ 1 (Prékopa [1971]); (x) Gamma distribution if the shape parameter is ≥ 1; (xi) chi-distribution if the number of degrees of freedom is ≥ 2; (xii) beta distribution if both shape parameters are ≥ 1; and (xiii) Weibull distribution if the shape parameter is ≥ 1.

All the above parameter restrictions are based on the fact that the exponent of a nonnegative quantity must be nonnegative in order that the function remains log-concave.

The following distributions are *not* log-concave for all parameters: (i) Student's t-distribution; (ii) Cauchy distribution; (iii) Pareto distribution; (iv) log-normal distribution; and (v) F-distribution.

Although the cumulative distribution function (c.d.f.) of all log-concave distributions is log-concave, the following non-log-concave distributions also have log-concave c.d.f.s: (i) log-normal distribution; (ii) Patero distribution; (iii) Weibull distribution when the shape parameter is < 1; and (iv) gamma distribution when the shape parameter is < 1.

Some useful properties of log-concave distributions are as follows:

(i) If the density (p.d.f.) is log-concave, so is its c.d.f.

(ii) If a multivariate density is log-concave, so is the marginal density over any subset of variables.

(iii) The sum of two independent log-concave random variables is log-concave, since the convolution of two log-concave functions is log-concave.

(iv) The product of two log-concave functions is log-concave, which means that the joint densities formed by multiplying two probability densities (e.g., normal gamma distribution, which always has a shape parameter ≥ 1) are log-concave. This property is extremely useful in general-purpose Gibbs sampling programs.

Example 8.6. In manufacturing projects, let $x_{\rm m} = x + v$, where $x_{\rm m}$ denotes the manufacturing yield, $x \in \mathbb{R}^n$ the nominal value of design parameters, $v \in \mathbb{R}^n$ the manufacturing errors with zero random value, and $S \subseteq \mathbb{R}^n$, the specs, i.e., acceptable values of $x_{\rm m}$. Then the *yield* $Y(x)$ is the probability that $x + v$ is in S, i.e., $Y(x) = \mathfrak{p}(x + v)$ is log-concave, if S is a convex set, and if the probability density of v is log-concave. ■

Example 8.7. Let $S = \{y \in \mathbb{R}^2 \,|\, y_1 \geq 1, y_2 \geq 1\}$, and let v_1, v_2 be independent, normal with $\sigma = 1$. Then, with $x + v \in S$, the yield is

$$Y(x) = \mathfrak{p}(x + v) = \frac{1}{2\pi}\Big(\int_{1-x_1}^{\infty} e^{-t^2/2}\,dt\Big)\Big(\int_{1-x_2}^{\infty} e^{-t^2/2}\,dt\Big), \tag{8.5.1}$$

where each factor on the right side of (8.5.1) is a normal distribution truncated at $1 - x_{1,2}$. Next, the maximum yield vs. cost is evaluated as follows: If the manufacturing cost $c = x_1 + 2x_2$, then the maximum yield for a given cost is given by

$$Y_{\rm max}(c) = \sup_{\substack{x_1+2x_2=c\\ x_1,x_2\geq 0}} Y(x), \tag{8.5.2}$$

where $Y(x)$ is log-concave, and

$$-\log Y_{\rm max}(c) = \inf_{\substack{x_1+2x_2=c\\ x_1,x_2\geq 0}} -\log Y(x_1, x_2). \tag{8.5.3}$$

The relation between the cost c and the maximum yield $Y_{\rm max}$ is presented in Figure 8.2, where the cost rises as the yield increases. ■

Let a density function f be defined as

$$f(x) \equiv f_\phi(x) = \exp\{\phi(x)\} = \exp\{-(-\phi(x))\}, \tag{8.5.4}$$

where ϕ is concave (and so $-\phi$ is convex). Let us call the class of all such densities f on $\mathbb{R}$ as the *class of log-concave densities* and denote this class by $\mathcal{P}_0 \equiv \mathcal{P}_{\text{log-concave}}$. The function $f \in \mathcal{P}_0$ is log-concave iff

(i) $\log f(tx + (1-t)y) \geq t \log f(x) + (1-t) \log f(y)$ for all $t \in [0,1]$ and for all $x, y \in \mathbb{R}$;

(ii) $f(tx + (1-t)y) \geq f(x)^t \cdot f(y)^{1-t}$;

(iii) $f((x+y)/2) \geq \sqrt{f(x)f(y)}$ (for $t = \frac{1}{2}$, and assuming f is measurable);

(iv) $f((x+y))^2 \geq f(x)f(y)$.

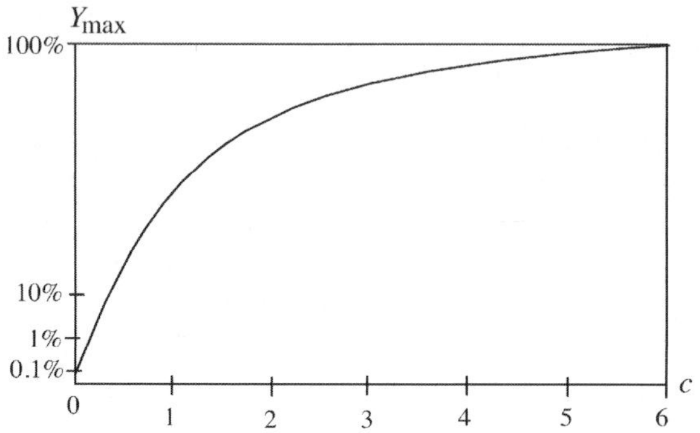

Figure 8.1 Cost vs. maximum yield.

Example 8.8. 1. Standard normal distribution: $f(x) = \dfrac{1}{\sqrt{2\pi}} e^{-x^2/2}$; then $-\log f(x) = \frac{1}{2}x^2 + \log\sqrt{2\pi}$, and $(-\log f(x))'' = 1$.

2. Laplace distribution: $f(x) = \frac{1}{2}e^{-|x|}$; then $-\log f(x) = |x| + \log 2$, $(-\log f(x))'' = 0$ for all $x \neq 0$.

3. Logistic distribution: $f(x) = \dfrac{e^x}{(1+e^x)^2}$; then $-log f(x) = -x + 2\log(1 + e^x)$, $(-\log f(x))'' = \dfrac{e^x}{(1+e^x)^2} = f(x)$. ■

8.6 Exercises

8.1. Prove that if P_1 and P_2 are log-concave probability measures, then the product measure $P_1 \times P_2$ is a log-concave probability measure.

PROOF. If a probability measure P in $\mathbb{R}^n$ assigns zero mass to every hyperplane in $\mathbb{R}^n$, then by (8.1.1), log-concavity of P holds if $P(t\mathbf{x} + (1-t)\mathbf{y}) \geq P(\mathbf{x})^t P(\mathbf{y})^{1-t}$, $0 < t < 1$, for all $\mathbf{x}, \mathbf{y} \in \mathbb{R}^n$. Let A, B denote two rectangular hyperplanes with sides parallel to the coordinate axes such that all $\mathbf{x} \in A$ and all $\mathbf{y} \in B$. Then, by the above inequality for these hyperplanes, we have $P(tA + (1-t)B) \geq P(A)^t P(B)^{1-t}$ for $0 < t < 1$. A similar argument applies to the product $P_1 \times P_2$.

8.2. Show that Condition B holds in Example 8.5. SOLUTION. For any i, $a_{r,s}^{(i)}$ is proportional to $\binom{i}{r}\binom{i}{s} - \binom{i-1}{r}\binom{i+1}{s}$, i.e., for part (a) of Condition B, for any given i, the increment term $a_{m+t,m-t}^{(i)} + a_{m-t,m+t}^{(i)}$ is proportional to $2\binom{i}{m+t}\binom{i}{m-t} - \binom{i-1}{m+t}\binom{i+1}{m-t} - \binom{i+1}{m+t}\binom{i-1}{m-t}$, which is positive for $t \leq T$ and negative for $t > T$ for some value of $T > 0$. Hence the partial sums $\sum\limits_{k=-t}^{t} a_{m+k,m-k}^{(i)}$ form a sequence which increases for $t \leq T$ and decreases thereafter. Then, using the identity $\sum_j \binom{a}{j}\binom{b}{r-j} = \binom{a+b}{r}$, we find that $\sum\limits_{k=-m}^{m} a_{m+k,m-k}^{(i)} = 0$, and thus the sequence of partial sums must be nonnegative for any t. A similar argument holds for part (b) of Condition B.

8.3. Show that the density function (p.d.f.) of which of the following probability distributions is log-concave, log-convex, or log-linear: (i) uniform distribution; (ii) standard normal distribution; (iii) logistic distribution; (iv) extreme-value distribution; (v) exponential distribution; (vi) Weibull distribution; (vii) power function distribution; (viii) Gamma distribution; (ix) chi-square distribution; (x) chi-distribution; (xi) beta distribution; and (xii) Student's t-distribution.

ANSWER. (i) Uniform distribution, defined on the interval $[0, 1]$, has density $f(x) = 1$, which is (weakly) log-concave.

(ii) Standard normal probability distribution has probability density $f(x) = \dfrac{1}{\sqrt{2\pi}} e^{-x^2/2}$, whence $(\log f(x))' = -x$ and $(\log f(x))'' = -1 < 0$. Thus, the density is log-concave.

(iii) Logistic distribution has density $f(x) = \dfrac{e^{-x}}{(1+e^{-x})^2}$, whence $(\log f(x))' = -1+2(1-F(x))$, and $(\log f(x))'' = -2f(x) < 0$; hence, $f(x)$ is log-concave.

(iv) Extreme-value distribution has density function $f(x) = \exp\{-e^{-x}\}$, giving $(\log f(x))'' = -e^{-x} < 0$; hence $f(x)$ is log-concave; (v) exponential distribution has density $f(x) = \lambda e^{-\lambda x}$, with $(\log f(x))'' = 0$, and $f'(x) < 0$ for all $x \in [0, \infty]$, and hence $f(x)$ is log-concave.

(vi) Weibull distribution with parameter c has density function $f(x) = cx^{c-1} e^{-x^c}$, $x \in (0, \infty)$. Also, $(\log f(x))'' = (1-c)x^{-2}(1+cx^c) \begin{cases} < 0 \text{ for } c > 1, \\ = 0 \text{ for } c = 1, \\ > 0 \text{ for } c < 1. \end{cases}$
Thus, $f(x)$ is (strictly) log-concave if $0 < c < 1$, log-linear if $c = 1$, and it is log-convex if $c > 1$.

(vii) Power function distribution has density function $f(x) = cx^{c-1} e^{-x^c}$,

$x \in (0, \infty)$. Also, $(\log f(x))'' = (1-c)x^{-2}(1+cx^c)\begin{cases} < 0 \text{ for } c > 1, \\ = 0 \text{ for } c = 1, \\ > 0 \text{ for } c < 1. \end{cases}$ Thus, the density function is (strictly) log-concave if $0 < c < 1$, log-linear if $c = 1$, and log-convex if $c > 1$.

(viii) Gamma distribution has density function $f(x) = \dfrac{x^{m-1}\theta^m\, e^{-x\theta}}{\Gamma(m)}$, $x \in (0, \infty), \theta > 0$ and $m > 0$. Then $(\log f(x))'' = \dfrac{1-m}{x^2}$. Thus, the density function is strictly log-concave for $m > 1$, but strictly log-convex for $m < 1$.

(viii) Chi-square distribution with n degrees of freedom is a gamma distribution with $\theta = 2$ and $m = n/2$. Since the sum of the squares of n independent standard normal random variables has a chi-square distribution with n degrees of freedom, and since the gamma distribution has a log-concave density function for $m \geq 1$, so the sum of the squares of two or more independent standard normal variables has a log-concave density function.

(ix) chi-distribution has density function $f(x) = \dfrac{x^{(n/2)-1}\, e^{-n/2} x^2}{2^{n/2}\Gamma(n/2)}, x > 0$, where n is a positive integer. Since $(\log f(x))'' = -\dfrac{n-1}{x^2} - n < 0$, the density function is log-concave.

(x) beta-distribution has density function $f(x) = \dfrac{x^{a-1}(1-x)^{b-1}}{B(a,b)}, x \in (0,1), a, b > 0$. Since $(\log f(x))'' = \dfrac{1-a}{x} + \dfrac{1-b}{x}$, the density function is log-concave if $a \geq 1$ and $b \geq 1$, and log-convex if $a < 1$ and $b < 1$. If $a < 1$ and $b > 1$, or if $a > 1$ and $b < 1$, then the density function is neither log-convex nor log-concave on $(0, 1)$.

(xi) Student's t-distribution is defined on the entire real line with density function

$$f(x) = \frac{(1 + x^2/n)^{-n+1/2}}{\sqrt{n}\, B(1/2, n/2)},$$

where $B(a, b)$ is the incomplete beta function and n is the number of degrees of freedom. Since $(\log f(x))'' = -(n+1)\dfrac{n - x^2}{(n+x^2)^2}$, the density function is log-concave on the central interval $[-\sqrt{n}, \sqrt{n}]$, and therefore, it is log-concave on this interval but log-convex on each of the outer intervals $[\infty, -\sqrt{n}]$ and $[\sqrt{n}, \infty]$. Thus, although this distribution is itself not log-concave, a truncate one on the interval $[-\sqrt{n}, \sqrt{n}]$ is log-concave.

9

Quadratic Programming

Quadratic programming is used to optimize functions $f(\mathbf{x})$, $\mathbf{x} \in \mathbb{R}^n$, of the form $\frac{1}{n}\mathbf{x}^T\mathbf{Q}\mathbf{x} + \mathbf{c}^T\mathbf{x}$ subject to equality and inequality constraints, where $n \geq 2$ is an integer, usually very large. We will discuss the iteration methods to solve such optimization problems.

9.1 Quadratic Programming

Quadratic programs (QP) are minimization programs that have a convex (quadratic) objective and a convex constraint set formed by linear constraints. The QP primal problem is defined as

$$\text{(P)}: \quad \text{Minimize } \tfrac{1}{2}\mathbf{x}^T\mathbf{Q}\mathbf{x} + \mathbf{c}^T\mathbf{x} \text{ subject to } \mathbf{A}\mathbf{x} \geq \mathbf{b}, \tag{9.1.1}$$

which is equivalent to $\mathbf{b}-\mathbf{A}\mathbf{x} \leq \mathbf{0}$, with any nonnegativity restrictions included in $\mathbf{A}\mathbf{x} \geq \mathbf{b}$, where $\mathbf{x} \in \mathbb{R}^n$; $\mathbf{Q}$ is an $n \times n$ real, symmetric and positive definite matrix; $\mathbf{A}$ is an $m \times n$ real matrix; and $\mathbf{b}, \mathbf{c} \in \mathbb{R}^m$ are each a column vector. Thus, we have a convex objective and linear constraints.

The Lagrangian is

$$L(\mathbf{x}, \boldsymbol{\lambda}) = f(\mathbf{x}) + \sum_{j=1}^{m} \lambda_j g_j(\mathbf{x}) = \tfrac{1}{2}\mathbf{x}^T\mathbf{Q}\mathbf{x} + \mathbf{c}^T\mathbf{x} + \lambda^T(\mathbf{b} - \mathbf{A}\mathbf{x}). \tag{9.1.2}$$

The dual problem is defined as $\max\limits_{\boldsymbol{\lambda}\geq 0} \big\{ \min\limits_{\mathbf{x}} L(\mathbf{x}, \boldsymbol{\lambda})\big\}$, that is,

$$\max_{\boldsymbol{\lambda}\geq 0} L(\mathbf{x}, \boldsymbol{\lambda}) \text{ subject to } \frac{\partial L(\mathbf{x}, \boldsymbol{\lambda})}{\partial \mathbf{x}} = \mathbf{0}. \tag{9.1.3}$$

This yields the dual problem

$$\text{(D)} \quad \text{Maximize } L(\mathbf{x}, \boldsymbol{\lambda}) = \tfrac{1}{2}\mathbf{x}^T\mathbf{Q}\mathbf{x} + \mathbf{c}^T\mathbf{x} + \boldsymbol{\lambda}^T(\mathbf{b} - \mathbf{A}\mathbf{x}), \tag{9.1.4}$$

subject to $L_{\mathbf{x}}(\mathbf{x}, \boldsymbol{\lambda}) = \mathbf{Q}\mathbf{x} + \mathbf{c} - \mathbf{A}^T\boldsymbol{\lambda} = \mathbf{0}$, $\boldsymbol{\lambda} \geq \mathbf{0}$. This dual constraint implies that $\mathbf{x}^T\left[\mathbf{Q}\mathbf{x} + \mathbf{c} - \mathbf{A}^T\boldsymbol{\lambda}\right] = \mathbf{x}^T\mathbf{0}$, where $L_{\mathbf{x}}$ denotes the first-order partial derivatives L with respect to $\mathbf{x}$, i.e.,

$$\mathbf{x}^T\mathbf{Q}\mathbf{x} + \mathbf{x}^T\mathbf{c} - \mathbf{x}^T\left(\mathbf{A}^T\boldsymbol{\lambda}\right) = \mathbf{x}^T\mathbf{Q}\mathbf{x} + \mathbf{x}^T\mathbf{c} - \boldsymbol{\lambda}^T\mathbf{A}\mathbf{x} = \mathbf{0}. \tag{9.1.5}$$

Using (9.1.4), the dual objective function can be rearranged as

$$\boldsymbol{\lambda}^T\mathbf{b} - \tfrac{1}{2}\mathbf{x}^T\mathbf{Q}\mathbf{x} + \left(\mathbf{x}^T\mathbf{Q}\mathbf{x} + \mathbf{x}^T\mathbf{c} - \boldsymbol{\lambda}^T\mathbf{A}\mathbf{x}\right) = \boldsymbol{\lambda}^T\mathbf{b} - \tfrac{1}{2}\mathbf{x}^T\mathbf{Q}\mathbf{x}. \tag{9.1.6}$$

Thus, the dual problem reduces to

$$\text{Minimize } \boldsymbol{\lambda}^T\mathbf{b} - \tfrac{1}{2}\mathbf{x}^T\mathbf{Q}\mathbf{x} \;\text{ subject to }\; \mathbf{Q}\mathbf{x} + \mathbf{c} - \mathbf{A}^T\boldsymbol{\lambda} = \mathbf{0},\ \boldsymbol{\lambda} \geq \mathbf{0}. \tag{9.1.7}$$

There are three particular cases:
CASE 1. If $\mathbf{Q} = \mathbf{0}$, then

$$\begin{aligned} &\text{(P) is} \quad \min\left\{\mathbf{c}^T\mathbf{x}\right\} \quad \text{subject to} \quad \mathbf{A}\mathbf{x} \geq \mathbf{b}, \\ &\text{(D) is} \quad \max\left\{\lambda^T\mathbf{b}\right\} \quad \text{subject to} \quad \mathbf{A}^T\lambda = \mathbf{c}, \end{aligned}$$

which is the regular linear programming pair.
CASE 2. If $\mathbf{Q} \neq \mathbf{0}$, then

(P) has n variables, and m linear inequality constraints, and

(D) has $(m+n)$ variables, n equality constraints, and m nonnegativity constraints.

CASE 3. If $\mathbf{Q}$ is nonsingular (i.e., $\mathbf{Q}^{-1}$ exists), then from the dual constraints we have

$$\mathbf{Q}\mathbf{x} + \mathbf{c} - \mathbf{A}^T\boldsymbol{\lambda} = \mathbf{0} \implies \mathbf{x} + \mathbf{Q}^{-1}\mathbf{c} - \mathbf{Q}^{-1}\mathbf{A}^T\boldsymbol{\lambda} = \mathbf{0},$$

i.e.,

$$\mathbf{x} = \mathbf{Q}^{-1}[\mathbf{A}^T\boldsymbol{\lambda} - \mathbf{c}]. \tag{9.1.8}$$

Thus, we may eliminate $\mathbf{x}$ altogether from the dual problem.

Given any two matrices $\mathbf{U}$ and $\mathbf{V}$, we will use the following four known results: (i) $[\mathbf{U}\mathbf{V}]^T = \mathbf{V}^T\mathbf{U}^T$; (ii) $[\mathbf{U}^T]^T = \mathbf{U}$; (iii) $\mathbf{U}^T\mathbf{V} = \mathbf{V}^T\mathbf{U}$ (assuming compatibility); and (iv) $\mathbf{Q}$ and $\mathbf{Q}^{-1}$ be symmetric and identical to their transposes. Then, substituting the value of $\mathbf{x}$ from (9.1.6) into the dual objective function and rearranging, we get

$$\boldsymbol{\lambda}^T\mathbf{b} - \tfrac{1}{2}\mathbf{x}^T\mathbf{Q}\mathbf{x} = \boldsymbol{\lambda}^T\mathbf{b} - \tfrac{1}{2}[\mathbf{Q}^{-1}(\mathbf{A}^T\boldsymbol{\lambda} - \mathbf{c}]^T \cdot \mathbf{Q} \cdot [\mathbf{Q}^{-1}(\mathbf{A}^T\boldsymbol{\lambda} - \mathbf{c})]$$

$$
\begin{aligned}
&= \mathbf{b}^T\boldsymbol{\lambda} - \tfrac{1}{2}\big(\mathbf{A}^T\boldsymbol{\lambda} - \mathbf{c}\big)^T \cdot \mathbf{Q}^{-1}\cdot\mathbf{Q}\cdot\mathbf{Q}^{-1}\cdot\big(\mathbf{A}^T\boldsymbol{\lambda} - \mathbf{c}\big) \quad \text{by (i) and (iv)}\\
&= \mathbf{b}^T\boldsymbol{\lambda} - \tfrac{1}{2}[(\mathbf{A}^T\boldsymbol{\lambda})^T - \mathbf{c}^T)]\cdot\mathbf{Q}^{-1}\cdot\big(\mathbf{A}^T\boldsymbol{\lambda} - \mathbf{c}\big)\\
&= \mathbf{b}^T\boldsymbol{\lambda} - \tfrac{1}{2}[\boldsymbol{\lambda}^T\mathbf{A}\mathbf{Q}^{-1} - \mathbf{c}^T\mathbf{Q}^{-1}]\cdot(\mathbf{A}^T\boldsymbol{\lambda} - \mathbf{c}) \text{ by (i) and (ii)}\\
&= \mathbf{b}^T\boldsymbol{\lambda} - \tfrac{1}{2}[\boldsymbol{\lambda}^T\mathbf{A}\mathbf{Q}^{-1}\mathbf{A}^T\boldsymbol{\lambda} + \mathbf{c}^T\mathbf{Q}^{-1}\mathbf{c} - \mathbf{c}^T\mathbf{Q}^{-1}\mathbf{A}^T\boldsymbol{\lambda} - \boldsymbol{\lambda}^T\mathbf{A}\mathbf{Q}^{-1}\mathbf{c}]\\
&= \mathbf{b}^T\boldsymbol{\lambda} - \tfrac{1}{2}[\boldsymbol{\lambda}^T\mathbf{A}\mathbf{Q}^{-1}\mathbf{A}^T\boldsymbol{\lambda} + \mathbf{c}^T\mathbf{Q}^{-1}\mathbf{c} - \mathbf{c}^T\mathbf{Q}^{-1}\mathbf{A}^T\boldsymbol{\lambda} - (\mathbf{A}\mathbf{Q}^{-1}\mathbf{c})^T\boldsymbol{\lambda}] \text{ by (iii)}\\
&= \mathbf{b}^T\boldsymbol{\lambda} - \tfrac{1}{2}[\boldsymbol{\lambda}^T\mathbf{A}\mathbf{Q}^{-1}\mathbf{A}^T\boldsymbol{\lambda} + \mathbf{c}^T\mathbf{Q}^{-1}\mathbf{c} - \mathbf{c}^T\mathbf{Q}^{-1}\mathbf{A}^T\boldsymbol{\lambda} - \mathbf{c}^T(\mathbf{Q}^{-1})^T\mathbf{A}^T\boldsymbol{\lambda}] \text{ by(i)}\\
&= \mathbf{b}^T\boldsymbol{\lambda} - \tfrac{1}{2}[\boldsymbol{\lambda}^T\mathbf{A}\mathbf{Q}^{-1}\mathbf{A}^T\boldsymbol{\lambda} + \mathbf{c}^T\mathbf{Q}^{-1}\mathbf{c} - 2\mathbf{c}^T\mathbf{Q}^{-1}\mathbf{A}^T\boldsymbol{\lambda}] \text{ by (iv)}\\
&= [\mathbf{b}^T + \mathbf{c}^T\mathbf{Q}^{-1}\mathbf{A}^T]\,\boldsymbol{\lambda} - \tfrac{1}{2}\,[\boldsymbol{\lambda}^T\,\mathbf{A}\mathbf{Q}^{-1}\mathbf{A}^T\,\boldsymbol{\lambda}] - \tfrac{1}{2}[\mathbf{c}^T\mathbf{Q}^{-1}\mathbf{c}]\\
&= \mathbf{u}^T\boldsymbol{\lambda} - \tfrac{1}{2}\,[\boldsymbol{\lambda}^T\mathbf{v}\,\boldsymbol{\lambda}] - \tfrac{1}{2}[\mathbf{c}^T\mathbf{Q}^{-1}\mathbf{c}],
\end{aligned}
$$

where

$$\mathbf{u} = \mathbf{b} + \mathbf{A}\mathbf{Q}^{-1}\mathbf{c}, \text{ and } \mathbf{v} = -\mathbf{A}\mathbf{Q}^{-1}\mathbf{A}^T.$$

Then for nonsingular $\mathbf{Q}$, the dual is

$$\max_{\boldsymbol{\lambda}\ge 0}\big\{\mathbf{u}^T\boldsymbol{\lambda} - \tfrac{1}{2}\boldsymbol{\lambda}^T\mathbf{v}\,\boldsymbol{\lambda} - \tfrac{1}{2}\,\mathbf{c}^T\mathbf{Q}^{-1}\mathbf{c}\big\}. \tag{9.1.9}$$

Example 9.1. Minimize $\frac{1}{2}x^2 + \frac{1}{2}y^2 - 4x - 4y$ subject to the constraints

$$0 \le x \le 1 \longrightarrow \begin{cases} -x \ge -1 \\ x \ge 0; \end{cases} \qquad 0 \le y \le 1 \Longrightarrow \begin{cases} -y \ge -1 \\ y \ge 0, \end{cases}$$

that is,

$$\text{Minimize } \tfrac{1}{2}\mathbf{x}^T\mathbf{Q}\mathbf{x} + \mathbf{c}^T\mathbf{x}, \quad \text{subject to} \quad \mathbf{A}\mathbf{x} \ge \mathbf{b},$$

where we will take $\mathbf{Q}$ as the identity matrix $\mathbf{I}$. Then

$$\mathbf{x} = \begin{bmatrix} x \\ y \end{bmatrix}, \ \mathbf{Q} = \begin{bmatrix} 1 & 0 \\ 0 & 1 \end{bmatrix}, \ \mathbf{c} = \begin{bmatrix} -4 \\ -4 \end{bmatrix}, \ \mathbf{A} = \begin{bmatrix} -1 & 0 \\ 0 & -1 \\ 1 & 0 \\ 0 & 1 \end{bmatrix}, \ \mathbf{b} = \begin{bmatrix} -1 \\ -1 \\ 0 \\ 0 \end{bmatrix}.$$

Since $\mathbf{Q}^{-1} = \mathbf{Q} = \mathbf{I}$, we get

$$\mathbf{u} = \mathbf{b} + \mathbf{A}\mathbf{Q}^{-1}\mathbf{c} = \begin{bmatrix} -1 \\ -1 \\ 0 \\ 0 \end{bmatrix} + \begin{bmatrix} -1 & 0 \\ 0 & -1 \\ 1 & 0 \\ 0 & 1 \end{bmatrix} \begin{bmatrix} 1 & 0 \\ 0 & 1 \end{bmatrix} \begin{bmatrix} -4 \\ -4 \end{bmatrix} = \begin{bmatrix} 3 \\ 3 \\ -4 \\ -4 \end{bmatrix},$$

$$\mathbf{v} = -\mathbf{A}\mathbf{Q}^{-1}\mathbf{A}^T = -\begin{bmatrix} -1 & 0 \\ 0 & -1 \\ 1 & 0 \\ 0 & 1 \end{bmatrix} \begin{bmatrix} 1 & 0 \\ 0 & 1 \end{bmatrix} \begin{bmatrix} -1 & 0 & 1 & 0 \\ 0 & -1 & 0 & 1 \end{bmatrix}$$

$$= \begin{bmatrix} -1 & 0 & 1 & 0 \\ 0 & -1 & 0 & 1 \\ 1 & 0 & -1 & 0 \\ 0 & 1 & 0 & -1 \end{bmatrix}.$$

Since the dual is given by (9.1.9), the optimal solution from (9.1.8) with the above computed values of $\mathbf{u}$ and $\mathbf{v}$ is

$$\mathbf{x}^* = \mathbf{Q}^{-1}[\mathbf{A}^T\boldsymbol{\lambda}^* - \mathbf{c}] = \begin{bmatrix} 1 & 0 \\ 0 & 1 \end{bmatrix} \left\{ \begin{bmatrix} -1 & 0 & 1 & 0 \\ 0 & -1 & 0 & 1 \end{bmatrix} \begin{bmatrix} \lambda_1^* \\ \lambda_2^* \\ \lambda_3^* \\ \lambda_4^* \end{bmatrix} - \begin{bmatrix} -4 \\ -4 \end{bmatrix} \right\}$$

$$= \begin{bmatrix} -\lambda_1^* + \lambda_3^* + 4 \\ -\lambda 2^* + \lambda_4^* + 4 \end{bmatrix}. \ \blacksquare \tag{9.1.10}$$

The values of $\boldsymbol{\lambda}$ could be determined from (9.1.10). However, these values are easily determined using the Hildreth-D'Epso method described in the next section. The values of $\boldsymbol{\lambda}^*$ so obtained are then used to determine the optimal value of $\mathbf{x}^*$. This example will continue after this method is explained.

9.2 Hildreth-D'Esopo Method

This method, which is an iterative method developed by Hildreth [1957] and D'Esopo [1959], will be applied to the dual (9.1.9). It follows the following three steps:

STEP 1. Start with an initial $\boldsymbol{\lambda}$, for example, $\boldsymbol{\lambda} = \mathbf{0}$.

STEP 2. For $j = 1, 2, \dots, m$, do the following: Search for the maximum in the direction parallel to the λ_j-axis by fixing λ_k, $k \neq j$, and solving $\dfrac{\partial \hat{L}}{\partial \lambda_j} = 0$ for λ_j. If this $\lambda_j < 0$, then fix $\lambda_j = 0$.

STEP 3. If λ from this iteration is the same as the $\boldsymbol{\lambda}$ from the previous iteration, STOP; else go to STEP 2.

The KKT conditions for the dual (9.1.9) are

$$\frac{\partial \hat{L}}{\partial \boldsymbol{\lambda}}(\boldsymbol{\lambda}) = \mathbf{u} + \mathbf{v}\,\boldsymbol{\lambda} = \mathbf{0}. \tag{9.2.1}$$

!!! *Some authors write this expression as* $\nabla_{\boldsymbol\lambda}\hat{L}(\boldsymbol\lambda) = \mathbf{u} + \mathbf{v}\,\boldsymbol\lambda$, *which is incorrect.*

Example 9.1, continued. We start with the KKT conditions (9.2.1), with

$$\mathbf{u} = \begin{bmatrix} 3 \\ 3 \\ -4 \\ -4 \end{bmatrix}, \quad \mathbf{v} = \begin{bmatrix} -1 & 0 & 1 & 0 \\ 0 & -1 & 0 & 1 \\ 1 & 0 & -1 & 0 \\ 0 & 1 & 0 & -1 \end{bmatrix},$$

which yields

$$\frac{\partial \hat{L}}{\partial \boldsymbol\lambda} = \begin{bmatrix} \partial\hat{L}/\partial\lambda_1 \\ \partial\hat{L}/\partial\lambda_2 \\ \partial\hat{L}/\partial\lambda_3 \\ \partial\hat{L}/\partial\lambda_4 \end{bmatrix} = \mathbf{u} + \mathbf{v}\,\boldsymbol\lambda = \begin{bmatrix} 3 \\ 3 \\ -4 \\ -4 \end{bmatrix} + \begin{bmatrix} -1 & 0 & 1 & 0 \\ 0 & -1 & 0 & 1 \\ 1 & 0 & -1 & 0 \\ 0 & 1 & 0 & -1 \end{bmatrix} \begin{bmatrix} \lambda_1 \\ \lambda_2 \\ \lambda_3 \\ \lambda_4 \end{bmatrix}$$
$$= \begin{bmatrix} 1 - \lambda_1 + \lambda_3 \\ 1 - \lambda_2 + \lambda_4 \\ -4 + \lambda_1 - \lambda_3 \\ -4 + \lambda_2 - \lambda_4 \end{bmatrix} = \begin{bmatrix} 0 \\ 0 \\ 0 \\ 0 \end{bmatrix}, \tag{9.2.2}$$

which must be solved to obtain the optimal $\boldsymbol\lambda^*$. This becomes simpler using the Hildreth-D'Espo method, which is as follows. We start with (9.2.1). Then

First iteration. Let $\boldsymbol\lambda = [0\ 0\ 0\ 0]^T$. Solve $\dfrac{\partial \hat{L}}{\partial \lambda_1} = 0$, keeping $\lambda_2 = 0 = \lambda_3 = \lambda_4$; we get $-\lambda_1 + \lambda_3 + 1 = 0$, which gives $\lambda_1 = 1$, thus $\boldsymbol\lambda = [1\ 0\ 0\ 0]^T$.

Next, solve $\dfrac{\partial \hat{L}}{\partial \lambda_2} = 0$, with $\lambda_1 = 1, \lambda_3 = 0 = \lambda_4$; we get $-\lambda_2 + \lambda_4 + 1 = 0$, yielding $\lambda_2 = 1$. Thus, $\boldsymbol\lambda = [1\ 1\ 0\ 0]^T$.

Next, solve $\dfrac{\partial \hat{L}}{\partial \lambda_3} = 0$, with $\lambda_1 = 1 = \lambda_2, \lambda_4 = 0$; we get $\lambda_1 - \lambda_3 - 4 = 0$, yielding $\lambda_3 = -3$. Fix $\lambda_3 = 0$. Thus, $\boldsymbol\lambda = [1\ 1\ 0\ 0]^T$.

Finally, solve $\dfrac{\partial \hat{L}}{\partial \lambda_4} = 0$, with $\lambda_1 = 1 = \lambda_2, \lambda_3 = 0$; we get $\lambda_2 - \lambda_4 - 4 = 0$, yielding $\lambda_4 = -3$. Fix $\lambda_4 = 0$. Thus, $\boldsymbol\lambda = [1\ 1\ 0\ 0]^T$.
End of First iteration.

Notice that $\boldsymbol\lambda$ has changed from $[0\ 0\ 0\ 0]^T$ to $[1\ 1\ 0\ 0]^T$. So we go to the second iteration.

Second iteration. It goes through the following four steps:

1. $-\lambda_1 + \lambda_3 + 1 = 0$ with $\lambda_2 = 1, \lambda_3 = 0 = \lambda_4$, yielding $\lambda_1 = 1$; thus, $\boldsymbol\lambda = [1\ 1\ 0\ 0]^T$.

2. $-\lambda_2 + \lambda_4 + 1 = 0$ with $\lambda_1 = 1, \lambda_3 = 0 = \lambda_4$, yielding $\lambda_2 = 1$; thus, $\boldsymbol{\lambda} = [1\ 1\ 0\ 0]^T$.
3. $\lambda_1 - \lambda_3 - 2 = 0$ with $\lambda_1 = 1 = \lambda_2, \lambda_4 = 0$, yielding $\lambda_3 = -3$, so fix $\lambda_3 = 0$; thus, $\boldsymbol{\lambda} = [1\ 1\ 0\ 0]^T$.
4. $\lambda_2 - \lambda_4 - 2 = 0$ with $\lambda_1 = 1 = \lambda_2, \lambda_3 = 0$, yielding $\lambda_4 = -3$, so fix $\lambda_4 = 0$; thus, $\boldsymbol{\lambda} = [1\ 1\ 0\ 0]^T$.

End of second iteration.

The iteration has converged since the value of $\boldsymbol{\lambda}$ remains the same for these two iterations. Thus, $\lambda_1^* = 1 = \lambda_2^*, \lambda_3^* = 0 = \lambda_4^*$. Then, using (9.1.8), the optimal primal vector $\mathbf{x}^*$ is given by

$$\mathbf{x}^* = \mathbf{Q}^{-1}[\mathbf{A}^T\boldsymbol{\lambda}^* - \mathbf{c}] = \begin{bmatrix} 1 & 0 \\ 0 & 1 \end{bmatrix} \left\{ \begin{bmatrix} -1 & 0 & 1 & 0 \\ 0 & -1 & 0 & 1 \end{bmatrix} \begin{bmatrix} \lambda_1 \\ \lambda_2 \\ \lambda_3 \\ \lambda_4 \end{bmatrix} - \begin{bmatrix} -4 \\ -4 \end{bmatrix} = \begin{bmatrix} 3 \\ 3 \end{bmatrix} . \blacksquare \right.$$

Finally, note that using these values of $\mathbf{x}$ the primal objective (9.1.1) yields

$$\tfrac{1}{2}\mathbf{x}^T\mathbf{Q}\mathbf{x} + \mathbf{c}^T\mathbf{x} = \tfrac{1}{2}\begin{bmatrix} 3 & 3 \end{bmatrix} \begin{bmatrix} 1 & 0 \\ 0 & 1 \end{bmatrix} \begin{bmatrix} 3 \\ 3 \end{bmatrix} + \begin{bmatrix} -4 & -4 \end{bmatrix} \begin{bmatrix} 3 \\ 3 \end{bmatrix} = -13. \blacksquare$$

9.3 Beale's Method

This method (Beale [1959]) is used for the quadratic programming problem

$$\text{Minimize } f(\mathbf{x}) = \left\{\tfrac{1}{n}\mathbf{x}^T\mathbf{Q}\mathbf{x} + \mathbf{c}^T\mathbf{x}\right\}, \text{ subject to } \mathbf{A}\mathbf{x} = \mathbf{b},\ \mathbf{x} \geq \mathbf{0}. \tag{9.3.1}$$

First, the vectors $\mathbf{x}$ and $\mathbf{c}$ and the matrices $\mathbf{A}$ and $\mathbf{Q}$ are partitioned into their basic and nonbasic partitions, which are denoted by the index B and N respectively. Thus, $\mathbf{A} = [\mathbf{A}^B \,|\, \mathbf{A}^N]$, where $\mathbf{A}^B$ has columns corresponding to the basic variables $\mathbf{x}_B$ and is nonsingular. Again, the constraint in (9.3.1) is expressed as $\mathbf{A}^B\mathbf{x}_B + \mathbf{A}^N\mathbf{x}_N = \mathbf{b}$. Thus, if $\mathbf{x}_N = \mathbf{0}$, then $\mathbf{x}_B = [\mathbf{A}^B]^{-1}\mathbf{b}$. In general, $\mathbf{x}_B = [\mathbf{A}^B]^{-1}\mathbf{b} - [\mathbf{A}^B]^{-1}\mathbf{A}^N\mathbf{x}_N$ for $\mathbf{x}$ that satisfies $\mathbf{A}\mathbf{x} = \mathbf{b}$.

The partitions of $\mathbf{c}$ and $\mathbf{Q}$ are

$$\mathbf{c} = [\mathbf{c}_B \,|\, \mathbf{c}_N]^T, \quad \mathbf{Q} = \begin{bmatrix} \mathbf{Q}_B^B & \mathbf{Q}_B^N \\ \mathbf{Q}_N^B & \mathbf{Q}_N^N \end{bmatrix}, \tag{9.3.2}$$

where $\mathbf{Q}_B^N = [\mathbf{Q}_N^B]^T$.

Example 9.2. Minimize

$$f(x, y, z) = \tfrac{1}{3}x^2 + \tfrac{1}{3}y^2 + z^2 + 2xy + 2xz + yz + 3x + y + z, \tag{9.3.3}$$

which is of the matrix form (9.3.1). The algebraic function $f(x,y,z)$ can be expressed in the matrix form as

$$\frac{1}{3}\underbrace{\begin{bmatrix} x & y & z\end{bmatrix}}_{\mathbf{x}^T}\underbrace{\begin{bmatrix}1&3&3\\3&1&1\\3&2&3\end{bmatrix}}_{\mathbf{Q}}\underbrace{\begin{bmatrix}x\\y\\z\end{bmatrix}}_{\mathbf{x}}+\underbrace{\begin{bmatrix}3&1&1\end{bmatrix}}_{\mathbf{c}^T}\underbrace{\begin{bmatrix}x\\y\\z\end{bmatrix}}_{\mathbf{x}}, \tag{9.3.4}$$

subject to

$$\left.\begin{array}{r} x+z=9\\ y+z=12\end{array}\right\} \Longrightarrow \underbrace{\begin{bmatrix}1&0&1\\0&1&1\end{bmatrix}}_{\mathbf{A}}\begin{bmatrix}x\\y\\z\end{bmatrix}=\underbrace{\begin{bmatrix}9\\12\end{bmatrix}}_{\mathbf{b}},\ \mathbf{x}\geq\mathbf{0}. \tag{9.3.5}$$

An initial choice $x=9, y=12, z=0$ makes the objective (9.3.3) equal to 330. Also, with this choice we have

$$\begin{aligned}
\mathbf{x}^0 &= \left[\mathbf{x}_B^0 \,|\, \mathbf{x}_N^0\right] = [x^0\ y^0 \,|\, z^0] = [9\ 12 \,|\, 0]^T,\\
\mathbf{A} &= \left[\mathbf{A}^B \,|\, \mathbf{A}^N\right] = \begin{bmatrix}1&0&|&1\\0&1&|&1\end{bmatrix},\\
\mathbf{c} &= \left[\mathbf{c}_B \,|\, \mathbf{c}_N\right] = [3\quad 1\quad |\quad 1]^T,\\
\mathbf{Q} &= \begin{bmatrix}\mathbf{Q}_B^B* & \mathbf{Q}_B^N\\ \mathbf{Q}_N^B & \mathbf{Q}_N^N\end{bmatrix} = \left[\begin{array}{cc|c}1&3&3\\3&1&1\\ \hline 3&2&3\end{array}\right].
\end{aligned}$$

Also

$$\begin{aligned}
\left(\mathbf{A}^B\right)^{-1}\mathbf{b} &= \begin{bmatrix}1&0\\0&1\end{bmatrix}^{-1}\begin{bmatrix}9\\12\end{bmatrix} = \begin{bmatrix}9\\12\end{bmatrix} = \mathbf{x}_B,\\
\left(\mathbf{A}^B\right)^{-1}\mathbf{A}^N\mathbf{x}_N &= \begin{bmatrix}1&0\\0&1\end{bmatrix}^{-1}\begin{bmatrix}1\\1\end{bmatrix}[0] = \begin{bmatrix}0\\0\end{bmatrix}.
\end{aligned}$$

The objective function (9.3.1) is

$$\begin{aligned}
f(\mathbf{x}) &= \tfrac{1}{3}\mathbf{x}^T\mathbf{Q}\mathbf{x}+\mathbf{c}^T\mathbf{x}\\
&= \tfrac{1}{3}\left[\mathbf{x}_B^T \quad | \quad \mathbf{x}_N^T\right]\begin{bmatrix}\mathbf{Q}_B^B & \mathbf{Q}_B^N\\ \mathbf{Q}_N^B & \mathbf{Q}_N^N\end{bmatrix}\begin{bmatrix}\mathbf{x}_B\\ \mathbf{x}_N\end{bmatrix} + \left[\mathbf{c}^T \quad | \quad \mathbf{c}^T\right]\begin{bmatrix}\mathbf{x}_B\\ \mathbf{x}_N\end{bmatrix}\\
&= \tfrac{1}{3}\left[\mathbf{x}_B^T \quad | \quad \mathbf{x}_N^T\right]\begin{bmatrix}\mathbf{Q}_B^B\,\mathbf{x}_B+\mathbf{Q}_B^N\,\mathbf{x}_N\\ \mathbf{Q}_N^B\,\mathbf{x}_B+\mathbf{Q}_N^N\,\mathbf{x}_N\end{bmatrix} + \mathbf{c}_B\mathbf{x}_B+\mathbf{c}_N\mathbf{x}_N\\
&= \tfrac{1}{3}\left[\mathbf{x}_B^T\mathbf{Q}_B^B\mathbf{x}_B+\mathbf{x}_B^T\mathbf{Q}_B^N\mathbf{x}_N+\mathbf{x}_N^T\mathbf{Q}_N^B\mathbf{x}_B+\mathbf{x}_N^T\mathbf{Q}_N^N\mathbf{x}_N\right]+\mathbf{c}_B\mathbf{x}_B+\mathbf{c}_N\mathbf{x}_N.
\end{aligned}$$

On substituting $\mathbf{x}_B = (\mathbf{A}^B)^{-1}\mathbf{b}$ and $(\mathbf{A}^B)^{-1}\mathbf{A}^N\mathbf{x}_N = \mathbf{0}$, we get

$$\begin{aligned}f(\mathbf{x}) = &\Big\{\mathbf{c}_B\big(\mathbf{A}^B\big)^{-1}\mathbf{b} + \tfrac{1}{3}[(\mathbf{A}^B)^{-1}\mathbf{b}]^T\mathbf{Q}_B^B(\mathbf{A}^B)^{-1}\mathbf{b}\Big\}\\ &+ \{[(\mathbf{A}^B)^{-1}\mathbf{b}]^T\mathbf{Q}_B^N - [(\mathbf{A}^B)^{-1}\mathbf{b}]^T\mathbf{Q}_B^B(\mathbf{A}^B)^{-1}\mathbf{A}^N + \mathbf{c}_N - \mathbf{c}_B^T(\mathbf{A}^B)^{-1}\mathbf{A}^N\}\\ &+ \tfrac{1}{3}\mathbf{x}_N^T\Big\{\mathbf{Q}_N^N - \mathbf{Q}_N^B(\mathbf{A}^B)^{-1}\mathbf{A}^N + [(\mathbf{A}^B)^{-1}\mathbf{A}^N]^T\mathbf{Q}_B^B(\mathbf{A}^B)^{-1}\mathbf{A}^N\Big\}\mathbf{x}_N.\end{aligned}$$

The objective function can also be written as

$$f(\mathbf{x}) = \mathbf{z} + \mathbf{p}^T + \tfrac{1}{3}\mathbf{x}_N^T\mathbf{R}\mathbf{x}_N, \tag{9.3.6}$$

where

$$\begin{aligned}\mathbf{z} &= \mathbf{c}_B\big(\mathbf{A}^B\big)^{-1}\mathbf{b} + \tfrac{1}{3}[(\mathbf{A}^B)^{-1}\mathbf{b}]^T\mathbf{Q}_B^B(\mathbf{A}^B)^{-1}\mathbf{b}\Big\},\\ \mathbf{p} &= [(\mathbf{A}^B)^{-1}\mathbf{b}]^T\mathbf{Q}_B^N - [(\mathbf{A}^B)^{-1}\mathbf{b}]^T\mathbf{Q}_B^B(\mathbf{A}^B)^{-1}\mathbf{A}^N + \mathbf{c}_N - \mathbf{c}_B^T(\mathbf{A}^B)^{-1}\mathbf{A}^N\},\\ \mathbf{R} &= \mathbf{Q}_N^N - \mathbf{Q}_N^B(\mathbf{A}^B)^{-1}\mathbf{A}^N + [(\mathbf{A}^B)^{-1}\mathbf{A}^N]^T\mathbf{Q}_B^B(\mathbf{A}^B)^{-1}\mathbf{A}^N.\end{aligned}$$

Thus, using KKT conditions we have

$$\begin{aligned}\frac{\partial f(\mathbf{x})}{\partial x_i} &= p_i + \sum_{k\in N} r_{ik}x_k \text{ (for } i \in N)\\ &= p_i \text{ (at } \mathbf{x}_N = \mathbf{0}).\end{aligned} \tag{9.3.7}$$

At this point we must choose any negative p_i and increase the corresponding value of x_i until
(i) one of the basic variables becomes zero (as in the simplex method); or
(ii) $\dfrac{\partial f(\mathbf{x})}{\partial x_i} = p_i + r_{ii}x_i = 0$, i.e., $x_i = \dfrac{-p_i}{r_{ii}}$. Note that this result is nonbasic but is a feasible solution.

Example 9.2, continued. We will use iteration method.

(First iteration:) We have

$$\mathbf{p} = [9 \quad 12]\begin{bmatrix}3\\1\end{bmatrix} - [9 \quad 12]\begin{bmatrix}1 & 3\\3 & 1\end{bmatrix}\begin{bmatrix}1 & 0\\0 & 1\end{bmatrix}\begin{bmatrix}1\\1\end{bmatrix} + [1] - [3 \quad 1]\begin{bmatrix}1\\1\end{bmatrix}\begin{bmatrix}1\\0\end{bmatrix} = [-51].$$

Thus, $\dfrac{\partial f}{\partial z} = -51$.

Note that we do not need $\mathbf{R}$, since $\mathbf{x}_N = \mathbf{0}$. Next, increase z from 0 until one of the basic variables goes to zero, i.e., by 9 units when $x = 0, y = 3, z = 9$. Then the new objective function is 84, as compared to 330 at the previous iteration.

(Second iteration) gives $\mathbf{x}_B = [y \quad z \mid x] = [3 \quad 9 \mid 0]$. This iteration continues to the next one until the two successive iterations give the same result. At that point the iterative process stops. ■

9.4 Wolfe's Method

This is an iterative procedure developed by Wolfe [1959] to solve maximization QP problem numerically with inequality constraints. If the QP problem is a minimization problem, it must be converted into a maximization problem by multiplying the constraints by -1. The inequality constraints are then converted into equality constraints by introducing slack variables. The QP problem is finally converted into a linear programming (LP) problem by introducing new variables $\mathbf{v} = \{v_1, v_2, \dots, v_n\}$ in $\mathbb{R}^n$, which is solved numerically by the simplex method.[1]

We will consider the following maximization QP problem in $\mathbb{R}^2$:

Maximize $f(\mathbf{x}) = \frac{1}{2}\mathbf{x}^T\mathbf{Q}\mathbf{x} + \mathbf{c}^T\mathbf{x}$ subject to the constraints $\mathbf{A}\mathbf{x} - \mathbf{b} \leq 0$, $\mathbf{x} \geq 0$,

which we rewrite as:

Maximize $\frac{1}{2} \sum\limits_{j,k=1}^{n} c_{jk}x_jx_k + \sum\limits_{j=1}^{n} c_jx_j$, where $c_{jk} = c_{kj}$, subject to the constraints $\sum\limits_{j=1}^{n} a_{ij}x_j \leq b_j, x_j \geq 0$ for $i = 1, 2, \dots, m; j = 1, 2, \dots, n$; and for all j, k, $b_i \geq 0$ for all $i = 1, 2, \dots, m$.

Also assume that the quadratic form $\sum\limits_{j,k=1}^{n} c_{jk}x_jx_k$ is negative semidefinite (NSD).

Wolfe's method is then developed using the following five steps:

STEP 1. Convert the inequality constraints to equations by introducing slack variables $\mathbf{s} = s_i^2$ in the ith constraint $(i = 1, 2, \dots, m)$ and the slack variable $\mathbf{t} = t_j^2$ in the jth non-negativity constraint $(j = 1, 2, \dots, n)$.

STEP 2. Construct the Lagrangian

$$L(\mathbf{x}, \mathbf{s}, \mathbf{t}, \boldsymbol{\lambda}, \boldsymbol{\mu}) = f(\mathbf{x}) - \sum_{i=1}^{m} \lambda_i \Big\{ \sum_{j=1}^{n} a_{ij}x_j - b_i + s_i^2 \Big\} - \sum_{j=1}^{n} \mu_j \{-x_j + t_j^2\},$$

where $\mathbf{x} = (x_1, x_2, \dots, x_n)$, $\mathbf{s} = (s_1^2, s_2^2, \dots, s_m^2)$, $\mathbf{t} = (t_1^2, t_2^2, \dots, t_n^2)$, $\boldsymbol{\lambda} = (\lambda_1, \lambda_2, \dots, \lambda_m)$, and $\boldsymbol{\mu} = (\mu_1, \mu_2, \dots, \mu_n)$. Next, equate to zero the first

[1] The *simplex method* is an algorithm for starting at some extreme feasible point and, by a sequence of exchanges, proceeding systematically to the other such points until a solution point is found. This is done in a way which steadily reduces the value of the linear function Z. The exchange process involved is essentially the same as used in matrix inversion. Details of this method can be found in Hildebrand [1974].

partial derivatives of L with respect to $\mathbf{x}, \mathbf{s}, \mathbf{t}, \boldsymbol{\lambda}, \boldsymbol{\mu}$, and obtain the KKT conditions.

STEP 3. Introduce the nonnegative vector $\mathbf{v} = [v_1, v_2, \dots , v_n]$ in the KKT conditions $c_j + \sum_{k=1}^{n} c_{jk}x_k - \sum_{i=1}^{m} a_{ij}\lambda_i + \mu_j = 0$ for $j = 1, 2, \dots , n$, and construct the objective function $Z = v_1 + v_2 + \cdots + v_n$.

STEP 4. Obtain the initial basic feasible solution for the following linear programming (LP) problem: $\min\{Z_v\} = v_1 + v_2 + \cdots + v_n$ subject to the constraints

$$\sum_{k=1}^{n} c_{jk}x_k - \sum_{i=1}^{m} a_{ij}\lambda_i + \mu_j = -c_j \text{ for } j = 1, 2, \dots , n,$$

$$\sum_{i=1}^{m} a_{ij}x_j + s_i^2 = b_i,$$

where $v_j, \lambda_j, \mu_j, x_j \geq 0$ for $i = 1, 2, \dots , m$; $j = 1, 2, \dots , n$, and satisfying the complementary slackness condition

$$\sum_{j=1}^{n} \mu_j x_j + \sum_{i=1}^{m} \lambda_i s_i^2 = 0; \quad \text{i.e., } \lambda_i s_i^2 = 0, \text{ and } \mu_j x_j = 0.$$

STEP 5. Apply the two-phase simplex method (Tableau Format)[2] to determine for the LP problem of Step 4 an optimum solution, which must satisfy the complementary slackness conditions.

The optimum solution obtained in Step 5 is the optimum solution of the required QP problem.

Example 9.3. (Singh [2012]) Maximize $Z = 8x + 10y - 2x^2 - x^2$ subject to the constraints $3x + 2y \leq 6$ and $x, y \geq 0$.

SOLUTION. Introduce all the constraints in the nonpositive form: $3x+2y \leq 6$, and $-x \leq 0, -y \leq 0$, and then introduce the slack variables so that $3x + 2y + s_1^2 = 6, -x + t_1^2 = 0$, and $-y + t_2^2 = 0$. Then the problem reduces to the LP problem:

Maximize $Z = 8x + 10y - 2x^2 - y^2$ subject to the constraints $3x + 2y + s_1^2 = 6, -x + t_1^2 = 0, -y + t_2^2 = 0$.
The Lagrangian is given by

$$\begin{aligned} L(x, y, \lambda_1, \mu_1, \mu_2, s_1, t_1, t_2) &= 8x + 10y - 2x^2 - y^2 + \lambda_1(6 - 3x - 2y - s_1^2) \\ &\quad + \mu_1(x - t_1^2) + \mu_2(y - t_2), \end{aligned}$$

[2] The algorithm and examples can be found in Bertsimas and Tsitsiklis [1997], Bertsekas [1999], and Cormen et al. [2001].

whence we obtain the KKT conditions

$$\frac{\partial L}{\partial x} = 8 - 4x - 3\lambda_1 + \mu_1 = 0, \quad \frac{\partial L}{\partial y} = 10 - 2y - 2\lambda_1 + \mu_2 = 0,$$
$$\frac{\partial L}{\partial \lambda_1} = 6 - 3x - 2y - s_1^2 = 0, \quad \frac{\partial L}{\partial \mu_1} = x - t_1^2 = 0, \quad \frac{\partial L}{\partial \mu_2} = y - t_2^2 = 0.$$

Next, we introduce the linearly independent variables v_1 and v_2, and we have the LP problem:

Maximize $Z = -v_1 - v_2$ subject to the constraints $4x + 3\lambda_1 - \mu_1 + v_1 = 8$, $3y + 2\lambda_1 - \mu_2 + v_2 = 10$, $3x + 2y + s_1^2 = 6$.

The details of the solution are given in the following four tables.

Table 1

BV	c_B	x_B	$x(0)$	$y(0)$	$\lambda_1(0)$	$\mu_1(0)$	$\mu_2(0)$	$v_1(-1)$	$v_2(-1)$	$s_1^2(0)$
v_1	-1	8	4	0	3	-1	0	1	0	0
v_2	-1	10	0	2	2	0	-1	0	1	0
s_1^2	0	6	3	2	0	0	0	0	0	1
	$Z = -18$		-4	-2	-5	1	1	0	0	0

Note that λ_1 cannot be the entering variable since s_1^2 is the basic variable such that $\lambda_1 s_1^2 = 0$. Hence, x is the entering variable since μ_1 is not basic; similarly, y can also be the entering variable since μ_2 is not basic. The min-ratio for v_1 and s_1^2, given by $(x_B/x(0)$, is $(8/4, 6/3) = (2, 2)$, which is a tie. So we must take y as the entering variable, since it has the min-ratio $(10/2, 6/2) = (5, 3)$. This leads to Table 2.

Table 2

BV	c_B	x_B	$x(0)$	$y(0)$	$\lambda_1(0)$	$\mu_1(0)$	$\mu_2(0)$	$v_1(-1)$	$v_2(-1)$	$s_1^2(0)$
v_1	-1	8	4	0	3	-1	0	1	0	0
v_2	-1	4	-3	0	2	0	-1	0	1	0
y	0	3	3/2	1	0	0	0	0	0	1/2
	$Z = -12$		-1	0	-5	1	1	0	0	0

Now λ_1 can enter since s_1^2 is not basic, and for v_1 and v_2 the min-ratio is $(x_B/y(0)) = (8/3, 4/2)$. This leaves the variable v_2, which leads to Table 3.

Table 3

BV	c_B	x_B	$x(0)$	$y(0)$	$\lambda_1(0)$	$\mu_1(0)$	$\mu_2(0)$	$v_1(-1)$	$v_2(-1)$	$s_1^2(0)$
v_1	-1	2	17/2	0	0	-1	3	1	-3	0
λ_1	0	2	$-3/2$	0	1	0	-1	0	1	0
y	0	3	3/2	1	0	0	0	0	0	1/2
	$Z=-2$		$-17/2$	0	0	1	-3	0	4	0

The min-ratio is $(4/17, 2)$. The final solution follows.

Table 4

BV	c_B	x_B	$x(0)$	$y(0)$	$\lambda_1(0)$	$\mu_1(0)$	$\mu_2(0)$	$v_1(-1)$	$v_2(-1)$	$s_1^2(0)$
x	0	4/17	1	0	0	$-2/17$	6/17	2/17	$-6/17$	0
λ_1	0	40/17	0	0	1	$-3/17$	8/17	20/17	$-6/17$	0
y	0	45/17	0	1	0	3/17	$-9/17$	$-3/17$	9/17	35/34
	$Z=0$		0	0	0	0	0	1	1	0

Hence, the optimum solution is $(x, y, \lambda_1) = (4/17, 45/17, 40/17)$, since $v_2 = 0 = v_1 = \mu_1 = \mu_2 = s_1^2$. This solution satisfies the conditions $\lambda_1 s_1^2 = 0, \mu_1 x = 0, \mu_2 t = 0$. The maximum value of Z is $\max\{Z\} = 6137/289$. ■

9.5 Exercises

9.1. For the following optimization problems, let $f(x)$ be the cost function and $g(x)$ the inequality constraint. Minimize $f(x)$ subject to $g(x) \le 0$, given:

(i) $f(x) = x, g(x) = |x|$ in a domain $D \subset \mathbb{R}$;

(ii) $f(x) = x^3, g(x) = -x + 1$ in a domain $D \subset \mathbb{R}$;

(iii) $f(x) = x^3, g(x) = -x + 1$ in a domain $D \subset \mathbb{R}_+$; and

(iv) $f(x) = x, g(x) = \begin{cases} -x-2 & \text{for } x \le -1, \\ x & \text{for } -1 \le x \le 1 \text{ in a domain } D \subset \mathbb{R}, \\ -x+2 & \text{for } x \ge 1. \end{cases}$

Also, in each problem plot the graph, state whether it is convex and if so, whether the Slater condition (§5.3.2) is satisfied, i.e., $g_i(x) \ge 0$ for all i.

ANS. (i) The problem is convex, $x^* = 0$ and $\lambda^* = 1$, and the Slater condition is not satisfied; (ii) the problem is not convex, $x^* = 1, \lambda^3 = 3$, and the Slater condition is not satisfied; (iii) the problem is convex, $x^* = 1, \lambda^* = 3$, and the Slater condition is satisfied; and (iv) the problem is not convex, $x^* = 2, \lambda^* = 1$, and the Slater condition is not satisfied.

9.2. (a) Minimize the distance between the origin and the convex region bounded by the constraints $x+y \geq 6$, $2x+y \geq 8$, and $x, y \geq 0$, and (b) verify that the KKT necessary conditions are satisfied at the point of minimum distance.

SOLUTION. Since minimizing the required distance is equivalent to minimizing the distance from the origin to the tangent of the circle that touches the given convex region, consider a circle $x^2 + y^2 = r^2$, and minimize r^2, or $f(x, y) = x^2+y^2$ subject to the constraints $x+y \geq 6$, $2x+y \geq 8$, and $x, y \geq 0$. The feasible region lies in the first quadrant $x \geq 0, y \geq 0$. In Figure 9.1, the lines $x + y = 6$ and $2x + y = 8$ are plotted, and the feasible region is the shaded region. We will determine a point (x, y) which gives a minimum value of $f(x, y) = x^2 + y^2$ subject to the given constraints.

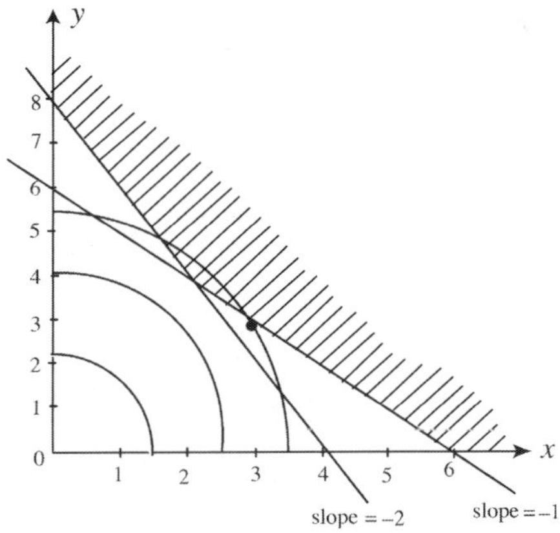

Figure 9.1 Feasible region.

The slope of the tangent to a circle $x^2 + y^2 = c^2$ is $\dfrac{dy}{dx} = -\dfrac{x}{y}$; the slope of the line $x + y = 6$ is -1, while that of the line $2x + y = 8$ is -2. Then

CASE 1. If the line $x+y = 6$ is tangent to the circle, then $\dfrac{dy}{dx} = -\dfrac{x}{y} = -1$, which gives $x = y$. Then solving $x + y = 6$ and $x = y$, we get $x = 3 = y$, i.e., this line touches the circle at $(3, 3)$.

CASE 2. If the line $2x + y = 8$ is tangent to the circle, then $\dfrac{dy}{dx} = -\dfrac{x}{y} =$

-2, which gives $x = 2y$. Then solving $2x + y = 8$ and $2x = y$, we get $x = 16/5, y = 8/5$, i.e., this line touches the circle at $(16/5, 8/5)$.

Of these two cases, since the point $(16/5, 8/5)$ lies outside and the point $(3,3)$ lies within the feasible region, we have $\min(f, y) = x^2 + y^2 = 18$, $x = 3, y = 3$.

(b) To verify KKT necessary conditions, let

$$\mathbf{x} = \begin{bmatrix} x \\ y \end{bmatrix}, \quad \text{and} \quad \boldsymbol{\lambda} = \begin{bmatrix} \lambda_1 \\ \lambda_2 \end{bmatrix}.$$

Then $f(\mathbf{x}) = x^2 + y^2, g(\mathbf{x}) = x + y - 6$, $h(\mathbf{x}) = 2x + y - 8$, and the KKT conditions are

$$L(\mathbf{x}, \boldsymbol{\lambda}) = f(\mathbf{x}) + \lambda_1(6 - x - y) + \lambda_2(8 - 2x - y).$$

subject to the constraints $g(\mathbf{x}) \ge 0$, $h(\mathbf{x}) \ge 0$, $\mathbf{x} \ge 0$, and $\mathbf{y} \ge 0$. Equating partial derivatives of $L(\mathbf{x}, \boldsymbol{\lambda})$ to zero we get

$$\frac{\partial L}{\partial x} = 2x - \lambda_1 - 2\lambda_2 = 0, \quad \frac{\partial L}{\partial y} = 2y - \lambda_1 - \lambda_2 = 0,$$
$$\frac{\partial L}{\partial \lambda_1} = 6 - x - y = 0, \quad \frac{\partial L}{\partial \lambda 2} = 8 - 2x - y = 0,$$

and solving these equations simultaneously we find at the point $(3,3)$ that $\lambda_1 = 6, \lambda_2 = 0$. Also, since $\lambda_1(6 - x - y) = 0$, $\lambda_2(8 - 2x - y) = 0$ at $(3,3)$, the KKT conditions are satisfied at the point $(3,3)$, and $\min f(x,y) = 18$ at this point.

9.3. Use the KKT conditions to minimize $\frac{1}{2}x^2 + \frac{1}{2}y^2 - 2x - 2y$ subject to the constraint

$$0 \le x \le 1 \Longrightarrow \begin{cases} -x \ge -1 \\ x \ge 0; \end{cases} \quad 0 \le y \le 1 \Longrightarrow \begin{cases} -y \ge -1 \\ y \ge 0. \end{cases}$$

ANS. $\mathbf{x}^* = [1 \quad 1]^T$.

9.4. Use KKT conditions to minimize $f(x,y) = x^2 + 4y^2 - 8x - 16y$ subject to the constraints $x + y \le 5$, $x \le 3, x \ge 0, y \ge 0$.

HINT. $\mathbf{Q} = \begin{bmatrix} 2 & 0 \\ 0 & 8 \end{bmatrix}$, $\mathbf{c}^T = \begin{bmatrix} -8 \\ -16 \end{bmatrix}$, $\mathbf{A} = \begin{bmatrix} 1 & 1 \\ 1 & 0 \end{bmatrix}$, and $\mathbf{b} = \begin{bmatrix} 5 \\ 3 \end{bmatrix}$. ANS. $\mathbf{x}^* = [3\ 2]^T$.

9.5. Use KKT conditions to minimize $f(x,y,z) = 2x^2 + y^2 + 4z^2$ subject to the constraints $x + 2y - z = 6$, $2x - 2y + 3z = 12$, $x, y, z \ge 0$.

HINT. $\mathbf{Q} = \begin{bmatrix} 4 & 0 & 0 \\ 0 & 2 & 2 \\ 0 & 0 & 8 \end{bmatrix}$, $\mathbf{A} = \begin{bmatrix} 1 & 2 & -1 \\ 2 & -2 & 3 \end{bmatrix}$, $\mathbf{c}^T = \begin{bmatrix} 0 \\ 0 \\ 0 \end{bmatrix}$, $\mathbf{b} = \begin{bmatrix} 6 \\ 12 \end{bmatrix}$.
ANS. $\mathbf{x}^* = [5.045\ 1.194\ 1.433]^T$.

9.6. Use Beale's method to solve the problem: min $\left\{\frac{1}{2}x^2+\frac{1}{2}y^2+z^2+2xy+2xz+yz+2x+y+z\right\}$ subject to the constraint $x+z=8$, $y+z=10$. HINT. $\mathbf{Q} = \begin{bmatrix} 1 & 2 & 2 \\ 2 & 1 & 1 \\ 2 & 1 & 2 \end{bmatrix}$, $\mathbf{c}^T = [2 \quad 1 \quad 1]$, $\mathbf{A} = \begin{bmatrix} 1 & 0 & 1 \\ 0 & 1 & 1 \end{bmatrix}$, and $\mathbf{b}^T = [8 \quad 10]$.

9.7. For the optimization problem: Minimize $\|\mathbf{Ax}-\mathbf{b}\|_2^2$ subject to $\mathbf{Gx} = \mathbf{h}$, (a) find the dual problem; (b) write the KKT conditions, and see if you can find a solution; (c) find the optimal multiplier vector and the optimal solution as a function of the optimal multiplier.

SOLUTION. (a) The Lagrangian is

$$L(\mathbf{x},\mathbf{c}) = \|\mathbf{Ax}-\mathbf{b}\|_2^2 + \mathbf{c}^T(\mathbf{Gx}-\mathbf{h}) = \mathbf{x}\mathbf{A}^T\mathbf{A}\mathbf{x} + (\mathbf{G}^T\mathbf{c} - 2\mathbf{A}^T\mathbf{b})^T\mathbf{x} - \mathbf{c}^T\mathbf{h}.$$

The dual function (9.2.1) is obtained by minimizing the above strictly convex function. Thus, we get

$$g(\mathbf{c}) = -\frac{1}{4}(\mathbf{G}^T\mathbf{c} - 2\mathbf{A}^T\mathbf{b})^T(\mathbf{A}^T\mathbf{A})^{-1}(\mathbf{G}^T\mathbf{c} - 2\mathbf{A}^T\mathbf{b}) - \mathbf{c}^T\mathbf{h}.$$

(b) The KKT conditions are: $2\mathbf{A}^T(\mathbf{Ax}^* - \mathbf{b}) + \mathbf{G}^T\mathbf{c}^* = 0$, and $\mathbf{Gx}^* = \mathbf{h}$. Since the problem is convex and satisfies the Slater condition, it is feasible and an optimal point $\mathbf{x}^*$ exists iff the KKT conditions have a solution. But since by the Weierstrass theorem[3] an optimal solution exists, so the KKT conditions must have a solution. Thus, from the first KKT condition, we get $\mathbf{x}^* = (\mathbf{A}^T\mathbf{A})^{-1}(\mathbf{A}^T\mathbf{b} - \frac{1}{2}\mathbf{G}^T\mathbf{c}^*)$, where from the second KKT condition we have $\mathbf{c}^* = -2(\mathbf{G}(\mathbf{A}^T\mathbf{A})\mathbf{G}^T)^{-1}(\mathbf{h} - \mathbf{G}(\mathbf{A}^T\mathbf{A})^{-1}\mathbf{A}^T\mathbf{b})$.

9.8. Consider the optimization problem in $\mathbb{R}_+^2$ with the objective function $f(\mathbf{x}) = [F_1(\mathbf{x}) \quad F_2(\mathbf{x})]^T$, where $F_1(\mathbf{x}) = x^2 + y^2$ and $F(\mathbf{x}) = (2x+5)^2$. Then (a) evaluate all Pareto optimal values and points in explicit expressions for both values and points using the scalarization method; and (b) solve the scalarization problem with either weight equal to zero; in both cases show that the solutions of the scalar problem are also Pareto optimal.

[3] This theorem states that a continuous function on a nonempty closed bounded set achieves a maximum and a minimum in this set. In fact, let $f : \mathbb{R}^n \mapsto \mathbb{R}$ be a continuous real-valued function, and let $M = \sup_{\mathbf{x}\in\mathbb{R}^n} \{f(\mathbf{x})\}$ and $m = \inf_{\mathbf{x}\in\mathbb{R}^n} \{f(\mathbf{x})\}$. Then there is a point $\mathbf{x}^M$ and a point $\mathbf{x}^m$ such that $f(\mathbf{x}^m) = m$ and $f(\mathbf{x}^M) = M$.

SOLUTION. Since the problem is convex, all Pareto optimal points can be obtained using the scalarization method with some weight vector $\boldsymbol{\lambda} \geq \mathbf{0}$. Thus, fix some $\lambda \geq \mathbf{0}$ and solve the problem: minimize $\{\lambda_1(x^2+y^2)+\lambda_2(2x+5)\}$. This problem is equivalent to minimizing $\{(\lambda_1+4\lambda_2)x^2+\lambda_1 y^2+20\lambda_2 x+25\lambda_2\}$. Any solution of this problem will give a Pareto optimal point and value. Since the cost function f is strictly convex, the corresponding Pareto optimal point is given by

$$\mathbf{x}^*(\lambda_1,\lambda_2)=\begin{bmatrix}\dfrac{-10\lambda_2}{\lambda_1+4\lambda_2}\\ 0\end{bmatrix},$$

which, by setting $\mu=\lambda_2/\lambda_1$, can be written as

$$\mathbf{x}^*(\mu)=\begin{bmatrix}\dfrac{-10\mu}{1+4\mu}\\ 0\end{bmatrix},$$

thus yielding

$$F_1^*(\mu)=\Big(\frac{-10\mu}{1+4\mu}\Big)^2, \quad \text{and} \quad F_2^*(\mu)=\Big(\frac{-20\mu}{1+4\mu}+5\Big)^2.$$

The remaining Pareto optimal points and values are calculated as follows: let $\mu\to 0$ and $\mu\to\infty$. As $\mu\to 0$, we get $\mathbf{x}^*=0$, which gives $f^*(\mathbf{x})=[0\ \ 25]$. Next, as $\mu\to\infty$, we get $\mathbf{x}^*=[-\frac{5}{2}\ \ 0]^T$ and $f^*(\mathbf{x})=[\frac{25}{4}\ \ 0]$, which corresponds to minimizing the error in the solution with the minimum norm, and this $\mathbf{x}^*$ is not necessarily a Pareto optimal point.

9.9. Prove that any local optimal point of a convex problem is (globally) optimal.

PROOF. Assume x is locally optimal and y is optimal with $f(x)<f(y)$. Since x is locally optimal, it means that there is an M such that if z is feasible and $\|z-x\|_2\leq M$, then $f(z)\geq f(x)$. Let $z=ty+(1-t)x$, where $t=\dfrac{M}{2\|y-x\|_2}$. Then $\|y-x\|_2>M$ for $0<t<\frac{1}{2}$. Since z is a convex combination of two feasible points, it is also feasible. Also, $\|z-x\|_2=M/2$, and thus, $f(z)\leq tf(x)+(1-t)f(x)+(1-t)f(y)<f(x)+(1-t)f(x)$, which contradicts the assumption that x is locally optimal. The result also holds in $\mathbb{R}^n$. ■

10

Optimal Control Theory

A Hamiltonian function is involved in dynamic programming of an objective function on the state variables in optimal control theory. A Hamiltonian is similar to a Lagrangian in concave programming and requires first-order conditions.

10.1 Hamiltonian

The Hamiltonian is the operator corresponding to the total energy of the system in most cases. Its spectrum is the set of possible outcomes when one measures the total energy of a system. Because of its close relation with the time-evolution of a system, it is very important in most formulations of quantum theory. On the other hand, the Hamiltonian in optimal control theory is distinct from its quantum mechanical definition. Pontryagin proved that a necessary condition for solving the optimal control problem is that the control should be chosen so as to minimize the Hamiltonian. This is known as *Pontryagin's minimum principle* which states that a control $u(t)$ is to be chosen so as to minimize the objective function

$$J(u) = \Psi(x(T)) + \int_0^T L(x, u, t)\, dt, \tag{10.1.1}$$

where $x(t)$ is the system state which evolves according to the state equations $\dot{x} = f(x, u, t)$, $x(0) = x_0, t \in [0, T]$, and the control must satisfy the constraints $a \leq u(t) \leq b, t \in [0, T]$.

10.2 Optimal Control

Problems in optimal control theory involve continuous time, a finite time horizon, and fixed endpoints. They are generally written as

$$\text{Maximize } J = \int_0^T f(x(t), y(t), t)\, dt, \tag{10.2.1}$$

subject to $\dot{x} = g(x(t), y(t), t)$, $x(0) = x_0$, $x(T) = x_T$, where J is the value of the functional to be optimized; $x(t)$ is the state variable which changes over time according to the differential equation set equal to zero in the constraint; $y(t)$ is the control variable whose value is selected or controlled to optimize J; t denotes time; and $\dot{x}$ denotes the time derivative of x, i.e., dx/dt. The solution of the optimal control problem (10.2.1) is obtained to set the limits of the optimal dynamic time path for the control variable $y(t)$.

The dynamic optimal control problems involve the *Hamiltonian function H* similar to the Lagrangian in static optimal control problems. The Hamiltonian is defined as

$$H(x(t), y(t), \lambda(t), t) = f(x(t), y(t), \lambda(t), t) + \lambda(t) g(x(t), y(t), t), \qquad (10.2.2)$$

where, unlike the static problems, the multiplier $\lambda(t)$, called the *costate variable*, is a function of t and estimates the marginal value or shadow price of the associate state variable $x(t)$. The method of solving problems of the type (10.2.1) is similar to that used for solving the static optimization problem involving the Lagrangian. Thus, assuming that the Hamiltonian is differentiable in y and strictly concave so that there is an interior solution, and not an endpoint solution, the necessary conditions for maximization are

$$\begin{aligned} &\text{(a)} \quad \frac{\partial H}{\partial y} = 0,\\ &\text{(b)} \quad \frac{\partial \lambda}{\partial t} = \dot{\lambda} = -\frac{\partial H}{\partial x},\\ &\text{(c)} \quad \frac{\partial x}{\partial t} = \dot{x} = \frac{\partial H}{\partial \lambda},\\ &\text{(d)} \quad x(0) = x_0,\\ &\text{(e)} \quad x(T) = x_T. \end{aligned} \qquad (10.2.3)$$

Conditions (a), (b) and (c) are known as the *maximum principle*, and conditions (d) and (e) as the *boundary conditions*; the two equations of motion in conditions (b) and (c) are called the *Hamiltonian system* or the *control system*.

For minimization, the objective functional can simply be multiplied by -1, as in concave programming. If the solution does not involve an end point, $\frac{\partial H}{\partial y}$ need not be equal to zero as in condition (a).

Example 10.1. Solve the following dynamic optimal control problem:

$$\text{Maximize} \int_0^4 (5x - 6y^2)\, dt \ \text{ subject to } \dot{x} = 6y, \ x(0) = 2, x(4) = 98.$$

The Hamiltonian is

$$H = 5x - 6y^2 + \lambda(6y).$$

Using conditions (10.2.1) we get from conditions (a)–(c):

$$\frac{\partial H}{\partial y} = -12y + 6\lambda = 0, \text{ which gives } y = 0.5\lambda, \tag{10.2.4a}$$

$$\dot{\lambda} = -\frac{\partial H}{\partial x} = -5, \tag{10.2.4b}$$

$$\dot{x} = \frac{\partial H}{\partial \lambda} = 6y, \tag{10.2.4c}$$

Next, we integrate (10.2.4b) to get:

$$\lambda = -5t + c_1. \tag{10.2.4d}$$

Also from (10.2.4a), (10.2.4c) and (10.2.4d) we find that $\dot{x} = 6(0.5\lambda) = 3\lambda = 3(-5t + c_1) = -15t + 3c_1$, which upon integration gives

$$x(t) = -7.5t^2 + 3c_1 t + c_2,$$

where the arbitrary constants c_1 and c_2 are determined using the boundary conditions $x(0) = 2, x(4) = 98$, giving $c_2 = 2, c_1 = 18$. Hence,

$$\begin{aligned} x(t) &= -7.5t^2 + 54t + 2 \quad \text{(state variable)}, \\ \lambda(t) &= -5t + 18 \quad \text{(costate variable)}, \\ y(t) &= 0.5\lambda = 0.5(-5t + 18) = -2.5t + 9 \quad \text{(control variable)}. \end{aligned}$$

The equation of motion $\dot{x} = 6y$ also gives the above control variable, since $-15t + 54 = 6y$ yields $y = -2.5t + 9$. Finally, at the endpoints we have $y(0) = 9, y(4) = -1$. Thus the optimal path of the control variable $y(t)$ is linear, starting at the point $(0, 9)$ and ending at the point $(4, -1)$, with a slope of $-10/4 = -2.5$. ■

10.2.1 Sufficient Conditions for Optimization. Assuming that the necessary conditions (10.2.3) for maximization in control theory are satisfied, the sufficiency conditions are satisfied if

(i) The objective functional $f(x(t), y(t), t)$ and the constraint $g(x(t), y(t), t)$ are both differentiable and both concave in x and y; and

(ii) $\lambda(t) = 0$ if the constraint is nonlinear in x and y. However, if the constraint is linear, λ may assume any sign.

Recall that linear functions are either concave or convex, but neither strictly concave nor strictly convex. In the case of nonlinear functions, the easiest test for joint concavity is the following *discriminant test*: Given the Hessian (discriminant of the second-order derivatives) of the function f,

$$\mathbf{H} = \begin{vmatrix} f_{xx} & f_{xy} \\ f_{yx} & f_{yy} \end{vmatrix}, \tag{10.2.5}$$

a function is *strictly concave* if the discriminant is negative definite, i.e., if

$$|H_1| = f_{xx} < 0 \quad \text{and} \quad |H_2| = |\mathbf{H}| > 0;$$

and simply concave if the discriminant is negative semidefinite, i.e., if

$$|H_1| = f_{xx} \leq 0 \quad \text{and} \quad |H_2| \geq |\mathbf{H}| \geq 0.$$

A negative definite discriminant implies a global maximum and is always sufficient for a maximum. A negative semidefinite discriminant indicates a local maximum and is sufficient for a maximum if the test is conducted for every possible ordering of the variables with similar results (see §A.5).

10.3 Free Endpoint

In the case of a free endpoint, say the upper endpoint, the optimal control problem becomes

$$\begin{aligned} &\text{Maximize } J = \int_0^T f(x(t), y(t), t)\, dt \\ &\quad \text{subject to } \dot{x} = g(x(t), y(t), t),\ x(0) = x_0,\ x(T) \text{ free.} \end{aligned} \tag{10.3.1}$$

Then, assuming that the Hamiltonian is differentiable in y and strictly concave so that there is an interior solution, the necessary conditions for maximization remain the same as in (10.2.1) except for condition (e) which becomes

$$\text{(e)} \quad \lambda(T) = 0. \tag{10.3.2}$$

Such a boundary condition is called the *transversality condition* for a free endpoint. The justification for condition (10.3.2) is as follows: If the value of $x(T)$ is free to vary, the constraint must also be free (i.e., nonbinding), and so the shadow price λ evaluated at $x = T$ must be zero.

Example 10.2. Solve the following optimal control problem with a free endpoint:

$$\text{Maximize } \int_0^3 (5x - y^2)\, dt \ \text{ subject to } \dot{x} = 4y,\ x(0) = 2, x(3) \text{ free.}$$

The Hamiltonian is $H = 5x - y^2 + \lambda(4y)$. Using conditions (10.2.1) we get from conditions (a)–(c):

$$\frac{\partial H}{\partial y} = -2y + 4\lambda = 0, \quad \text{which gives} \quad y = 2\lambda, \tag{10.3.3a}$$

$$\dot{\lambda} = -\frac{\partial H}{\partial x} = -5, \tag{10.3.3b}$$

$$\dot{x} = \frac{\partial H}{\partial \lambda} = 4y, \tag{10.3.3c}$$

Then (10.3.3b) gives

$$\lambda(t) = -5t + c_1. \tag{10.3.3d}$$

The constant c_1 is determined using the boundary condition $\lambda(3) = 0$ in (10.3.3d), which gives $c_1 = 15$. From (10.3.3a), (10.3.3c) and (10.3.3d), we get $\dot{x} = -40t + 120$, which upon integration gives

$$x(t) = -20t^2 + 120t + c_2,$$

where c_2 is determined using condition (d) $x(0) = 2$, giving $c_2 = 2$. Thus,

$$x(t) = -20t^2 + 120t + 2.$$

From (10.3.3a) we get the control variable

$$y(t) = 2\lambda = -10t + 30,$$

with the endpoint values $y(0) = 30$ and $y(3) = 0$. Thus, the optimal path of the control variable is linear, from point $(0, 30)$ to $(3, 0)$, with the slope of $-30/3 = -10$. ■

10.4 Inequality Constraints at the Endpoints

In the case when the terminal value of the state variable $x(t)$ is subject to an inequality constraint of the type $x(T) \geq x_{\min}$, the optimal value $x^*(T)$ may be chosen freely as long as it does not violate the value set by the constraint $x_{\min}$. Thus, if $x^*(T) > x_{\min}$, the constraint becomes nonbinding like the case of a free endpoint, and we can take $\lambda(T) = 0$ when $x^*(T) > x_{\min}$. In such a case, conditions (a) through (d) of (10.2.3) remain the same, but the condition (e) is replaced by $\lambda(T) = 0$, as in the case of a free endpoint. However, if $x^*(T) < x_{\min}$, the constraint is binding and the optimal solution will require setting $x(T) = x_{\min}$, i.e., a fixed-end problem with

$$\lambda(T) \geq 0 \quad \text{when } x^*(T) = x_{\min}. \tag{10.4.1}$$

The endpoint problems are sometimes reduced to a single statement

$$\lambda(T) \geq 0, \quad x(T) \geq x_{\min}, \quad [x(T) - x_{\min}]\,\lambda(T) = 0, \tag{10.4.2}$$

which is similar to the KKT condition. In practice, however, the problems with inequality constraints are easy to solve if we follow the following three steps:

STEP 1. Solve the problem as if it were a free endpoint problem.

STEP 2. If the optimal value of the state variable $x(t)$ is greater than the minimum required by the endpoint condition, i.e., if $x^{(}T) \geq x_{\min}$, then the correct solution has been found.

STEP 3. If $x^*(T) < x_{\min}$, set the terminal endpoint equal to the value of the constraint, $x(T) = x_{\min}$, and solve as a fixed endpoint problem.

Example 10.3. Solve

$$\text{Maximize } \int_0^3 (5x - y^2)\, dt \text{ subject to } \dot{x} = 4y,\ x(0) = 2, x(3) \geq 180.$$

First, solve it as an unconstrained problem with a free endpoint. From Example 10.2, we have:

$$x(t) = -40t^2 + 120t + 2, \quad \text{which gives } x(3) = 182 > 180.$$

Since the free endpoint solution satisfies the terminal endpoint constraint $X(T) \geq 180$, the constraint is not binding and we thus have a proper solution, where from Example 10.2, the control variable $y(t) = -10t + 30$. ■

Example 10.4. Solve the same problem as in Example 10.3 but with the new boundary conditions: $x(0) = 5$ and $x(4) \geq 190$. First, we will use the complementary slackness condition to find the solution by assuming that $x(1) - S^2 = 190$, where S is called the *slackness variable*. There are two cases to consider:

CASE 1. $\lambda = 0$: Then $-2y = 0$, or $y = 0$. Also, $\dot{x} = 4y = 0$ gives $x(t) = a_1$, which using the initial condition $x(0) = 2$, gives $a_1 = 2$. Then the terminal condition $x(1) = S^2 + 190$ gives $S^2 + 190 = 2$, or $S^2 = -188$, which is infeasible.

CASE 2. $S = 0$: Then the first two steps are the same as in Example 10.3 solved as a free endpoint problem. The maximum principle gives

$$y = 2\lambda, \quad \dot{\lambda} = -5, \quad \dot{x} = -40t + 8c_1,$$

thus, giving $\lambda(t) = -5t + c_1$, $x(t) = -20t^2 + 8c_1 t + c_2$. Now the new boundary conditions are $x(0) = 5$ and $x(4) = 190$, which yield $c_2 = 5$ and $c_1 = 15.2$. Hence, $\lambda(t) = -5t + 15.2$, $x(t) = -20t^2 + 121.6t + 5$, and $y(t) = -10t + 30.4$. ■

10.5 Discounted Optimal Control

Optimal control problems involving discounting are expressed as follows:

$$\begin{gathered}\text{Maximize } J = \int_0^T e^{-pt} f(x(t), y(t), t)\, dt \\ \text{subject to } \dot{x} = g(x(t), y(t), t),\ x(0) = x_0,\ x(T) \text{ free.}\end{gathered} \tag{10.5.1}$$

The Hamiltonian for this problem is

$$H = e^{-pt}\, f(x(t), y(t), t)\, dt + \lambda(t) g(x(t), y(t), t)\, dt.$$

If we set $\mu(t) = e^{pt}\, \lambda(t)$, then the Hamiltonian is modified to

$$H_c = e^{pt}\, H = f(x(t), y(t), t)\, dt + \mu(t) g(x(t), y(t), t)\, dt \tag{10.5.2}$$

where H_c is called the *current value Hamiltonian.* The above optimization problem can be solved by the above method for a free endpoint condition, where H is replaced by H_c and λ by μ. Then, $\dot{\lambda} = -\dfrac{\partial H}{\partial x} = -\dfrac{\partial H_c}{\partial x}\, e^{-pt}$. But since $la(t) = \mu(t)\, er^{-pt}$, we get by differentiating, $\dot{\lambda} = \dot{\mu}\, e^{-pt} - p\mu\, e^{-pt}$. By equating these two expressions for $\dot{\lambda}$ and canceling the term e^{-pt}, condition (b) in (10.2.3) is replaced by $\dot{\mu} = p\mu - \dfrac{\partial H_c}{\partial x}$. The boundary conditions (d) and (e) are similarly adjusted. Thus, assuming an interior solution exists, the necessary conditions for the current-value Hamiltonian H_c are

$$\begin{aligned}
&\text{(a)} \quad \frac{\partial H_c}{\partial y} = 0, \\
&\text{(b)} \quad \frac{\partial \mu}{\partial t} = \dot{\mu} = -\frac{\partial H_c}{\partial x}, \\
&\text{(c)} \quad \frac{\partial x}{\partial t} = \dot{x} = \frac{\partial H_c}{\partial \mu}, \\
&\text{(d)} \quad x(0) = x_0, \\
&\text{(e)} \quad \mu(T)\, e^{-pt} = 0.
\end{aligned} \tag{10.5.3}$$

If the solution does not involve an endpoint, then $\dfrac{\partial H_c}{\partial y} \neq 0$ in condition (a), but H_c must still be maximized with respect to y. Since $H_c = H\, e^{pt}$, the value of y that will maximize H_c will also maximize H_c since e^{pt}, being independent of y, is treated like a constant when maximizing with respect to y. The sufficiency conditions, depending on the sign of the Hessian $|\mathbf{H}|$, remain the same for the discounted optimal control problem.

Example 10.5. Maximize $\displaystyle\int_0^3 e^{-0.02t}(x - 2x^2 - 5y^2)\, dt$, subject to $\dot{x} = y - 0.5x$, $x(0) = 90$, and $x(3)$ free. SOLUTION. First, we check the sufficient conditions to ensure that this problem has a global maximum. The Hessian

$$|\mathbf{H}| = \begin{vmatrix} f_{xx} & f_{xy} \\ f_{yx} & f_{yy} \end{vmatrix} = \begin{vmatrix} -4 & 0 \\ 0 & -10 \end{vmatrix} = 40 > 0,$$

and the first principal minor $|H_1| = -4 < 0$, while the second principal minor $|H_2| = |\mathbf{H}| > 0$, which imply that the Hessian is negative definite. Thus, f is concave in both x and y, and $g = y - 4x$ is linear, so conditions for a global maximum are satisfied.

Now, the current-valued Hamiltonian is

$$H_c = x - 2x^2 - 5y^2 + \mu(y - 0.5x).$$

Applying the maximum principle (10.5.3) and using $p = 0.02$, we get

$$\begin{aligned}
\frac{\partial H_c}{\partial y} &= 10y + \mu = 0, \quad \text{which gives} \quad y = 0.1\mu, \\
\dot{\mu} = p\mu - \frac{\partial H_c}{\partial x} &= 0.02\mu - (1 - 4x - 0.5\mu) = 0.52\mu + 4x - 1, \\
\dot{x} = \frac{\partial H_c}{\partial \mu} &= y - 0.5x = 0.1\mu - 0.5x.
\end{aligned}$$

To solve for μ and x, the last two equations above are written in the matrix form $\mathbf{Y} = \mathbf{AY} + \mathbf{B}$ as

$$\begin{bmatrix} \dot{\mu} \\ \dot{x} \end{bmatrix} = \begin{bmatrix} 0.52 & 4 \\ 0.1 & -0.5 \end{bmatrix} \begin{bmatrix} \mu \\ x \end{bmatrix} + \begin{bmatrix} -1 \\ 0 \end{bmatrix}.$$

The characteristic equation for this system of equations is

$$|\mathbf{A} - r\mathbf{I}| = \begin{vmatrix} 0.52 - r & 4 \\ 0.1 & -0.5 - r \end{vmatrix} = 0,$$

where, using formula (A.20) with λ replaced by r so it does not conflict with the Lagrange multiplier λ, the characteristic roots are

$$r_{1,2} = \frac{0.02 \pm \sqrt{(0.02)^2 - 4(-0.66)}}{2} = 0.82245, -0.80245.$$

For $r_1 = 0.82245$, the eigenvector $\mathbf{y}_e^1$ is determined by solving the equation

$$\begin{bmatrix} 0.52 - 0.82245 & 4 \\ 0.1 & -0.5 - 0.82245 \end{bmatrix} \begin{bmatrix} c_1 \\ c_2 \end{bmatrix} = \begin{bmatrix} -0.30245 & 4 \\ 0.1 & 1.32245 \end{bmatrix} \begin{bmatrix} c_1 \\ c_2 \end{bmatrix} = 0,$$

which gives $-0.30245c_1 + 4c_2 = 0$, or $c_1 = 3.306c_2$, so that

$$\mathbf{y}_e^1 = \begin{bmatrix} 3.306 \\ 1 \end{bmatrix} k_1 \, e^{0.82245t}.$$

For $r_2 = -0.80245$, the eigenvector $\mathbf{y}_e^2$ is obtained by solving

$$\begin{bmatrix} 0.52 + 0.80245 & 4 \\ 0.1 & -0.5 + 0.80245 \end{bmatrix} \begin{bmatrix} c_1 \\ c_2 \end{bmatrix} = \begin{bmatrix} 1.32245 & 4 \\ 0.1 & 0.30245 \end{bmatrix} \begin{bmatrix} c_1 \\ c_2 \end{bmatrix} = 0,$$

which gives $1.32245c_1 + 4c_2 = 0$, or $c_1 = -3.0247c_2$, so that

$$\mathbf{y}_e^2 = \begin{bmatrix} -3.0247 \\ 1 \end{bmatrix} k_2\, e^{-0.80245t}.$$

The particular solution is given by $\mathbf{Y}^* = -\mathbf{A}^{-1}\mathbf{B}$, or using (A.10), by

$$\mathbf{Y} = \begin{bmatrix} \mu^* \\ x^* \end{bmatrix} = -\frac{1}{-0.66}\begin{bmatrix} -0.5 & -6 \\ -0.1 & 0.52 \end{bmatrix}\begin{bmatrix} -1 \\ 0 \end{bmatrix} = \begin{bmatrix} 0.758 \\ 0.152 \end{bmatrix}.$$

Adding the complementary and particular solutions we get

$$\begin{aligned} \mu(t) &= 3.306k_1\, e^{0.82245t} - 3.0247k_2\, e^{-0.80247t} + 0.758, \\ x(t) &= k_1\, e^{0.82245t} + k_2\, e^{-0.80247t} + 0.152. \end{aligned}$$

Now, we apply the boundary conditions: Since $\mu(T)\, e^{-pt} = 0$ at the free endpoint, we get at $T = 3$,

$$\mu(3)\, e^{-0.02(3)} = 0,$$

or

$$[3.306k_1\, e^{0.82245(3)} - 3.0247k_2\, e^{-0.80247(3)} + 0.758]\, e^{-0.06} = 0,$$

or

$$37.033k_1 - 0.256k_2 + 0.713 = 0. \tag{10.5.4}$$

Also, at $T = 0$, $x(0) = 90$, so we have

$$k_1 + k_2 + 0.152 = 90. \tag{10.5.5}$$

Solving (10.5.4) and (10.5.5) simultaneously by Cramer's rule, we get $k_1 = 0.59$, and $k_2 = 89.25$. Hence,

$$\begin{aligned} \mu(t) &= 1.95\, e^{0.82245t} - 269.94\, e^{-0.80247t} + 0.758, \quad \text{costate variable}, \\ x(t) &= 0.59\, e^{0.82245t} + 89.25\, e^{-0.80247t} + 0.152, \quad \text{state variable}, \end{aligned}$$

and

$$y(t) = 0.1\mu(t) = 0.195\, e^{0.82245t} - 26.994\, e^{-0.80247t} + 0.0758, \quad \text{control variable.}\ \blacksquare$$

10.6 Exercises

FIXED ENDPOINT:

10.1. Maximize $\int_0^2 (3x - 2y^2)\,dt$ subject to $\dot{x} = 8y$, $x(0) = 5$, and $x(2)$ free. HINT. Hamiltonian $H = 3x - 2y^2 + \lambda(8y)$; $\dot{x} = 16\lambda$, $x(t) = -24t^2 + 96t + 5$ (state variable); $y(t) = -6t + 12$ (control variable). The optimal path of the control variable is linear starting at $(0, 12)$ and ending at $(2, 0)$ with slope -6.

FREE ENDPOINTS:

10.2. Solve the optimal control problem with a free endpoint:

$$\text{Maximize } \int_0^4 (5x - 2y^2)\,dt \text{ subject to } \dot{x} = 8y,\ x(0) = 2, x(4) \text{ free.}$$

SOLUTION. We have $\dfrac{\partial H}{\partial y} = -4y + 8\lambda = 0$, which gives $y = 0.5\lambda$, and $\dot{\lambda} = -\dfrac{\partial H}{\partial x} = -5$, $\dot{x} = \dfrac{\partial H}{\partial \lambda} = 8y$. Then $\lambda(t) = -5t + c_1$, where $c_1 = 20$ using the boundary condition $\lambda(4) = 0$. Also, since $\dot{x} = -80t + 16c_1$, we get on integrating, $x(t) = -40t^2 + 320t + c_2$, where $c_2 = 2$ using condition $x(0) = 2$. Thus, $x(t) = -40t^2 + 320t + 2$, and then $y(t) = 0.5\lambda = -2.5t + 10$, with the endpoint values $y(0) = 10$ and $y(4) = 0$. Thus, the optimal path of the control variable is linear, from point $(0, 10)$ to $(4, 0)$, with the slope of $-10/4 = -2.5$.

10.3. Maximize $\displaystyle\int_0^2 (9x - 12y^2)\,dt$, $\dot{x} = 18y$, $x(0) = 5$, $x(3)$ free. SOLUTION. First we check if the sufficiency condition is met. The Hessian $|\mathbf{H}| = \begin{vmatrix} f_{xx} & f_{xy} \\ f_{yx} & F_{yy} \end{vmatrix} = \begin{vmatrix} 9 & 0 \\ 0 & -24 \end{vmatrix} = -24 < 0$, and the first principal minor $|H_1| = 9 > 0$, while the second principal minor $|H_2| = |\mathbf{H}| > 0$, which imply that the Hessian is negative definite. Thus, f is concave in both x and y, and $g = 8y$ is linear, so conditions for a global maximum are satisfied.

The Hamiltonian is

$$H = 9x - 12y^2 + \lambda(18y).$$

Then we have

$$\begin{aligned}
&\frac{\partial H}{\partial y} = -24y + 18y, \quad \text{which gives} \quad y = 0.75\lambda,\\
&\dot{\lambda} = -\frac{\partial H}{\partial x} = -9,\\
&\dot{x} = \frac{\partial H}{\partial \lambda} = 18y = 18(0.75\lambda) = 13.5\lambda.
\end{aligned}$$

Integrating the first of the last two equations, we get $\lambda(t) = -9t + c_1$. Then $\dot{x} = 13.5(-9t + c_1) = -121.5t + 13.5c_1$, which on integrating gives

$$x(t) = -60.5t^2 + 13.5c_1 t + c_2.$$

Using the transversality condition $\lambda(3) = 0$, we find that $c_1 = 27$, and thus, the costate variable is $\lambda(t) = -9t + 27$. Condition $x(0) = 5$ gives $c_2 = 5$, and the state variable is

$$x(t) = -60.5t^2 + 364.5t + 5,$$

and the control variable is given by

$$y(t) = 0.7\lambda(t) = -12t + 20.25.$$

10.4. Maximize $\int_0^2 (3y - y^2 - x - 2x^2)\,dt$, $\dot{x} = x + 2y$, $x(0) = 6$, $x(2)$ free.

SOLUTION. First we check if the sufficiency condition is met. The Hessian

$$|\mathbf{H}| = \begin{vmatrix} f_{xx} & f_{xy} \\ f_{yx} & F_{yy} \end{vmatrix} = \begin{vmatrix} -4 & 0 \\ 0 & -2 \end{vmatrix} = 8 > 0,$$

and the first principal minor $|H_1| = -4 < 0$, while the second principal minor $|H_2| = |\mathbf{H}| > 0$, which imply that the Hessian is negative definite. Thus, f is concave in both x and y, and $g = x + 2y$ is linear, so conditions for a global maximum are satisfied.

The Hamiltonian is

$$H = 3y - y^2 - x - 2x^2 + \lambda(x + 2y).$$

Then we have

$$\begin{aligned} \frac{\partial H}{\partial y} &= 3 - 2y + 2\lambda, \quad \text{which gives} \quad y = \lambda + 1.5, \\ \dot{\lambda} &= -\frac{\partial H}{\partial x} = 1 + 4x - \lambda, \\ \dot{x} &= \frac{\partial H}{\partial \lambda} = x + 2y = x + \lambda + 1.5\lambda. \end{aligned}$$

The last two equations in matrix form $\mathbf{Y} = \mathbf{AX} + \mathbf{B}$ are

$$\begin{bmatrix} \dot{\lambda} \\ \dot{x} \end{bmatrix} = \begin{bmatrix} -1 & 4 \\ 1 & 1 \end{bmatrix} \begin{bmatrix} \lambda \\ x \end{bmatrix} + \begin{bmatrix} 1 \\ 1.5 \end{bmatrix}.$$

The characteristic roots are given by ($|\mathbf{A}| = -5, \operatorname{tr}\mathbf{A} = 0$)

$$r_{1,2} = \frac{0 \pm \sqrt{(0)^2 - 4(-5)}}{2} = \pm 2.236.$$

The eigenvector $\mathbf{y}_e^1$ for the root $r_1 = 2.236$ is obtained from

$$\begin{bmatrix} -1-2.236 & 4 \\ 1 & 1-2.236 \end{bmatrix} \begin{bmatrix} c_1 \\ c_2 \end{bmatrix} = \begin{bmatrix} -3.236 & 4 \\ 1 & -1.236 \end{bmatrix} \begin{bmatrix} c_1 \\ c_2 \end{bmatrix} = 0,$$

or $-3.236c_1 + 4c_2 = 0$, which gives $c_1 = 1.236c_2$. Thus,

$$\mathbf{y}_e^1 = \begin{bmatrix} 1.236 \\ 1 \end{bmatrix} k_1\, e^{2.236t}.$$

The eigenvector $\mathbf{y}_e^2$ for the root $r_2 = -2.236$ is obtained from

$$\begin{bmatrix} -1+2.236 & 4 \\ 1 & 1+2.236 \end{bmatrix} \begin{bmatrix} c_1 \\ c_2 \end{bmatrix} = \begin{bmatrix} 1.236 & 4 \\ 1 & 3.236 \end{bmatrix} \begin{bmatrix} c_1 \\ c_2 \end{bmatrix} = 0,$$

or $1.236c_1 + 4c_2 = 0$, which gives $c_1 = -3.236c_2$. Thus,

$$\mathbf{y}_e^2 = \begin{bmatrix} -3.236 \\ 1 \end{bmatrix} k_2\, e^{-2.236t}.$$

For the particular solution we get

$$\begin{bmatrix} \lambda \\ x \end{bmatrix} = -\frac{1}{(-5)} \begin{bmatrix} 1 & -4 \\ -1 & -1 \end{bmatrix} \begin{bmatrix} 1 \\ 1.5 \end{bmatrix} = \begin{bmatrix} -1 \\ -0.5 \end{bmatrix},$$

which gives

$$\begin{aligned} \lambda(t) &= 1.236k_1e^{2.236t} + 3.236k_2e^{-2.236t} - 1, \\ x(t) &= k_1e^{2.236t} + k_2e^{-2.236t} - 0.5. \end{aligned} \tag{9.6.1}$$

The transversality condition $\lambda(2) = 0$ and the initial condition $x(0) = 6$, when applied to the above two equations, respectively, give

$$\begin{aligned} 108k_1 - 0.37k_2 &= 1, \\ 87.532k_1 + 0.115k_2 &= 6.5, \end{aligned}$$

which when solved simultaneously, e.g., by Cramer's rule, give $k_1 = 0.056$ and $k_2 = 13.73$. Hence, from Eq (9.6.1) we get

$$\begin{aligned} \lambda(t) &= 0.67e^{2.236t} - 44.43e^{-2.236t} - 1, \quad \text{costate variable} \\ x(t) &= 0.05e^{2.236t} + 13.73e^{-2.236t} - 0.5, \quad \text{state variable} \\ y(t) &= \lambda(t) + 1.5 = 0.67e^{2.236t} - 44.43e^{-2.236t} + 0.5, \quad \text{control variable.} \end{aligned}$$

10.5. Maximize $\int_0^4 (5x - 2y^2)\, dt$ subject to $\dot{x} = 8y,\ x(0) = 2, x(4) \geq 800.$

SOLUTION. From Exercise 10.2 we have

$$\lambda(t) = -5t + 20,$$
$$x(t) = -40t^2 + 320t + 2,$$
$$y(t) = -2.5t + 10.$$

Now we evaluate $x(t)$ at $x = 4$: $x(4) = 642 < 800$, so the constraint is violated. Thus, we redo this problem as a new fixed endpoint conditions: $x(0) = 2$ and $x(4) = 800$. Then from Exercise 10.2, since $\lambda(t) = -5t + c_1$, and $x(t) = -40t^2 + 16c_1 t + c_2$, we apply the new endpoint conditions, and get $x(0) = c_2 = 2$, and $x(4) = -640 + 64c_1 + 2 = 800$, giving $c_1 = 22.47$. Hence,

$$\lambda(t) = -5t + 22.47, \quad \text{costate variable}$$
$$x(t) = -40t^2 + 359.52 + 2, \quad \text{state variable}$$
$$y(t) = -2.5t + 11.26, \quad \text{control variable.}$$

10.6. Solve the same problem as in Exercise 9.2 but with the new boundary conditions: $x(0) = 5$ and $x(4) \geq 650$. The first two steps are the same as in Example k.1 solved as a free endpoint problem. The maximum principle gives

$$y = 2\lambda, \quad \dot{\lambda} = -5, \quad \dot{x} = -80t^2 + 16c_1,$$

thus, giving $\lambda(t) = -5t + c_1$, $x(t) = -40t^2 + 16c_1 t + c_2$. Now the new boundary conditions are $x(0) = 2$ and $x(4) = 650$, which yield $c_2 = 2$ and $c_1 = 20.125$. Hence, $\lambda(t) = -5t + 20.125$, $x(t) = -40t^2 + 322t + 2$, and $y(t) = -2.5t + 20.125$.

10.7. Solve

$$\text{Maximize} \int_0^4 (5x - 2y^2)\, dt \ \text{ subject to } \dot{x} = 8y,\ x(0) = 2, x(4) \geq 620.$$

First, solve it as an unconstrained problem with a free endpoint. From Exercise 10.5, we have:

$$x(t) = -40t^2 + 320t + 2, \quad \text{which gives } x(4) = 642 > 620.$$

Since the free endpoint solution satisfies the terminal endpoint constraint $X(T) \geq 620$, the constraint is not binding and we thus have a proper solution, where from Example 9.2, the control variable $y(t) = -10t + 40$.

10.8. Maximize $\int_0^1 (8x + 3y - 2y^2)\, dt$ subject to $\dot{x} = 8y,\ x(0) = 9, x(1) \geq 90.$

SOLUTION. The Hamiltonian is

$$H = 8x + 3y - 2y^2 + \lambda(8y).$$

Then, using conditions (10.2.3) we have

$$\begin{aligned}
&\frac{\partial H}{\partial y} = 3 - 4y + 8\lambda = 0, \quad \text{which gives } y = 2\lambda + 1.5, \\
&\dot{\lambda} = -\frac{\partial H}{\partial x} = -8, \\
&\dot{x} = \frac{\partial H}{\partial \lambda} = 8y = 16\lambda + 12.
\end{aligned} \tag{9.6.2}$$

First, we will use the complementary slackness condition to find the solution by assuming that $x(1) - S^2 = 90$. There are two cases to consider:

CASE 1. $\lambda = 0$: Then $3 - 4y = 0$, or $y = 0.75$. Also, $\dot{x} = 12$ gives $x(t) = 12t + a_1$, which using the initial condition $x(0) = 9$, gives $a_1 = 9$. Then the terminal condition $x(1) = S^2 + 90$ gives $12 + 9 - S^2 = 90$, or $S^2 = -69$, which is infeasible.

CASE 2. $S = 0$: Then integrating the last two equations in (9.6.2), we get

$$\begin{aligned}
&\lambda(t) = -8t + c_1, \\
&\dot{x} = 16(-8t + c_1) + 12 = -128t + 16c_1 + 12.
\end{aligned} \tag{9.6.3}$$

The last equation on integration gives

$$x(t) = -64t^2 + 16c_1 + 12t + c_2.$$

Using the initial condition $x(0) = 9$, we get $c_2 = 9$. Also, using the transversality condition $\lambda(1) = 0$, we get $c_1 = 8$. Thus,

$$\begin{aligned}
\lambda(t) &= -8t + 8, \\
x(t) &= -64t^2 = 140t + 9.
\end{aligned}$$

Now, to see if this solution is acceptable, we check $x(1) = 85 < 90$. So the terminal constraint is violated. Thus, in this situation we solve the problem with a fixed endpoint condition $x(1) = 90$. Then from Eqs (9.6.3), condition $x(0) = 9$ gives $c_2 = 9$, and the new constraint $x(1) = 90$ gives $c_1 = 9.0625$. Hence,

$$\begin{aligned}
&\lambda(t) = -8t + 9.0625, \quad \text{costate variable} \\
&x(t) = -64t^2 + 15t + 9, \quad \text{state variable} \\
&y(t) = 2\lambda(t) + 1.5 = -16t + 19.625, \quad \text{control variable.}
\end{aligned}$$

10.9. Maximize $\int_0^3 e^{-0.04t}\,(xy - x^2 - y^2)\,dt$ subject to $\dot{x} = x + 2y$, $x(0) = 130.2$, and $x(3)$ free. SOLUTION. First, we check the sufficient conditions to ensure that this problem has a global maximum. The Hessian

$$|\mathbf{H}| = \begin{vmatrix} f_{xx} & f_{xy} \\ f_{yx} & f_{yy} \end{vmatrix} = \begin{vmatrix} -2 & 0 \\ 0 & -2 \end{vmatrix} = 4 > 0,$$

and the first principal minor $|H_1| = -2 < 0$, while the second principal minor $|H_2| = |\mathbf{H}| > 0$, which imply that the Hessian is negative definite. Thus, f is concave in both x and y, and $g = x + 2y$ is linear, so conditions for a global maximum are satisfied.

Now, the current-valued Hamiltonian is

$$H_c = xy - x^2 - y^2 + \mu(x + 2y).$$

Applying the maximum principle (9.5.3) and using $p = 0.04$, we get

$$\begin{aligned} \frac{\partial H_c}{\partial y} &= x - 2y + 2\mu = 0, \quad \text{which gives} \quad y = 0.5(x + 2\mu), \\ \dot{\mu} = p\mu - \frac{\partial H_c}{\partial x} &= 0.04\mu - (y - 2x + \mu) = -0.96\mu + 1.5x, \\ \dot{x} = \frac{\partial H_c}{\partial \mu} &= x + 2y = 2x + 2\mu. \end{aligned}$$

To solve for μ and x, the last two equations above are written in the matrix form $\mathbf{Y} = \mathbf{AY} + \mathbf{B}$ as

$$\begin{bmatrix} \dot{\mu} \\ \dot{x} \end{bmatrix} = \begin{bmatrix} -1.96 & 1.5 \\ 2 & 2 \end{bmatrix} \begin{bmatrix} \mu \\ x \end{bmatrix} + \begin{bmatrix} 0 \\ 0 \end{bmatrix}.$$

The characteristic equation for this system of equations is

$$|\mathbf{A} - r\mathbf{I}| = \begin{vmatrix} -1.96 - r & 1.5 \\ 2 & 2 - r \end{vmatrix} = 0,$$

where, using formula (A.20), the characteristic roots are

$$r_{1,2} = \frac{0.04 \pm \sqrt{(0.04)^2 - 4(-0.692)}}{2} = 2.65, -2.61.$$

For $r_1 = 2.65$, the eigenvector $\mathbf{y}_e^1$ is determined by solving the equation

$$\begin{bmatrix} -1.96 - 2.65 & 1.5 \\ 2 & 2 - 2.65 \end{bmatrix} \begin{bmatrix} c_1 \\ c_2 \end{bmatrix} = \begin{bmatrix} -4.61 & 1.5 \\ 2 & -0.65 \end{bmatrix} \begin{bmatrix} c_1 \\ c_2 \end{bmatrix} = 0,$$

which gives $-4.61c_1 + 1.5c_2 = 0$, or $c_2 = 3.073c_1$, so that

$$\mathbf{y}_e^1 = \begin{bmatrix} 1 \\ 3.073 \end{bmatrix} k_1\, e^{2.65t}.$$

For $r_2 = -2.61$, the eigenvector $\mathbf{y}_e^2$ is obtained by solving

$$\begin{bmatrix} -1.96 + 2.61 & 1.5 \\ 2 & 2 + 2.61 \end{bmatrix} \begin{bmatrix} c_1 \\ c_2 \end{bmatrix} = \begin{bmatrix} 0.65 & 1.5 \\ 2 & 4.61 \end{bmatrix} \begin{bmatrix} c_1 \\ c_2 \end{bmatrix} = 0,$$

which gives $0.65c_1 + 1.5c_2 = 0$, or $c_1 = -2.308c_2$, so that

$$\mathbf{y}_e^2 = \begin{bmatrix} -2.308 \\ 1 \end{bmatrix} k_2\, e^{-2.61t}.$$

The particular solution is given by $\mathbf{Y}^* = -\mathbf{A}^{-1}\mathbf{B}$, where $\mathbf{B} = \{\mathbf{0}\}$. Thus, adding the complementary and particular solutions we get

$$\mathbf{Y}^* = \begin{bmatrix} \mu \\ x \end{bmatrix} = -\frac{1}{-6.92} \begin{bmatrix} 2 & -2 \\ -1.5 & 1.96 \end{bmatrix} \begin{bmatrix} 0 \\ 0 \end{bmatrix} = \begin{bmatrix} 0 \\ 0 \end{bmatrix}.$$

Then

$$\begin{aligned} \mu(t) &= k_1\, e^{2.65t} - 2.308k_2\, e^{-2.61t}, \\ x(t) &= 3.073k_1\, e^{2.65t} + k_2\, e^{-2.61t}. \end{aligned}$$

Now, we apply the boundary conditions: at the free endpoint $\mu(3)\, e^{-0.04t} = 0$, or $\mu(3)e^{-0.04(3)} = \mu(3)\, e^{-0.12} = 0$, which gives

$$13.599k_1 - 0.162k_2 = 0.$$

Also, $x(0) = 130.2$ gives

$$43.495k_1 + 0.0735k_2 = 130.2.$$

Solving these two equations simultaneously by Cramer's rule, we get $k_1 = 2.62$, and $k_2 = 220.0589$. Hence,

$$\begin{aligned} \mu(t) &= 2.62\, e^{2.65t} - 507.89\, e^{-2.61t}, && \text{costate variable,} \\ x(t) &= 8.05\, e^{2.65t} + 220.06\, e^{-2.61t}, && \text{state variable,} \end{aligned}$$

and the control variable is

$$y(t) = 0.5(x + 2\mu(t)) = 6.645\, e^{2.65t} - 397.56\, e^{-2.61t}.$$

10.10. Maximize $\int_0^1 e^{-0.07t}\,(8x + 3y + xy - 2x^2 - 0.8y^2)\,dt$ subject to $\dot{x} = x + 4y$, $x(0) = 91$, and $x(1)$ free. SOLUTION. First, we check the sufficient conditions to ensure that this problem has a global maximum. The Hessian

$$|\mathbf{H}| = \begin{vmatrix} f_{xx} & f_{xy} \\ f_{yx} & f_{yy} \end{vmatrix} = \begin{vmatrix} -4 & 1 \\ 1 & -1.6 \end{vmatrix} = 5.6 > 0,$$

and the first principal minor $|H_1| = -4 < 0$, while the second principal minor $|H_2| = |\mathbf{H}| > 0$, which imply that the Hessian is negative definite. Thus, f is concave in both x and y, and $g = x + 4y$ is linear, so conditions for a global maximum are satisfied.

Now, the current-valued Hamiltonian is

$$H_c = 8x + 3y + xy - 2x^2 - 0.8y^2 + \mu(x + 4y).$$

Applying the maximum principle (9.5.3) and using $p = 0.04$, we get

$$\frac{\partial H_c}{\partial y} = 3 + x - 1.6y + 4\mu = 0, \quad \text{which gives} \quad y = 0.625x + 2.5\mu + 1.875,$$

$$\dot{\mu} = p\mu - \frac{\partial H_c}{\partial x} = 0.07\mu - (8 + y - 4x + \mu) = -1.57\mu + 3.375x - 9.875,$$

$$\dot{x} = \frac{\partial H_c}{\partial \mu} = x + 4y = 10\mu + 3.5x + 7.5.$$

To solve for μ and x, the last two equations above are written in the matrix form $\mathbf{Y} = \mathbf{AY} + \mathbf{B}$ as

$$\begin{bmatrix} \dot{\mu} \\ \dot{x} \end{bmatrix} = \begin{bmatrix} -1.57 & 3.375 \\ 10 & 3.5 \end{bmatrix} \begin{bmatrix} \mu \\ x \end{bmatrix} + \begin{bmatrix} -9.875 \\ 7.5 \end{bmatrix}.$$

The characteristic equation for this system of equations is

$$|\mathbf{A} - r\mathbf{I}| = \begin{vmatrix} -1.57 - r & 3.375 \\ 10 & 3.5 - r \end{vmatrix} = 0,$$

where, using formula (A.20), the characteristic roots are

$$r_{1,2} = \frac{0.07 \pm \sqrt{(0.07)^2 - 4(-39.245)}}{2} = 6.299, -6.229.$$

For $r_1 = 6.299$, the eigenvector $\mathbf{y}_e^1$ is determined by solving the equation

$$\begin{bmatrix} -1.57 - 6.299 & 3.375 \\ 10 & 3.5 - 6.299 \end{bmatrix} \begin{bmatrix} c_1 \\ c_2 \end{bmatrix} = \begin{bmatrix} -7.869 & 3.375 \\ 10 & 3.5 - 6.299 \end{bmatrix} \begin{bmatrix} c_1 \\ c_2 \end{bmatrix} = 0,$$

which gives $-7.869c_1 + 3.375c_2 = 0$, or $c_2 = 2.3315c_1$, so that

$$\mathbf{y}_e^1 = \begin{bmatrix} 1 \\ 2.3315 \end{bmatrix} k_1\, e^{6.299t}.$$

For $r_2 = -6.229$, the eigenvector $\mathbf{y}_e^2$ is obtained by solving

$$\begin{bmatrix} -1.75 + 6.229 & 3.375 \\ 10 & 3.5 + 6.229 \end{bmatrix} \begin{bmatrix} c_1 \\ c_2 \end{bmatrix} = \begin{bmatrix} 4.659 & 3.375 \\ 10 & 9.729 \end{bmatrix} \begin{bmatrix} c_1 \\ c_2 \end{bmatrix} = 0,$$

which gives $4.659c_1 + 3.375c_2 = 0$, or $c_1 = -0.724c_2$, so that

$$\mathbf{y}_e^2 = \begin{bmatrix} --0.724 \\ 1 \end{bmatrix} k_2\, e^{-6.229t}.$$

The particular solution is given by $\mathbf{Y}^* = -\mathbf{A}^{-1}\mathbf{B}$. Thus, by adding the complementary and particular solutions we get

$$\mathbf{Y}^* = \begin{bmatrix} \mu \\ x \end{bmatrix} = -\frac{1}{-39.245} \begin{bmatrix} 3.5 & -3.375 \\ -10 & -1.57 \end{bmatrix} + \begin{bmatrix} -59.875 \\ 78.975 \end{bmatrix}.$$

Then

$$\begin{aligned} \mu(t) &= k_1\, e^{6.299t} - 0.724k_2\, e^{-6.229t} - 59.875, \\ x(t) &= 2.3315k_1\, e^{6.299t} + k_2\, e^{-6.229t} + 78.975. \end{aligned}$$

Now, we apply the boundary conditions: at the free endpoint $\mu(1)\, e^{-0.07t)} = 0$, which gives

$$505.946k_1 - 0.0013k_2 = 55.684.$$

Also, $x(0) = 91$ gives

$$2.3315k_1 + k_2 = 12.025.$$

Solving these two equations simultaneously by Cramer's rule, we get $k_1 = 0.11$, and $k_2 = 11.77$. Hence,

$$\begin{aligned} \mu(t) &= 0.11\, e^{6.299t} - 8.52\, e^{-6.229t} - 59.875, \quad \text{costate variable}, \\ x(t) &= 0.256\, e^{6.299t} + 11.77\, e^{-6.229t} + 78.975, \quad \text{state variable}, \end{aligned}$$

and the control variable is

$$y(t) = 0.675x + 2.5\mu + 1.875 = 0.435\, e^{6.299t} - 18.29\, e^{-6.229t} - 168.03.$$

11

Demands

In microeconomics there are three well-known and highly used demands. They are the Marshallian, the Hicksian, and the Walrasian demands, which deal with what consumers will buy in different situations so as to maximize their profit. Some useful results like Shephard's lemma and the Slutsky equation are introduced, and so-called Giffen and Veblen goods are discussed. We will first introduce Shephard's lemma, and then analyze the above-mentioned three demands.

11.1 Shephard's Lemma

This lemma is used in microeconomics in applications to the theory of the firm and consumer choices. For more details, see Varian [1992].

Shephard's lemma states that the demand h for a particular good j, for a given level of utility u and given price p, is equal to the derivative of the expenditure function e with respect to the price of the relevant good, i.e.,

$$h_j(p, u) = \frac{\partial e(p, u)}{\partial p_j}. \tag{11.1.1}$$

In the theory of the firm, this lemma has a similar form for the conditional factor demand $c(w, y)$ for each input factor (or good) j:

$$x_j(w, y) = \frac{\partial c(w, y)}{\partial w_j}. \tag{11.1.2}$$

PROOF. We will consider (11.1.1) only for the two-good case. The general case and proof of (11.1.2) are analogous. The expenditure function $e(p_1, p_2, u)$ is the minimand of the constrained optimization problem, and thus, using the Lagrange multiplier method, the Lagrangian is given by

$$L(p_1, p_2, u) = p_1 x_1 + p_2 x_2 + \lambda(u - U(x_1, x_2)), \tag{11.1.3}$$

where $U(x_1, x_2)$ is the prescribed constraint. The derivatives of the minimand

$e(p_1, p_2, u)$ with respect to the parameters $p_j, j = 1, 2$, are given by

$$\frac{\partial e}{\partial p_j} = \frac{\partial L}{\partial p_j} = x_j^h,\ j = 1, 2,$$

where x_j^h is the minimizer (i.e., the Hicksian demand function for good j, $j = 1, 2$). ■

11.2 Marshallian Demand

Named after Alfred Marshall (British, 1842-1924, author of *Principles of Economics* 1890), the Marshallian demand in microeconomics specifies what the consumer would buy in each price and income or wealth situation, assuming it perfectly solves the utility maximization problem. It is compared with *Walrasian demand* (named after Léon Walras), also known as *uncompensated demand*, since the original Marshallian analysis ignored wealth effects. Using the utility maximization problem, there are m commodities with price vector p and choice vector x. The consumer has income i, and a set of affordable packages

$$b(p, i) = \{x : \langle p, x\rangle \leq i\}, \tag{11.2.1}$$

where $\langle p, x\rangle$ is the inner product of price and quantity vectors. The consumer has a utility function $u : \mathbb{R}_+^m \mapsto \mathbb{R}$. Then the consumer's Marshallian demand correspondence is defined uniquely by

$$x^*(p, i) = \arg\{\max_{x\in b(p,i)} \{u(x)\}\}, \tag{11.2.2}$$

which is a homogeneous function of degree zero, i.e., for every constant $a > 0$,

$$x^*(a\,p, a\,i) = x^*(p, i). \tag{11.2.3}$$

Suppose p and i are measured in dollars. When $a = 100$, ap and $a\,i$ are exactly the same quantities measured in cents. Obviously, changing the units of measurement does not affect the demand.

The following examples deal with two goods, 1 and 2.

1. The utility function has the Cobb-Douglas form: $u(x_1, x_2) = x_1^\alpha x_2^\beta$; then the constrained optimization problem leads to the Marshallian demand function

$$x^*(p_1, p_2, i) = \Big(\frac{\alpha\, i}{(\alpha+\beta)p_1}, \frac{\beta\, i}{(\alpha+\beta)p_2}\Big). \tag{11.2.4}$$

2. The utility function is a CES utility function:

$$u(x_1, x_2) = \Big(\frac{x_1^\delta}{\delta} + \frac{x_2^\delta}{\delta}\Big)^{1/\delta}. \tag{11.2.5}$$

Then

$$x^*(p_1, p_2, i) = \Big(\frac{i\,p_1^{\varepsilon-1}}{p_1^{\varepsilon-1} + p_2^{\varepsilon-1}}, \frac{i\,p_2^{\varepsilon-1}}{p_1^{\varepsilon-1} + p_2^{\varepsilon-1}}\Big), \quad \varepsilon = \frac{\delta}{\delta - 1}. \tag{11.2.6}$$

In both case, the preferences are strictly convex, the demand is unique, and the demand function is continuous.
3. The utility function has the linear form $u(x_1, x_2) = x_1 + x_2$, which is weakly convex, and in fact the demand is not unique: when $p_1 = p_2$, the consumer may divide his income in arbitrary ratios between goods 1 and 2 and get the same utility.
4. The utility function exhibits a non-diminishing marginal rate of substitution

$$u(x_1, x_2) = (x_1^\alpha + x_2^\alpha), \quad \alpha > 1. \tag{11.2.7}$$

The utility function is concave, and the demand is not continuous: When $p_1 < p_2$, the consumer demands only good 1, and when $p_1 > p_2$, the consumer demands only good 2, and when $p_1 = p_2$ the demand correspondence contains two distinct bundles, either buy only good 1 or buy only good 2.

Example 11.1. A consumer's preferences are represented by the utility function

$$U(x, y) = x + y + 2y^{1/2},$$

where good 1 is the numéraire with price $p_1 = 1$, good 2 has price p_2, and the consumer's income is m. (i) Find the consumer's Marshallian demands for both goods as a function of p_1, p_2 and m, and the corner solutions, if any; (ii) use the solution in part (i) and the relevant homogeneity property of Marshallian demands to determine the consumer's demands for goods 1 and 2 for arbitrary non-negative prices p_1, p_2 and income m; (iii) find the consumer's Hicksian demand (§11.2) functions $h_i(p_1, p_2, u), i = 1, 2$; (iv) find the consumer's expenditure function $e(p_1, p_2, u)$; and (v) verify if Shephard's lemma applies in this case.

SOLUTION. (i) The marginal rate of substitution between good 2 and good 1 is $1 + y^{-1/2} > 1$. At an interior solution, we must have $p_2 = 1 + y^{-1/2}$, which is possible only if $p_2 > 1$. If the consumer chooses positive amounts of both goods, the Marshallian demands are given by

$$y(1, p_2, m) = \Big(\frac{1}{p_2 - 1}\Big)^2, \quad \text{and} \quad x(1, p_2, m) = m - p_2\Big(\frac{1}{p_2 - 1}\Big)^2. \tag{11.2.8}$$

The consumption of good 1 is positive iff $p_2 > 1$ and $m > p_2\Big(\dfrac{1}{p_2 - 1}\Big)^2$. If the consumption of good 1 is zero, then $y(1, p_2, m) = m/p_2$ and $x(1, p_2, m) = 0$.

The consumption of good 2 is always positive, because the marginal rate of substitution tends to infinity as $y \to 0$.

(ii) Since the demand is homogeneous of degree zero in prices and income, we have

$$x_i(p_1, p_2, m) = x_i\Big(1, \frac{p_2}{p_1}, \frac{m}{p_1}\Big), \quad i = 1, 2. \tag{11.2.9}$$

Then from part (i) we find that at an interior solution

$$\begin{aligned} y(p_1, p_2, m) &= \Big(\frac{1}{(p_2/p_1) - 1}\Big)^2 = \Big(\frac{p_1}{p_2 - p_1}\Big)^2, \\ x(p_1, p_2, m) &= \frac{m}{p_1} - \frac{p_2}{p_1}\Big(\frac{1}{(p_2/p_1) - 1}\Big)^2 = \frac{m}{p_1} - \frac{p_1 p_2}{p_2 - p_1}. \end{aligned} \tag{11.2.10}$$

(iii) At an interior solution,

$$\begin{aligned} h_2(p_1, p_2, u) &= \Big(\frac{p_1}{p_2 - p_1}\Big)^2, \\ h_1(p_1, p_2, u) &= u - h_2(p_1, p_2, u) - 2\sqrt{h_2(p_1, p_2, m)} \\ &= u - \Big(\frac{p_1}{p_2 - p_1}\Big)^2 - 2\Big(\frac{p_1}{p_2 - p_1}\Big) \\ &= u - \Big[\Big(\frac{p_1}{p_2 - p_1} + 1\Big)^2\Big] = u + 1 = \Big(\frac{p_1}{p_2 - p_1}\Big)^2. \end{aligned} \tag{11.2.11}$$

Thus, there is an interior solution iff

$$u + 1 > \Big(\frac{p_1}{p_2 - p_1}\Big)^2.$$

If

$$u + 1 < \Big(\frac{p_1}{p_2 - p_1}\Big)^2,$$

then $h_1(p_1, p_2, u) = 0$, and $u = h_2(p_1, p - 2, u) + 2\sqrt{h_2(p_1, p_2, u)}$. The right-hand side of the last equation is a strictly increasing function of h_2, which means that for any prescribed (p_1, p_2, u) there is a unique solution for $h_2(p_1, p_2, u)$. However, such a solution does not have a simple closed form representation.

(iv) We have

$$\begin{aligned} e(p_1, p_2, u) &= p_1 h(p_1, p_2, u) + p_2 h_2(p_1, p_2, u) \\ &= p_1\Big(u - \Big(\frac{p_1}{p_2 - p_1} + 1\Big)^2 + 1\Big) + p_2\Big(\frac{p_1}{p_2 - p_1}\Big)^2 \quad \text{using part (iii)}, \\ &= p_1(u + 1) + \frac{p_2 p_1^2 - p_1 p_2^2}{(p_2 - p_1)^2} = p_1(u + 1) - \frac{p_1 p_2}{p_2 - p_1}. \end{aligned} \tag{11.2.12}$$

(v) First, in the case of interior solutions, we have, using Shephard's lemma,

$$\frac{\partial e(p_1, p_2, u)}{\partial p_i} = h_i(p_1, p_2, u), \quad i = 1, 2.$$

Next, differentiating (11.2.12) with respect to p_2, we get

$$\begin{aligned}\frac{\partial e(p_1, p_2, u)}{\partial p_2} &= \frac{-(p_2 - p_1)p_1 + p_1 p_2}{(p_2 - p_1)^2} = \frac{p_1^2}{(p_2 - p_1)^2} \\ &= h_2(p_1, p_2, u) \quad \text{from part (iii)},\end{aligned}$$

which verifies that Shephard's lemma applies to good 2. Again, differentiating (11.2.12) with respect to p_1, we get

$$\begin{aligned}\frac{\partial e(p_1, p_2, u)}{\partial p_1} &= u + 1 - \frac{(p_2 - p_1)p_2 + p_1 p_2 - 2}{(p_2 - p_1)^2} \\ &= u + 1 - \Big(\frac{p_2}{p_2 - p_1}\Big)^2 \\ &= h_1(p_1, p_2, u) \quad \text{from part (iii)},\end{aligned}$$

which verifies that Shephard's lemma holds for good 1.

11.3 Hicksian Demand

Named after John Richard Hicks (British neo-Keynesian economist, 1904-1989, NL 1972), in microeconomics, a consumer's *Hicksian demand correspondence* is the demand of a consumer over a bundle of goods that minimizes their expenditure while delivering a fixed level of utility. It is defined as follows: Given a utility function $u : \mathbb{R}^n_+ \mapsto \mathbb{R}$, the *Hicksian demand correspondence* $h^* : \mathbb{R}^n_{++} \times u(\mathbb{R} + +^n) \mapsto \mathbb{R}^n_+$ is defined by

$$h^*(p, \bar{u}) = \arg\big\{ \min_{x \in \mathbb{R}^n_+} p \cdot x \big\} \quad \text{subject to} \;\; u(x) \geq \bar{u}.$$

If this correspondence is a function, then it is called the *Hicksian demand function* or *compensated demand function.* Mathematically,

$$h(p, \bar{u}) = \arg\Big\{ \min_x \Big\{ \sum_j p_j x_j \Big\}\Big\} \text{ subject to } u(x) \geq \bar{u}. \tag{11.3.1}$$

Hicksian demand functions are often convenient for mathematical manipulation because they do not require income or wealth to be represented. Also, the function to be minimized is linear in x_j, which makes the optimization problem simpler. However, the Marshallian demand functions of the form

$x(p, w)$ that describe demand given by prices p and income w are easier to observe directly. The two are related by

$$h(p, u) = x(p, e(p, u)), \tag{11.3.3}$$

where $e(p, u)$ is the expenditure function (the function that gives the minimum wealth required to get to a utility level), and by

$$h(p, v(p, w)) = x(p, w), \tag{11.3.3}$$

where $v(p, w)$ is the indirect utility function (which gives the utility level of having a given wealth under a fixed price regime). Their derivatives are related by the Slutsky equation (see §11.4).

Whereas the Hicksian demand comes from the expenditure minimization problem, the Marshallian demand comes from the utility maximization problem. Thus, the two problems are mathematical duos, and hence the duality theorem provides a method of proving the above relationships.

The Hicksian demand function is intimately related to the expenditure function. If the consumer's utility function $u(x)$ is locally nonsatiated and strictly convex, then $h(p, u) = \nabla_p e(p, u)$. If the consumer's utility function $u(x)$ is continuous and represents a locally nonsatiated preference relation, then the Hicksian demand correspondence $h(p, u)$ satisfies the following properties:

1. Homogeneity of degree zero in p: For all $a > 0$, $h(ap, u) = h(p, u)$. This is because the same x that minimizes $\sum_j p_j x_j$ also minimizes $\sum_j a p_j x_j$ subject to the same constraint.
2. No excess demand: The constraint $u(x) \geq \bar{u}$ holds with strict equality, $u(x) = \bar{u}$. This follows from the continuity of the utility function. Informally, they could simply spend less until utility becomes exactly $\bar{u}$.
3. Hicksian demand finds the cheapest consumption bundle that achieves a given utility level.

11.4 Slutsky Equation

The Slutsky equation (or Slutsky identity) relates changes in the Marshallian (uncompensated) demand to changes in Hicksian (compensated) demand, just to maintain a fixed level of utility. Thus, this equation decomposes the change in demand for good j in response to the change in price of good k, and is defined by

$$\frac{\partial x_j(p, w)}{\partial p_k} = \frac{\partial h_j(p, u)}{\partial p_k} - \frac{\partial x_j(p, w)}{\partial w} x_k(p, w), \tag{11.4.1}$$

where $h(p, u)$ is the Hicksian demand and $x(p, w)$ is the Marshallian demand, both at the price levels p (vector), wealth level (or income level) w, and fixed utility u by maximizing the utility at the original price and income vectors, formally given by $v(p, w)$. The right-hand side of Eq (11.4.1) represents the change in demand for good j holding utility fixed at u minus the quantity of good k demanded, multiplied by the change in demand for good j when wealth changes. Thus, the first term on the right-hand side of Eq (11.4.1) represents the substitution effect, and the second term the income effect. Since utility cannot be observed, the substitution effect is not directly observable, but it can be calculated using the other two terms in Eq (11.4.1).

To derive Eq (11.4.1), use the identity $h_j(p, u) = x_j(p, e(p, u))$, where $e(p, u)$ is the expenditure function, and u is the utility obtained by maximizing utility for a given p and w. Then differentiating $h_j(p, u)$ partially with respect to p_k, we get

$$\frac{\partial h_j(p,u)}{\partial p_k} = \frac{\partial x_j(p, e(p,u))}{\partial p_k} + \frac{\partial x_j(p, e(p,u))}{\partial e(p,u)} \cdot \frac{\partial e(p,u)}{\partial p_k}. \tag{11.4.2}$$

Then, using the fact that $\dfrac{\partial e(p,u)}{\partial p_k} = h_k(p,u)$ by Shephard's lemma, and that $h_k(p, u) = h_k(p, v(p, w)) = x_k(p, w)$ at the optimum, where $v(p, w)$ is the indirect utility function, and substituting these results into Eq (11.4.2), we obtain Eq (11.4.1).

The Slutsky equation (11.4.1) shows that the change in the demand for a good, caused by a price change, can be explained by the following two effects: (i) a substitution effect that results in a change in the relative prices of two goods, and (ii) an income effect that results in a change in the consumer's purchasing power. For more details, see Nicholson [1978].

The Slutsky equation (11.4.1) can also be written as

$$\frac{\partial h_j(p,u)}{\partial p_k} = \frac{\partial x_j(p,w)}{\partial p_k} + \frac{\partial x_j(p,w)}{\partial w} x_k(p,w), \tag{11.4.3}$$

or in matrix form as

$$D_p h(p,u) = \underbrace{D_p x(p,w)}_{n\times n} + \underbrace{D_w x(p,e)}_{n\times 1} \underbrace{x(p,e)^T}_{1\times n}. \tag{11.4.4}$$

Another Proof. Take any (p, w, u), and recall that $h(p, u) = x(p, w)$ and $e(p, u) = w$. Now, differentiate $h_j(p, u) = x_j(p, e(p, u))$ with respect to p_j:

$$\begin{aligned}
\frac{\partial h_j(p,u)}{\partial p_k} &= \frac{\partial x_j(p, e(p,u))}{\partial p_k} + \frac{\partial x_j(p, e(p,u))}{\partial w} \frac{\partial e(p,u)}{\partial p_j} \\
&= \frac{\partial x_j(p, e(p,u))}{\partial p_k} + \frac{\partial x_j(p, e(p,u))}{\partial w} h_j(p,u) \\
&= \frac{\partial x_j(p,w)}{\partial p_k} + \frac{\partial x_j(p,w)}{\partial w} x_j(p,w). \blacksquare
\end{aligned}$$

The formula (11.4.1) signifies both substitution effect and income effect as follows:

$$\frac{\partial x_j(p,w)}{\partial p_k} = \underbrace{\frac{\partial h_j(p,u)}{\partial p_k}}_{\text{substitution effect}} - \underbrace{\frac{\partial x_j(p,w)}{\partial w} x_k(p,w)}_{\text{income effect}}. \tag{11.4.5}$$

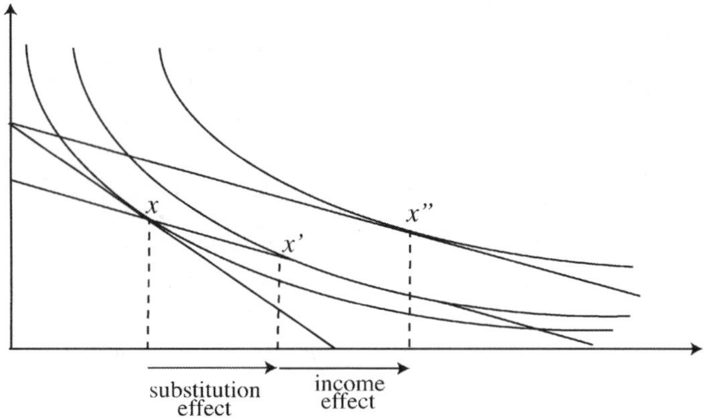

Figure 11.1 Normal goods.

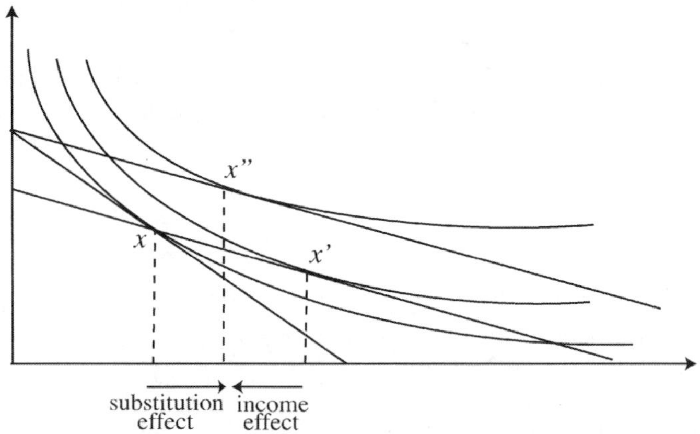

Figure 11.2 Inferior goods.

According to Slutsky, the substitution effect, which is the change from x to x', always moves opposite to the price effect, which is the change from x' to x''. Thus, the substitution effect is negative since the change in demand due to the substitution effect is opposite from the price change. On the other hand, the income effect may be negative or positive depending on whether the good is inferior or not. Hence, in the case of normal goods, since both the substitution and income effects increase demand when their own price decreases, the demand curve slopes downward (Figure 11.1). But in the case of inferior good the income effect and the substitution effect are in the opposite direction (Figure 11.2).

Example 11.2. (1). Let $p = (p_1, p_2)$ be original prices, and $x = (x_1, x_2)$ be the original demand. Let p_1 be decreased to p_1'. While the initial income was $i = p_1x_1 + p_2x_2$, now the consumer needs to buy only $p_1'x_1 + p_2x_2 = i'$, thus, the consumer's income becomes $i - i' = (p_1 - p_1')x_1$. Thus, at the new price, (i) if less income is needed than before to buy the original choice, then real income has increased; and (ii) if more income than before is needed to buy the original choice, then the real income has decreased.

(2). To determine the changes in quantities demanded when the consumer's income is adjusted so that at new prices he/she can just buy the original choice, let (i) the change be from x to x' (known as the *pure substitution effect*), and (ii) the subsequent change from x' to x'' (known as the *pure income effect*). For example, suppose that the demand function for milk is $x_1 = 10 + \dfrac{i}{10p_1}$. If initially, $i = 120$ and $p_1 = 3$, then $x_1 = 10 + \dfrac{120}{10 \times 3} =$ 14 units of milk. Thus, $14 \times 3 = 42$ is spent on milk, and the remaining $120 - 42 = 78$ is spent on other goods. Now, suppose p_1 has decreased to $p_1' = 2$. Then how much income is needed to buy the initial bundle? Obviously, $78 + 2 \times 14 = 106$. Then the consumer will buy $x_1' = 10 + \dfrac{106}{10} = 15.3$ units of milk with that money. Thus, $15.3 - 14 = 1.3$ is the substitution effect. Next, what is the income effect? Obviously, $x_1'' = 10 + \dfrac{120}{10 \times 2} = 16$ will be the eventual consumption, giving a change of $16 - 14 = 2$ units. Hence, the $2 - 13 = 0.7$ is the income effect.

The substitution effect always moves opposite to the price effect. Thus, we say: the substitution effect is *negative* when the change in demand due to the substitution effect is opposite to the price change. However, the income effect may be *negative* or *positive* depending on whether the good is inferior or not. ■

11.4.1 Giffen Goods. If the income effect and the substitution effect are in opposite directions and if the income effect is larger than the substitution effect, then a price decrease lessens the demand. Such goods are called *Giffen goods*. For a consumer who consumes multiple goods, the Giffen goods are inferior goods (Figure 11.3).

In fact, the above definition of Giffen good implies that

$$\frac{\partial x_i^c(p, M)}{\partial p_i} > 0. \tag{11.4.6}$$

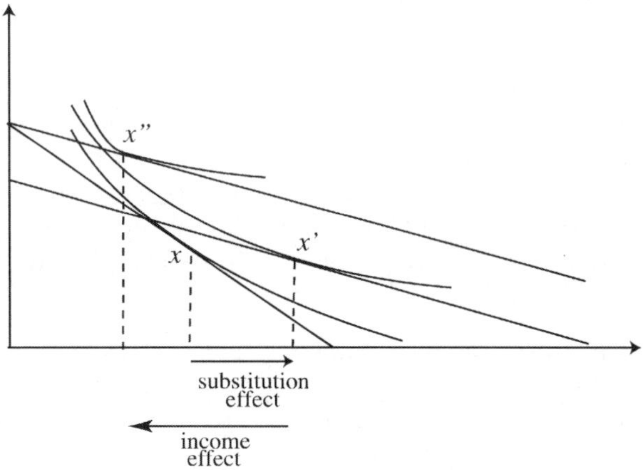

Figure 11.3 Giffen goods.

Using the Slutsky equation for the own-price effect of good i, the inequality (11.4.6) implies that

$$\frac{\partial x_i^c(p,\bar{u})}{\partial p_i} - x_i \frac{\partial x_i^c(p,M)}{\partial M} > 0,$$

or

$$\frac{\partial x_i^c(p,\bar{u})}{\partial p_i} > x_i \frac{\partial x_i^c(p,M)}{\partial M}. \tag{11.4.7}$$

Also,

$$\frac{\partial x_i^c(p,\bar{u})}{\partial p_i} < 0, \tag{11.4.8}$$

so from the above section we know that for inequality (11.4.7) to hold we must at least have

$$\frac{\partial x_i^c(p,M)}{\partial M} < 0. \tag{11.4.9}$$

But this is the definition of an inferior good, which means that every Giffen good is also inferior. On the other hand, not every inferior good is a Giffen good because the right-hand side of the inequality (11.4.7) could be negative but larger than the left-hand side (in absolute value). In such a case the own-price effect would be negative, which means that for a good to be a Giffen good, the income effect times demand, $x_i \frac{\partial x_i^c(p,M)}{\partial M}$, must be smaller (or larger in absolute vale) than the substitution effect $\frac{\partial x_i^c(p,\bar{u})}{\partial p_i}$, which is inequality (11.4.7).

Finally, note that a Giffen good faces an upward sloping demand curve because the income effect dominates the substitution effect, which means that quantity demand increases as the price rises. However, a good cannot always have an upward sloping demand curve, because the consumer eventually runs out of money. At some point the rising price of the Giffen good surpasses the consumer's budget, and a price increase will lower the amount of the good the consumer is able to buy. This means that at higher enough prices, the demand curve will start sloping downward (Figure 11.4, where A marks the point at which the good surpasses consumer's budget; and the 'Giffen good range' represents the range where only such good is consumed).

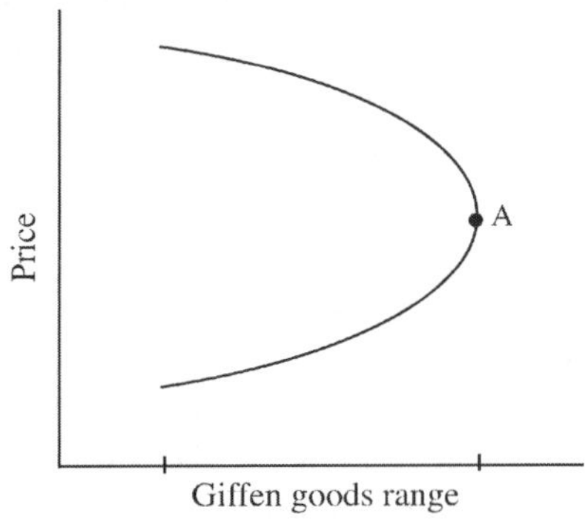

Figure 11.4 Giffen goods and consumer's budget.

11.4.2 Veblen Goods. (Veblen [1899]) These goods are types of luxury goods for which the quantity demanded increases as the price increases, which appears to be in contradiction to the *Law of Demand*. Consumers prefer more of a good as its price rises, resulting in an upward sloping demand curve. These goods are, for example, personal goods such as wines, jewelry, designer handbags and luxury cars, and are in demand simply because the high prices asked for them. They make a desirable status symbol as conspicuous consumption and leisure. This phenomenon is also known as the *Veblen effect*, where goods are desired even if they become over-priced. A corollary is that a decrease in their price decreases their demand. This effect is known as the *snob effect*, the *bandwagon effect*, the *network effect*, the *hot-hand fallacy*, and the *common law of business balance*. None of these effects, however, can predict what will happen to the actual quantity of goods demanded as prices change. The actual effect on quantity demanded will depend on the range of other available goods, their prices and substitutions for consumed goods.

11.5 Walrasian Demand

Consider the utility maximization problem

$$\max_{x \in X} u(x) \text{ subject to } p \cdot x \leq w, \;\; x \in B(p, w). \tag{11.5.1}$$

The *Walrasian Demand correspondence* is defined as the set of solutions $x(p,w) \subset X$ of the maximization problem (11.5.1) given $p \gg 0$ and $w \geq 0$. The set $x(p,w)$ is not empty for any such (p,w) if u is continuous. An important property of $x(p,w)$ is the so-called *Walrasian demand.*

For the definition of a *locally nonsatiated function*, see Appendix F.

Theorem 11.1. *Suppose u is a continuous, locally nonsatiated function, and let $X = \mathbb{R}^n_+$. Then the Walrasian demand correspondence $x : \mathbb{R}^n_{++} \times \mathbb{R}_+ \mapsto \mathbb{R}^n_+$ has the following four properties:*

(i) Homogeneity of degree 0: $x(\alpha p, \alpha w) = x(p,w)$ *for any* $\alpha > 0$ *and* (p,w);

(ii) Walras's law: $p \cdot x' = w$ *for any* $x' \in x(p,w)$ *and* (p,w);

(iii) Convexity: $x(p,w)$ *is convex for any* (p,w) *if* u *is quasi-concave; and*

(iv) Continuity: x *is upper semicontinuous.*

PROOF. Property (i) follows from the definition and (11.5.1); for (ii) use local satiation; (iii) is obvious; and (iv) follows from the maximum theorem. ■

NOTES. The function $x(p,w)$ is a single point if u is strictly quasi-concave, and is continuous if it is single valued.

The function $x(p,w)$ can be obtained for each (p,w) by applying the KKT conditions if u is differentiable; these conditions are necessary in the case of constrained optimization, and are sufficient if u is quasi-concave.

Suppose u is locally nonsatiated and the optimal solution for problem (11.5.1) is an interior solution. Then the KKT conditions are

$$\frac{\partial u(x)}{\partial x} - \lambda p = 0,\ p \cdot x = w. \tag{11.5.2}$$

Boundary solution: For some k, $\dfrac{\partial u(x)}{\partial x_k} - \lambda p_k = 0$ should be replaced by

$$\frac{\partial u(x)}{\partial x_k} - \lambda p_k \begin{cases} \leq 0 \text{ if } x_k = 0, \\ = 0 \text{ if } x_k > 0. \end{cases} \tag{11.5.3}$$

Example 11.3. For a boundary solution, let

$$DU_{x_1}(x_1,x_2) - \lambda p_1 \leq 0, \quad Du_{x_2}(x_1,x_2) - \lambda p_2 = 0.$$

This equation can be written as

$$\nabla u(x) - \lambda p + \mu \begin{bmatrix} 1 \\ 0 \end{bmatrix} = 0, \quad \text{for some } \mu \geq 0.$$

Specifically, take $u(x) = \sqrt{x_1} + x_2$. Assume an interior solution for x_1. Then the KKT conditions become

$$\frac{1}{2\sqrt{x_1}} - \lambda p_1 = 0, \quad 1 - \lambda p_2 \begin{cases} \leq 0 \text{ if } x_2 = 0 \\ = 0 \text{ if } x_2 > 0 \end{cases}; \quad p \cdot x = w.$$

Then the solution is

(i) $x_1(p,w) = \dfrac{p_2^2}{4p_1^2}$, $x_2(p,w) = \dfrac{w}{p_2} - \dfrac{p_2}{4p_1}$, $\lambda(p,w) = 1$ when $4p_1 w > p_2^2$;

(ii) $x_1(p,w) = \dfrac{w}{p_1}$, $x_2(p,w) = 0$, $\lambda(p,w) = \dfrac{1}{2\sqrt{p_1 w}}$ when $4p_1 w \le p_2^2$.

Note that there is no income effect on x_1, i.e., x_1 is independent of w as long as $4p_2 w > p_2^2$. ■

For any $(p,w) \in \mathbb{R}^n_{++} \times \mathbb{R}_+$, the *indirect utility function* is defined as

$$v(p,w) := u(x'),$$

where $x' \in x(p,w)$.

Theorem 11.2. *The indirect utility function $v(p,w)$ satisfies the following properties: (i) it is homogeneous of degree 0; (ii) it is nonincreasing in p_k for any k and strictly increasing in w; (iii) it is quasi-convex; and (iv) it is continuous.*

PROOF. Proofs of (i) and (ii) are obvious; (iv) follows from the maximum theorem. For the proof of (iii), suppose $\max\{v(p',w'), v(p'',w'') \le \bar{v}$ for any $(p',w'),(p'',w'') \in \mathbb{R}^n_{++} \times \mathbb{R}_+$ and $\bar{v} \in \mathbb{R}$. Then for any $t \in [0,1]$ and any $x \in B(tp' + (1-t)p'', tw' + (1-t)w")$, either $x \in B(p',w')$ or $x \in B(p'',w'')$ must hold. Thus,

$$v(tp' + (1-t)p'', tw' + (1-t)w'') \le \max\{v(p',w'), v(p'',w'') \le \bar{v}. \blacksquare \tag{11.5.4}$$

Example 11.4. Let the Cobb-Douglas utility function be

$$u(x) = \sum_{j=1}^{n} \alpha_j \log x_j,\ \alpha_j \ge 0,\ \sum_{j=1} n\alpha_j = 1, \tag{11.5.5}$$

where α_j is the fraction of the expense for good j. Then the Walrasian demand is $x_j(p,w) = \dfrac{\alpha_j w}{p_j}$. Hence,

$$v(p,w) = \log w + \sum_{j=1}^{n} \alpha_j(\log \alpha_j - \log p_j). \blacksquare$$

Example 11.5. Let the quasi-linear utility function be

$$v(p,w) = \sqrt{x_1(p,w)} + x_2(p,w) = \frac{p_2}{4p_1} + \frac{w}{p_2},$$

where we assume an interior solution. Then the indirect utility function is of the form $v(p, w) = a(p) + w$. ■

Example 11.6. (Labor supply) Consider the following simple labor decision problem:

$$\max_{q,\tau \geq 0}\{(1-t)\log q + t\log\tau\} \text{ subject to } pq + w\tau \leq wT + P,\ \tau \leq T,$$

where q is the amount of consumed good, p is the price, T is the total time available, τ is the time spent for 'leisure,' which determines $h = T - \tau$ hours of work, w is wage, so wh is labor income, and P is nonlabor income. Since the utility function is Cobb-Douglas, the Walrasian demand is

$$q(p, w, P) = \frac{(1-t)(wT + P)}{p}, \ \tau(p, w, P) = \frac{\tau(wT + P)}{w} \quad \text{for } \tau < T.$$

In the case when $\tau \leq T$, the consumer does not participate in the labor market ($\tau(p, w, P) = t$) and spends all labor income to purchase goods ($q(p, w, P) = P/p$). For $\tau < T$, the indirect utility function is

$$v(p, w, wT + P) = K + \log(wT + P) - t\log p - (1-t)\log w,$$

where K is a constant. ■

11.6 Cost Minimization

For each $p \gg 0$ and $u \in \mathbb{R}$, consider the problem

$$\min_{x \in X}\{p \cdot x\} \quad \text{subject to } u(x) \geq \bar{u}. \tag{11.6.1}$$

In other words, this problem asks as to what is the cheapest way to get utility as high as $\bar{u}$. Let the Hicksian demand correspondence[1] $h(p, \bar{u})$ be the set of solutions to the problem (11.6.1). Assuming local nonsatiation, the constraint can be modified as follows: if u is continuous and the set $U \equiv \{x \in X : u(x) \geq \bar{u}\}$ is not empty, then choose any x' such that $u(x') > \bar{u}$, and choose any $\bar{x} \in \mathbb{R}_+$ such that $p'_j\bar{x} \geq p' \cdot x'$ for all j and all p' in a neighborhood of $p \gg 0$. This replaces the constraint by a compact set. Then the cost minimizing solution of problem (11.6.1) is the same locally with respect to $(p, \bar{u})$. Moreover, the set $h(p, \bar{u})$ is not empty in the neighborhood of $(p, \bar{u}) \in \mathbb{R}^n_{++} \times U$.

Theorem 11.6. (Hicksian demand) *Let u be continuous and locally nonsatiated, and let $X = \mathbb{R}^n_+$. Then the set h, known as the Hicksian correspondence and defined by $h : \mathbb{R}^n_{++} \times U \mapsto \mathbb{R}^n_+$ (i) is homogeneous of degree 0 in*

[1] This correspondence is also used in welfare analysis.

p; (ii) achieves $\bar{u}$ exactly (i.e., $u(x') = \bar{u}$ for any $x' \in h(p, \bar{u})$ if $\bar{u} \geq u(0)$); (iii) is convex for any given $(p, \bar{u})$ if u is quasi-concave; and (iv) is upper semicontinuous.

The set $h(p, u)$ reduces to a point if u is strictly quasi-concave.

PROOF. (i), (ii), and (iii) are easy. To prove (iv), we cannot apply the maximum theorem as we did in the proof of Theorem 11.1 on the Walrasian demand, because in the present case the set is not locally bounded. ∎

11.7 Expenditure Function

The *expenditure function* $e(p, \bar{u})$ is defined as $e(p, \bar{u}) := p \cdot x'$ for any $x' \in h(p, \bar{u})$. The following result holds:

Theorem 11.7. *Let u be continuous and locally nonsatiated, and let $X = \mathbb{R}^n_+$. Then the expenditure function $e : \mathbb{R}^n_{++} \times U \mapsto \mathbb{R}$ is (i) homogeneous of degree 1 in p; (ii) nondecreasing in p_j for any j and strictly increasing in $\bar{u}$ for $\bar{u} > u(0)$; (iii) concave in p; and (iv) continuous.*

Hence, we find that utility maximization and cost minimization go hand-in-hand together; one implies the other. The following result holds.

Theorem 11.8. (Mas-Colell et al. [1995: 156]; Nicholson [1978: 90-93]) *Let u be continuous and locally nonsatiated, and let $X = \mathbb{R}^n_+$. Then (i) if $x^* \in x(p, w)$ for given $p \gg 0$ and $w \geq 0$, then $x^* \in h(p, v(p, w))$ and $e(p, v(p, w)) = w$; and (ii) if $x^* \in h(p, \bar{u})$ for given $p \gg 0$ and $\bar{u} \geq u(0)$, then $x^* \in x(p, e(p, \bar{u}))$ and $v(p, e(p, \bar{u})) = \bar{u}$.*

PROOF. Suppose utility maximization does not imply cost minimization, i.e., there exists an $x' \in \mathbb{R}^n_+$ that satisfies $u(x') \geq u(x^*)$ and $p \cdot x' < p \cdot x^*$ (which, by Walras's law, equals w). Then by local nonsatiation, there also exists an $x'' \in \mathbb{R}^n_+$ that satisfies $u(x'' > u(x^*)$ and $p \cdot x'' < w$. But this contradicts utility maximization. Hence, $x^* \in h(p, v(p, w))$ and $e(p, v(p, w)) = p \cdot x^* = w$.

Next, suppose cost minimization does not imply utility maximization, i.e., there exists an $x' \in \mathbb{R}^n_+$ that satisfies $u(x') \geq u(x^*) \geq \bar{u}$ and $p \cdot x' \leq p \cdot x^*$. Note that $p \cdot x' > 0$ because $\bar{u} \geq u(0)$. Let $x^t := t0 = (1 - t)x' \in X$ for $t \in [0, 1]$. Then $u(x^t) > u(x^*)$ and $p \cdot x^t < p \cdot x^*$ because $p \cdot x' > 0$ for small t. But this contradicts cost minimization. Hence, $x^* \in x(p, e(p, w))$ and $v(p, v(p, \bar{u})) = u(x^*) = \bar{u}$ since $u > u(0)$. ∎

Example 11.7. The Walrasian demand curve of good j is downward sloping (i.e., $\dfrac{\partial x_j(p, w)}{\partial p_j} < 0$) if it is a *normal good* (i.e., $\dfrac{\partial x_j(p, w)}{\partial w} \geq 0$). If good j is an *inferior good* (i.e., $\dfrac{\partial x_j(p, w)}{\partial w} < 0$), then x_j can be a *Giffen good* (i.e., $\dfrac{\partial x_j(p, w)}{\partial p_j} > 0$). ∎

The right-hand side of formula (11.4.3) is known as the *Slutsky matrix* or *substitution matrix*, denoted by $S(p,w)$, which for all $(p,w) \gg 0$ and $\bar{u} = v(p,w)$, has the following properties: (i) $S(p,w) = D_p^2 e(p,\bar{u})$, (ii) $S(p,w)$ is negative semidefinite, (iii) $S(p,w)$ is a symmetric matrix, and (iv) $S(p,w)p = 0$.

To prove these properties, note that (i) follows from the previous theorem; (ii) and (iii) hold, since $e(p,\bar{u})$ is concave and twice continuously differentiable; and (iv) follows since $h(p,\bar{u})$ is homogeneous of degree 0 in p, or $x(p,w)$ is homogeneous of degree 0 in (p,w), and Walras's law applies.

11.8 Applications

LABOR SUPPLY. We will apply a Slutsky equation to the labor/leisure decision problem. Let the income vector i be given by $i = wT + P$. Then, since w affects the labor τ, we get from (11.4.3),

$$\frac{\partial \tau(p,w,i)}{\partial w} = \frac{\partial \tau(p,w,i)}{\partial w} + \frac{\partial \tau(p,w,i)}{\partial i} T. \tag{11.8.1}$$

If we differentiate $\tau(p,w,i) = \tau(p, we(p,w,\bar{u}))$ with respect to $w(\tau(p,w,\bar{u}))$, we obtain

$$\frac{\partial \tau(p,w,i)}{\partial w} = \frac{\partial \tau(p,w,i)}{\partial w} + \frac{\partial \tau(p,w,i)}{\partial i} \tau(p,w,i). \tag{11.8.2}$$

Hence,

$$\begin{aligned} \frac{\partial \tau(p,w,i)}{\partial w} &= \underbrace{\frac{\partial \tau(p,w,\bar{u})}{\partial w}}_{\text{substitution effect}} - \underbrace{\frac{\partial \tau(p,w,i)}{\partial i}}_{\text{income effect I}} \tau(p,w,i) + \underbrace{\frac{\partial \tau(p,w,i)}{\partial i}}_{\text{income effect II}} T \\ &= \frac{\partial \tau(p,w,\bar{u})}{\partial w} + \frac{\partial \tau(p,w,i)}{\partial i} \big(T - \tau(p,w,i)\big). \end{aligned} \tag{11.8.3}$$

For the labor supply $h = T - \tau$, the relation (11.8.3) becomes

$$\frac{\partial \tau(p,w,i)}{\partial w} = \frac{\partial h(p,w,\bar{u})}{\partial w} - \frac{\partial h(p,w,i)}{\partial i} h(p,w,i), \tag{11.8.4}$$

which is called *Roy's identity*. This identity can also be written as follows: For all $(p,w) \gg 0$,

$$x(p,w) = -\frac{1}{D_w v(p,w)} \nabla_p v(p,w). \tag{11.8.5}$$

PROOF. For any $(p,w) \gg 0$, and $\bar{u} = v(p,w)$, we have $v(p, e(p, \text{varu})) = \bar{u}$, which upon differentiation gives

$$\begin{aligned} &\nabla_p v(p, e(p,\bar{u})) + D_w v(p, e(p,\bar{u}))\, \nabla_p e(p,\bar{u}) = 0, \\ &\nabla_p v(p, e(p,\bar{u})) + D_w v(p, e(p,\bar{u}))\, h(p,\bar{u}) = 0 \\ &\nabla_p v(p,w) + D_w v(p,w) x(p,w) = 0, \end{aligned} \tag{11.8.6}$$

which on rearranging yields (11.8.5). ∎

Notes: (1) If u is differentiable and locally nonsatiated, and if $X = \mathbb{R}^n_+$, then all previous theorems are applicable.

(2) If $\bar{u} > u(0)$ and $w > 0$, and u is quasi-concave, and prices p and demands h_j are strictly positive, then Walrasian demand and Hicksian demand are characterized by the following KKT conditions:

For Walrasian demand: $\nabla u(x) - \lambda p = 0$, and $w - p \cdot x = 0$.

For Hicksian demand: $p - \lambda \nabla u(x) = 0$ and $u(x) = \bar{u} = 0$.

(3) The implicit function theorem (see Appendix E) implies that $x(p, w)$ is a C^1 function if the derivative

$$\begin{bmatrix} D^2u(x) & -p \\ -p^T & 0 \end{bmatrix}$$

with respect to (x, λ) is a full rank matrix, i.e., we must show that

$$\begin{bmatrix} D^2u(x) & -\dfrac{1}{\lambda}\, Du(x)^T \\ -\dfrac{1}{\lambda}\, Du(x) & 0 \end{bmatrix} \tag{11.8.7}$$

is full rank. This is satisfied when u is differentiable strictly quasi-concave. (Use the following definition):

$u : X\ (\subset \mathbb{R}^n) \mapsto \mathbb{R}$ is *differentiable strictly quasi-concave* if

$\Delta x^T D^2 u(x) \Delta x < 0$ for any $\Delta x (\neq 0) \in \mathbb{R}^n$ such that $Du(x)\Delta x = 0$.

11.9 Exercises

11.1. A consumer has preferences represented by the utility function $u(x_1, x_2) = x_1 + x_2 + 2x_2^{1/2}$, where good 1 is the numéraire and price $p_1 = 1$. The price of good 2 is p_2, and the consumer's income is m. (a) Find this consumer's Marshallian demand for goods 1 and 2 as a function of p_2 and m, accounting for corner solutions, if any; (b) using the result of part (a) and the relevant homogeneity property of Marshallian demand, find this consumer's demand for goods 1 and 2 for arbitrary nonnegative prices p_1, p_2 and income m; (c) find the consumer's Hicksian demand functions $h_1(p_1, p_2 u)$ and $h_2(p_1, p_2 u)$; (d) find this consumer's expenditure function $e(p_1, p_2, u)$; and (e) verify that Shephard's lemma applies in this case (part (d)).

Ans. (a) The marginal rate of substitution between good 1 and good 2 is $1 + x_2^{-1/2} > 1$. At an interior solution, we must have $p_2 = 1 + x_2^{-1/2}$, which is possible only if $p_2 > 1$. If the consumer chooses positive amounts of both goods, the Marshallian demands are given by

$$x_2(1, p_2, m) = \Big(\frac{1}{p_2 - 1}\Big)^2, \quad x_1(1, p_2, m) = m - p_2\Big(\frac{1}{p_2 - 1}\Big)^2. \tag{11.9.1}$$

There is positive consumption of good 1 iff $p_2 > 1$ and $m > p_2\Big(\frac{1}{p_2-1}\Big)^2$. If the consumption of good 1 is zero, then $x_1(1,p_2,m) = 0$ and $x_2(1,p_2,m) = m/p_2$. Thus the consumption of good 2 is always positive since the marginal rate of substitution approaches infinity as x_2 approaches zero.

(b) Since demand is homogeneous of degree zero in prices and income, we have

$$x_i(p_1,p_2,m) = x_i\Big(1, \frac{p_2}{p_1}, \frac{m}{p_1}\Big). \tag{11.9.2}$$

Then, using the result of part (a), at an interior solution we have

$$\begin{aligned} x_2(p_1,p_2,m) &= \Big(\frac{1}{(p_2/p_1)-1}\Big)^2 = \Big(\frac{p_2}{p_2-p_1}\Big)^2, \\ x_1(p_1,p_2,m) &= \frac{m}{p_1} - \frac{p_2}{p_1}\Big(\frac{1}{(p_2/p_1)-1}\Big)^2 = \frac{m}{p_1} - \Big(\frac{p_1p_2}{p_2-p_1}\Big)^2. \end{aligned} \tag{11.9.3}$$

(c) An interior solution,

$$h_2(p_1,p_2,u) = \Big(\frac{p_1}{p_2-p_1}\Big)^2, \tag{11.9.4}$$

and

$$\begin{aligned} h_1(p_1,p_2,u) &= u - h_2(p_1,p_2,u) - 2h_2(p_1,p_2,u)^{1/2} \\ &= u - \Big(\frac{p_1}{p_2-p_1}\Big)^2 - 2\Big(\frac{p_1}{p_2-p_1}\Big) \\ &= u - \Big(\Big(\frac{p_1}{p_2-p_1}+1\Big)^2 - 1\Big) \\ &= u + 1 - \Big(\frac{p_2}{p_2-p_1}\Big)^2. \end{aligned} \tag{11.9.5}$$

Thus, there will be an interior solution iff

$$u + 1 > \Big(\frac{p_2}{p_2-p_1}\Big)^2. \tag{11.9.7}$$

If

$$u + 1 < \Big(\frac{p_2}{p_2-p_1}\Big)^2, \tag{11.9.7}$$

then $h_2(p_1,p_2,u) = 0$, and $u = h_2(p_1,p_2,u) + 2\sqrt{h_2(p_1,p_2,u)}$. The right-hand side of the second equation is a strictly increasing function of h_2, i.e., for any specified (p_1,p_2,u), there is a unique solution for $h_2(p_1,p_2,u)$., but this solution cannot be expressed in a simple closed form.

(d) $e(p_1, p_2, u) = p_1h(p_1, p_2, u) + p_2h_2(p_1, p_2, u)$. Using result of part (c), this expression can be written for an interior solution as

$$\begin{aligned} e(p_1, p_2, u) &= p_1\left(u - \left(\frac{p_2}{p_2 - p_1}\right)^2 + 1\right) + p_2\left(\frac{p_1}{p_2 - p_1}\right)^2 \\ &= p_1(u+1) + \frac{p_2p_1^2 - p_1p_2^2}{(p_2 - p_1)^2} \\ &= p_1(u+1) - \frac{p_1p_2}{p_2 - p_1}. \end{aligned} \tag{11.9.8}$$

(e) In the case of interior solutions, by Shephard's lemma we have

$$\frac{\partial e(p_1, p_2, u)}{\partial p_i} = h_i(p_1, p_2, u). \tag{11.9.9}$$

Differentiating (11.9.8) with respect to p_2 we get

$$\begin{aligned} \frac{\partial e(p_1, p_2, u)}{\partial p_2} &= \frac{-(p_2 - p_1)p_1 + p_1p - 2}{(p_2 - p_1)^2} \\ &= \frac{p_2}{(p_2 - p_1)^2} = h_2(p_1, p_2, u) \quad \text{from (11.9.4).} \end{aligned} \tag{11.9.10}$$

Thus, Shephard's lemma applies to good 2. Again, differentiating (11.9.8) with respect to p_1, we get

$$\begin{aligned} \frac{\partial e(p_1, p_2, u)}{\partial p_1} &= u_1 + 1 - \frac{(p_2 - p_1)p_2 + p_1p_2}{(p_2 - p_1)^2} \\ &= u + 1 - \left(\frac{p_2}{p_2 - p_1}\right)^2 = h_1(p_1, p_2, u) \quad \text{from (11.9.5).} \end{aligned} \tag{11.9.11}$$

Thus, Shephard's lemma applies to good 1.

11.2. Use the Cobb-Douglas utility function $u(x_1, x_2) = x_1^{0.35}x_2^{0.65}$, let initially $p_1 = 2, p_2 = 1$, $i = 100$, and finally $p_1' = 1, p_2 = 1, i = 100$. Then,

initially: $x_1 = 0.35\,\frac{i}{p_1} = 0.35\,\frac{100}{0.2} = 17.5,\ x_2 = 0.65\frac{100}{1} = 65;$

finally: $x_1'' = 0.35\frac{i}{p_1'} = 0.35\frac{100}{1} = 35,\ x_2'' = 0.65\frac{100}{1} = 65.$

(a) How much money is needed to buy the initial bundle with new prices?
ANS. $i' = p_1'x_1 + p_2x_2 = 1 \times 17.5 + 1 \times 65 = 82.5$.

(b) What is the demand with i' and new prices?

ANS. Initially $x_1' = 0.35\frac{i'}{p_1} = 0.35\frac{82.5}{1} = 28.875,\ x_2 = 0.65\frac{82.5}{1} = 53.625.$

Hence, substitution effect is $= x_1' - x_1 = 28.875 - 17.5 = 11.375$, and income effect is equal to $x_1'' = x_1' = 35 - 28.875 = 6.125$. Note that these two effects add up to $11.375 + 6.125 = 17.5$.

11.3. An electric company is setting up a power plant in a foreign country, and it has to plan its capacity. The peak-period demand for power is given by $p_1 = 400 - q_1$, and the off-peak demand is given by $p_2 = 380 - q_2$. The variable cost is 20 per unit (paid in both markets) and capacity costs 10 per unit which is only paid once and is used in both periods.
(i) Write out the Lagrangian and KKT conditions for this problem.
(ii) Find the optimal outputs and capacity for this problem.
(iii) How much of the capacity is paid for by each market (i.e., what are the values of λ_1 and λ_2)?
(iv) Now suppose the capacity cost is 30 cents per unit (paid only once). Find the quantities, capacity, and how much of the capacity is paid for by each market (i.e., λ_1 and λ_2).

11.4. Show that the Walrasian and Hicksian demands are equal. HINT. (i) In both demands the consumption bundles that maximize utility are the same as the consumption bundles which maximize expenditure, provided the constraints of the two 'match up'. (ii) Both demands must coincide when computed according to the same prices, income, and utility. (iii) The proposition implies that the expenditure function $e(p, \bar{u}(p, w)) = w$ and $\bar{u}(p, e(p, \bar{u})) = \bar{u}$, so for a fixed price vector p, the quantities $e(p, \cdot)$ and $\bar{u}(p, \cdot)$ are inverses of each other.

12

Black-Scholes Model

Although far from being perfect, the Black-Scholes model is still useful. It demands a prerequisite of partial differential equations and the Laplace transform.

12.1 Black-Scholes Equation

The Black-Scholes equation is a one-dimensional diffusion equation, which describes the price of the option over time, and is defined as

$$\frac{\partial V}{\partial t} + \frac{1}{2}\sigma^2 S^2 \frac{\partial^2 V}{\partial S^2} + rS\frac{\partial V}{\partial S} - rV = 0, \tag{12.1.1}$$

where V denotes the value of the option, r the rate of interest, and S the asset at time t. The derivation of this equation is as follows: The Black-Scholes model based on return has two components: (i) $\mu\, dt$, which is predictable and deterministic, where μ is the drift, and (ii) $\sigma\, dX$, which is a random contribution to the return dS/S, where σ is the volatility of the asset S at time t. For each interval dt, the quantity dX is a sample drawn from the normal distribution $N(0, (\sqrt{dt})^2)$, which when multiplied by σ produces the term $\sigma\, dX$. The value of the parameters σ and μ are estimated from the historical data. Thus, we obtain the following *stochastic differential equation*:

$$\frac{dS}{S} = \mu\, dt + \sigma\, dX. \tag{12.1.2}$$

Three particular cases are as follows:

CASE 1. If $\sigma = 0$, the behavior of the asset price is given by the equation $dS/S = \mu\, dt$, which, with the initial condition $S(0) = S_0$, has the solution $S = S_0\, e^{\mu t}$. Thus, the asset price is totally deterministic.

CASE 2. Eq (12.1.2) represents a random walk, and it cannot be solved to give a deterministic solution for the share price S, but it gives probabilistic information about the behavior of S.

CASE 3. Eq (12.1.2) can be regarded as a scheme for constructing the time series that may be realized by the share prices (see Exercise 12.1 for a discrete model).

The discrete model presented in Exercise 12.1 (Figure 12.9.8) with a finite time interval in each step is not practicable because the large amount of data involved would be unmanageable. Hence, we will develop a continuous model by taking the limit as $dt \to 0$. This requires Itô's lemma which is a version of Taylor's series for functions of random variable.

12.1.1 Itô's Lemma. Assume that X_t is a drift-diffusion process that satisfies the SDE

$$dX_t = \mu_t\, dt + \sigma_t\, dB_t,$$

where B_t is a Wiener process. If $f(t, x)$ is a C^2-function, its Taylor's series expansion is

$$df = \frac{\partial f}{\partial t} + \frac{\partial f}{\partial x}\, dx + \frac{1}{2}\frac{\partial^2 f}{\partial x^2}\, dx^2 + \cdots .$$

Substituting X_t for x, and therefore, $\mu_t\, dt + \sigma_t\, dB_t$ for dx gives

$$df = \frac{\partial f}{\partial t}\, dt + \frac{\partial f}{\partial x}(\mu_t\, dt + \sigma_t\, dB_t) + \frac{1}{2}\frac{\partial^2 f}{\partial x^2}(\mu_t^2\, dt^2 + 2\mu_t\sigma_t\, dt\, dB_t + \sigma_t^2\, dB_t^2) + \cdots .$$

In the limit as $dt \to 0$, the terms dt^2 and $dt\, dB_t$ tend to zero faster than dB_t^2, which is $O(dt)$. Setting the dt^2 and $dt\, dB_t$ terms to zero, substituting dt for dB_t^2, and collecting the dt and dB_t terms, we obtain

$$df = \Big(\frac{\partial f}{\partial t} + \mu_t \frac{\partial f}{\partial x} + \frac{1}{2}\sigma_t^2 \frac{\partial^2 f}{\partial x^2}\Big)\, dt + \sigma_t \frac{\partial f}{\partial x}\, dB_t. \tag{12.1.3}$$

This is Itô's lemma.

12.1.2 Derivation of Black-Scholes Equation. To derive the Black-Scholes equation (1.1), suppose that a stock price follows a geometric Brownian motion given by the stochastic differential equation $dS = S(\mu\, dt + \sigma\, dB)$. Then, if the value of an option at time t is $f(t, S_t)$, by Itô's lemma (12.1.3) we get

$$df(t, S_t) = \Big(\frac{\partial f}{\partial t} + \frac{1}{2}(S_t\sigma)^2 \frac{\partial^2 f}{\partial S^2}\Big)\, dt + \frac{\partial f}{\partial S}\, dS_t.$$

The term $\dfrac{\partial}{\partial S}\, dS$ represents the change in value in time dt of the trading strategy consisting of holding an amount $\dfrac{\partial f}{\partial S}$ of the stock. If this trading strategy is followed, and any cash held is assumed to grow at the risk-free rate r, then the total value V of this portfolio satisfies the SDE

$$dV_t = r\Big(V_t - \frac{\partial f}{\partial S}\, S_t\Big)\, dt + \frac{\partial f}{\partial S}\, dS_t.$$

This strategy replicates the option if $V = f(t, S)$. Combining these options we get the Black-Scholes equation

$$\frac{\partial f}{\partial t} + \frac{1}{2}\sigma^2 S^2 \frac{\partial^2 f}{\partial S^2} + rS\frac{\partial f}{\partial S} - rf = 0, \tag{12.1.4}$$

which is Eq (12.1.1) with $f = V$.

12.2 Solution of Black-Scholes Equation

The solution is determined by first transforming the Black-Scholes equation (12.1.1) into one-dimensional diffusion equation (heat equation), and then solving this equation by different methods.

12.2.1 Transformation. We will use the following transformations to convert the Black-Scholes partial differential equation (12.1.1) into the heat equation. First, we convert the spot price to log-moneyness and the time to one-half of the total variance, to remove the S and S^2 terms from the Black-Scholes equation; thus, we set

$$\begin{aligned} x &= \ln \frac{S}{K} \quad \text{so that } S = K\,e^x, \\ \tau &= \frac{1}{2}\sigma^2 (T-t) \quad \text{so that } t = T - \frac{2\tau}{\sigma^2}, \\ U(x,t) &= \frac{1}{K}V(S,t) = \frac{1}{K}V(Ke^x, T - \frac{2\tau}{\sigma^2}). \end{aligned} \tag{12.2.1}$$

Next, we apply the chain rule to the partial derivatives in the Black-Scholes equation:

$$\begin{aligned} \frac{\partial V}{\partial t} &= K\frac{\partial U}{\partial \tau}\frac{\partial \tau}{\partial t} = -\frac{1}{2}K\sigma^2\frac{\partial U}{\partial \tau}, \\ \frac{\partial V}{\partial S} &= K\frac{\partial U}{\partial x}\frac{\partial x}{\partial S} = \frac{K}{S}\frac{\partial U}{\partial x} = e^{-x}\frac{\partial U}{\partial x}, \\ \frac{\partial^2 V}{\partial S^2} &= -\frac{K}{S^2}\frac{\partial U}{\partial x} + \frac{K}{S}\frac{\partial}{\partial S}\Big(\frac{\partial U}{\partial x}\Big) \\ &= -\frac{K}{S^2}\frac{\partial U}{\partial x} + \frac{K}{S}\frac{\partial}{\partial x}\Big(\frac{\partial U}{\partial x}\Big)\frac{\partial x}{\partial S} \\ &= -\frac{K}{S^2}\frac{\partial U}{\partial x} + \frac{K}{S^2}\frac{\partial^2 U}{\partial x^2} = \frac{e^{-2x}}{K}\Big(\frac{\partial^2 U}{\partial x^2} - \frac{\partial U}{\partial x}\Big). \end{aligned}$$

Substitution for the partials in the Black-Scholes equation (12.1.1) yields

$$-\frac{1}{2}K\sigma^2\frac{\partial U}{\partial \tau} + rK\frac{\partial U}{\partial x} + \frac{1}{2}\sigma^2 K\Big(\frac{\partial^2 U}{\partial x^2} - \frac{\partial U}{\partial x}\Big) - rU = 0,$$

which simplifies to

$$-\frac{\partial U}{\partial \tau}+(k-1)\frac{\partial U}{\partial x}+\frac{\partial^2 U}{\partial x^2}-kU=0, \tag{12.2.2}$$

where $k=\dfrac{2r}{\sigma^2}$. Notice that the coefficients of Eq (12.2.2) are independent of x and τ. The boundary condition for V is $V(S_T,T)=(S_T-K)^+$. Now, from Eq (12.2.1), $x=\ln\{S_T/K\}\equiv x_T$ when $t=T$ and $S_t=S_T$, and $\tau=0$. Then the boundary condition for U is

$$U_0(x_T)=U(x_T,0)=\frac{1}{K}V(S_T-K)^+=\frac{1}{K}(Ke^{xT}-K)^+=(e^{x_T}-1)^+.$$

Lastly, we set

$$W(x,T)=e^{\alpha x+\beta^2\tau}\,U(x,t), \tag{12.2.3}$$

where $\alpha=\frac{1}{2}(k-1)$ and $\beta=\frac{1}{2}(k+1)$, $(k=2r/\sigma^2)$. The transformation (12.2.3) converts Eq (12.2.2) into the heat equation; details are as follows:

$$\begin{aligned}
\frac{\partial U}{\partial \tau}&=e^{-\alpha x-\beta^2\tau}\Big(\frac{\partial W}{\partial \tau}-\beta^2 W(x,\tau)\Big),\\
\frac{\partial U}{\partial x}&=e^{-\alpha x-\beta^2\tau}\Big(\frac{\partial W}{\partial x}-\alpha W(x,\tau)\Big),\\
\frac{\partial^2 U}{\partial x^2}&=e^{-\alpha x-\beta^2\tau}\Big(\alpha^2 W(x,\tau)-2\alpha\frac{\partial W}{\partial x}+\frac{\partial^2 W}{\partial x^2}\Big).
\end{aligned}$$

Substituting these derivatives into Eq (12.2.2) we obtain

$$\begin{aligned}
\beta^2 W(x,\tau)-\frac{\partial W}{\partial \tau}+(k-1)\Big(-\alpha W(x,\tau)+\frac{\partial W}{\partial x}\Big)&\\
+\alpha W(x,\tau)=2\alpha\frac{\partial W}{\partial x}+\frac{\partial^2 W}{\partial x^2}-kW(x,\tau)=0,&
\end{aligned}$$

which simplifies to the heat equation

$$\frac{\partial W}{\partial \tau}=\frac{\partial^2 W}{\partial x^2}. \tag{12.2.4}$$

The boundary condition for $W(x,\tau)$ is obtained from (12.2.3) as

$$\begin{aligned}
W_0(x_T)&=W(x_T,0)=e^{\alpha x_T}\,U(x_T,0)\\
&=\Big(e^{(\alpha+1)x_T}-e^{\alpha x_T}\Big)^+=\Big(e^{\beta x_T}-e^{\alpha x_T}\Big)^+,
\end{aligned} \tag{12.2.5}$$

since $\beta = \alpha + 1$. Notice that the transformation from V to W is given by

$$V(S,t) = \frac{1}{K}\, e^{-\alpha x - \beta^2 \tau} W(x,\tau). \tag{12.2.6}$$

12.2.2. Solution of the Heat Equation. The solution of the heat equation (12.2.4) subject to the boundary condition (12.2.5) is

$$\begin{aligned} W(x,\tau) &= \frac{1}{2\sqrt{\pi\tau}} \int_{-\infty}^{\infty} e^{-(x-y)^2/(4\tau)} W_0(y)\, dy \\ &= \frac{1}{2\sqrt{\pi\tau}} \int_{-\infty}^{\infty} e^{-(x-y)^2/(4\tau)} (e^{\beta y} - e^{\alpha y})^+ \, dy, \end{aligned} \tag{12.2.7}$$

where the Green's function $G(x,t,y)$ is given by

$$G(x,t,y) = \frac{1}{2\sqrt{\pi\tau}}\, e^{-(x-y)^2/(4\tau)}. \tag{12.2.8}$$

For details, see Kythe [2011:§6.1.2]. The graphs of this Green's function G for some values of $t > 0$ are shown in Figure 12.1, which shows normal distributions.

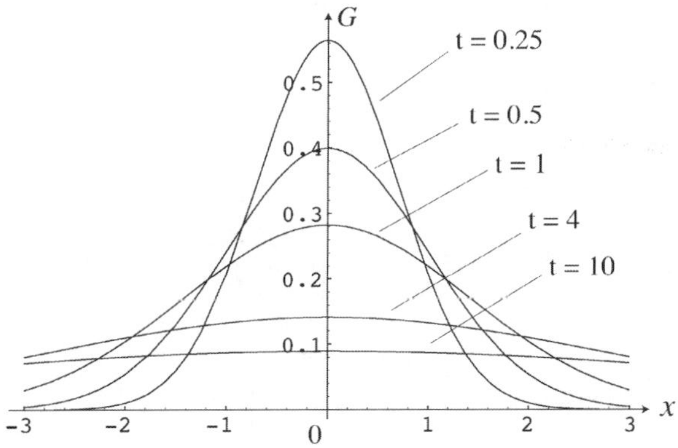

Figure 12.1 Green's function for 1-D diffusion operator.

12.2.3 Black-Scholes Call Price. The solution of the heat equation (12.2.4) subject to the boundary condition (12.2.5) is obtained from Eq (12.2.8) as

$$\begin{aligned} W(x,\tau) &= \frac{1}{2\sqrt{\pi\tau}} \int_{-\infty}^{\infty} e^{-(x-y)^2/(4\tau)} W_0(y)\, dy \\ &= \frac{1}{2\sqrt{\pi\tau}} \int_{-\infty}^{\infty} e^{-(x-y)^2/(4\tau)} (e^{\beta y} - e^{\alpha y})^+ \, dy. \end{aligned} \tag{12.2.9}$$

Now, set $z = (y-x)/\sqrt{2\tau}$, so that $y = \sqrt{2\tau}\, z + x$, giving $dy = \sqrt{2\tau}\, dz$. Then

$$W(x,\tau) = \frac{1}{\sqrt{2\pi}} \int_{-\infty}^{\infty} \exp\Big\{-\frac{1}{2}z^2\Big\} \exp\Big\{\beta(\sqrt{2\tau}\, z + x) - \alpha(\sqrt{2\tau}\, z + x)^+\Big\}\, dz. \tag{12.2.10}$$

Note the integral in (12.2.10) is nonzero only when the second exponent term is positive, i.e., when $\beta(\sqrt{2\tau}\, z + x) > \alpha(\sqrt{2\tau}\, z + x)$, or $z > -x/\sqrt{2\pi}$. Let us write Eq (12.2.10) as

$$W(x,\tau) = I_1 - I_2,$$

where

$$\begin{aligned} I_1 &= \frac{1}{\sqrt{2\pi}} \int_{-x/\sqrt{2\pi}}^{\infty} \exp\Big\{-\frac{1}{2}z^2\Big\} \exp\Big\{\beta(\sqrt{2\tau}\, z + x)\Big\}\, dz, \\ I_2 &= \frac{1}{\sqrt{2\pi}} \int_{-x/\sqrt{2\pi}}^{\infty} \exp\Big\{-\frac{1}{2}z^2\Big\} \exp\Big\{\alpha(\sqrt{2\tau}\, z + x)\Big\}\, dz. \end{aligned}$$

Completing the square in the integrand in I_1, we get $-\frac{1}{2}z^2 + \beta\sqrt{2\tau}z + \beta x = -\frac{1}{2}(z - \beta\sqrt{2\tau})^2 + \beta x + \beta^2\tau$, and thus,

$$I_1 = \frac{1}{\sqrt{2\pi}}\, e^{\beta x + \beta^2\tau} \int_{-x/\sqrt{2\pi}}^{\infty} e^{-\frac{1}{2}(z-\beta\sqrt{2\tau})^2}\, dz.$$

Now set $u = z - \beta\sqrt{2\tau}$. Then

$$\begin{aligned} I_1 &= \frac{1}{\sqrt{2\pi}}\, e^{\beta x + \beta^2\tau} \int_{-x/\sqrt{2\pi}-\beta\sqrt{2\tau}}^{\infty} e^{-\frac{1}{2}u^2}\, du \\ &= e^{\beta x + \beta^2\tau}\, \Phi\Big(\frac{x}{\sqrt{2\tau}} + \beta\sqrt{2\tau}\Big), \end{aligned}$$

where Φ is the c.d.f. of a normal random variable. The I_2 is similar, so replacing β by α, we get

$$I_2 = e^{\alpha x + \alpha^2\tau}\, \Phi\Big(\frac{x}{\sqrt{2\tau}} + \alpha\sqrt{2\tau}\Big).$$

Recall that $x = \ln(S/K), k = 2r/\sigma^2, \alpha = \frac{1}{2}(k-1) = \frac{r-\sigma^2/2}{\sigma^2}, \beta = \frac{1}{2}(k+1) = \frac{r+\sigma^2/2}{\sigma^2}$, and $\tau = \frac{1}{2}\sigma^2(T-t)$. Thus,

$$\begin{aligned} \frac{x}{\sqrt{2\tau}} + \beta\sqrt{2\tau} &= \frac{\ln\frac{S}{K} + (r+\sigma^2/2)(T-t)}{\sigma(T-t)} \equiv d_1, \\ \frac{x}{\sqrt{2\tau}} + \alpha\sqrt{2\tau} &= d_1 - \sigma\sqrt{T-t} \equiv d_2, \end{aligned}$$

and then the integrals I_1 and I_2 are given by

$$I_1 = e^{\beta x+\beta^2\tau}\,\Phi(d_1), \quad \text{and} \quad I_2 = e^{\alpha x+\alpha^2\tau}\,\Phi(d_2).$$

Hence, the solution is

$$W(x,\tau) = I_1 - I_2 = e^{\beta x+\beta^2\tau}\,\Phi(d_1) - e^{\alpha x+\alpha^2\tau}\,\Phi(d_2). \tag{12.2.11}$$

Finally, to obtain the solution for the call price $V(S_t,t)$, we use Eq (12.2.6) in Eq (12.2.11) and obtain

$$V(S,t) = K\,e^{-\alpha x-\beta^2\tau}\,W(x,\tau) = K\,e^{-\alpha x-\beta^2\tau}\,(I_1 - I_2). \tag{12.2.12}$$

The first and the second integrals in Eq (12.2.12) are

$$\begin{aligned} &K\,e^{-\alpha x-\beta^2\tau}\,e^{\beta x+\beta^2\tau}\Phi(d_1) = K\,e^{(\beta-\alpha)x}\,\Phi(d_1) = S\Phi(d_1), \quad \text{since } \beta-\alpha=1,\\ &K\,e^{-\alpha x-\beta^2\tau}\,e^{\alpha x+\alpha^2\tau}\Phi(d_2) = K\,e^{(\alpha^2-\beta^2)\tau}\,\Phi(d_2) = Ke^{-r(T-t)}\Phi(d_2), \end{aligned} \tag{12.2.13}$$

since $\beta^2 - \alpha^2 = 2r/\sigma^2$. Then combining both terms in (12.2.13) we get the Black-Scholes call price $C(S_t,K,T)$ in Eq (12.6.28).

12.2.4 Some Finance Terms and Arbitrage

1. The price of a contingent claim can be determined using risk-neutral probability, like martingales, change of measure, etc.
2. Examples of contingent claims are call options and put options.
3. A call option gives the holder the right (but not the obligation) to buy a specified item for an agreed upon price at an agreed upon time.
4. Arbitrage is simply (risk-free) free money. An arbitrage argument says that there should be no (risk-free) free money. Arbitrage can determine prices.
5. How does one use arbitrage to price a claim? Try to replicate the claim with stocks and bonds.
6. Stocks and bonds are called securities.
7. A contingent claim, f, is *replicable* if we can construct a portfolio Π such that
   a. the values of Π and f are the same under all circumstances, and
   b. Π is self-financing, i.e., as time passes, we only shift money around within the portfolio, but we do not put any more in (or take any out).
8. Π is called the *replicating portfolio* (of f).

Example 12.1. Suppose the price of gold today is \$2000 per ounce and the risk-free interest is 3%. Suppose you do not want gold today (because it is out of fashion), but you do want it in 6 months (when, of course, it will be suddenly very popular). Therefore, you buy a *forward contract* F_0, which says that you will receive gold in 6 months. You have locked in a price today for something you will buy in 6 months. Then how much should you pay for this wonderful opportunity?

Suppose the forward contract costs \$2500. Then you should go to the bank, and borrow \$2000 at 3% interest, and use this money to buy some gold right now. Then *short* (sell) the forward (to a sucker). In 6 months, the following things will happen:

(i) You sell your gold for \$2500.
(ii) You pay back your bank loan with your newly received funds.
(iii) You are left with $\$2500 - \$2000 \cdot e^{0.5(0.3)} = \$176.33$, which is a lot of free money.

What if the forward contract F_0 is selling for less than $\$200 \cdot e^{0.5(0.3)}$? Assume that you have gold lying around, and you sell your gold today and get \$2000, because you know the trick. Next, you put this \$2000 in the bank. Finally, you *go long* (buy) the forward contract. Then what happens after 6 months?

You take money out of the bank, which is now $\$2000 \cdot e^{0.5(0.3)}$.
Use it to buy your gold back for F_0.
Now you have your gold back, and $\$2000 \cdot e^{0.5(0.3)} - F_0$. Since this number is positive, you are happy.

Hence, arbitrage sets the price of the forward contract to be $\$2000 \cdot e^{0.5(0.3)}$. If the price is anything else, there is risk-free free money to be made. This is true of all forward contracts on an asset with no storage costs and which do not pay dividends, assuming that the interest rates are constant. Generally, any replicable claim has the same price as its self-financing replicating portfolio. ■

12.2.5 Self-Financing Portfolio. We have seen that forward contracts are simple to price, mostly because of the linearity of payoffs at maturity. On the other hand, options are difficult to price, because the payoff at maturity has a kink (a small problem or flaw due to randomness). However, a self-financing portfolio can be constructed using Itô's lemma. Thus, if $dS_t = S_t\mu\, dt + S_5\sigma\, dX$, and $f : (S_t, t) \mapsto \mathbb{R}$, we find that if $dx = (dS_t, dt)'$, then

$$\begin{aligned} df = (\nabla f, dx) &= \frac{\partial f}{\partial S}\, dS + \frac{\partial f}{\partial t}\, dt \\ &= \left(\frac{\partial f}{\partial S}\, S_t\mu + \frac{\partial f}{\partial t}\right) dt + \frac{\partial f}{\partial S} S_t\sigma\, dX. \end{aligned}$$

Since dX behaves like $\sqrt{dt}$, we may take $(dX)^2 = dt$. Then $\frac{1}{2}\frac{\partial^2 f}{\partial S^2}(dS)^2 =$

$\frac{1}{2}\frac{\partial^2 f}{\partial S^2}S_t^2\sigma^2\,dt$ up to first order. Hence, by Itô's lemma

$$df = \Big(\frac{\partial f}{\partial S}S_t\mu + \frac{\partial f}{\partial t} + \frac{\partial^2 f}{\partial S^2}S_t^2\sigma^2\Big)dt + \frac{\partial f}{\partial S}S_t\sigma\,dX, \tag{12.2.14}$$

where $dX = \sqrt{dt}$, as defined above. Notice that the only randomness in df is the dX term. Thus, we can contract a portfolio that eliminates the random part, and the rest we can easily control.

First, we will rely on the discrete version of (12.2.14). Since we want to price a contingent claim, or derivative, a simple method is to set

$$\Pi = \begin{cases} -1 & \text{for derivative,} \\ \Delta & \text{for shares,} \end{cases}$$

where $\Delta \equiv \frac{\partial f}{\partial S}$. For a small change δt in time, the corresponding change in Π is $\delta\Pi = -\delta f + \Delta\,\delta S$. The discrete version of (12.2.14) gives

$$\delta\Pi = \Big(-\frac{\partial f}{\partial t} - \frac{1}{2}\frac{\partial^2 f}{\partial S^2}\sigma^2 S_t^2\Big)\,\delta t, \tag{12.2.15}$$

which implies that the portfolio is risk-less (no uncertainty), and then by arbitrage argument we must have $\delta\Pi = r\Pi\,\delta t$, or

$$\Big(-\frac{\partial f}{\partial t} - \frac{1}{2}\frac{\partial^2 f}{\partial S^2}\sigma^2 S_t^2\Big)\,\delta t = r(-f + \Delta S)\,\delta t,$$

which yields

$$\Big(\frac{\partial f}{\partial t} + \frac{1}{2}\frac{\partial^2 f}{\partial S^2}\sigma^2 S_t^2 + r\Delta S\Big)\,\delta t = rf\,\delta t,$$

or

$$\frac{\partial f}{\partial t} + \frac{1}{2}\frac{\partial^2 f}{\partial S^2}\sigma^2 S_t^2 + r\Delta S = rf,$$

which yields the *Black-Scholes-Merton* partial differential equation

$$\frac{\partial f}{\partial t} + \frac{1}{2}\frac{\partial^2 f}{\partial S^2}\sigma^2 S_t^2 + r\frac{\partial f}{\partial S}S - rf = 0, \tag{12.2.16}$$

with known Cauchy data $f(S_t, T)$, which are initial conditions (at $t = 0$) on S and S_t. Thus, any function f that satisfies Eq (12.2.16) denotes the price of some theoretical contingent claim, and every contingent claim must satisfy Eq (12.2.16).

A solution of Eq (12.2.16) with boundary conditions depicting a European call option with strike K, i.e., with $f(S,T) = \max\{S - K, 0\}$, we obtain the Black-Scholes price of the European call option. Let c denote the Black-Scholes price of a European call option on a stock with no dividend, i.e.,

$$c \equiv c(K, r, S_t, t, T, \sigma) = S_t N(d_1) - K\,e^{-r(T-t)} N(d_2), \tag{12.2.17}$$

where N is the cumulative distribution function of the standard normal variable, $N = N(0, 1)$), and

$$\begin{aligned} d_1 &= \frac{\ln(S_t/K) + (r + \frac{1}{2}\sigma^2)(T-t)}{\sigma\sqrt{T-t}}, \\ d_2 &= d_1 - \sigma\sqrt{T-t}. \end{aligned} \tag{12.2.18}$$

Some properties of the Black-Scholes price of c are as follows:

1. If S_t is very large, c should be priced like a forward contract, since $c \approx S_t - Ke^{-r(T-t)}$ if S_t is large, which is the price of a forward contract.

2. If σ is very small, the payoff is expected to be

$$c \approx \max\{S_t\,e^{r(T-t)} - K, 0\}. \tag{12.2.19}$$

These two properties make the benchmark test.

3. c is an increasing function of σ.

4. $\dfrac{\partial c}{\partial S} = N(d_1)$. This is used to estimate $\Delta = \dfrac{\partial f}{\partial S}$, to be used in the replicating portfolio of c.

Notice that the price determined by risk-neutral expectation is the same as the price determined by solving the Black-Scholes equation.

12.2.6 Implied Volatility. Next, we will discuss the implied volatility, and where Black-Scholes goes wrong. Remember that prices are not set by the Black-Scholes options price. It is the markets that set prices, and according to some economists they set prices nearly perfectly. Therefore, go to the market to see what a call option on a certain underlying is selling for at this moment, i.e., at $t = 0$. Observe K, r, S_t, T, but remember that we cannot observe σ. So we solve for σ using (12.2.17), which is easy since the Black-Scholes call option price is monotonic in σ. The number we get is called the *implied volatility*.

If we check market data for different strike prices K with every thing else being equal, we get different implied volatilities. In fact, what we get is called a *volatility smile* or a *volatility skew* depending on its shape. This problem is due to our assumption that σ is an intrinsic property of the underlying. It should not vary with K.

Example 12.2. The prices for European call and put options on the QQQ (a NASDAQ 100 composite) for October 11, 2016 and expiration dates in October 17 and November 28, 2016, are presented in Table 12.1.

Table 12.1 Prices for European Call and Put Options

	Calls		Puts	
Strike	Oct	Nov	Oct	Nov
34	3.9	4.1	0.05	0.25
35	2.8	3.2	0.05	0.35
36	1.85	2.35	0.1	0.55
37	1	1.65	0.25	0.85
38	0.35	1.05	0.6	1.25
39	0.1	0.6	1.4	1.9
40	0.05	0.35	2.35	2.6

From this table, we notice that
$S_0 = 37.73$ (price at the closing of October 17, 2016),
$T - t = 42/365 = 0.1151$,
$r = 0.83$,
$q = 0.18$.
This data give the implied volatility as given in Table 12.2.

Table 12.2 Implied Volatility

Strike	October Call	November Call
34	0.3230	0.29
35	0.2592	0.2493
36	0.2455	0.2369
37	0.2455	0.2369
38	0.2279	0.2198
39	0.2156	0.2279
40	0.2181	0.2206

The above data is plotted in Figure 12.2 with the strike price on the x-axis and implied volatility on the y-axis.

Volatility smiles also occur with commodities. It is found that σ not only varies with the strike price, it also depends on whether a call or a put is being priced. Moreover, implied volatility varies with the expiration of the option. Thus, Black-Scholes is used to show that it lacks certain features. This model could be enlarged. Some suggestions are: (i) assume volatility is stochastic, i.e., let $\sigma = \mu_\sigma\, dt + \hat{\sigma}\, dX$; (ii) assume volatility is local, i.e., $\sigma = \sigma(S,t)$; (iii) assume the process that is underlying follows a jump-diffusion process; and (iv) assume interest rates are, at the very least, nonconstant. However, no improvement of the Black-Scholes model is available so far.

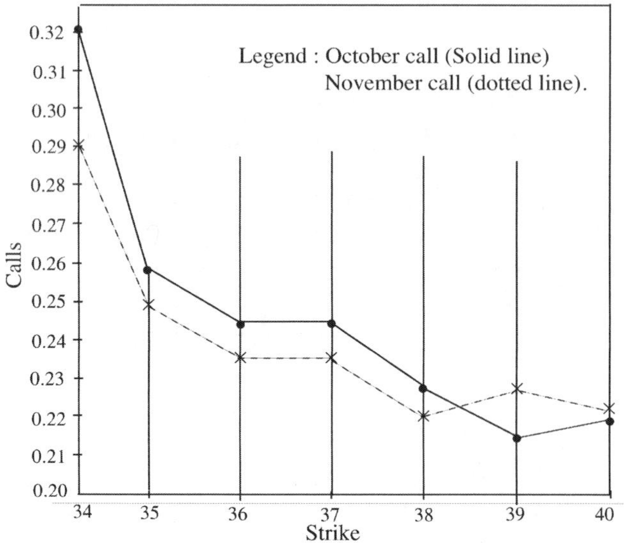

Figure 12.2 Implied volatility, October call.

12.3 Black-Scholes Formula

The Black-Scholes formula is used to calculate the price of the European put and call options. This price remains consistent with the Black-Scholes equation (12.1.1), since the formula is obtained by solving this equation using the terminal and boundary conditions. Thus, for example, the value of a call option for a non-dividend-paying underlying stock in terms of the Black-Scholes parameters is

$$C(S,t) = N(d_1)S - N(d_2)Ke^{-r(T-t)}, \tag{12.3.1}$$

where $N(\cdot)$ is the cumulative distribution function of the standard normal distribution, S is the spot price of the underlying asset, K is the strike price, $T-t$ is the time to maturity, r is the risk-free rate (annual, expressed as continuously compounding), σ is the volatility of returns of the underlying asset, and

$$d_1 = \frac{1}{\sigma\sqrt{T-t}}\Big\{\ln\Big(\frac{S}{K}\Big) + \Big(r + \frac{\sigma^2}{2}\Big)(T-t)\Big\}, \tag{12.3.2}$$

$$d_2 = d_1 - \sigma\sqrt{T-t}. \tag{12.3.3}$$

The price of the corresponding put option based on put-call parity is given by

$$P(S,t) = Ke^{-t(T-t)} - S + C(S,t) = N(-d_2)ke^{-r(T-t)} - N(-d_1)S. \quad (12.3.4)$$

An alternative formulation of the Black-Scholes formula, obtained by introducing some new variables, is

$$C(F,\tau) = D\big(N(d_+)F - N(d_-)K\big), \quad (12.3.5)$$

where

$$\begin{aligned} d_\pm &= \frac{1}{\sigma\sqrt{\tau}}\Big\{\ln\Big(\frac{F}{K}\Big) \pm \frac{1}{2}\sigma^2\tau\Big\}, \\ d_\pm &= d_\mp \pm \sigma\sqrt{\tau}, \end{aligned} \quad (12.3.6)$$

where $\tau = T - t$ is the time to expiry (i.e., remaining time, backwards time), $D = e^{-r\tau}$ is the discount factor, $F = e^{r\tau}S = S/D$ is the forward price of the underlying asset, $S = DF$, and $d_+ = d_1$ and $D_- = d_2$. The formula (12.3.5) is a special case of the so-called Black-76 formula. Thus, if a put-call parity, defined by $C - P = D(F - K) = S - DK$, is given, then the price of a put option is

$$P(F,\tau) = D\big(N(-d_-)K - N(-d_+)F\big). \quad (12.3.7)$$

The Black-Scholes formula (12.3.5) is a difference of two terms, sum of which is equal to the value of the binary call options. According to Nielsen [1993], this formula can be interpreted in terms of the $N(d_\pm)$ (and a fortiori $d_\pm$) terms as follows: it allows us to decompose a call option into the difference of two binary options, which are simply an asset-or-nothing call minus a cash-or-nothing call, where a call option exchanges cash for an asset at expiry, while an asset-or-nothing call just yields the asset (with no cash in exchange) and a cash-or-nothing call just yields cash (with no asset in exchange).

Next, we can rewrite formula (12.3.1) as

$$C = D\big(N(d_+)F - N(d_-)K\big). \quad (12.3.8)$$

This formula is made up of the difference of two parts: $DN(d_+)F$, and $DN(d_-)K$, where the first part is the present value of an asset-or-nothing call while the second is the present value of a cash-or-nothing call; the factor D in each is for discounting, because the expiry date is in the future, and removing it changes the present value to the future value (i.e., value at expiry). In simple terms, $N(d_+)F$ is the future value of an asset-or-nothing call and $N(d_-)K$ is the future value of a cash-or-nothing call. However, in risk-neutral terms, they are the expected value of the asset and the expected value of the cash, respectively, in the risk-neutral measure.

An obviously incorrect interpretation ensues if the $N(d_+)F$ term is regarded simply as the product of the probability of the option expiring in money $N(d_+)$ and the value of the underlying at expiry F, while the $N(d_-)K$ term is the product of the probability of the option expiring in money $N(d_-)$ and the value of the cash at expiry K. This is because as either both binaries expire in the money or both expire out of money, i.e., either cash is exchanged for asset or it is not, the probabilities $N(d_+)$ and $N(d_-)$ are not equal. In fact, the quantities $d_\pm$ are measures of moneyness (in standard deviation), while $N(d_-)$ are the probabilities of expiring ITM (percent moneyness). Thus, the interpretation of the cash option, $N(d_-)K$, is correct, since the value of the cash is independent of movements of the underlying, and therefore, it can be interpreted simply as 'probability times value.'

On the other hand, the product $N(d_+)F$ is more complicated, since, according to Nielsen [1993], the probability of expiring in the money and the value of the asset at expiry are not independent. In fact, the value of the asset at expiry is variable in terms of cash, but is constant in terms of the asset itself (i.e., it is a fixed quantity of the asset). Thus, these quantities are independent only if one changes stock (numéraire) to the asset rather than cash.

In formula (12.3.1), if S replaces the forward F in $d_\pm$ instead of the $\frac{1}{2}\sigma^2$ term, there is the term $(r \pm \frac{1}{2}\sigma^2)$, which can be interpreted as a drift factor in the risk-neutral measure for appropriate numéraire. The reason for the use of the factor $\frac{1}{2}\sigma^2$ is to account for the difference between the median and mean of the log-normal distribution, if d_- is used for moneyness rather than the standardized moneyness $m = \dfrac{1}{\sigma\sqrt{\tau}} \ln\left(\dfrac{F}{K}\right)$. The same factor is found in Itô's lemma on the geometric Brownian motion. Another reason for the incorrectness of the naive interpretation of replacing $N(d_+)$ by $N(d_-)$ in formula (12.3.5) is that it will yield a negative value for out-of-the-money call options.

Thus, the terms $N(d_1)$ and $N(d_2)$ represent, respectively, the probabilities of the option expiring in-the-money under the exponential martingale probability measure for stock and the equivalent martingale probability measure for risk-free asset. The risk-neutral probability for a finite stock price S_T is defined by

$$p(S,T) = \frac{N'(d_2(S_T))}{S_T\sigma\sqrt{T}}, \quad S_T \in (0,\infty), \tag{12.3.9}$$

where N' is the standard normal probability density function, and $d_2 = d_2(K)$ is defined in (12.3.3). Note that the term $N(d_2)$ represents the probability that the call will be exercised under the assumption that the asset drift is risk-free. On the other hand, the term $N(d_1)$ has no simple probability interpretation, but $SN(d_1)$ represents the present value, under risk-free interest rate, of the expected asset price at expiration provided the asset price at expiry is above the exercise price.

The solution of Eq (12.1.1), when discounted appropriately, is a martingale. What this means is that the option price is the expected value of the discounted payoff of the option. Any computation of the option price using this expectation is the risk neutrality approach and can be performed without any knowledge of the theory of partial differential equations, since the expectation of the option payoff is not done under the real world probability measure, but an artificial risk-free measure, which is different from the real world measure. For details, see Hull [2008; 307-309].

12.4 Use of Greek Letters

The letters of the Greek alphabet, commonly known as the *Greeks*, are important not only in the mathematical theory of finance but also in active trading. They measure the sensitivity of the value of a derivative or a portfolio to changes in parameter value(s) while the other measures are held fixed. Mathematically, they are partial derivatives of the price with respect to the parameter values. For example, the Greek letter *gamma* (γ) is a partial derivative of another Greek letter *delta* (δ). Financial institutions typically set (risk) limit values for each of the *Greeks* that their traders must *not* exceed. The Greek *delta* is most important since it usually confers the largest risk. For example, many traders who speculate and follow a *delta*-neutral hedging approach will zero their *delta* at the end of the day.

The *Greeks* for Black-Scholes can be obtained by differentiation of the Black-Scholes formula (12.3.1) or (12.3.5) (see Chen et al. [2010]). Note that it is clear from the formulae that the *gamma* and *vega*[1] are both the same value for call and put options. This can be seen directly from put-call parity, since the difference of a put and a call is a forward, which is linear in S and independent of σ; so a forward has zero *gamma* and zero *vega*. Recall that N' is the standard normal probability density function.

In practice, some sensitivities are usually quoted in scaled-down terms, to match the scale of likely changes in the parameters. For example, *rho* (ρ) is often reported divided by 10,000 (1 basis point rate change), *vega* by 100 (1 vol point change), and *theta* (θ) by 365 or 252 (1 day decay based on either calendar days or trading days per year).

12.5 Log-normal Distribution

12.5.1 Log-normal p.d.f and c.d.f. First, note that if a random variable $Y \in \mathbb{R}$ follows the normal distribution with mean μ and variance σ^2, then $X = e^Y$ follows the log-normal distribution with mean

$$E[X] = e^{\mu+\frac{1}{2}\sigma^2}, \tag{12.5.1}$$

[1] *vega* is not a letter in the Greek alphabet; it arises from reading the Greek letter ν as 'v'.

and variance

$$\mathrm{Var}[X] = \left(e^{\sigma^2} - 1\right) e^{2\mu+\sigma^2}. \tag{12.5.2}$$

The p.d.f for X is

$$dF_X(x) = \frac{1}{\sigma x\sqrt{2\pi}} \exp\Big\{ -\frac{1}{2}\Big(\frac{\ln x - \mu}{\sigma}\Big)^2\Big\}, \tag{12.5.3}$$

and the c.d.f is

$$F_X(x) = \Phi\Big(\frac{\ln x - \mu}{\sigma}\Big), \tag{12.5.4}$$

where

$$\Phi(y) = \frac{1}{\sqrt{2\pi}} \int_{-\infty}^{y} e^{-t^2/2}\, dt$$

is the standard normal c.d.f.

12.5.2 Log-normal Conditional Expected Value. The expected value of X conditional on $X > x$ is $L_X(K) = E[X|X > x]$. Using Eq (12.5.3), we find that $L_X(K)$ for the log-normal distribution is

$$L_X(K) = \int_K^\infty \frac{1}{\sigma\sqrt{2\pi}} \exp\Big\{ -\frac{1}{2}\left(\frac{\ln x - \mu}{\sigma}\right)^2 \Big\}\, dx. \tag{12.5.5}$$

With the change of variables $y = \ln x$, so that $x = e^y, dx = e^y\, dy$, and the Jacobian e^y, we get from (12.5.5)

$$L_X(k) = \int_{\in K}^\infty \frac{e^y}{\sigma\sqrt{2\pi}} \exp\Big\{ -\frac{1}{2}\left(\frac{\ln x - \mu}{\sigma}\right)^2 \Big\}\, dy. \tag{12.5.6}$$

The exponent in (12.5.6), after completing the square and combining terms, becomes

$$-\frac{1}{2\sigma^2}(y^2 - 2y\mu + \mu^2 - 2\sigma^2 y) = -\frac{1}{2\sigma^2}\left(y - (\mu + \sigma^2)\right)^2 + \mu + \tfrac{1}{2}\sigma^2,$$

and Eq (12.5.6) reduces to

$$L_X(K) = \exp\{\mu + \tfrac{1}{2}\sigma^2\}\frac{1}{\sigma} \int_{\ln K}^\infty \frac{1}{\sigma\sqrt{2\pi}} \exp\Big\{ -\frac{1}{2}\Big(\frac{y - (\mu + \sigma^2)}{\sigma}\Big)^2\Big\}\, dy. \tag{12.5.7}$$

Now, for the random variable X with p.d.f. $f_X(x)$ and c.d.f. $F_X(x)$, and the scale-location transformation $Y = \sigma X + \mu$, it is easy to show that the Jacobian is $1/\sigma$, the p.d.f for Y is $f_Y(y) = \dfrac{1}{\sigma} f_X\big(\dfrac{y-\mu}{\sigma}\big)$, and c.d.f is

$F_Y(y) = F_X\big(\frac{y-\mu}{\sigma}\big)$. Hence, the integral of Eq (12.5.7) involves the scale-location transformation of the standard normal c.d.f. Since $\Phi(-x) = 1-\Phi(x)$, we get (Hogg and Kulgman [1984])

$$L_X(K) = \exp\{\mu + \tfrac{1}{2}\sigma^2\}\Phi\Big(\frac{-\ln K + \mu + \sigma^2}{\sigma}\Big). \tag{12.5.8}$$

12.6 Black-Scholes Call Price

Let $C(S_t, K, T)$ be the t-time price of a European call option. Then we have

$$C(S_t, K, T) = S_t\Phi(d_1) - e^{-r\tau}K\Phi(d_2), \tag{12.6.1}$$

where

$$\begin{aligned} d_1 &= \frac{\ln(S_t/K) + (r + \frac{1}{2}\sigma^2)\tau}{\sigma\sqrt{\tau}}, \\ d_2 &= d_1 - \sigma\sqrt{\tau}, \\ \Phi(y) &= \frac{1}{\sqrt{2\pi}}\int_{-\infty}^{y} e^{-t^2/2}\,dt, \end{aligned} \tag{12.6.2}$$

and $\Phi(y)$ is the standard normal c.d.f.

12.6.1 Black-Scholes Economy. There are two assets: a risky stock S, and a risk-less bond B. These assets are derived by the stochastic differential equations

$$dS_t = \mu S_t\,dt + \sigma S_t\,dX_t, \tag{12.6.3}$$

$$dB_t = r_t B_t\,dt. \tag{12.6.4}$$

The value of the bond at time zero is $B_0 = 1$, and that of the stock is S_0. This model is valid under certain market assumptions, for which see Hull [2008]. By Itô's lemma, the value V_t of a derivative written on the stock follows the diffusion equation

$$\begin{aligned} dV_t &= \frac{\partial V}{\partial t} + \frac{\partial V}{\partial S}\,dS + \frac{1}{2}\frac{\partial^2 V}{\partial S^2}(dS)^2 \\ &= \frac{\partial V}{\partial t} + \frac{\partial V}{\partial S}\,dS + \frac{1}{2}\frac{\partial^2 V}{\partial S^2}\sigma^2 S^2\,dt \\ &= \Big(\frac{\partial V}{\partial t} + \mu S_t\frac{\partial V}{\partial S}\,dS + \frac{1}{2}\sigma^2 S_t^2\frac{\partial^2 V}{\partial S^2}\Big)\,dt + \Big(\sigma S_t\frac{\partial V}{\partial S}\Big)\,dX_t. \end{aligned} \tag{12.6.5}$$

There are four different methods to derive Eq (12.6.1): (i) By straightforward integration; (ii) by applying the Feynman-Kac theorem, (iii) by transforming the Black-Scholes equation into the heat equation, for which a solution is known (this was the original method used by Black and Scholes [1973]; see §12.2.3), and (iv) using the Capital Asset Pricing Model (CAPM). We will discuss these methods in the sequel.

With constant interest rate r, the time t price of a European call option on a non-dividend paying stock when its spot pricc is S_t and with strike K and time to maturity $\tau = T - t$ is given by

$$C(S_t, K, T) = e^{-r\tau} E^{\mathcal{Q}}[(S_T - K)^+ | \mathcal{F}_t], \tag{12.6.6}$$

which can be evaluated to give Eq (12.6.1), but rewritten here for convenience as

$$\begin{aligned} C(S_t, K, T) &= S_t \Phi(d_1) = K e^{-r\tau} \Phi(d_2), \\ d_1 &= \frac{\ln \dfrac{S_t}{K} + (r + \dfrac{1}{2}\sigma^2)\tau}{\sigma\sqrt{\tau}}, \\ d_2 &= d_1 - \sigma\sqrt{\tau} = \frac{\ln \dfrac{S_t}{K} + (r - \dfrac{1}{2}\sigma^2)\tau}{\sigma\sqrt{\tau}}. \end{aligned} \tag{12.6.7}$$

To find a measure $\mathcal{Q}$ such that under this measure the discounted stock price that uses B_t is a martingale, let

$$dS_t = r_t S_t \, dt + \sigma S_t \, dW_t^{\mathcal{Q}}, \quad \text{where } W_t^{\mathcal{Q}} = W_t + \frac{\mu - r_t}{\sigma} t. \tag{12.6.8}$$

12.6.2 Black-Scholes under a Different Numéraire. The principle behind the 'pricing by arbitrage' is that if the market is complete, we can find a portfolio that replicates the derivative at all times, and we can find an equivalent martingale measure (EMM) $\mathcal{N}$ such that the discounted stock price is a martingale. Moreover, $\mathcal{N}$ determines the unique numéraire N_t that discounts the stock price. The time t value $V(S_t, t)$ of the derivative with payoff $V(S_T, T)$ at time T discounted by the numéraire N_t is

$$V(S - t, t) = N_t E^{\mathcal{N}} \left[\frac{V(S_T, T)}{N_T} | \mathcal{F}_t \right]. \tag{12.6.9}$$

Recall that the bond $B_t = e^{r\tau}$ serves as the numéraire, and since r is deterministic, we can take $N_T = e^{rT}$ out of the expectation with $V(S_T, T) = (S_T - K)^+$ we can write

$$V(S_t, t) = e^{-r(T-t)} E^{\mathcal{N}}[(S_T - K)^+ | \mathcal{F}_t],$$

which is Eq (12.6.6) for the call price.

Now we will use the stock price S_t as the numéraire and recover the Black-Scholes call price. We start with the stock price process in Eq (12.6.8) under the measure $\mathcal{Q}$ and with a constant interest rate

$$dS_t = rS_t\,dt + \sigma S_t\,dW_t^{\mathcal{Q}}. \tag{12.6.10}$$

The related bond price is defined by $\tilde{B} = B/S$. Then by Itô's lemma we get the process

$$dB_t = \sigma^2 B_t\,dt - \sigma B_t\,dW_t^{\mathcal{Q}}. \tag{12.6.11}$$

The measure $\mathcal{Q}$ turns $\tilde{S} = S/B$ into a martingale, and not into $\tilde{B}$. The measure $\mathcal{P}$ that turns $\tilde{B}$ into a martingale is

$$W_t^{\mathcal{P}} = W_t^{\mathcal{Q}} - \sigma t, \tag{12.6.12}$$

so that we find that $d\tilde{B}_t = -\sigma\tilde{B}_t\,dW_t^{\mathcal{P}}$ is a martingale under $\mathcal{P}$.

The value of the European call is determined by using $N_t = S_t$ as the numéraire together with the payoff function $V(S_T, T) = (S_T - K)^+$ in evaluating Eq (12.6.9) as

$$V(S_t, t) = S_t E^{\mathcal{P}}\left[\frac{(S_T - K)^+}{S_T}|\mathcal{F}_t\right] = S_t E^{\mathcal{P}}[(1 - KZ_T)|\mathcal{F}_t], \tag{12.6.13}$$

where $Z_t = 1/S_t$. To evaluate $V(S_t, t)$ we need the distribution for Z_T. The process for $Z = 1/S$ is obtained using Itô's lemma in Eq (12.6.10) and change of measure in Eq (12.6.12), i.e.,

$$dZ_t = (-r + \sigma^2)Z_t\,dt - \sigma Z_t\,dW_t^{\mathcal{Q}} = -rZ_t\,dt - \sigma Z_t\,dW_t^{\mathcal{P}}. \tag{12.6.14}$$

Thus, to solve for Z_t define $Y_t = \ln Z_t$ and apply Itô's lemma again, yielding

$$dY_t = -(r + \frac{1}{2}\sigma^2)\,dt - \sigma\,dW_t^{\mathcal{P}}. \tag{12.6.15}$$

This equation after integration yields the solution

$$Y_T - Y_t = -(r + \frac{1}{2}\sigma^2)(T - t) - \sigma(W_T^{\mathcal{P}} - W_t^{\mathcal{P}}),$$

so that the solution for Z_T is

$$Z_T = \exp\Big\{\ln Z_t - (r + \frac{1}{2}\sigma^2)(T - t) - \sigma(W_T^{\mathcal{P}} - W_t^{\mathcal{P}})\Big\}. \tag{12.6.16}$$

Note that since $W_T^{\mathcal{P}} - W_t^{\mathcal{P}}$ is identical in distribution to $W_\tau^{\mathcal{P}}$, where $\tau = T - t$ is the time to maturity, and since $W_\tau^{\mathcal{P}}$ follows the normal distribution with zero mean and variance $\sigma^2\tau$, the exponent in Eq (12.6.16) also follows the normal distribution with mean $\ln Z_t - (r + \frac{1}{2}\sigma^2)\tau = -\ln S_t - (r + \frac{1}{2}\sigma^2)\tau \equiv u$ and variance $\sigma^2\tau \equiv v$. This implies that Z_T follows the log-normal distribution with mean $e^{u+v/2}$ and variance $(e^v - 1)\, e^{2u+v}$. Note that the factor $(1 - KZ_T)^+$ in the expectation of Eq (12.6.13) is nonzero for $Z_T < 1/K$. Hence, we can rewrite this expectation as two integrals

$$E^{\mathcal{P}}[(1 - KZ_T)|\mathcal{F}_t] = \int_{-\infty}^{1/k} dF_{Z_T} - K\int_{-\infty}^{1/K} Z_T\, dF_{Z_T} \equiv I_1 - I_2, \quad (12.6.17)$$

where F_{Z_T} is the c.d.f of Z_T defined in Eq (12.6.16). Using the definition of $L_{Z_T}(x)$ from Eq (12.5.8) for I_1, we find that

$$\begin{aligned}
I_1 &= F_{Z_T}\left(\frac{1}{k}\right) = \Phi\Big(\frac{\ln\frac{1}{K} - u}{v}\Big) \\
&= \Phi\Big(\frac{-\ln K + \ln S_t + (r + \frac{1}{2}\sigma^2)\tau}{\sigma\sqrt{\tau}}\Big) = \Phi(d_1), \\
I_2 &= K\left[\int_{-\infty}^{\infty} Z_T\, dF_{Z_T} - \int_{1/K}^{\infty} Z_T\, dF_{Z_T}\right] \\
&= K\left[E^{\mathcal{P}}[Z_T] - L_{Z_T}\left(\frac{1}{k}\right)\right] \\
&= K\left[e^{u+v/2} - e^{u+v/2}\Phi\Big(\frac{-\ln\dfrac{1}{K} + u + v}{\sqrt{v}}\Big)\right] \\
&= Ke^{u+v/2}\left[1 - \Phi\Big(\frac{-\ln\dfrac{S_t}{K} - (r - \frac{1}{2}\sigma^2)\tau}{\sigma\sqrt{\tau}}\Big)\right] \\
&= \frac{K}{S}\, e^{-r\tau}\Phi(d_2), \quad \text{since } 1 - \Phi(-d_2) = \Phi(d_2).
\end{aligned} \quad (12.6.18)$$

Substituting these values of I_1 and I_2 into Eq (12.6.13) we get

$$\begin{aligned}
V(S_t, t) &= S_t E^{\mathcal{P}}[(1 - KZ_T)|\mathcal{F}_t] = S_t[I_1 - I_2] \\
&= S_t\Phi(d_1) - Ke^{-r\tau}\Phi(d_2),
\end{aligned} \quad (12.6.19)$$

which is the Black-Scholes call price in Eq (12.6.1).

12.6.3 Black-Scholes by Direct Integration. The European call price $C(S_t, K, T)$ is a discounted time t expected value of $(S_t - K)^+$ under the

EMM II when the interest rates are constant. Thus, from Eq (12.6.6)

$$\begin{aligned} C(S_t, K, T) &= e^{-r\tau} E^{\mathcal{Q}}[(S_T - K)^+ | \mathcal{F}_t] \\ &= e^{-r\tau} \int_K^\infty (S_T - K)\, dF(S_T) \\ &= e^{-r\tau} \int_K^\infty S_T\, dF(S_T) - e^{-r\tau} K \int_K^\infty dF(S_T). \end{aligned} \tag{12.6.20}$$

To evaluate these two integrals, we will use the results derived in §12.6.2, that under $\mathcal{Q}$ and at time t the terminal stock price S_T follows the log-normal distribution with mean $\ln S_t + \left(r - \frac{1}{2}\sigma^2\right)\tau$ and variance $\sigma^2\tau$, where $\tau = T - t$ denotes the time to maturity.

The first integral in the last line of Eq (12.6.20) uses the conditional expectation of S_T, given that $S_T > K$, thus

$$\int_K^\infty S_T\, dF(S_T) = E^{\mathcal{Q}}[S_T | S_T > K] = L_{S_T}(K), \tag{12.6.21}$$

where the conditional expectation from Eq (12.5.8) is given by

$$\begin{aligned} L_{S_T} &= \exp\left\{ \ln S_t + \left(r - \frac{1}{2}\sigma^2\right)\tau + \frac{1}{2}\sigma^2\tau \right\} \\ &\quad \times \Phi\Big(\frac{-\ln K + \ln S_t + (r - \frac{1}{2}\sigma^2)\tau + \sigma^2\tau}{\sigma\sqrt{\tau}} \Big) \\ &= S_t\, e^{r\tau} \Phi(d_1), \end{aligned} \tag{12.6.22}$$

and thus, the first integral in the last line of Eq (12.6.20) is $S_t\Phi(d_1)$. Next, using Eq (12.5.4), the second integral in the last line of Eq (12.6.20) can be written as

$$\begin{aligned} e^{-r\tau} K \int_K^\infty dF(S_T) &= e^{-r\tau} K[1 - F(K)] \\ &= e^{-r\tau} K\Big[1 - \Phi\Big(\frac{\ln K - \ln S_t - (r - \frac{1}{2}\sigma^2)\tau}{\sigma\sqrt{\tau}} \Big)\Big] \\ &= e^{-r\tau} K[1 - \Phi(-d_2)] = e^{-r\tau} K \Phi(d_2). \end{aligned}$$

Hence, combining these two terms we obtain Eq (12.6.1) for the European call price.

12.6.4 Feynman-Kac Theorem. First, we will discuss this theorem.

Theorem 12.1. (Feynman-Kac theorem) Suppose that x_t follows the process

$$dx_t = \mu(x_t, t)\, dt + \sigma(x_t, t)], dW_t^{\mathcal{Q}}, \tag{12.6.23}$$

and suppose that the differentiable function $V = V(x_t, t)$ satisfies the partial differential equation

$$\frac{\partial V}{\partial t} + \mu(x_t, t)\frac{\partial V}{\partial x} + \frac{1}{2}\sigma(x_t, t)^2 \frac{\partial^2 V}{\partial x^2} - r(t, x)V(x_t, t) = 0, \tag{12.6.24}$$

with boundary condition $V(x_T, T)$. Then this equation has the solution

$$V(x_t, t) = E^{\mathcal{Q}}[\exp\Big\{-\int_t^T r(S_u, u)\, du\Big\} V(x_T, T)\big|\mathcal{F}_t]. \tag{12.6.25}$$

In this equation the time-t expectation is defined with respect to the same measure $\mathcal{Q}$ under which the stochastic part of Eq (12.6.23) defines Brownian motion.

In order to apply the Feynman-Kac theorem to the Black-Scholes call price, note that the value $V_t = V(S_t, t)$ of a European call option written at time t with strike price K and constant rates of interest r satisfies the Black-Scholes equation

$$\frac{\partial V}{\partial t} + rS_t \frac{\partial V}{\partial S} + \frac{1}{2}\sigma^2 S_t^2 \frac{\partial^2 V}{\partial S^2} - rV_t = 0, \tag{12.6.26}$$

with boundary condition $V(S_T, T) = (S_T - K)^+$.[1] Eq (12.6.26) is the same as Eq (12.6.24) for $x_t = S_t, \mu(x_t, t) = rS_t$, and $\sigma(x_t, t) = \sigma S_t$. Thus, we can apply the Feynman-Kac theorem so that the value of the European call is given by

$$\begin{aligned} V(S_t, t) &= E^{\mathcal{Q}}[\exp\Big\{-\int_t^T r(X_u, u)\, du\Big\} V(S_T, T)|\mathcal{F}_t] \\ &= e^{-r\tau}\, E^{\mathcal{Q}}[(S_T - K)^+ \big|\mathcal{F}_t]. \end{aligned} \tag{12.6.27}$$

This equation is the same as Eq (12.6.7). Hence, the expectation in Eq (12.6.27) can be evaluated exactly in the same way as above in §12.6.3, and thus, we obtain the call price in Eq (12.6.1).

12.6.5 CAPM. The Capital Asset Pricing Model (CAPM) is based on the assumption that the expected return r_i of a security i in excess of the risk-free rate r is

$$E[r_i] - r = \beta_i(E[r_M] - r),$$

[1] See www.frouah.com for the derivation of Eq (12.6.26).

where r_M denotes the return on the market, and the security's *beta* is given by

$$\beta_i = \frac{\mathrm{Cov}[r_i, r_M]}{\mathrm{Var}[r_M]}.$$

12.6.6 CAPM for Assets. During the time increment dt, the expected stock price return $E[r_S\, dt]$ is $E\left[\frac{dS_t}{S_t}\right]$, where S_t satisfies the diffusion equation (12.6.3). Then the expected return is

$$E\left[\frac{dS_t}{S_t}\right] = r\, dt + \beta_S(E[r_M] - r)\, dt. \tag{12.6.28}$$

Similarly, the expected return on the derivative $E[r_V\, dt]$, where V_t satisfies the diffusion equation (12.6.5), is

$$E\left[\frac{dV_t}{V_t}\right] = r\, dt + \beta_V(E[r_M] - r)\, dt. \tag{12.6.29}$$

If we divide both sides of Eq (12.6.4) by V_t, we get

$$\frac{dV_t}{V_t} = \frac{1}{V_t}\Big(\frac{\partial V}{\partial t} + \frac{1}{2}\sigma^2 S_t^2 \frac{\partial^2 V}{\partial S^2}\Big)\, dt + \frac{\partial V}{\partial S}\frac{dS_t}{S_t}\frac{S_t}{V_t},$$

or

$$r_V\, dt = \frac{1}{V_t}\Big(\frac{\partial V}{\partial t} + \frac{1}{2}\sigma^2 S_t^2 \frac{\partial^2 V}{\partial S^2}\Big)\, dt + \frac{\partial V}{\partial S}\frac{S_t}{V_t} r_S\, dt. \tag{12.6.30}$$

If we drop dt from both sides and take covariance of r_V and R_M, and note that only the second term on the right-hand side of Eq (12.6.30) is stochastic, we get

$$\mathrm{Cov}[r_V, r_M] = \frac{\partial V}{\partial S}\frac{S_t}{V_t}\,\mathrm{Cov}[r_S, r_M], \tag{12.6.31}$$

which yields the relation between the *beta* of the derivative β_V and the *beta* of the stock β_S as

$$\beta_V = \Big(\frac{\partial V}{\partial S}\frac{S_t}{V_t}\Big)\beta_S,$$

which is same relation as in Black-Scholes [1973, Eq(15)]. Next, multiplying Eq (12.6.29) by V_t we obtain

$$\begin{aligned} E[dV_t] &= rV_t\, dt + V_t\beta_V(E[r_M] - r)\, dt \\ &= rV_t\, dt + \frac{\partial V}{\partial S} S_t\beta_S(E[r_M] - r)\, dt, \end{aligned} \tag{12.6.32}$$

which is same relation as in Black-Scholes [1973, Eq(18)]. Now, taking expectations of the second line in Eq (12.6.5), and substituting for $E[dS_t]$ from Eq (12.6.29), we obtain

$$E[dV_t] = \frac{\partial V}{\partial t}\,dt + \frac{\partial V}{\partial S}[rS_t\,dt + S_t\beta_S(E[r_M - r])\,dt] + \frac{1}{2}\frac{\partial^2 V}{\partial S^2}\sigma^2 S^2\,dt. \quad (12.6.33)$$

On equating Eqs (12.6.32) and (12.6.33), dropping dt from both sides, and canceling terms in β_S, we get the Black-Scholes equation (12.6.26). Hence, we have obtained the Black-Scholes call price by using the Feynman-Kac theorem exactly the same way as in §12.6.4 and solving the integral as in §12.6.3.

12.7 Dividends

The Black-Scholes call price in Eq (12.6.1) is for a call written on a non-dividend-paying stock. There are two ways to incorporate dividends into the call price: (i) by assuming that the stock pays a continuous dividend yield q, or (ii) by assuming that the stock pays dividends in lump sum or 'lumpy' dividends.

12.7.1 Continuous Dividends. Assume that the dividend yield q is constant so that the stockholder receives an amount $qS_t\,dt$ of dividend in the time increment dt. After the dividend is paid out, the value of the stock drops by the dividend amount. In other words, without the dividend yield, the value of the stock increases by $rS_t\,dt$, but with the dividend yield, the value of the stock increases by $rS_t\,dt - qS_t\,dt = (r-q)S_t\,dt$. Thus, the expected return becomes $r-q$ instead of r, which means that the risk-neutral process for S_t satisfies Eq (12.6.8) but with drift $r-q$ instead of r we have

$$dS_t = (r-q)S_t\,dt + \sigma S_t\,dW_t^Q. \quad (12.7.1)$$

Following the same derivation method as in §12.6.2, Eq (12.7.1) has the solution

$$S_T = S_t \exp\Big\{(r - q - \tfrac{1}{2}\sigma^2)\tau + \sigma W_\tau^Q\Big\}, \quad \tau = T - t. \quad (12.7.2)$$

Thus, S_T follows the log-normal distribution with mean $S_t e^{(r-q)\tau}$ and variance $S_t^2 e^{2(r-q)\tau}(e^{\sigma^2\tau} - 1)$. Then, proceeding as in Eq (12.6.20), the call price is

$$C(S_t, K, T) = e^{-r\tau} L_{S_T}(K) - e^{-r\tau}(1 - F(K)), \quad (12.7.3)$$

where the conditional expectation $L_{S_T}(K)$ from Eq (12.5.8) becomes

$$\begin{aligned} L_{S_T}(K) &= \exp\Big\{\ln S_t + (r - q - \frac{1}{2}\sigma^2)\tau + \frac{1}{2}\sigma^2\tau\Big\} \\ &\quad \times \Phi\Big(\frac{-\ln K + \ln S_t + (r - q - \frac{1}{2}\sigma^2)\tau + \sigma^2\tau}{\sigma\sqrt{\tau}}\Big) \\ &= S_t\, e^{(r-q)\tau}\,\Phi(d_1), \end{aligned} \quad (12.7.4)$$

where

$$d_1 = \frac{\ln \dfrac{S_t}{K} + \left(r - q + \dfrac{1}{2}\sigma^2\right)\tau}{\sigma\sqrt{\tau}}.$$

Using Eq (12.5.4), the second term in Eq (12.6.20) becomes

$$\begin{aligned} e^{-r\tau}K[1 - F(K)] &= e^{-r\tau}\Big[1 - \Phi\Big(\frac{\ln K - \ln S_t - \left(r - q - \frac{1}{2}\sigma^2\right)\tau}{\sigma\sqrt{\tau}}\Big)\Big] \\ &= e^{-r\tau}K\Phi(d_2), \end{aligned} \tag{12.7.5}$$

where $d_2 = d_1\sigma\sqrt{\tau}$, as before. Then substituting Eqs (12.7.4) and (12.7.5) into Eq (12.7.2), we obtain the Black-Scholes price of a European call written on a stock that pays continuous dividends as

$$C(S_t, K, T) = S_t e^{-q\tau}\,\Phi(d_1) - e^{-r\tau}\,K\Phi(d_2). \tag{12.7.6}$$

Notice that the only modification is that the current value of the stock price is decreased by $e^{-q\tau}$, and the return on the stock is decreased from r to $r - q$.

12.7.2 Lumpy Dividends. The concept is the same as above, except that the current value of the stock price is decreased by the dividends, though not continuously.

12.8 Solutions of SDEs

12.8.1 Stock Price. Recall that S_t is driven by the diffusion in Eq (12.6.3). Now, apply Itô's lemma to $\ln S_t$, and get

$$d\ln S_t = (\mu - \tfrac{1}{2}\sigma^2)\,dt + \sigma\,dW_t, \tag{12.8.1}$$

which, upon integration from 0 to t, gives

$$\int_0^t d\ln S_u = \int_0^t (\mu - \tfrac{1}{2}\sigma^2)\,du + \sigma\int_0^t dW_u,$$

so that

$$\ln S_t - \ln S_0 = (\mu - \tfrac{1}{2}\sigma^2)\,t + \sigma W_t, \quad \text{since } W_0 = 0.$$

Hence, the solution of the SDE is

$$S_t = S_0 \exp\Big\{(\mu - \tfrac{1}{2}\sigma^2)t + \sigma W_t\Big\}. \tag{12.8.2}$$

Since W_t is distributed normal $N(0, t)$ with zero mean and variance t, we find that $\ln S_t$ follows the normal distribution with mean $\ln S_0 + (\mu - \frac{1}{2}\sigma^2)t$

and variance $\sigma^2 t$. Hence, in view of (12.5.1) and (12.5.2), S_t follows the log-normal distribution with mean $S_0\, e^{\mu t}$ and variance $S_0^2 e^{2\mu t}(e^{\sigma^2 t} - 1)$. We can also integrate Eq (12.8.1) from t to T and obtain

$$S_T = S_t \exp\left\{(\mu - \tfrac{1}{2}\sigma^2)\tau + \sigma(W_T - W_t)\right\},$$

which is similar to Eq (12.8.2), and thus, S_T follows the log-normal distribution with mean $S_t e^{\mu\tau}$ and variance $S_t^2 e^{2\mu\tau}(e^{\sigma^2 \tau} - 1)$.

12.8.2 Bond Price. Applying Itô's lemma to the function $\ln B_t$, we find that this function follows the SDE

$$d \ln B_t = r_t\, dt.$$

Integrating this equation from 0 to t we get

$$d \ln B_t - d\text{Ln}B_0 = \int_0^t r_u\, du,$$

which yields the solution of this SDE as

$$B_t = \exp\left\{\int_0^t r_u\, du\right\}, \quad \text{since } B_0 = 1.$$

Note that when interest rates are constant, $r_t = r$ and $B_t = e^{rt}$. Thus, integrating from t to T we get the solution

$$B_{t,T} = \exp\left\{\int_t^T r_u\, du\right\} \Longrightarrow B_{t,T} = e^{r\tau} \text{ for constant interest rates.}$$

12.8.3 Discounted Stock Price as Martingale. To find a measure $\mathcal{Q}$ such that under this measure the discounted stock price that uses B_t is a martingale, let

$$dS_t = r_t S_t\, dt + \sigma S_t\, dW^{\mathcal{Q}}, \quad \text{where } W_t^{\mathcal{Q}} = W_t + \frac{\mu - r_t}{\sigma}\, t. \tag{12.8.3}$$

We know that under $\mathcal{Q}$, at time $t = 0$, the stock price S_t follows the log-normal distribution with mean $S_0\, e^{r_t t}$ and variance $S_0^2\, e^{2r_t T}\big(e^{\sigma^2 t} - 1\big)$; however S_t is not a martingale. If we use B_t as the numéraire, the discounted stock price $\tilde{S}_t = S_t/B + t$ will be a martingale. Applying Itô's lemma to $\tilde{S}_t$, we get the SDE

$$d\tilde{S}_t = \frac{\partial \tilde{S}_t}{\partial B}\, dB_t + \frac{\partial \tilde{S}_t}{\partial S}\, dS_t, \tag{12.8.4}$$

since all terms involving second-order derivatives are zero. Expanding Eq (12.8.4) we get the SDE

$$\begin{aligned} d\tilde{S}_t &= -\frac{S_t}{B_t^2}\,dB_t + \frac{1}{B_t}\,dS_t \\ &= -\frac{S_t}{B_t^2}\,r_t B_t\,dt + \frac{1}{B_t}\Big(r_t S_t\,dt + \sigma S_t\,dW_t^{\mathcal{Q}}\Big) \\ &= \sigma \tilde{S}_t\,dW_t^{\mathcal{Q}} \end{aligned}$$

The solution of this SDE is

$$\tilde{S}_t = \tilde{S}_0 \,\exp\Big\{ -\frac{1}{2}\sigma^2 t + \sigma W_t^{\mathcal{Q}} \Big\}. \tag{12.8.5}$$

Thus, $\ln \tilde{S}_t$ follows the normal distribution with mean $\ln \tilde{S}_0 - \frac{1}{2}\sigma^2 t$ and variance $\sigma^2 t$. For a proof that $\tilde{S}$ is a martingale, see Exercise 12.5.

12.8.4 Summary. Apply Itô's lemma to the stock price $dS_t = \mu S_t\,dt + \sigma S_t\,dW_t$ and the bond prince $dB_t = r_t B_t\,dt$, we obtain the processes for $\ln S_t$ and $\ln B_t$:

$$\begin{aligned} d\ln S_t &= \Big(\mu - \frac{1}{2}\sigma^2\Big)\,dt + \sigma\,dW_t, \\ d\ln B_t &= r_t\,dt. \end{aligned}$$

Solving these equations, we get

$$S_t = S_0\,\exp\Big\{ \Big(\mu - \frac{1}{2}\sigma^2\Big)t + \sigma W_t \Big\}, \quad \text{and} \quad B_t = \exp\Big\{ \int_0^t r_s\,ds \Big\}.$$

If we apply a change of measure to get the stock price under the risk-neutral measure $\mathcal{Q}$, we have

$$dS_t = rS_t + \sigma S_t\,dW_t^{\mathcal{Q}} \Longrightarrow S_t = S_0 \exp\Big\{ \Big(r - \frac{1}{2}\sigma^2\Big)t + \sigma W_t^{\mathcal{Q}} \Big\}.$$

Since S_t is not a martingale under $\mathcal{Q}$, we discount S_t by B_t to obtain $\tilde{S}_t = S_t/B_t$ and

$$d\tilde{S}_t = \sigma \tilde{S}_t\,dW_t^{\mathcal{Q}} \Longrightarrow \tilde{S}_t = \tilde{S}_0 \exp\Big\{ -\frac{1}{2}\sigma^2 t + \sigma W_t^{\mathcal{Q}} \Big\}.$$

Thus, $\tilde{S}_t$ is a martingale under $\mathcal{Q}$.

The distributions of the above processes are summarized in Table 12.3, where $\tilde{S} = S/B$. Also note that the logarithm of the stock price is normally distributed.

Table 12.3 Distributions

Stochastic Process	Log-normal Distribution† Mean	Variance	Process a Martingale
$dS = \mu S\,dt + \sigma S\,dW$	$S_t\,e^{\mu\tau}$	$S_t^2 e^{2\mu\tau}\big(e^{\sigma^2\tau} - 1\big)$	no
$dS = rS\,dt + \sigma S\,dW^{\mathbb{Q}}$	$S_t\,e^{r\tau}$	$S_t^2 e^{2r\tau}\big(e^{\sigma^2\tau} - 1\big)$	no
$d\tilde{S} = \mu\tilde{S}\,dt + \sigma S\,dW^{\mathbb{Q}}$	$\tilde{S}_t$	$\tilde{S}_t^2\big(e^{\sigma^2\tau} - 1\big)$	yes
$d\tilde{S} = (\mu - r)\tilde{S}\,dt + \sigma\tilde{S}\,dW^{\mathbb{Q}}$	$S_t\,e^{(\mu-r)\tau}$	$S_t^2 e^{2(\mu-r)\tau}\big(e^{\sigma^2\tau} - 1\big)$	no

† This distribution is for $S_T|\mathcal{F}_t$.

12.9 Exercises

12.1. Let the price S of a share today be \$ 12.00. We will construct a time series for the share prices over four intervals if $\mu = 0.35, \sigma = 0.26, dt = 1/252$, where 252 is the number of trade days in a year. We will determine dX at each step from the normal distribution with mean 0 and standard deviation $1/\sqrt{252} \approx 0.063$, i.e., $N\big(0, 1/\sqrt{252}\big) = N(0, 0.063)$.

STEP 1. At time $t = 0$ we have $S_0 = 12$. We choose a value for dX from $N(0, 0.063)$, so take $dX = -0.05$. Then Eq (12.1.2) gives

$$\frac{dS}{12} = \frac{0.35}{252} + 0.26(-0.05) = -0.0116 \Longrightarrow dS = 12(-0.0116) = -0.14,$$

thus, $S_1 = 12 - 0.14 = 11.86$.

STEP 2. $S_1 = 11.86$; choose $dX = 0.15$. Then

$$\frac{dS_1}{11.86} = \frac{0.35}{252} + 0.26(0.15) = 0.04 \Longrightarrow dS = 11.86(0.04) = 0.48,$$

thus, $S_2 = 11.86 + 0.48 = 12.34$.

STEP 3. $S_2 = 12.34$; choose $dX = 0.09$. Then

$$\frac{dS_2}{12.34} = \frac{0.35}{252} + 0.26(0.09) = 0.025 \Longrightarrow dS = 12.34(0.025) = 0.31,$$

thus, $S_2 = 12.34 + 0.31 = 12.65$.

STEP 4. $S_3 = 12.65$; choose $dX = 0.12$. Then

$$\frac{dS_3}{12.65} = \frac{0.35}{252} + 0.26(0.12) = 0.0325 \Longrightarrow dS = 12.65(0.0325) = 0.41,$$

thus, $S_2 = 12.65 + 0.41 = 13.06$.

12.2. Derive Eq (12.1.2).

SOLUTION. Let the return from a stock be defined discretely as

$$\frac{\delta S_t}{S_t} = \mu\,\delta t + \sigma\sqrt{\delta t}. \tag{12.9.1}$$

To find the continuous version of Eq (12.9.1) we will use Brownian motion which is a stochastic process, i.e., a set of random variables $\{X_t\}, t \geq 0$ such that (i) the mapping $t \mapsto X_t$ is a.s. continuous, (ii) the process has stationary, independent increments, and (iii) the increment $X_{t+s} - X_t$ is normally distributed with variance t. Let us consider the stochastic process $\{X_t^n\}, t \geq 0, n \geq 1$, defined by

$$X_t^n = \frac{1}{\sqrt{n}} \sum_{1 \leq j \leq \lfloor nt \rfloor} \varepsilon_j, \tag{12.9.2}$$

where $\varepsilon_1, \varepsilon_2, \dots$ denote a sequence of independent standard normal random variables such that $\varepsilon_j \sim N(0,1)$. Thus, X_t is a random walk that takes a new step every $1/n$ units of time. Note that by the central limit theorem[1]

$$\frac{1}{\sqrt{\lfloor nt \rfloor}} \sum_{1 \leq j \leq \lfloor nt \rfloor} \varepsilon_j \tag{12.9.3}$$

converges (in distribution) to a standard normal random variable Z. Then using (12.9.3), we have

$$X_t^n = \frac{\sqrt{\lfloor nt \rfloor}}{\sqrt{n}} \frac{1}{\sqrt{\lfloor nt \rfloor}} \sum_{1 \leq j \leq \lfloor nt \rfloor} \varepsilon_j.$$

Now, since $\lim\limits_{n \to \infty} \dfrac{\sqrt{\lfloor nt \rfloor}}{\sqrt{n}} = \sqrt{t}$, we get in the limit $X_t = \sqrt{t}\,Z$. Moreover, for $s, t \in \{0, 1/n, 2/n, \dots\}$ we have

$$X_{t+s}^n - X_t^n = \frac{1}{\sqrt{n}} \sum_{1 \leq j \leq n(t+s)} \varepsilon_j - \frac{1}{\sqrt{n}} \sum_{1 \leq j \leq nt} \varepsilon_j = \frac{1}{\sqrt{n}} \sum_{nt+1 \leq j \leq n(t+s)} \varepsilon_j.$$

Since $X_{t+s}^n - X_t^n \to N(0, s)$, we get $X_{t+s} - X_t = \sqrt{s}Z$. Thus, dX_t behaves like $\sqrt{dt}$. Hence, we obtain the continuous time analog of Eq (12.9.1) as

$$\frac{dS_t}{S_t} = \mu\,dt + \sigma\,dX, \tag{12.9.4}$$

[1] This theorem states that, given certain conditions, the arithmetic mean of a sufficiently large number of iterates of independent random variables, each with a well-defined finite expected value and finite variance, will be approximately normally distributed, regardless of underlying distribution.

which is Eq (12.1.2).

12.3. Derive the Black-Scholes equation for a contingent claim $f(S_t, t)$ with a risk-free self-financing portfolio Π.

SOLUTION. A self-financing portfolio can be constructed using Itô's lemma. Thus, if $dS_t = S_t\mu\, dt + S_5\sigma\, dX$, and $f : (S_t, t) \mapsto \mathbb{R}$, we find that if $dx = (dS_t, dt)'$, then

$$\begin{aligned} df = (\nabla f, dx) &= \frac{\partial f}{\partial S}\, dS + \frac{\partial f}{\partial t}\, dt \\ &= \Big(\frac{\partial f}{\partial S}\, S_t\mu + \frac{\partial f}{\partial t}\Big)\, dt + \frac{\partial f}{\partial S} S_t\sigma\, dX. \end{aligned}$$

Since dX behaves like $\sqrt{dt}$, we may take $(dX)^2 = dt$. Then $\frac{1}{2}\frac{\partial^2 f}{\partial S^2}(dS)^2 = \frac{1}{2}\frac{\partial^2 f}{\partial S^2} S_t^2\sigma^2\, dt$ up to first order. Hence, by Itô's lemma

$$df = \Big(\frac{\partial f}{\partial S} S_t\mu + \frac{\partial f}{\partial t} + \frac{\partial^2 f}{\partial S^2} S_t^2\sigma^2\Big) dt + \frac{\partial f}{\partial S} S_t\sigma\, dX, \tag{12.9.5}$$

where dX is defined in (12.9.4). Notice that the only randomness in df is the dX term. Thus, we can construct a portfolio that eliminates the random part, and the rest we can easily control.

First, we will rely on the discrete version of (12.9.4) and (12.9.5). Since we want to price a contingent claim, or derivative, a simple method is to set

$$\Pi = \begin{cases} -1 & \text{for derivative,} \\ \Delta & \text{for shares,} \end{cases}$$

where $\Delta \equiv \dfrac{\partial f}{\partial S}$. For a small change δt in time, the corresponding change in Π is $\delta\Pi = -\delta f + \Delta\, \delta S$. The discrete version of (12.9.4) and (12.9.5) give

$$\delta\Pi = \Big(-\frac{\partial f}{\partial t} - \frac{1}{2}\frac{\partial^2 f}{\partial S^2}\sigma^2 S_t^2\Big)\, \delta t, \tag{12.9.6}$$

which implies that the portfolio is risk-less (no uncertainty), and then by arbitrage argument we must have $\delta\Pi = r\Pi\, \delta t$, or

$$\Big(-\frac{\partial f}{\partial t} - \frac{1}{2}\frac{\partial^2 f}{\partial S^2}\sigma^2 S_t^2\Big)\, \delta t = r(-f + \Delta S)\, \delta t,$$

which yields

$$\Big(\frac{\partial f}{\partial t} + \frac{1}{2}\frac{\partial^2 f}{\partial S^2}\sigma^2 S_t^2 + r\Delta S\Big)\, \delta t = r f\, \delta t,$$

or

$$\frac{\partial f}{\partial t} + \frac{1}{2}\frac{\partial^2 f}{\partial S^2}\sigma^2 S_t^2 + r\Delta S = rf,$$

which yields the *Black-Scholes-Merton* partial differential equation

$$\frac{\partial f}{\partial t} + \frac{1}{2}\frac{\partial^2 f}{\partial S^2}\sigma^2 S_t^2 + r\frac{\partial f}{\partial S}S - rf = 0, \tag{12.9.7}$$

with known Cauchy data $f(S_t, T)$. Thus, any function f that satisfies Eq (12.9.7) denotes the price of some theoretical contingent claim, and every contingent claim must satisfy Eq (12.9.7).

12.4. Prove that $\tilde{S}_t$, defined in Eq (12.8.5), is a martingale under $\mathcal{Q}$.

SOLUTION. Consider the expectation under $\mathcal{Q}$ for $s < t$

$$\begin{aligned} E^{\mathcal{Q}}[\tilde{S}_t|F_s] &= \tilde{S}_0 \exp\{-\frac{1}{2}\sigma^2 t\}E^{\mathcal{Q}}[\exp\{\sigma W_t^{\mathcal{Q}}\}|\mathcal{F}_s] \\ &= \tilde{S}_0 \exp\Big\{-\frac{1}{2}\sigma^2 t + \sigma W_s^{\mathcal{Q}}\Big\}E^{\mathcal{Q}}[\exp\Big\{\sigma(W_t^{\mathcal{Q}} - W_s^{\mathcal{Q}})\Big\}|\mathcal{F}_s]. \end{aligned}$$

Note that at time s the quantity $W_t^{\mathcal{Q}} - W_s^{\mathcal{Q}}$ is distributed as $N(0, t-s)$, which is the same as the distribution to $W_{t-s}^{\mathcal{Q}}$ at time zero. Hence, we have

$$E^{\mathcal{Q}}[\tilde{S}_t|\mathcal{F}_s] = \tilde{S}_0 \exp\Big\{-\frac{1}{2}\sigma^2 t + \sigma W_s^{\mathcal{Q}}\Big\}\, E^{\mathcal{Q}}[\exp\Big\{\sigma W_{t-s}^{\mathcal{Q}}\Big\}|\mathcal{F}_0].$$

Now, the moment generating function (mgf) of a random variable X with normal distribution $N(\mu, \sigma^2)$ is

$$E[e^{\phi X}] = \exp\Big\{\mu\phi + \frac{1}{2}\sigma^2\phi^2\Big\}.$$

Since $W_{t-s}^{\mathcal{Q}}$ under $\mathcal{Q}$ is $\mathcal{Q}$-Brownian motion distributed as $N(0, t-s)$, the mfg of $W_{t-s}^{\mathcal{Q}}$ is $E^{\mathcal{Q}}[\exp\Big\{\sigma W_{t-s}^{\mathcal{Q}}\Big\}] = \exp\big\{\frac{1}{2}\sigma^2(t-s)\big\}$, where σ is the same as ϕ. Thus, we have

$$\begin{aligned} E^{\mathcal{Q}}[\tilde{S}_t|\mathcal{F}_t] &= \tilde{S}_0 \exp\Big\{-\frac{1}{2}\sigma^2 t + \sigma W_s^{\mathcal{Q}}\Big\}\exp\Big\{\frac{1}{2}\sigma^2(t-s)\Big\} \\ &= \tilde{S}_0 \exp\Big\{-\frac{1}{2}\sigma^2 s + \sigma W_s^{\mathcal{Q}}\Big\} = \tilde{S}_s, \end{aligned}$$

which shows that $\tilde{S}_t$ is a $\mathcal{Q}$-martingale.

Note that (i) pricing a European call option under Black-Scholes uses the fact that under $\mathcal{Q}$, at time t, the terminal stock price at expiry, S_T, follows the normal distribution with mean $S_t\,e^{r\tau}$ and variance $S_t^2\,e^{2r\tau}\big(e^{\sigma^2\tau}-1\big)$ when the interest rate r_t is the constant value r; and (ii) under the original measure, the process for $\tilde{S}_t$ is $d\tilde{S}_t = (\mu - r)\tilde{S}_t\,dt + \sigma\tilde{S}_t\,dW_t$, which is clearly not a martingale.

12.5. Show that the process Y defined by $Y_t = t^2 W_t^3, t \geq 0$ satisfies the SDE

$$dY_t = \Big(2\frac{Y_t}{t} + 3(t^4 Y_t)^{1/3}\Big)\,dt + 3(tY_t)^{2/3}\,dW_t,\ Y_0 = 0.$$

SOLUTION. Obviously, $Y_=0$. The function $f(t,x) = t^2x^3$ is in C^2, and so by Itô's lemma

$$\begin{aligned} dY_t &= \frac{\partial f}{\partial t}(t,W_t)\,dt + \frac{\partial f}{\partial x}(t,W_t)\,dW_t + \frac{1}{2}\frac{\partial^2 f}{\partial x^2}(t,W_t)\,d(W,W)_t \\ &= 2tW_t^3\,dt + 3t^2W_t^2\,dW_t + \frac{1}{2}6t^2W_t\,dt \\ &= (2tW_t^3 + 3t^2W_t)\,dt + 3t^2W_t^2\,dW_t. \end{aligned}$$

The result is obtained since $2tW_t^3 = 2Y_t/t, 3t^2W_t = 3(t^4Y_t)^{1/3}$, and $3t^2W_t^2 = 3(tY_t)^{2/3}$.

12.6. Show that the process X given by $X_t = e^{W_t+t/2} + e^{W_t-t/2}, t \geq 0$, satisfies the SDE

$$dX_t = X_t\,dW_t + e^{W_t+t/2}\,dt,\ X_0 = 2.$$

SOLUTION. Since $X_t = e^{W_t-t/2}\,(e^t+1),\ t \geq 0$, set $Z_t = e^{W_t-t/2}$ and $Y_t = (e^t+1), t \geq 0$. Then $Z = E[W]$ satisfies the SDE $dZ_t = Z_t\,dW_t$, and Y satisfies the SDE $dY_t = e^t\,dt$. Since Z is a continuous semi-martingale and Y is continuous martingale of bounded variation, we have $\langle Z, Y\rangle \equiv [Z,Y] \equiv 0$ (P– a.s.). Hence, by the product rule we get

$$\begin{aligned} dX_t &= Y_t\,dZ_t + Z_t\,dY_t + \langle Z, Y\rangle_t = Y_tZ_t\,dW_t + Z_te^t\,dt \\ &= X_t\,dW_t + e^{W_t+t/2}\,dt. \end{aligned}$$

Also, $X_0 = e^0 + e^0 = 2$.

12.7. Derive the Black-Scholes Equation (12.1.1). SOLUTION. Given a continuous and continuously differentiable function f over an interval I, its Taylor's series at a point $x = a \in I$ is

$$f(x) = \sum_{n=0}^{\infty} \frac{f^{(n)}(a)}{n!}\,(x-a)^n. \tag{12.9.8}$$

If this series converges to f on I, then

$$f(x) = f(a) + \sum_{n=1}^{\infty} \frac{f^{(n)}(a)}{n!}(x-a)^n,$$

or

$$f(x) - f(a) = \sum_{n=1}^{\infty} \frac{f^{(n)}(a)}{n!}(x-a)^n. \tag{12.9.9}$$

If we replace x by $x + \Delta x$ and a by x, then (12.9.9) gives

$$\Delta f = f(x+\Delta x) - f(x) = \sum_{n=1}^{\infty} \frac{f^{(n)}(x)}{n!}(\Delta x)^n = f'(x) + \frac{f''(x)}{2!}(\Delta x)^2 + \cdots. \tag{12.9.10}$$

Now, let $t \to 0$ in Eq (12.1.2), i.e., we have: with probability 1, $(dX)^2 \to dt$ as $t \to 0$. Suppose that $f(S)$ is a function of the asset price S. If we change S by a small amount dS, then by (12.9.10) we have

$$df = \frac{df}{dS}dS + \frac{1}{2}\frac{d^2 f}{dS^2}(dS)^2 + \cdots. \tag{12.9.11}$$

By (12.1.2)

$$\begin{aligned} dS &= S(\mu\, dt + \sigma\, dS), \\ (dS)^2 &= S^2(\mu^2 (dt)^2 + 2\mu\sigma\, dt\, dX + \sigma^2 (dX)^2). \end{aligned} \tag{12.9.12}$$

Since, with probability 1, $(dX)^2 \to dt$ as $dt \to 0$, the term $S^2\sigma^2(dX)^2$ dominates the expression for $(dS)^2$ in (12.9.12) as dt becomes small. Hence, we will retain only this term and use $S^2\sigma^2\, dt$ as an approximation in (12.9.11) for $(dS)^2$ as $dt \to 0$, thus giving Itô's lemma:

$$\begin{aligned} df &= \frac{df}{dS}\, dS + \frac{1}{2}\frac{d^2 S}{dt^2}(S^2\sigma^2\, dt) = \frac{df}{dS}(S\mu\, dt + \sigma\, dX) + \frac{d^2 S}{dt^2}(S^2\sigma^2\, dt) \\ &= \sigma S \frac{df}{dS}\, dX + \left(\mu S \frac{df}{dS} + \frac{1}{2}\sigma^2 S^2 \frac{d^2 S}{dt^2}\right) dt. \end{aligned} \tag{12.9.13}$$

Itô's lemma relates a small change in a function of a random variable to a small change in the variable itself, as it contains a deterministic component dt and a random component dX. We will, however, need the following multivariate version of Itô's lemma: If f is a function of two variables S, t, then

$$df = \sigma S \frac{\partial f}{\partial S}\, dX + \left(\mu S \frac{\partial f}{\partial S} + \frac{1}{2}\sigma^2 S^2 \frac{\partial^2 f}{\partial S^2} + \frac{\partial f}{\partial t}\right) dt. \tag{12.9.14}$$

We will now derive the Black-Scholes equation (12.1.1). Let $V(S,t)$, which is called $C(S,t)$ for a call and $P(S,t)$ for a put, denote the value of an option, and be r the interest rate. Using Itô's lemma (12.9.14) we have

$$dV = \sigma S \frac{\partial V}{\partial S}\, dX + \Big(\mu S \frac{\partial V}{\partial S} + \frac{1}{2}\sigma^2 S^2 \frac{\partial^2 V}{\partial S^2} + \frac{\partial V}{\partial t}\Big)\, dt. \tag{12.9.15}$$

To derive Eq (12.9.15), consider a portfolio containing one option and $-\Delta$ units of the underlying stock. Then the value of the portfolio is $\Pi = V - \Delta S$, which gives $d\Pi = dV - \delta\, dS$, and

$$\begin{aligned} d\Pi &= \sigma S \frac{\partial V}{\partial S}\, dX + \Big(\mu S \frac{\partial V}{\partial S} + \frac{1}{2}\sigma^2 S^2 \frac{\partial^2 V}{\partial S^2} + \frac{\partial V}{\partial t}\Big)\, dt - \Delta S \mu\, dt - \Delta D \sigma\, dX \\ &= \sigma S\Big(\frac{\partial V}{\partial S} - \Delta\Big)\, dX + \Big(\mu S \frac{\partial V}{\partial S} + \frac{1}{2}\sigma^2 S^2 \frac{\partial^2 V}{\partial S^2} + \frac{\partial V}{\partial t} - \mu\, \Delta S\Big)\, dt \\ &= \Big(\frac{1}{2}\sigma^2 S^2 \frac{\partial^2 V}{\partial S^2} + \frac{\partial V}{\partial t}\Big)\, dt, \quad \text{choosing } \Delta = \frac{\partial V}{\partial S}. \end{aligned} \tag{12.9.16}$$

Now, if Π was invested in risk-less assets, it would have a growth of $r\Pi\, dt$ during the interval dt. Then, using (12.9.16), we should get for a fair price

$$r\Pi\, dt = \Big(\frac{1}{2}\sigma^2 S^2 \frac{\partial^2 V}{\partial S^2} + \frac{\partial V}{\partial t}\Big)\, dt \Longrightarrow r\Big(V - \frac{\partial V}{\partial S}\, S\Big) = \frac{1}{2}\sigma^2 S^2 \frac{\partial^2 V}{\partial S^2} + \frac{\partial V}{\partial t},$$

which yields the Black-Scholes equation (12.1.1).

A

Probability Topics

A.1 Definition of Probability Measure

(i) $0 \leq P(A) \leq 1$ for each event A;
(ii) $P(\Omega) = 1$;
(iii) $A \cap B = \emptyset \Longrightarrow P(A \cup B) = P(A) + P(B)$; and
(iv) $A_i \cap A_j = \emptyset \Longrightarrow P\Big(\bigcup_{i=1}^{\infty} A_i\Big) = \sum_{i=1}^{\infty} P(A_i)$.

Property (iv) is equivalent to $A_i \supset A_{i+1}$, $\bigcap_{i=1}^{\infty} A_i = \emptyset \Longrightarrow \lim_{n\to\infty} P(A_n) = 0$.

A.2 Probability Laws (Note: $\neg A$ is used for not-A)

(i) $P(\neg A) = 1 - P(A)$;
(ii) $P(A \cup B) = P(A) + P(B) - P(A \cap B)$;
(iii) $P(A \cup B \cup C) = P(A) + P(B) + P(C) - P(A \cap B) - P(A \cap C)$
$-P(B \cap C) + P(A \cap B \cap C)$;
(iv) $P(B \backslash A) = P(B) - P(A \cap B)$;
(v) $P(A \triangle B) = P(A) + P(B) - 2P(A \cap B)$;
(vi) $P\Big(\bigcup_{i=1}^{n} A_i\Big) \leq \sum_{i=1}^{n} P(A_i)$ (*Boole's inequality*);
(vii) $P\Big(\bigcap_{i=1}^{n} A_i\Big) \geq 1 - \sum_{i=1}^{n} (1 - P(A_i))$ (*Bonferroni's inequality*).

A.3 Conditional Probability and Independent Events

(i) $P(A \,|\, B) = \dfrac{P(A \cap B)}{P(B)}$, $P(B \,|\, A)) = \dfrac{P(A \cap B)}{P(A)}$;
(ii) $P(A \cap B) = P(A) \cdot P(B \,|\, A)$;
(iii) $P(A \cap B \cap C) = P(A) \cdot P(B \,|\, A) \cdot P(C \,|\, A \cap B)$;
(iv) $P(A_1 \cap \ldots \cap A_n) = P(A_1)P(A_2 \,|\, A_1)P(A_3 \,|\, A_1 \cap A_2) \cdots$
$P(A_n \,|\, A_1 \cap \cdots \cap A_{n-1})$ (see Figure A.1 for $n = 4$);

(v) $P(A_i \mid B) = \dfrac{P(A_i)P(B \mid A_i)}{\sum\limits_{i=1}^{n} P(A_i)P(B \mid A_i)}$ (*Bayes' formula*);

(vi) $P(B) = \sum\limits_{i=1}^{n} P(A_i)P(B \mid A_i)$ (*Total probability formula*)

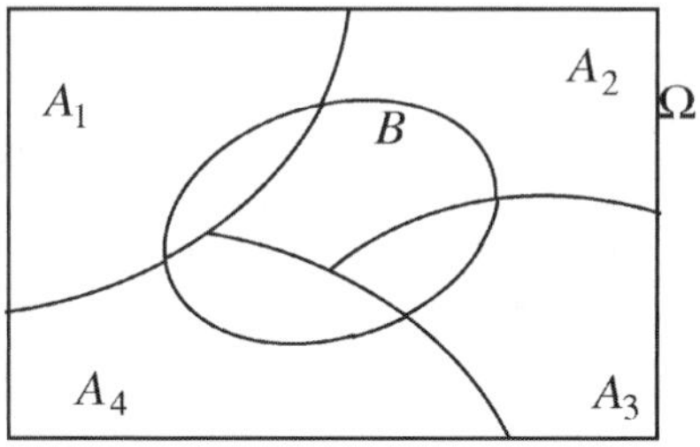

Figure A.1 Conditional probability and independent events.

For *independent* events the following hold:

(i) $P(A \cap B) = P(A)P(B)$.

(ii) $P(B \mid A) = P(B), \quad P(A \mid B) = P(A)$.

(iii) $P(A \cap B \cap C) = P(A)P(B)P(C)$.

(iv) $P(A_1 \cap A_2 \cap \cdots \cap A_n) = P(A_1)P(A_2) \cdots P(A_n)$.

(v) If A and B are independent, then $(\neg A, B), (A, \neg B)$ and $(\neg A, \neg B)$ are pairs of independent events.

A.4 Random (Stochastic) Variables

Discrete random variable X and continuous random variable X are presented in Figure A.2(a) and (b). Let $f(X)$ denote their probability function.

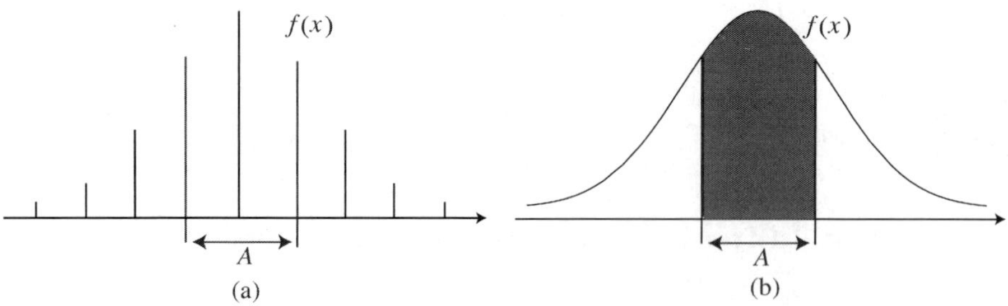

Figure A.2 (a) Discrete, and (b) continuous random variable X.

Then, using the following notation: $P(X \in A)$ for probability; $E[x]$ or μ for expectation; $F(x)$ for distribution function; $\mathrm{Var}[X]$ or σ^2 for variance; $H(x)$ for entropy; DRV for discrete random variable; CRV for continuous random variable, we have

	DRV X	CRV X
$P(X \in A) =$	$\sum_{x \in A} f(x)$	$\int_A f(x)\,dx$
$E[X] \equiv \mu =$	$\sum_{x \in A} x f(x)$	$\int_A x f(x)\,dx$
$F(x) =$	$P(X \leq x) = \sum_{t \leq x} f(t)$	$\int_{-\infty}^{\infty} f(t)\,dt$
$\mathrm{Var}[X] \equiv \sigma^2 =$	$E[(X-\mu)^2]$ $= \sum_{x \in \Omega} (x-\mu)^2 f(x)$	$E[(X-\mu)^2]$ $= \int_\Omega (x-\mu)^2 f(x)\,dx$
$H(x) =$	$E[\ln(1/f(X))]$ $= -\sum_{x \in \Omega} f(x)(\ln f(x))$	$E[\ln(1/f(X))]$ $= -\int_\Omega f(x)\,\ln f(x)\,dx$

Note that if $y = \ln x$, then $y' = \frac{1}{x}$, $x > 0$. In general, if $y = \log_a x$, then $y' = \frac{1}{x \ln a}$, $a > 1$.

Expectations : $E[aX] = aE[X], \quad E[X+Y] = E[X] + E[Y]$.

For discrete random variable : $E[g((X)] = \sum_{x \in \Omega} g(x) f_X(x)$.
For continuous random variable : $E[g((X)] = \int_\Omega g(x) f_X(x)\,dx$.

Variances : (i) $\mathrm{Var}[aX] = a^2 \mathrm{Var}[X]$; (ii) $\mathrm{Var}[X] = E[X^2] - (E[X])^2$ *(Steiner's theorem)*; and (iii) $\mathrm{Var}[X+Y] = \mathrm{Var}[X] + \mathrm{Var}[Y] + 2\mathrm{Cov}[X,Y]$.

Chebyshev inequality : (i) $P(|X| \geq a) \leq E[X^2]/a^2$; (ii) $P(|X-\mu| \geq a) \leq \mathrm{Var}[X]/a^2$; and (iii) $P(|X-\mu| \geq k\sigma) \leq 1/k^2$.

A.5 Bivariate (Two-dimensional) Random Variable.

The discrete and continuous variable (X, Y) is defined below and presented in Figure A3.

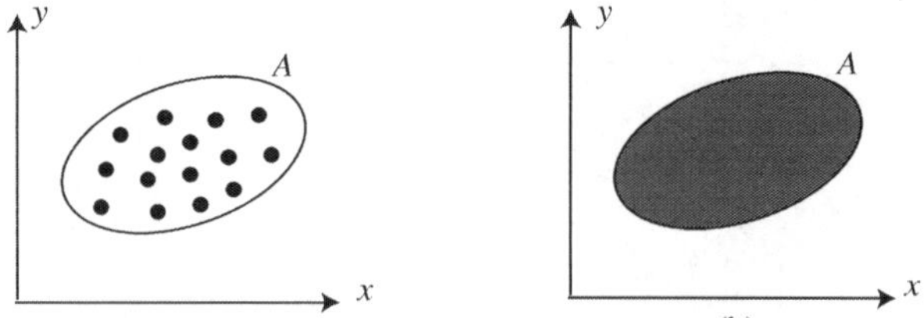

(a) (b)

Figure A.3 (a) Discrete, and (b) continuous random variables X, Y.

	Discrete Variable (X, Y)	Continuous Variable (X, Y)
$P((X,Y) \in A) =$	$\sum\sum_{(x,y)\in A} f(x,y)$	$= \iint_A f(x,y)\,dx\,dy$
$F(x,y) = P(X \leq x, Y \leq y) =$	$\sum_{u\leq x}\sum_{v\leq y} f(u,v)$	$= \int_{-\infty}^{x}\int_{-\infty}^{y} f(u,v)\,du\,dv$
$f_X(x) =$	$\sum_y f(x,y)$	$= \int_{-\infty}^{\infty} f(x,y)\,dy$
$f_Y(y) =$	$\sum_x F(x,y)$	$= \int_{-\infty}^{\infty} f(x,y)\,dx$
$E[g[X,Y]] =$	$\sum_x\sum_y g(x,y)f(x,y)$	$= \iint_{-\infty}^{\infty} g(x,y)f(x,y)\,dx\,dy$

In this Table $f_X(x)$ denotes the marginal distribution, and $E[g[X,Y]]$ the expectation.

Schwarz inequality: $(E[XY])^2 \leq E[X^2]E[Y^2]$.

For independent random variables X and Y,
$f(x,y) = f_X(x)f_Y(y)$.
$F(x,y) = F_X(x)F_Y(y)$.
$E[XY] = E[X] + E[Y]$.
$\text{Var}[X+Y] = \text{Var}[X] + \text{Var}[Y]$.

For independent random variables X and Y,
$f_{X+Y}(x) = \int_{-\infty}^{\infty} f_X(t)f_Y(x-t)\,dt$.

For independent random variables X and Y $(X \geq 0, Y \geq 0)$,
$f_{X+Y}(x) = \int_0^x f_X(t)f_Y(x-t)\,dt$.

COVARIANCE :

$\text{Cov}[X,Y] = E[(X-\mu_1)(Y-\mu_2)],\ E[X] = \mu_1, E[Y] = \mu_2$.
$\text{Cov}[X,X] = \text{Var}[X]$.
$\text{Cov}[X,Y] = E[XY] - E[X]E[Y] = \text{Cov}[Y,X]$.
X and Y independent $\Longrightarrow \text{Cov}[X,Y] = 0$.

$\text{Var}[X+Y] = \text{Var}[X] + \text{Var}[Y] + 2\text{Cov}[X,Y].$

CORRELATION COEFFICIENT ρ :

$\rho = \dfrac{\text{Cov}[X,Y]}{\sqrt{\text{Var}[X]\text{Var}[Y]}}, \quad -1 \le \rho \le 1,$

X and Y independent $\Longrightarrow \rho = 0,$

$\text{Var}[X+Y] = \text{Var}[X] + \text{Var}[Y] + 2\rho\sqrt{\text{Var}[X]\text{Var}[Y]}.$

A.6 Moments

The *kth central moment* μ_k is defined as $\mu_k = E[(X-\mu)^k]$.
The *skewness* γ_1 and *kurtosis* γ_2 are defined as

$$\gamma_1 = \mu_3/\sigma^3, \quad \gamma_2 = (\mu_4/\sigma^4) - 3.$$

For the $N(\mu,\sigma)$ normal distribution
$\mu_{2k+1} = 0, \quad \mu_2 = \sigma^{2k}(2k-1)!!, \quad \gamma_1 = \gamma_2 = 0.$

A.7 Convergence

Convergence in probability:

$\lim_{n\to\infty} X_n = X \Longleftrightarrow \lim_{n\to\infty} P(|X_n - X| > \varepsilon) = 0$ for each $\varepsilon > 0$.

Convergence almost surely:

$\text{plim}_{n\to\infty} X_n = xX \Longleftrightarrow P\left(\lim_{n\to\infty} X_n = X\right) = 1.$

Convergence in distribution:

$X_n \mapsto X \Longleftrightarrow \lim_{n\to\infty} P(X_n \le x) = P(X \le x)$ for each x such that $P(X \le x)$ is continuous in x.

Convergence in mean:

$\text{l.i.m.}_{n\to\infty} X_n = X \Longleftrightarrow \lim_{n\to\infty} E[|X_n - X|^2] = 0.$

For these kinds of convergence we have:

$$\left\{\begin{array}{l} \text{plim}_{n\to\infty} X_n = X \\ \text{l.i.m.}_{n\to\infty} X_n = X \end{array}\right. \Longleftrightarrow \text{limp}_{n\to\infty} X_n = X \Longrightarrow X_n \mapsto X$$

in distribution, where plim is the limit in probability, and *l.i.m.* is the limit in mean.

A.8 Central Limit Theorem

The central limit theorem (CLT) used in probability theory establishes that in most cases when the independent variables are added, their sum tends toward a normal distribution (known as the *bell curve*) even if the original variables themselves are not normally distributed. In other words, under certain conditions the arithmetic mean of a sufficiently large number of iterates of independent random variables, each with a finite expected value and finite variance, will be approximately normally distributed, regardless of the distribution used. A simple example is that if one flips a coin many times, the probability of getting a given number of heads in a series of flips must follow a normal curve, with mean equal to half of the total number of flips in each series. More details are available in Rice [1995].

Let $\{X_1, \dots, X_n\}$ denote a sequence of independent and identically distributed random variables drawn from distributions of expected values by μ and finite variance by σ^2. Consider the sample average

$$S_n = \frac{X_1 + \cdots + X_n}{n}$$

of such random variables. Then by the law of large numbers, this sample average converges in probability and almost surely to the expected value μ as $n \to \infty$. The classical central limit theorem determines the size and the distributional form of the deterministic number μ during the convergence. This theorem states that as n gets larger, the distribution of the difference $\sqrt{n}\,(S_n - \mu)$ approximates the normal distribution with mean zero and variance σ^2. For very large n the distribution of S_n becomes close to the normal distribution with mean μ and variance σ^2/n.

Theorem A.1. (Lindeberg-Lévy CLT) *Suppose $\{X_1, \dots, X_n\}$ is a sequence of independent and identically distributed random variables with $E[X_i] = \mu$ and $Var[X_i] = \sigma^2 < \infty$. Then as $n \to \infty$, the random variables $\sqrt{n}\,(Sn - \mu)$ converge in distribution to a normal $N(0, \sigma^2)$, i,e.,*

$$\sqrt{n}\Big[\Big(\frac{1}{n}\sum_{i=1}^{n} X_i\Big) - \mu\Big] \longrightarrow N(0, \sigma^2).$$

For proof, see Billingsley [1995:357].

B

Differentiation of Operators

Let X and Y denote normed linear spaces over a field F which may be $\mathbb{R}$, or $\mathbb{R}^n$. The mappings defined by $f : \mathbb{R}^n \mapsto \mathbb{R}$ may, in general, not be linear. Let $L(X, Y)$ denote the class of all linear operators from X into Y, and let $B(X, Y)$ denote the class of all bounded linear operators.

B.1 Gateaux Differential

Let $\mathbf{x}_0 \in X$ be a fixed point and let $f : X \mapsto Y$. If there exists a function $\delta f(\mathbf{x}_0, \cdot) : X \mapsto Y$ such that

$$\lim_{t \to 0} \left\| \frac{f(\mathbf{x}_0 + t\mathbf{h}) - f(\mathbf{x}_0)}{t} - \delta f(\mathbf{x}_0, \mathbf{h}) \right\| = 0, \tag{B.1}$$

where $t \in F$ for all $\mathbf{h} \in X$, then f is said to be *Gateaux differentiable* at $\mathbf{x}_0$, and $\delta f(f_0, \mathbf{h})$ is called the *Gateaux differential* of f at $\mathbf{x}_0$ with increment $\mathbf{h}$. The Gateaux differential is also called the *weak differential* of f or the G-differential of f.

B.2 Fréchet Derivative

Let $\mathbf{x}_0 \in X$ be a fixed point, and let $f : X \mapsto Y$. If there exists a bounded linear operator $F(\mathbf{x}_0) \in B(X, Y)$ such that

$$\lim_{\|\mathbf{h}\| \to 0} \left\| f(\mathbf{x}_0 + \mathbf{h}) - f(\mathbf{x}_0) - F(\mathbf{x}_0)\mathbf{h} \right\| = 0, \tag{B.2}$$

where $\mathbf{h} \in X$, then f is said to be *Fréchet differentiable* at $\mathbf{x}_0$, and $F(\mathbf{x}_0)$ is called the *Fréchet derivative* of f at $\mathbf{x}_0$.

The following theorem establishes that Fréchet differentiability implies Gateaux differentiability.

Theorem B.1. *Let $f : X \mapsto Y$, and let $\mathbf{x}_0 \in X$ be a fixed point. If f is Fréchet differentiable at $\mathbf{x}_0$, then f is Gateaux differentiable, and the Gateaux differential is given by*

$$\delta f(\mathbf{x}_0, \mathbf{h}) = f'(\mathbf{x}_0)\mathbf{h} \quad \text{for all } \mathbf{h} \in X. \tag{B.3}$$

PROOF. The proof follows from the definition (B.2). Suppose that $F(\mathbf{x}_0) = f'(\mathbf{x}-))$, let $\varepsilon > 0$, and let $\mathbf{h} \in X$. Then there exists a $\delta > 0$ such that

$$\frac{1}{\|t\mathbf{h}\|}\big\|f(\mathbf{x}_0+\mathbf{h}) - f(\mathbf{x}_0) - F(\mathbf{x}_0)\mathbf{h}\big\| < \varepsilon\,\|\mathbf{h}\|,$$

provided that $\|t\mathbf{h}\| < \delta$ if $t\mathbf{h} \neq 0$. But this implies that

$$\Big\|\frac{f(\mathbf{x}_0+t\mathbf{h}) - f(\mathbf{x}_0)}{t} - F(\mathbf{x}_0)\mathbf{h}\Big\| < \varepsilon, \tag{B.4}$$

provided that $|t| < \delta\|\mathbf{h}\|$. ■

Example B.1. Let $X = \mathbb{R}^n$ and let $f : \mathbb{R}^n \mapsto \mathbb{R}$. Let $\mathbf{x} = (x_1, \dots, x_n) \in \mathbb{R}^n$, and $\mathbf{h} = (h_1, \dots, h_n) \in \mathbb{R}^n$. If f has continuous partial derivatives with respect to x_i, $i = 1, \dots, n$, then the Fréchet differential of f is given by

$$\delta f(\mathbf{x}, \mathbf{h}) = \frac{\partial f(\mathbf{x})}{\partial x_1} h_1 + \cdots + \frac{\partial f(\mathbf{x})}{\partial x_n} h_n.$$

For a fixed $\mathbf{x}_0 \in \mathbb{R}^n$, define the bounded linear functional $f(\mathbf{x}_0)$ on $\mathbb{R}^n$ by

$$f(\mathbf{x}_0)\ \mathbf{h} = \sum_{i=1}^{n} \frac{\partial f(\mathbf{x})}{\partial x_i} h_i\Big|_{\mathbf{x}=\mathbf{x}_0} \quad \text{for } \mathbf{h} \in \mathbb{R}^n.$$

This means that $f(\mathbf{x}_0)$ is the Fréchet derivative of f at $\mathbf{x}_0$, and we denote it by $Df(\mathbf{x}_0)$:

$$Df(\mathbf{x}_0) = \Big[\frac{\partial f(\mathbf{x}_0)}{\partial x_1} \ \ \dots \ \ \frac{\partial f(\mathbf{x}_0)}{\partial x_n}\Big]. \tag{B.5}$$

In $\mathbb{R}$ the derivative $Df(\mathbf{x})$ becomes $f'(x)$.

Example B.2. Consider the line $3x + 2y = 6$ in $\mathbb{R}$ with slope $-3/2$. In $\mathbb{R}^2$, the function $f(,y) = 3x + 2y - 6$ has the gradient $\nabla f = 3\mathbf{i} + 2\mathbf{j}$. Then $f'(x) = -\dfrac{\partial f/\partial x}{\partial f/\partial y} = -3/2$ is the slope of the plane of f. The gradient ∇f is the vector starting at the origin, passing through the pane, and ending at the point $(3, 2)$ (see Figure B.1). ■

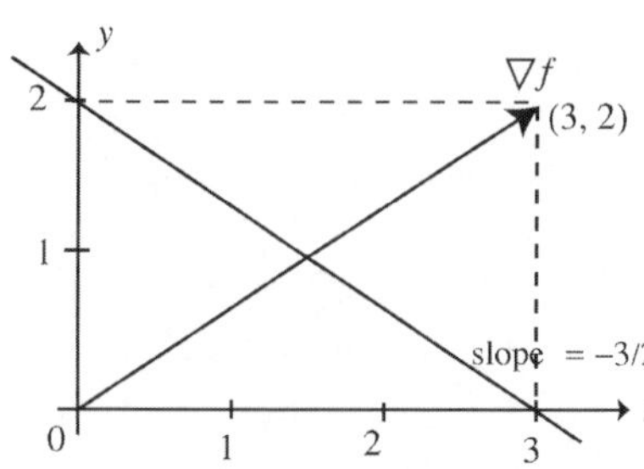

Figure B.1 Slope and gradient.

The fact still remains that the gradient is defined as $\nabla f(\mathbf{x}) = Df(\mathbf{x})\,[\mathbf{e}]$, i.e.,

$$\begin{aligned}\nabla f(\mathbf{x}) &= \frac{\partial f(\mathbf{x})}{\partial x_1}\mathbf{e}_1 + \cdots + \frac{\partial f(\mathbf{x})}{\partial x_n}\mathbf{e}_n \\ &= \begin{bmatrix} \frac{\partial f(\mathbf{x})}{\partial x_1} & \cdots & \frac{\partial f(\mathbf{x})}{\partial x_n} \end{bmatrix} \begin{bmatrix} \mathbf{e}_1 \\ \vdots \\ \mathbf{e}_n \end{bmatrix} = \begin{bmatrix} \frac{\partial f(\mathbf{x})}{\partial x_1} & \cdots & \frac{\partial f(\mathbf{x})}{\partial x_n} \end{bmatrix} [\mathbf{e}]^T,\end{aligned} \tag{B.6}$$

and the vector $[\mathbf{e}]^T$ cannot just be removed from this definition unless ∇f occurs in an equation that satisfies Condition A of Theorem 2.18.

According to Smith [1985:71], using the scalar product of $\nabla f(\mathbf{x})$ and the increment $\mathbf{h}$, the gradient is defined implicitly by

$$\langle \nabla f(\mathbf{x}), \mathbf{h} \rangle = \lim \frac{f(\mathbf{x} + t\,\mathbf{h}) - f(\mathbf{x})}{t} \quad \text{as } \mathbf{h} \to 0. \tag{B.7}$$

If this limit exists for all directions $\mathbf{h}$ in an inner product space, then $\nabla f(x)$ is simply described as the gradient of f at $\mathbf{x}$. In the case of Fréchet differentiation, the gradient, if it exists, can be expressed as

$$f(\mathbf{x} + t\,\mathbf{h}) = f(\mathbf{x}) + \langle \nabla f, \mathbf{h} \rangle + o(\|\mathbf{h}\|), \tag{B.8}$$

provided that $\|\mathbf{h}\| \to 0$ in a subspace containing $f(\mathbf{x})$ and $f(\mathbf{x}){+}\mathbf{h}$. Practically, the gradient is usually defined by

$$\langle \nabla f(\mathbf{x}), \mathbf{h}) \rangle = \frac{d}{d\mathbf{x}} f(\mathbf{x} + \mathbf{h})\big|_{\mathbf{h}=0}. \tag{B.8}$$

Example B.3. Smith [1985:78]) Let $f : \mathbb{R}^n \mapsto \mathbb{R}$ have continuous second-order partial derivatives. The gradient of f is given by

$$\nabla f = \begin{bmatrix} \frac{\partial f}{\partial x_1} & \cdots & \frac{\partial f}{\partial x_n} \end{bmatrix} \tag{B.9}$$

with respect to the usual inner product. The second Gateaux variation is

$$d^2 f(\mathbf{x}, \mathbf{h}) = \frac{d^2}{dt^2} \int_0^1 f(x_1 + th_1 + \cdots + x_n + th_n)\, dt\big|_{t=0} = \langle \mathbf{H}, \mathbf{h} \rangle, \tag{B.10}$$

where $\mathbf{h} = (h_1, \dots, h_n)$, and $\mathbf{H} = \dfrac{\partial^2 f(x_1, \dots, x_n)}{\partial x_i \partial x_j}$ is the Hessian of f. Then $\langle \mathbf{H}\mathbf{h}, \mathbf{h} \rangle = \mathbf{h}^T \mathbf{H}\mathbf{h}$. ■

It is obvious from (B.8) and (B.9) that gradient ∇f is expressed as either equal or equivalent to the first partial derivatives of f, using words like 'practically' or 'usually.' But the question still remains about the validity of such a conclusion. The second variation (B.10) is, however, correct.

B.3 Taylor's Series Approximations

Taylor's series approximations are presented in summation form. In $\mathbb{R}$, the Taylor's series approximation about $x = a$ is

$$f(x) = f(a) + f'(a)(x-a) + \frac{f''(a)}{2!}(x-a)^2 + \cdots + \frac{f^{(n-1)}(a)}{(n-1)!}(x-a)^{(n-1)} + R_n,$$

where $R_n = \dfrac{f^n(\xi)\,(x-a)^n}{n!}$ is the remainder after n terms, and $a < \xi < x$. If $R_n = 0$, the infinite series is called the Taylor series about $x = a$.

In $\mathbb{R}^2$, let $\mathbf{x} = (x, y)$ and $\mathbf{a} = (a, b)$. Then the second-order approximation of Taylor's series expansion is

$$\begin{aligned} f(x,y) &\approx f(a,b) + f_x(a,b)(x-a) + f_y(a,b)(y-b) \\ &\quad + \frac{1}{2!}\Big[f_{xx}(a,b)(x-a)^2 + 2f_{xy}(a,b)(x-a)(y-b) + f_{yy}(a,b)(y-b)^2\Big], \end{aligned}$$

or

$$f(\mathbf{x}) \approx f(\mathbf{a}) + \sum_{i=1}^{2} f_i(\mathbf{a})(x_i - a_i) + \frac{1}{2!}\Big[\sum_{i=1}^{2} f_{ij}(\mathbf{a})(x_i - a_i)(x_j - a_j)\Big],$$

where $f_i = \dfrac{\partial f}{\partial x_i}$ and $f_{ij} = \dfrac{\partial}{\partial x_i}\Big(\dfrac{\partial f}{\partial x_j}\Big)$ for $i, j = 1, 2$.

In $\mathbb{R}^n$, where $\mathbf{x} = (x_1, \dots, x_n)$ and $\mathbf{a} = (a_1, \dots, a_n)$, the second-order approximation of Taylor's series expansion is

$$f(\mathbf{x}) \approx f(\mathbf{a}) + \sum_{i=1}^{n} f_i(\mathbf{a})(x_i - a_i) + \frac{1}{2!}\Big[\sum_{i,j=1}^{n} f_{ij}(\mathbf{a})(x_i - a_i)(x_j - a_j)\Big], \tag{B.11}$$

where $f_i = \dfrac{\partial f}{\partial x_i}$ and $f_{ij} = \dfrac{\partial}{\partial x_i}\Big(\dfrac{\partial f}{\partial x_j}\Big)$ $\Big[= \dfrac{\partial}{\partial x_j}\Big(\dfrac{\partial f}{\partial x_i}\Big)$, if f is continuous$\Big]$ for $i, j = 1, 2, \dots, n$. Note that the first two terms in (B.11) give the first-order approximation of the Taylor's series expansion at $\mathbf{x} = \mathbf{a}$ in $\mathbb{R}^n$.

C

Distributions

C.1 Definitions

We will follow the convention of denoting a random variable by an upper case letter, e.g., X, and using the corresponding lower case letter, e.g., x, for a particular value of that variable.

A real-valued function $F(x)$ is called a (univariate) *cumulative distribution function* (c.d.f.), or simply a *distribution function*, or *distribution*, if (i) $F(x)$ is nondecreasing, i.e., $F(x_1) \leq F(x_2)$ for $x_1 \leq x_2$; (ii) $F(x)$ is everywhere continuous from the right, i.e., $F(x) = \lim_{\varepsilon \to 0+} F(x+\varepsilon)$; and (iii) $F(-\infty) = 0, F'(\infty) = 1$.

The function $F(x)$ describes probability of the event: $X \leq x$, i.e., the probability $p\{X \leq x\} = f(x)$, which describes the c.d.f. of X. There are two principal types of distributions: *discrete* and *continuous*.

DISCRETE DISTRIBUTIONS. They are characterized by the random variable X taking on an enumerable number of values $\dots, x_{-1}, x_0, x_1, \dots$ with point probabilities $p_n = P\{X = x_n\} \geq 0$ which is subject only to the restriction $\sum_n p_n = 1$. In this case the distribution is written as

$$F(x) = P\{X \leq x\} = \sum_{x_n \leq x} p_n, \tag{C.1}$$

where the summation is taken over all values of x for which $x_n \leq x$. The set $\{x_n\}$ of values for which $p_n > 0$ is called *the domain* of the random variable X.

A discrete distribution of a random variable is called a *lattice distribution* if there exist numbers a and $b \neq 0$ such that every possible value of X can be represented in the form $a + nb$, where n takes only integer values.

CONTINUOUS DISTRIBUTIONS They are characterized by $F(x)$ being absolutely continuous. Thus, $F(x)$ possesses a derivative $F'(x) = f(x)$, and the c.d.f. can be written as

$$F(x) = P\{X \le x\} = \int_{-\infty}^{x} f(t)\,dt. \tag{C.2}$$

The derivative $f(x)$ is called the *probability density function* (p.d.f.) or *frequency function*, and the values of x for which $f(x) > 0$ make up the domain of the random variable X.

C.2 Some Common Distributions

1. Uniform Distribution: It is defined on the interval $[0,1]$; density (p.d.f.) $f(x) = 1$; c.d.f. $F(x) = x$; difference $d(x) = \int_0^x \frac{\bar{F}(t)}{F(x)}\,d(t) = x/2$; failure rate $r(x) = f(x)/\bar{F}(x) = 1/(1-x)$; mean residual lifetime $\text{mrl}(x) = \int_x^1 \frac{\bar{F}(t)}{\bar{F}(x)}\,dt = (1-x)/2$. Thus, the density function is (weakly) log-concave; $d(x)$ is monotone increasing: $r(x)$ is monotone increasing; and $\text{mrl}(x)$ is monotone decreasing. These properties follow from the log-concavity of f.

2. Normal Distribution. The p.d.f. is $f(x) = \dfrac{1}{\sigma\sqrt{2\pi}} e^{-\frac{1}{2}((x-a)/n)^2}$, $-\infty < x < \infty$, where $-\infty < m < \infty$, $0 < \sigma < m$, and m the mean and σ^2 the variance. Since the normal c.d.f. does not have a closed form, it is very difficult to determine directly whether the c.d.f. is log-concave, or to determine where the functions $d(x)$, $r(x)$, and $\text{mrl}(x)$ are monotone. The standard normal probability distribution has probability density $f(x) = \dfrac{1}{\sqrt{2\pi}} e^{-x^2/2}$. Then $(\ln f(x))' = -x$ and $(\ln f(x))'' = -1 < 0$. Thus, the normal distribution has log-concave density. The normed normal distribution has $f(x) = \phi(x) = \dfrac{1}{\sqrt{2\pi}}\, e^{-x^2/2}$, $x \in (0,1)$, with expectation $\mu = 0$, and variance $\sigma^2 = 1$.

3. Log-normal Distribution. If Y is a log normal distribution, then $Y = e^X$, where X is a normal distribution. Since the normal distribution has a log-concave c.d.f., by Theorem 8.2.8 the log normal distribution has a concave c.d.f. By Theorem 8.2.7, the difference function $d(x)$ is increasing for a log normally distributed variable. The log normal density function is $f(x) = \dfrac{1}{x\sqrt{2\pi}}\, e^{-(\ln x)^2/2}$, which, unlike the normal distribution, is not log-concave; however, since $(\ln f(x))'' = \dfrac{\ln x}{x^2}$, so $f(x)$ is log-concave on the interval $(0,1)$ and log-convex on the interval $(1,\infty)$. The failure rate is neither monotone increasing nor monotone decreasing (Patel, et al. [1976]). Further, the function $\text{mrl}(x)$ is not monotone, but increasing for small values and decreasing for large values of x (Muth [1975]). These last two statements have been verified only by numerical computation and computer graphics.

4. Mirror-image of Log-normal Distribution. It has a monotone decreasing $d(x)$ for some values of x and increasing for others. The support is the set of all negative real numbers. Unlike the case of the mirror-image Pareto distribution, we cannot calculate closed form expressions for the c.d.f. or the $d(x)$ function for this distribution. But it is known that both these functions are non-monotone because the failure rate and the mean residual lifetime function are non-monotone.

5. Logistic Distribution. c.d.f. $F(x) = \dfrac{1}{1+e^{-x}}$; density $f(x) = \dfrac{e^{-x}}{(1+e^{-x})^2}$. Also, we have $(\ln f(x))' = -1 + 2(1 - F(x))$, and $(\ln f(x))'' = -2f(x) < 0$; hence, this distribution has log-concave density.

6. Extreme-value Distribution. Density function is $f(x) = \exp\{-e^{-x}\}$, giving $(\ln f(x))'' = -e^{-x} < 0$; hence this distribution had log-concave density. This distribution arises as the limit as $n \to \infty$ of the greatest value among n independent random variables.

7. chi-Square Distribution with n Degrees of Freedom. It is a gamma distribution with $\theta = 2$ and $m = n/2$. Since the sum of the squares of n independent standard normal random variables has a chi-square distribution with n degrees of freedom, and since the gamma distribution has a log-concave density function for $m \geq 1$, the sum of the squares of two or more independent standard normal variables has a log-concave density function.

8. chi Distribution. Its support is $\{x : x > 0\}$; density function

$$f(x) = \frac{x^{(n/2)-1}\, e^{-n/2} x^2}{2^{n/2}\Gamma(n/2)},$$

where n is a positive integer. If a random variable X has a chi-distribution with n degrees of freedom, then $\sqrt{X/n}$ has a chi-distribution with parameter n. The sample standard deviation from the sum of n independent standard normal variables has a chi-distribution. Since $(\ln f(x))'' = -\dfrac{n-1}{x^2} - n < 0$, the chi-distribution has a log-concave density function. The chi-distribution with $n = 2$ is known as the *Rayleigh distribution*, and with $n = 3$, as the *Maxwell distribution*.

9. Exponential Distribution. Its support is $[0, \infty]$; density function $f(x) = \lambda e^{-\lambda x}$; c.d.f. $F(x) = 1 - e^{-\lambda x}$. The density function is log-linear, and hence log-concave, with $(\ln f(x))'' = 0$, and $f'(x) < 0$, and $F'(x) > 0$ for all x. Thus, $f(x)/F(x)$ is monotone decreasing, and hence f is strictly log-concave. From Theorem 8.2.6, we conclude that $d(x)$ is a monotone increasing function.

Barlow and Proschan [1981] have noted that this is the only distribution for which the failure rate $r(x)$ and the mean residual lifetime $\mathrm{mrl}(x)$ are constant: $r(x) = f(x)/\bar{F}(x) = \lambda$ and $\mathrm{mrl}(x) = \int_x^h \bar{F}(t)\,dt = 1/\lambda$. If the lifetime of an object has an exponential distribution, then it does not 'wear out', i.e., the probability of failure and the expected remaining lifetime remain constant so long as the object 'survives.'

10. Laplace Distribution. It has density function $f(x) = \dfrac{\lambda}{2} e^{-\lambda|x|}$, where $\lambda > 0$; c.d.f. is

$$f(x) = \begin{cases} \frac{1}{2}\, e^{\lambda x} & \text{if } x < 0, \\ 1 - \frac{1}{2}\, e^{-\lambda x} & \text{if } x \geq 0. \end{cases}$$

The density function is sometimes known as the *double exponential*, since it is proportional to the exponential density for positive x and to the mirror-image of the exponential distribution for negative x. Also, $\ln f(x) = \lambda|x|$, which is clearly a concave function, although its derivative $(\ln f(x))'$ does not exist at $x = 0$.

11. Weibull Distribution with Parameter $c > 0$. The density function is $f(x) = cx^{c-1}\, e^{-x^c}$, $x \in (0, \infty)$. Also, $(\ln f(x))'' = (1-c)x^{-2}(1 + cx^c)\begin{cases} < 0 \text{ for } c > 1, \\ = 0 \text{ for } c = 1, \\ > 0 \text{ for } c < 1. \end{cases}$ Thus, the density function is (strictly) log-concave if $0 < c < 1$, log-linear if $c = 1$, and log-convex if $c > 1$. Further, the reliability function $\bar{F}(x) = 1 - F(x) = e^{-x^c}$, giving $(\ln \bar{F}(x))'' = -c(c-1)x^{c-2}$ which is positive for $c < 1$ and nonpositive for $c \geq 1$. Thus, the reliability function is log-concave for $c \geq 1$ and log-convex for $c < 1$. For this distribution with $0 < c < 1$ the failure rate is a decreasing function of age.

12. Power Function Distribution. Its c.d.f. is $F(x) = x^\beta$, with support $(0, 1)$; density function $f(x) = \beta x^{\beta-1}$, giving $(\ln f(x))'' = (1-\beta)x^{-2}$, so that the density function is log-concave if $\beta \geq 1$ and log-convex if $0 < \beta < 1$. This distribution has a log-concave c.d.f. for all positive β, since $(\ln F(x))'' = -\beta x^{-2} < 0$. The difference function is $d(x) = \int_l^x \dfrac{F(t)}{F(x)}\, dt = \dfrac{x}{\beta + 1}$; thus, $d(x)$ is monotone increasing for all $\beta \geq 0$, because log-concavity of $F(x)$ implies that $d(x)$ is monotone increasing. Moreover, the reliability function $\bar{F}(x) = 1 - x^\beta$, giving $(\ln \bar{F}(x))'' = \dfrac{\beta x^{\beta-2}(1 - \beta - x^\beta)}{(1 - x^\beta)^2}$, which has the same sign as $1 - \beta - x^\beta$; thus, this expression is positive for x near zero and negative for x near 1. Hence, the reliability function is neither log-concave nor log-convex on $(0, 1)$. The right-side integral of the reliability function $R(x) = \dfrac{\beta - x^\beta + 1}{1 + \beta}$, which is neither log-concave nor log-convex.

13. Gamma Distribution. Its density function is $f(x) = \dfrac{x^{m-1}\theta^m\, e^{-x\theta}}{\Gamma(m)}$, $x \in (0, \infty), \theta > 0$ and $m > 0$. Then $(\ln f(x))'' = \dfrac{1-m}{x^2}$. Thus, the density function is strictly log-concave for $m > 1$, but strictly log-convex for $m < 1$, and in this case $f'(x) < 0$ for all $x > 0$. Therefore, the c.d.f. is log-concave, and by Theorem 8.7, the left-side integral of the c.d.f. is log-concave. Barlow and Proschan [1981: 75] have shown that for $m < 1$, the failure rate is a monotone decreasing function of age, implying that the reliability function is log-convex.

14. beta Distribution. Its density function is $f(x) = \dfrac{x^{a-1}(1-x)^{b-1}}{B(a,b)}, x \in (0,1), a, b > 0$. Since $(\ln f(x))'' = \dfrac{1-a}{x} + \dfrac{1-b}{x}$, the density function is log-concave if $a \geq 1$ and $b \geq 1$, and log-convex if $a < 1$ and $b < 1$. If $a < 1$ and $b > 1$, or if $a > 1$ and $b < 1$, then the density function is neither log-convex nor log-concave on $(0, 1)$. The log-concavity of the c.d.f. and reliability function for $a > 1$ and $b < 1$, or for $a < 1$ and $b > 1$ are not known in general, but only in the following two special cases: (i) $a = b = \frac{1}{2}$, in which case the distribution is known as the *arcsin distribution*. Closed form expressions for the c.d.f. and the reliability functions are known but they are not simple. Their plots show that the density function and the reliability function are neither log-concave nor log-convex. (ii) $a = 2, b = \frac{1}{2}$, in which case the closed-form expressions when plotted show that the c.d.f. and its integrals are neither log-concave nor log-convex, but the reliability function and its integrals are log-convex.

15. Pareto Distribution. It is defined over the nonnegative real numbers, has c.d.f. $F(x) = 1 - x^{\beta}$, $\beta > 0$; density function $f(x) = \beta x^{-\beta-1}$; then $(\ln f(x))'' = \dfrac{\beta+1}{x^2} > 0$, implying that the density function is log-convex rather than log-concave. But since $f'(x) < 0$ for all x, the c.d.f. $F(x)$ is log-concave. This distribution is a simple example for which both failure rate and mean residual lifetime behave 'unusually.' The reliability function is $\bar{F}(x) = x^{-\beta}$, thus $(\ln \bar{F}(x))'' = \beta/x^2 > 0$, implying that the reliability function is log-convex rather than log-concave, and therefore, the failure rate is a decreasing function of x. Also, since $R(x) = \int_x^{\infty} F(t)\, dt$, the function $R(x)$ converges iff $\beta > 1$, in which case $R(x) = \dfrac{1}{\beta - 1}\, x^{1-\beta}$. Then $(\ln R(x))'' = \dfrac{\beta - 1}{x^2} > 0$, and therefore, $R(x)$ is log-concave and $\mathrm{mrl}(x)$ is a decreasing function of x.

16. Mirror-image of Pareto Distributions. Its support is $(-\infty, -1)$; c.d.f. $F(x) = (-x)^{-\beta}, \beta > 0$. For $\beta > 1$, the left integral of the c.d.f. $G(x) = \int_{-\infty}^{x} F(t)\, dt$ converges and $G(x) = (\beta-1)^{-1}(-x)^{1-\beta}$; then $d(x) = G(x)/G'(x)$ $= G(x)/F(x) = \dfrac{x}{1-\beta}$, and $d'(x) = \dfrac{1}{1-\beta} < 0$. Thus, the function $d(x)$ is decreasing, and the mean residual lifetime function $\mathrm{mrl}(x)$ is increasing.

17. Student's t-Distribution. It is defined on the entire real line with density function

$$f(x) = \frac{(1 + x^2/n)^{-n+1/2}}{\sqrt{n}\, B(1/2, n/2)},$$

where $B(a, b)$ is the incomplete beta function and n is the number of degrees of freedom. Then since $(\ln f(x))'' = -(n+1)\dfrac{n - x^2}{(n + x^2)^2}$, the density function is log-concave on the central interval $[-\sqrt{n}, \sqrt{n}]$, and therefore, it is log-concave on this interval but log-convex on each of the outer intervals $[\infty, -\sqrt{n}]$ and $[\sqrt{n}, \infty]$. Thus, although this distribution is itself not log-concave, a truncate one on the interval $[-\sqrt{n}, \sqrt{n}]$ is log-concave. There does not exist any proof for the log-concavity or log-convexity of the c.d.f. function, but numerical computations, using the program *gauss*, show that the c.d.f. is neither log-concave nor log-convex for the cases $n = 1, 2, 3, 4$ and 24. Since this distribution is symmetric, the log reliability function is the mirror-image of the log of the log c.d.f., and hence, the c.d.f. is neither log-concave nor log-convex, and so is the reliability function.

18. Cauchy Distribution. It is a Student's t distribution with one degree of freedom, and is equal to the ratio of two independent standard normal random variables. The density function is $f(x) = \dfrac{1}{\pi(1+x^2)}$, and c.d.f. $F(x) = \frac{1}{2} + \dfrac{\arctan(x)}{\pi}$; then $(\ln f(x))'' = -2\dfrac{x^2-1}{(x_2+1))2}$ which is negative for $|x| < 1$ and positive for $|x| > 1$. Thus, the density function is neither log-concave nor log-convex. Since the integral $\int_{-\infty}^{x} F(t)\,dt$ does not converge, the function G is not well-defined.

19. *F*-Distribution. It has support as the set of positive real numbers. It has two integer-valued parameters m_1 and m_2, known as 'degrees of freedom.' The density function is

$$f(x) = cx^{(m_1/2)-1}(1 + (m_1/m_2)x)^{-(m_1+m_2)/2},$$

where c is a constant that depends only on m_1 and m_2. This distribution arises in statistical applications as the distribution of the ratio of two independent chi-square distributions with m_1 and m_2 degrees of freedom. Since $(\ln f(x))'' = -(m_1/2 - 1)/x^2 + (m_1/m_2)^2(m_1 + m_2)/2(1 + m_1/m_2 x)^{-2}$, the density function is log-convex if $m_1 \leq 2$; but since $(\ln f(x))''$ is positive or negative depending on whether x is greater than or less than

$$\frac{m_2\sqrt{\dfrac{m_1-2}{m+1+m_2}}}{1-\sqrt{\dfrac{m_1-2}{m+1+m_2}}},$$

this function is neither log-concave nor log-convex.

D

Laplace Transforms

The technique of integral transforms, and in particular of Laplace transforms, is a powerful tool for the solution of linear ordinary or partial differential equations. A function $f(x)$ may be transformed by the formula

$$F(s) = \int_a^b f(x)K(s,x)\,dx,$$

where $F(s)$, provided it exists, is the integral transform of $f(x)$, s is the variable of the transform, and $K(s,x)$ is known as the *kernel* of the transform. An integral transform is a linear transformation, which, when applied to a linear initial or boundary value problem, reduces the number of independent variables by one for each application of the integral transform. Thus, a partial differential equation can be reduced to an algebraic equation by repeated application of integral transforms. The algebraic problem is generally easy to solve for the function $F(s)$, and the solution of the problem is obtained if we can determine the function $f(x)$ from $F(s)$ by some inversion formula.

D.1 Notation

The Laplace transform is defined as

$$\mathcal{L}\{f(t)\} \equiv F(s) = \bar{f}(s) = \int_0^\infty f(t)e^{-st}\,dt, \tag{D.1}$$

and its inverse is

$$\mathcal{L}^{-1}\{F(s)\} \equiv f(t) = \frac{1}{2\pi i}\int_{c-i\infty}^{c+i\infty} F(s)e^{st}\,ds, \tag{D.2}$$

where s is the variable of the transform, which in general is a complex variable. Note that the Laplace transform $F(s)$ exists for $s > \alpha$, if the function $f(t)$ is piecewise continuous in every finite closed interval $0 \le t \le b$ $(b > 0)$, and $f(t)$ is of exponential order α, i.e., there exist α, M, and $t_0 > 0$ such that $e^{-\alpha t}\,|f(t)| < M$ for $t > t_0$.

Two basic properties of the Laplace transforms are:

(i) Convolution Theorem:

$$\mathcal{L}^{-1}\left\{G(s)F(s)\right\} = \int_0^t f(t-u)g(u)du = \int_0^t f(u)g(t-u)du.$$

(ii) If $\mathcal{L}\left\{f(x,t)\right\} = F(x,s)$, then

$$\mathcal{L}\left\{\frac{\partial f(x,t)}{\partial x}\right\} = \frac{\partial F(x,s)}{\partial x}, \quad \text{and} \quad \mathcal{L}^{-1}\left\{\frac{\partial F(x,s)}{\partial x}\right\} = \frac{\partial f(x,t)}{\partial x}.$$

The second property is very useful; it is based on the Leibniz rule, which states that if $g(x,t)$ is an integrable function of t for each value of x, and the partial derivative $\dfrac{\partial g(x,t)}{\partial x}$ exists and is continuous in the region under consideration, and if $f(x) = \int_a^b g(x,t)dt$, then $f'(x) = \displaystyle\int_a^b \frac{\partial g(x,t)}{\partial x} dt$.

A table of some useful Laplace transform pairs is given at the end of this Appendix (Table D.1, p. 340); larger tables are available in reference books on mathematical formulas, e.g., in Abramowitz and Stegun [1972].

We will now explain certain techniques to derive Laplace transforms from known transform pairs.

Example D.1. Consider (see formula 19, Table D.1)

$$\mathcal{L}\left\{e^{at}\right\} = \frac{1}{s-a}. \tag{D.3}$$

Differentiating both sides with respect to a, we get

$$\mathcal{L}\left\{te^{at}\right\} = \frac{1}{(s-a)^2}, \tag{D.4}$$

and repeating this differentiation n times, we find that

$$\mathcal{L}\left\{t^n e^{at}\right\} = \frac{n!}{(s-a)^{n+1}}. \tag{D.5}$$

Next, replacing a by ib, choosing an appropriate n, and comparing the real and imaginary parts on both sides, we get the Laplace transforms of functions $t^n \cos bt$ and $t^n \sin bt$, and then combining with the Laplace transform of e^{at}, we obtain the Laplace transforms of functions $t^n e^{at} \cos bt$ and $t^n e^{at} \sin bt$. For example, if we choose $n = 2$, then we have

$$\mathcal{L}\left\{t^2 e^{at}\right\} = \frac{2!}{(s-a)^3}.$$

Now letting $a = ib$, we get

$$\mathcal{L}\left\{t^2 e^{ibt}\right\} = \frac{2!}{(s-ib)^3},$$

which yields

$$\mathcal{L}\left\{t^2(\cos bt + i\sin bt)\right\} = \frac{2(s+ib)^3}{(s^2+b^2)^3} = \frac{2(s^3+3is^2b-3sb^2-ib^3)}{(s^2+b^2)^3},$$

and equating the real and imaginary parts of this equality, we obtain

$$\mathcal{L}\left\{t^2\cos bt\right\} = \frac{2(s^3-3sb^2)}{(s^2+b^2)^3}, \tag{D.6}$$

and

$$\mathcal{L}\left\{t^2\sin bt\right\} = \frac{2(3s^2b-b^3)}{(s^2+b^2)^3}. \tag{D.7}$$

The Laplace transforms of $\mathcal{L}\left\{e^{at}t^2\cos bt\right\}$ and $\mathcal{L}e^{at}\left\{t^2\sin bt\right\}$ can then be easily obtained. ■

Example D.2. Consider

$$\mathcal{L}^{-1}\left\{\frac{e^{-a\sqrt{s}}}{s}\right\} = \operatorname{erfc}\frac{a}{2\sqrt{t}}, \tag{D.8}$$

where

$$\operatorname{erf}(x) = \frac{2}{\sqrt{\pi}}\int_0^x e^{-u^2}\,du, \quad \text{and} \quad \operatorname{erfc}(x) = 1-\operatorname{erf}(x) = \frac{2}{\sqrt{\pi}}\int_x^\infty e^{-u^2}\,du.$$

We can derive certain Laplace inverses by differentiating and integrating (D.8). Thus,

$$\mathcal{L}^{-1}\left\{\frac{e^{-a\sqrt{s}}}{\sqrt{s}}\right\} = \frac{1}{\sqrt{\pi t}}e^{-a^2/4t} \tag{D.9}$$

is obtained after differentiating (D.8) with respect to a and canceling out the negative sign on both sides.

Although the usual method of deriving (D.9) is by contour integration (see Exercise D.10), or by using the Laplace inverse of $\dfrac{e^{-a\sqrt{s}}}{s}$, an interesting method is given in Exercise D.1.

Next, we obtain $\mathcal{L}^{-1}\left\{e^{-a\sqrt{s}}\right\} = \dfrac{a}{2\sqrt{\pi t^3}}e^{-a^2/4t}$ which is obtained by differentiating (D.9) with respect to a and canceling out the negative sign. ■

Example D.3. If we integrate the formula $\mathcal{L}^{-1}\left\{\dfrac{e^{-a\sqrt{s}}}{s}\right\} = \operatorname{erfc}\dfrac{a}{2\sqrt{t}}$ with respect to a from 0 to a, we get

$$\int_0^a \mathcal{L}^{-1}\left\{\frac{e^{-x\sqrt{s}}}{s}\right\} dx = \int_0^a \operatorname{erfc}\frac{x}{2\sqrt{t}} dx.$$

Then, after changing the order of integration and the Laplace inversion and carrying out the integration on the left side, we get

$$\int_0^a \mathcal{L}^{-1}\left\{\frac{e^{-x\sqrt{s}}}{s}\right\} dx = \mathcal{L}^{-1}(s^{-3/2} - s^{-3/2}e^{-a\sqrt{s}}), \tag{D.10}$$

while the right side yields

$$\begin{aligned}\int_0^a \operatorname{erfc}\frac{x}{2\sqrt{t}}\, dx &= \left[x \operatorname{erfc}\frac{x}{2\sqrt{t}}\right]_0^a + \frac{1}{\sqrt{\pi t}}\int_0^a x\, e^{-x^2/4t}\, dx \\ &= a \operatorname{erfc}\frac{a}{2\sqrt{t}} - 2\sqrt{\frac{t}{\pi}} e^{-a^2/4t} + 2\sqrt{\frac{t}{\pi}}.\end{aligned}$$

Since $\mathcal{L}^{-1}\left\{s^{-3/2}\right\} = 2\sqrt{\dfrac{t}{\pi}}$, we get

$$\mathcal{L}^{-1}\left\{s^{-3/2}e^{-a\sqrt{s}}\right\} = 2\sqrt{\frac{t}{\pi}} e^{-a^2/4t} - a \operatorname{erfc}\frac{a}{2\sqrt{t}}. \;\blacksquare \tag{D.11}$$

Example D.4. To evaluate $\mathcal{L}^{-1}\left\{\dfrac{e^{-a\sqrt{s+c}}}{s}\right\}$, we know from (D.9) that

$$\mathcal{L}^{-1}\left\{e^{-a\sqrt{s}}\right\} = \frac{a}{2\sqrt{\pi t^3}}\, e^{-a^2/4t}.$$

Hence, using formula 2, Table D.1,

$$\mathcal{L}^{-1}\left\{e^{-a\sqrt{s+c}}\right\} = \frac{a}{2\sqrt{\pi t^3}} e^{-ct-a^2/4t},$$

which, in view of the convolution theorem (Property (i)) with $F(s) = \dfrac{1}{s}$ and $G(s) = e^{-a\sqrt{s+c}}$, yields

$$\mathcal{L}^{-1}\left\{\frac{e^{-a\sqrt{s+c}}}{s}\right\} = \int_0^t \frac{a}{2\sqrt{\pi u^3}}\, e^{-cu-a^2/4u} du. \tag{D.12}$$

Since

$$\frac{a}{2\sqrt{u^3}} = \frac{a}{4\sqrt{u^3}} + \frac{1}{2}\sqrt{\frac{c}{u}} + \frac{a}{4\sqrt{u^3}} - \frac{1}{2}\sqrt{\frac{c}{u}},$$

and

$$cu + \frac{a^2}{4u} = \left(\sqrt{cu} + \frac{a}{2\sqrt{u}}\right)^2 - a\sqrt{c} = \left(\sqrt{cu} - \frac{a}{2\sqrt{u}}\right)^2 + a\sqrt{c},$$

we define $x = \dfrac{a}{2\sqrt{u}} + \sqrt{cu}$ and $y = \dfrac{a}{2\sqrt{u}} - \sqrt{cu}$, and use the notation

$$\frac{a}{2\sqrt{t}} + \sqrt{ct} = x_1, \quad \text{and} \quad \frac{a}{2\sqrt{t}} - \sqrt{ct} = y_1.$$

Then the integral on the right side of (D.12) becomes

$$\frac{1}{\sqrt{\pi}}\left[e^{a\sqrt{c}}\int_{x_1}^{\infty} e^{-x^2}\,dx + e^{-a\sqrt{c}}\int_{y_1}^{\infty} e^{-y^2}\,dy\right]$$
$$= \frac{1}{2}\left[e^{a\sqrt{c}}\operatorname{erfc}\left(\frac{a}{2\sqrt{t}} + \sqrt{ct}\right) + e^{-a\sqrt{c}}\operatorname{erfc}\left(\frac{a}{2\sqrt{t}} - \sqrt{ct}\right)\right].$$

Hence,

$$\mathcal{L}^{-1}\left\{\frac{e^{-a\sqrt{s+c}}}{s}\right\} = \frac{1}{2}\Big[e^{a\sqrt{c}}\operatorname{erfc}\left(\frac{a}{2\sqrt{t}} + \sqrt{ct}\right) + e^{-a\sqrt{c}}\operatorname{erfc}\left(\frac{a}{2\sqrt{t}} - \sqrt{ct}\right)\Big]. \blacksquare \tag{D.13}$$

We state a useful theorem without proof.

Theorem D.1. *If $G(s) = \sum\limits_{k=1}^{\infty} G_k(s)$ is uniformly convergent series, then*

$$\mathcal{L}^{-1}\{G(s)\} = g(t) = \sum_{k=1}^{\infty} g_k(t), \tag{D.14}$$

where $\mathcal{L}^{-1}\{G_k(s)\} = g_k(t)$.

Example D.5. As an application of Theorem D.1, since

$$\begin{aligned} &\mathcal{L}^{-1}\left\{s^{-3/2}e^{-1/s}\right\} \\ &= \mathcal{L}^{-1}\left\{\frac{1}{s^{3/2}}\left[1 - \frac{1}{s} + \frac{1}{2!s^2} - \frac{1}{3!s^3} + \cdots + (-1)^n\frac{1}{n!s^n} + \cdots\right]\right\} \\ &= \mathcal{L}^{-1}\sum_{0}^{\infty}\frac{(-1)^n}{n!\,s^{n+3/2}} = \frac{1}{\sqrt{\pi}}\sum_{0}^{\infty}\frac{(-1)^n(2\sqrt{t})^{2n+1}}{(2n+1)!} = \frac{1}{\sqrt{\pi}}\sin(2\sqrt{t}), \end{aligned} \tag{D.15}$$

we find that this result and formula 7, Table D.1 give

$$\mathcal{L}^{-1}\left\{s^{-1/2}e^{-1/s}\right\} = \frac{1}{\sqrt{\pi t}}\cos(2\sqrt{t}). \blacksquare \tag{D.16}$$

Example D.6. Consider a semi-infinite medium bounded by $0 \leqq x \leqq \infty$, $-\infty < y,\ z < \infty$, which has an initial zero temperature, while its face $x = 0$ is maintained at a time-dependent temperature $f(t)$. The problem is to find the temperature for $t > 0$. By applying the Laplace transform to the heat conduction equation $kT_{xx} = T_t$, we get $\overline{T}_{xx} = \dfrac{s}{k}\overline{T}$, where $\overline{T} = \mathcal{L}\{T\}$. The solution of this equation is

$$\overline{T} = Ae^{mx} + Be^{-mx}, \tag{D.17}$$

where $m = \sqrt{s/k}$. Since $\overline{T}$ remains bounded as $x \to \infty$, we find that $A = 0$. The boundary condition at $x = 0$ in the transform domain yields $B = \bar{f}(s)$, where $\bar{f}(s)$ is the Laplace transform of $f(t)$. Thus, the solution in the transform domain is

$$\overline{T} = \bar{f}(s)\, e^{-mx}.$$

To carry out the inversion, we use the convolution property and Example D.2 and get

$$T = \int_0^t \frac{x\, e^{-x^2/4k\tau}}{2\tau\sqrt{\pi k\tau}}\, f(t-\tau)\, d\tau.$$

If $\bar{f}(s) = 1$, then the solution for T reduces to

$$T = \frac{x\, e^{-x^2/4kt}}{2t\sqrt{\pi kt}}.$$

This solution is the fundamental solution for the heat conduction equation for the half-space. In the special case when $f(t) = T_0$, the solution is given by

$$T = T_0 \operatorname{erfc}\left(\frac{x}{2\sqrt{kt}}\right). \blacksquare$$

Example D.7. Consider an infinite slab bounded by $0 \le x \le l$, $-\infty < y,\ z < \infty$, with initial zero temperature. The face $x = 0$ is maintained at a constant temperature T_0, and the face $x = l$ is maintained at zero temperature. The problem is to find the temperature inside the slab for $t > 0$. Proceeding as in the above example, the solution in the transform domain is given by Eq (D.17). Applying the boundary conditions in the transform domain we get

$$A + B = \frac{T_0}{s}, \quad \text{and} \quad A\, e^{ml} + B\, e^{-ml} = 0.$$

These two equations yield

$$B = \frac{T_0\, e^{ml}}{2s \sinh ml}, \qquad \text{and} \qquad A = \frac{T_0}{s} - B.$$

Substituting these values into $\overline{T}$, given by (D.17), and simplifying, we find that

$$\overline{T} = \frac{T_0 \sinh m(l-x)}{s \sinh ml},$$

Rewriting this solution as

$$\overline{T} = \frac{T_0}{s}\, e^{-ml} \left(e^{m(l-x)} - e^{-m(l-x)}\right) \left(1 - e^{-2ml}\right)^{-1},$$

and expanding the last factor by the binomial theorem, we get

$$\begin{aligned}\overline{T} &= \frac{T_0}{s}\, e^{-ml} \left(e^{m(l-x)} - e^{-m(l-x)}\right) \sum_0^\infty e^{-2nml} \\ &= \frac{T_0}{s} \sum_0^\infty \left(e^{-m(2nl+x)} - e^{-m[(2n+2)l-x]}\right),\end{aligned}$$

which, on inversion, yields

$$T = T_0 \sum_0^\infty \left[\operatorname{erf}\Big(\frac{2(n+1)l - x}{2\sqrt{kt}}\Big) - \operatorname{erf}\Big(\frac{2nl + x}{2\sqrt{kt}}\Big)\right].$$

Alternatively, we can use the Cauchy residue theorem and obtain the solution in terms of the Fourier series. Thus,

$$\begin{aligned}T &= \sum \text{ residues of } \frac{T_0 e^{st} \sinh m(l-x)}{s \sinh ml} \\ &= T_0 \Big[1 - \frac{x}{l} - \sum_1^\infty \frac{2}{n\pi}\, e^{-n^2\pi^2 kt/l^2} \sin\left(n\pi x/l\right)\Big]. \blacksquare\end{aligned}$$

Theorem D.2. (Inversion Theorem) *If $F(s)$ is the Laplace transform of $f(t)$, then*

$$f(t) = \frac{1}{2\pi i} \int_{c-i\infty}^{c+i\infty} F(s)\, e^{st}\, ds, \tag{D.18}$$

where $F(s)$ is of order $O(s^{-k})$, where $k > 0$.

To prove this theorem, we first state and prove a lemma.

Lemma D.1. *If $f(z)$ is analytic and of order $O(z^{-k})$ in the half-plane $\Re\{z\} > \gamma$, where γ and $k > 0$ are real constants, then*

$$f(z_0) = \frac{1}{2\pi i} \lim_{\beta\to\infty} \int_{\gamma-i\beta}^{\gamma+i\beta} \frac{f(z)}{z_0 - z}\, dz, \quad \Re\{z_0\} > \gamma. \tag{D.19}$$

PROOF. Consider the rectangle in Figure D.1. Choose $\beta > |\gamma|$ and such that z_0 lies inside this rectangle. By the Cauchy integral formula, we have

$$\int_\Gamma \frac{f(z)}{z - z_0}\, dz = 2\pi i f(z_0), \tag{D.20}$$

where Γ is the contour ABCDA. Let S denote the contour ABCD, then

$$\int_\Gamma \frac{f(z)}{z - z_0}\, dz = \int_{DA} \frac{f(z)}{z - z_0}\, dz + \int_S \frac{f(z)}{z - z_0}\, dz.$$

Since

$$\int_{DA} \frac{f(z)}{z - z_0}\, dz = -\int_{AD} \frac{f(z)}{z - z_0}\, dz,$$

we get from (D.20)

$$-\int_{\gamma-i\beta}^{\gamma+i\beta} \frac{f(z)}{z - z_0}\, dz + \int_S \frac{f(z)}{z - z_0}\, dz = 2\pi i f(z_0). \tag{D.21}$$

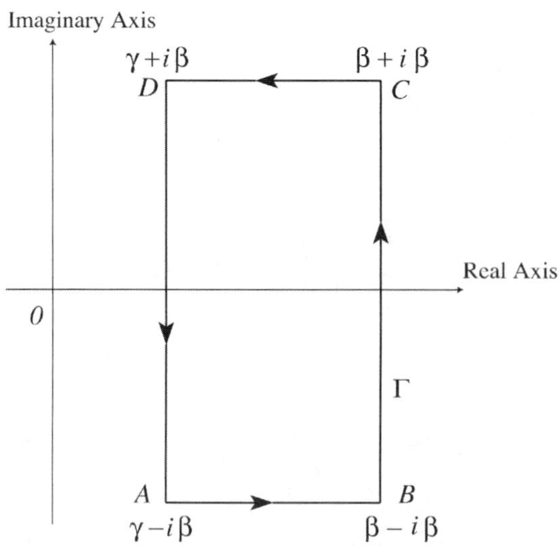

Figure D.1 Rectangular contour.

Now, consider $\int_S \frac{f(z)}{z-z_0}\,dz$ as $\beta \to \infty$. Obviously, $\beta \to \infty$ implies that $|z| \to \infty$ on S. Thus $|z| \geqq \beta$ for points on S. If we take β large enough so that $\beta > 2|z_0|$, then $|z_0| < \frac{1}{2}\beta \leqq \frac{1}{2}|z|$, or $\left|\frac{z_0}{z}\right| < \frac{1}{2}$ implies that $\left|1 - \frac{z_0}{z}\right| \geqq 1 - \left|\frac{z_0}{z}\right| > \frac{1}{2}$. Noting that $|f(z)| < M|z|^{-k}$ for large z, we get

$$\left|\frac{f(z)}{z-z_0}\right| = \left|\frac{f(z)}{z}\frac{1}{\left(1-\frac{z_0}{z}\right)}\right| \leqq \frac{M}{\left|z^{k+1}\left(1-\frac{z_0}{z}\right)\right|} \leqq \frac{2M}{\beta^{k+1}}.$$

It now follows that

$$\begin{aligned}\left|\int_S \frac{f(z)}{z-z_0}\,dz\right| &< \frac{2M}{\beta^{k+1}}\int_S |dz| = \frac{2M}{\beta^{k+1}}\,(\text{length of } S)\\ &= \frac{2M}{\beta^k}\left(\frac{4\beta-2\gamma}{\beta}\right) = \frac{2M}{\beta^k}\left(4-\frac{2\gamma}{\beta}\right).\end{aligned}$$

Thus, $\lim_{\beta\to\infty}\int_S \frac{f(z)}{z-z_0}\,dz = 0$. Hence, from (D.21),

$$-\int_{\gamma-i\infty}^{\gamma+i\infty} \frac{f(z)}{z-z_0}\,dz = 2\pi i f(z_0),$$

or

$$F(s) = \frac{1}{2\pi i}\int_{\gamma-i\infty}^{\gamma+i\infty} \frac{F(z)}{s-z}\,dz. \blacksquare \tag{D.22}$$

The proof of Theorem 6.2 for the Laplace transform now becomes elementary. By taking the Laplace inverse of both sides of Eq (D.22), we have

$$f(t) = L^{-1}\{F(s)\} = \frac{1}{2\pi i}\int_{\gamma-i\infty}^{\gamma+i\infty} L^{-1}\left\{\frac{F(z)}{s-z}\right\}dz = \frac{1}{2\pi i}\int_{\gamma-i\infty}^{\gamma+i\infty} F(z)e^{zt}\,dz. \blacksquare \tag{D.23}$$

Lemma D.2. *If* $|f(z)| < CR^{-k}$, $z = Re^{i\theta}, -\pi \leqq \theta \leqq \pi, R > R_0$, *where* $R_0, C,$ *and* k *are constants, then* $\int_\Gamma e^{zt} f(z)\,dz \to 0$ *as* $R \to \infty$, *provided* $t > 0$, *where* Γ *is the arc* $BB'CA'A$, *and* R *is the radius of the circular arc* *with chord AB (Figure D.2).*

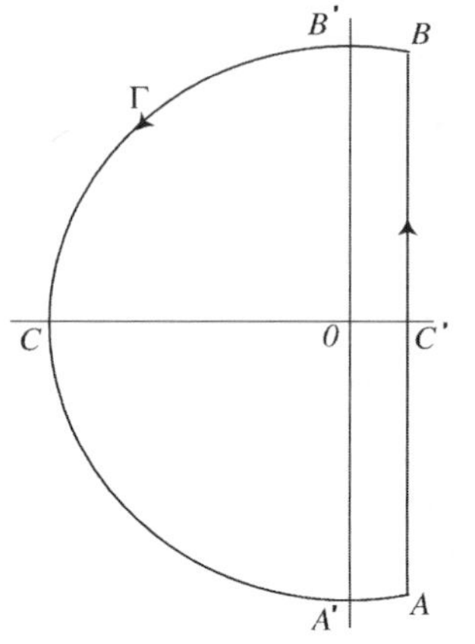

Figure D.2 Contour Γ.

PROOF. Consider the integral over the arc BB'. Let the angle BOC' be denoted by α. On BB' we have $z = Re^{i\theta}$, where θ varies from α to $\pi/2$, $\alpha = \cos^{-1}(\gamma/R)$, and $\gamma = OC'$. Then we get

$$\begin{aligned}\left|\int_{BB'} e^{zt} f(z)\,dz\right| &< \int_\alpha^{\pi/2} \left|CR^{-k} e^{Rte^{i\theta}} Rie^{i\theta}\right| d\theta \\ &= CR^{-k+1} \int_\alpha^{\pi/2} \left|e^{Rt\cos\theta}\right| d\theta \leqq CR^{-k+1} \int_\alpha^{\pi/2} \left|e^{\gamma t}\right| d\theta \\ &= CR^{-k+1}(\pi/2 - \alpha)e^{\gamma t} \\ &= CR^{-k+1}\, e^{\gamma t} \sin^{-1}\frac{\gamma}{R} \to 0 \quad \text{as } R \to \infty.\end{aligned}$$

Similarly, $\displaystyle\int_{A'A} e^{zt} f(z)\,dz \to 0$ as $R \to \infty$.

Let us now consider the integral over the arc $B'CA'$. By following the above procedure, we get

$$\begin{aligned}\left|\int_{B'CA'} e^{zt} f(z)\,dz\right| &< CR^{-k+1} \int_{\pi/2}^{3\pi/2} \left|e^{Rt\cos\theta}\right| d\theta \\ &= CR^{-k+1} \int_0^\pi e^{-Rt\sin\phi} d\phi \quad \text{where } \theta = \pi/2 + \phi \\ &= 2CR^{-k+1} \int_0^{\pi/2} e^{-Rt\sin\phi} d\phi \le 2CR^{-k+1} \int_0^{\pi/2} e^{-2Rt\phi/\pi}\, d\phi \\ &= \frac{\pi CR^{-k}}{t}(1 - e^{-Rt}) \to 0 \quad \text{as } R \to \infty.\end{aligned}$$

Hence, $\int_{\Gamma} e^{zt} f(z)\, dz \to 0$ as $R \to \infty$, provided that $t > 0$. ■

The justification to use the inequality $e^{-RT \sin\phi} \le e^{-2RT\phi/\pi}$ in the above penultimate step is as follows: The function $g(\phi) = \sin\phi - 2\phi/\pi \ge 0$ for $0 \le \phi \le \pi/2$, and with $g(0) = 0 = g(\pi/2)$ has only one critical point at $\phi = \cos^{-1}(2/\pi)$, which gives a maximum.

This result enables us to convert the integral $\dfrac{1}{2\pi i} \displaystyle\int_{\gamma - i\beta}^{\gamma + i\beta} F(z)\, e^{zt}\, dz$ into an integral over the contour $(-\Gamma)$.

D.2 Laplace$_2$-Transform

The Laplace$_2$-Transform (also denoted by $\mathcal{L}_2$-transform), which is a Laplace-type integral transform, was introduced by Yurekli and Sadek [1991]. It is defined as

$$\mathcal{L}_2\{f(t); s\} = \int_0^\infty t\, e^{-s^2t^2} f(t)\, dt. \tag{D.24}$$

A useful property of the $\mathcal{L}_2$-transform is as follows:

If f is a class C^n function, i.e., if $f, f', \dots, f^{(n-1)}$ are all continuous functions with a piecewise continuous derivative $f^{(n)}$ on the interval $t \geqq 0$ and if all functions are of exponential order $e^{c^2t^2}$ as $t \to \infty$ for some constant c, then for $n = 1, 2, \dots$

$$\begin{aligned} \mathcal{L}_2\{\delta_t^n f(t); s\} = 2^n s^{2n} \mathcal{L}_2\{f(t); s\} - 2^{n-1} s^{2(s-1)} f(0^+) \\ - 2^{n-2} s^{2(n-2)} (\delta_t f)(0^+) - \cdots - (\delta_t^{n-1} f)(0^+). \end{aligned} \tag{D.25}$$

For proof, see Yurekli and Sadek [1991]. ■

Lemma D.3. (Inversion formula) Let $F(\sqrt{s})$ be an analytic function of s (assuming that $s = 0$ is not a branch point) except at a finite number of poles each of which lies to the left of the vertical line $\Re\{s\} = c$, and if $F(\sqrt{s}) \to 0$ as $s \to \infty$ through the left-plane $\Re\{s\} \leqq c$, and if $\mathcal{L}_2\{f(t); s\} \equiv F(s)$ (see Figure D.3), then

$$\begin{aligned} \mathcal{L}_2^{-1}\{F(s)\} = f(t) &= \frac{1}{2\pi i} \int_{c-i\infty}^{c+i\infty} 2F(\sqrt{s}\,)\, e^{st^2}\, ds \\ &= \sum_{k=1}^{m} \big[\mathrm{Res}\,\{2F(\sqrt{s})\, e^{st^2}\}, s = s_k\big]. \end{aligned} \tag{D.26}$$

PROOF. Aghili, Ansari and Sedgi [2007].

Lemma D.4. (Convolution theorem) If $F(s), G(s)$ is the $\mathcal{L}_2$-transforms of the functions $f(t)$ and $g(t)$, respectively, then

$$F(s)G(s) = \mathcal{L}_2\{f \star g\} = \mathcal{L}_2\Big\{ \int_0^t x g(x) f(\sqrt{t^2 - x^2})\, dx \Big\}. \tag{D.27}$$

PROOF. Using the definition (D.24) for $F(s)$ and $G(s)$, we get

$$\begin{aligned} F(S)G(s) &= \int_0^\infty y\, e^{-s^2y^2} f(y)\,dy \int_0^\infty x\, e^{-s^2x^2} g(x)\,dx \\ &= \int_0^\infty \int_0^\infty yx\, e^{-s^2(x^2+y^2)} f(y)g(x)\,dx\,dy \\ &= \int_0^\infty t\, e^{-s^2t^2}\,dt \int_0^t xg(x)f(\sqrt{t^2-y^2})\,dx \\ &= \int_0^\infty t\, e^{-s^2t^2} \Big\{ \int_0^t xg(x)f(\sqrt{t^2-y^2})\,dx \Big\}\,dt \\ &= \mathcal{L}_2 \Big\{ \int_0^t xg(x)f(\sqrt{t^2-x^2})\,dx \Big\}, \end{aligned}$$

where we have set $x^2 + y^2 = t^2$, so that $y\,dy = t\,dt$, holding x as constant, and $y = \sqrt{(t^2 - x^2)}$. ■

A generalization of this theorem is the Efros theorem, also known as the *generalized product theorem*; for another version, see §13.1.

Theorem D.3. (Efros Theorem) Assuming that $\Phi(s)$ and $q(s)$ are analytic, let $\mathcal{L}_2\{\Phi(t,\tau)\} = \Phi(s)\tau e^{-\tau^2 q^2(s)}$. Then

$$\mathcal{L}_2 \Big\{ \int_0^\infty f(\tau)\Phi(t,\tau)\,d\tau \Big\} = F(q(s))\Phi(s), \tag{D.28}$$

where $\mathcal{L}_2\{f(t)\} = F(s)$.

PROOF. Using the definition (D.24) of the $\mathcal{L}_2$-transform, and changing the order of integration, we get

$$\begin{aligned} &\mathcal{L}_2 \Big\{ \int_0^\infty f(\tau)\Phi(t,\tau)\,d\tau \Big\} = \int_0^\infty t\, e^{-s^2t^2} \Big\{ \int_0^\infty f(\tau)\Phi(t,\tau)\,d\tau \Big\}\,dt \\ &\quad = \int_0^\infty f(\tau)\,dt \Big\{ \int_0^\infty te^{-s^2t^2}\Phi(t,\tau)\,d\tau \Big\} = \Phi(s) \int_0^\infty f(\tau)\tau e^{-\tau^2q^2(s)}\,d\tau \\ &\quad = \Phi(s)F(q(s)). \ ■ \end{aligned}$$

More details about $\mathcal{L}_2$-transform can also be found in Yurekli and Wilson [2002], [2003].

D.3 Exercises

D.1. The formula (D.9) can be obtained as follows (Churchill [1972]): Define $\dfrac{e^{-a\sqrt{s}}}{\sqrt{s}} = x$ and $e^{-a\sqrt{s}} = z$. Then $y' = \dfrac{dy}{ds} = -\dfrac{1}{2s^{3/2}}\, e^{-a\sqrt{s}} - \dfrac{a}{2s}\, e^{-a\sqrt{s}}$, which yields $2sy'+y+az=0$. Similarly, $z' = -\frac{a}{2\sqrt{s}}e^{-a\sqrt{s}}$ yields $2z'+ay=0$. Taking the inverse transform of these equations, we get

$$aG - F - 2tF' = 0, \quad \text{and} \quad aF - 2tG = 0,$$

where $\mathcal{L}^{-1}\{y\} = F(t)$ and $\mathcal{L}^{-1}\{z\} = G(t)$. From these two equations in F and G, we get $F' = \frac{1}{2t}\left(\frac{a^2F}{2t} - F\right)$, whose solution is $F = \frac{A}{\sqrt{t}}\, e^{-a^2/4t}$, which gives $G = \frac{aA}{2\sqrt{t^3}}\, e^{-a^2/4t}$. Note that if $a=0$, then $y = \frac{1}{\sqrt{s}}$, and $F(t) = \frac{1}{\sqrt{\pi t}}$ implies that $A = \frac{1}{\sqrt{\pi}}$. Hence,

$$F(t) = \frac{1}{\sqrt{\pi t}}\, e^{-a^2/4t}, \quad G = \frac{a}{\sqrt{\pi t^3}}\, e^{-a^2/4t}.$$

Then we integrate $\mathcal{L}^{-1}\left\{\frac{e^{-a\sqrt{s}}}{\sqrt{s}}\right\} = \frac{1}{\sqrt{\pi t}}e^{-a^2/4t}$ with respect to a from 0 to a and obtain $\mathcal{L}^{-1}\left\{\frac{e^{-a\sqrt{s}}}{s}\right\}$, using $\mathcal{L}\left\{\frac{1}{\sqrt{t}}\right\} = \sqrt{\frac{\pi}{s}}$ (formula 12, Table D.1).

D.2. Find (a) $\mathcal{L}^{-1}\left\{\frac{\cosh a\sqrt{s}}{s\cosh b\sqrt{s}}\right\}$, and (b) $\mathcal{L}^{-1}\left\{\frac{\sinh a\sqrt{s}}{\sinh b\sqrt{s}}\right\}$, $b>a>0$.

HINT. Use $\cosh x = (e^x + e^{-x})/2$, $\sinh x = (e^x - e^{-x})/2$, and $(1+z)^{-1} = \sum_{n=0}^{\infty}(-1)^n z^n$.

ANS. (a) $\sum_{n=0}^{\infty}(-1)^n\left\{\operatorname{erfc}\left(\frac{(2n+1)b-a}{2\sqrt{t}}\right) + \operatorname{erfc}\left(\frac{(2n+1)b+a}{2\sqrt{t}}\right)\right\}$,

(b) $\sum_{n=0}^{\infty}\left[\frac{(2n+1)b-a}{\sqrt{4\pi t^3}}\, e^{-[(2n+1)b-a]^2/(4t)} - \frac{(2n+1)b+a}{\sqrt{4\pi t^3}}\, e^{-[(2n+1)b+a]^2/(4t)}\right]$.

D.3. Show that $\mathcal{L}\{t^n\} = \frac{\Gamma(n+!)}{s^{n+1}}$, where $\Gamma(x)$ is the gamma function.

ANS. We have $\mathcal{L}\{t^n\} = \int_0^\infty t^n e^{-st}\, dt = \frac{1}{s^{n+1}}\int_0^\infty x^p e^{-x}\, dx = \frac{\Gamma(n+!)}{s^{n+1}}$, where we have set $st = x$. ■

D.4. Solve the partial differential equation $u_{tt} = u_{xx}$, with the initial conditions $u(x,0) = -\frac{(1-x)^2}{2}$, $u_t(x,0) = 0$, and the boundary conditions $u_x(0,t) = 1$ and $u_x(1,t) = 0$.

ANS. $u = -\frac{1}{2}t^2 - \frac{(1-x)^2}{2}$.

D.5. Solve the partial differential equation $u_t = u_{xx}$, with the initial condition $u(x,0) = 0$ and the boundary conditions $u_x(0,t) = 0$ and $u(1,t) = 1$.

HINT. The solution in the transform domain is $\bar{u} = \dfrac{\cosh x\sqrt{s}}{s \cosh \sqrt{s}}$.
Find two different inverses of this solution, by expanding the solution in a series of the type shown in Example D.7 and by the residue theorem.

ANS. $u = \sum_0^\infty (-1)^n \left[\operatorname{erfc} \dfrac{2n+1-x}{2\sqrt{t}} + \operatorname{erfc} \dfrac{2n+1+x}{2\sqrt{t}} \right]$, or

$$u = 1 - \sum_0^\infty (-1)^n \frac{4 \cos(2n+1)\pi x/2}{(2n+1)\pi}\, e^{-(2n+1)^2\pi^2 t/4}.$$

D.6. Solve in the transform domain the partial differential equation $u_t = u_{zz} + k u_{tzz}$, given that $u(z,0) = 0$, and $u(0,t) = u_0$, $\lim\limits_{z\to\infty} u(z,t) = 0$ for $t > 0$. Expand the solution in the transform domain in the form $\bar{u} = \dfrac{u_0}{s}\, e^{-z\sqrt{s}}\big(1 + \text{powers of } k\big)$. Invert the first two terms of this expansion.

ANS. $\dfrac{u(z,t)}{u_0} = 1 - \dfrac{1}{\pi} \displaystyle\int_0^\lambda \frac{1}{x} e^{-xt} \sin \sqrt{\frac{\lambda x}{\lambda - x}}\, dx$, where $\dfrac{1}{\lambda} = k$.

D.7. Using the Laplace transform method, solve the partial differential equation $u_t = u_{xx}$, with the initial condition $u(x,0) = 0$ and the boundary conditions $u_x(0,t) = 0$, and $u_x(1,t) = 1$.

HINT. The solution in the transform domain is $\bar{u} = \dfrac{\cosh x\sqrt{s}}{s^{3/2} \sinh \sqrt{s}}$. Find two different inverses of this solution, by expanding the solution in a series of the type shown in Example D.7 and by the residue theorem.

ANS.
$$u = \sum_{n=0}^\infty \Big\{ 2\sqrt{t/\pi}\, \Big(e^{-(2n+1-x)^2/4t} + e^{-(2n+1+x)^2/4t} \Big) - (2n+1-x) \operatorname{erfc} \frac{2n+1-x}{2\sqrt{t}} - (2n+1+x) \operatorname{erfc} \frac{2n+1+x}{2\sqrt{t}} \Big\},$$

and

$$u = \frac{x^2}{2} + t - \frac{1}{6} - \sum_{n=1}^\infty \frac{2(-1)^n}{n^2\pi^2}\, e^{-n^2\pi^2 t} \cos n\pi x.$$

D.8. Use contour integration to evaluate $\mathcal{L}^{-1}\left\{ \dfrac{e^{-a\sqrt{s}}}{s} \right\}$.

SOLUTION. Using the Laplace inversion formula (D.2), we have

$$f(t) = \frac{1}{2\pi i} \int_{c-i\infty}^{c+i\infty} \frac{e^{-a\sqrt{s}}}{s}\, e^{st}\, ds. \tag{D.29}$$

Consider the Bromwich contour $MABC_1CDL$ (Figure D.3). Then by Cauchy's theorem we get

$$I = \int_{c-i\infty}^{c+i\infty} \frac{e^{-a\sqrt{s}}}{s} e^{st}\, ds = \int_{LD} F(s)\, ds + \int_{DC} F(s)\, ds$$
$$+ \int_{C_1} F(s)\, ds + \int_{BA} F(s)\, ds + \int_{AM} F(s) ds.$$

As established in Lemma D.2, we have

$$\int_{LD} F(s)\, ds + \int_{AM} F(s)\, ds = 0,$$

where $F(s) = \dfrac{e^{-a\sqrt{s}}}{s}\, e^{st}$.

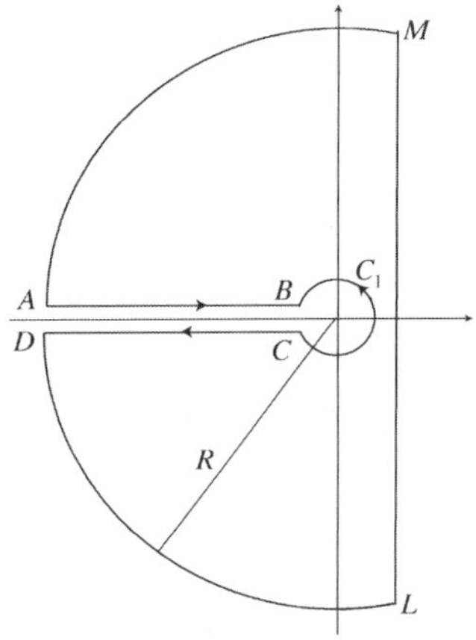

Figure D.3 Bromwich contour.

The integral over the circle C_1 is easily shown to be equal to $2\pi i$. This is done by taking the radius to be ε and substituting $s = \varepsilon e^{i\theta}$. On BA, $s = u\, e^{i\pi}$, and

$$\begin{aligned} I_{BA} &= \int_{\varepsilon\to 0}^{R\to\infty} \frac{1}{u\, e^{i\pi}} e^{-a\sqrt{u}e^{i\pi/2 + ute^{i\pi}}}\, e^{i\pi}\, du \\ &= \int_0^\infty \frac{1}{u}\, e^{-ia\sqrt{u}-ut}\, du = \int_0^\infty \frac{1}{u} e^{-ut}(\cos a\sqrt{u} - i \sin a\sqrt{u})\, du \\ &= 2\int_0^\infty \frac{1}{v} e^{-v^2 t}(\cos av - i\sin av)\, dv, \end{aligned}$$

where $u = v^2$. Similarly, $\int_{CD} = -2\int_0^\infty \dfrac{1}{v} e^{-v^2 t}(\cos av + i \sin av)\, dv$. Hence,

$$\int_{CD} + \int_{BA} = -4i \int_0^\infty \frac{1}{v} e^{-v^2 t} \sin av\, dv.$$

In order to evaluate the integral $\int_0^\infty \frac{1}{v}\, e^{-v^2 t} \sin av\, dv$, we consider the integral $\int_0^\infty e^{-v^2 t} \cos av\, dv$. Then

$$\begin{aligned}\int_0^\infty e^{-v^2 t} \cos av\, dv &= \Re\Big\{\int_0^\infty e^{-v^2 t+iav}\, dv\Big\}\\ &= \Re\Big\{e^{-a^2/4t}\int_0^\infty e^{-(v\sqrt{t}-ia/2\sqrt{t})^2}\, dv\Big\}\\ &= \Re\Big\{\frac{e^{-a^2/4t}}{\sqrt{t}}\int_{-ia/2\sqrt{t}}^\infty e^{-w^2}\, dw\Big\}, \quad \text{where } w = v\sqrt{t} - ia/2\sqrt{t}\\ &= \Re\Big\{\frac{e^{-a^2/4t}}{\sqrt{t}}\Big[\int_0^\infty e^{-w^2}\, dw + \int_{-ia/2\sqrt{t}}^0 e^{-w^2}\, dw\Big]\Big\}.\end{aligned}$$

Hence,

$$\int_0^\infty e^{-v^2 t} \cos av\, dv = \frac{\sqrt{\pi}\, e^{-a^2/4t}}{2\sqrt{t}}.$$

Integrating both sides of this equation with respect to a from 0 to a, we get

$$\int_0^\infty \frac{1}{v}\, e^{-v^2 t} \sin av\, dv = \sqrt{\frac{\pi}{4t}}\int_0^a e^{-x^2/4t}\, dx = \frac{\pi}{2}\, \operatorname{erf}\frac{a}{2\sqrt{t}}.$$

Thus

$$\mathcal{L}^{-1}\Big\{\frac{e^{-a\sqrt{s}}}{s}\Big\} = \frac{1}{2\pi i}\Big[2\pi i - 4i\frac{\pi}{2}\, \operatorname{erf}\frac{a}{2\sqrt{t}}\Big] = \operatorname{erfc}\frac{a}{2\sqrt{t}}. \ \blacksquare \tag{D.30}$$

D.9. Determine $f(x,t) \equiv \mathcal{L}_2^{-1}\Big\{\dfrac{e^{-x\sqrt{s^2+a^2}}}{2s^2(s^2-b)}\Big\}$.

SOLUTION. Let $F(s) = \dfrac{e^{-x\sqrt{s^2+a^2}}}{2s^2(s^2-b)}$, which gives $2F(\sqrt{s}) = \dfrac{e^{-x\sqrt{s+a^2}}}{s(s-b)}$. If we denote $x\sqrt{s+a^2}$ by z and use $e^{-z} = \dfrac{2}{\sqrt{\pi}}\int_0^\infty e^{-y^2-z^2/(4y^2)}\, dy$, we find that

$$\begin{aligned} f(x,t) &= \frac{1}{2\pi i}\int_{c-i\infty}^{c+i\infty} \frac{e^{-x\sqrt{s+a^2}}}{s(s-b)}\, e^{st^2}\, ds\\ &= \frac{1}{2\pi i}\int_{c-i\infty}^{c+i\infty} \frac{1}{s(s-b)}\Big(\int_0^\infty e^{-y^2-(x^2(s+a^2))/(4y^2)}\, dy\Big)\, e^{st^2}\, ds\\ &= \int_0^\infty \frac{e^{-y^2-a^2x^2/(4y^2)}}{s}\, dy\Big(\frac{1}{2\pi i}\int_{c-i\infty}^{c+i\infty} \frac{e^{-x^2/(4y^2)-t^2)s}}{s-b}\, ds\Big)\end{aligned}$$

$$= \frac{1}{b}\int_0^\infty e^{-y^2-a^2x^2/(4y^2)}\left(e^{b(t-x^2/(4y^2)+t^2)}-1\right)H(t-x^2/(4y^2)+t^2)\,dy$$
$$= \frac{1}{b}\int_A^\infty e^{-y^2-(a^2+1)x^2/(4y^2)+b(t+t^2)}\,dy - \frac{1}{b}\int_A^\infty e^{-y^2-a^2x^2/(4y^2)}\,dy,$$

where $A=\frac{1}{2}\frac{x}{\sqrt{t+t^2}}$, and in the last two lines above we have used the formulas $\mathcal{L}^{-1}\Big\{\frac{e^{-as}}{s}\Big\}=H(t-a)$ and $\mathcal{L}^{-1}\Big\{\frac{e^{-as}}{s-b}\Big\}=e^{b(t-a)}\,H(t-a)$ (see Table D.1).

D.10. To show that $\mathcal{L}_2\{\sin(t\tau)\}=\dfrac{\pi}{4s^3}t\,e^{-t^2/(4s^2)}$. HINT. Use the definition (D.24).

D.11. Use the $\mathcal{L}_2$-transform to solve the singular integral equation

$$\frac{2}{\pi}\int_0^\infty f(y)\sin(ty)\,dy=\operatorname{erf}\Big(\frac{t}{2a}\Big),\quad a\in\mathbb{R}.$$

SOLUTION. Using Example D.8, we find that

$$\frac{2}{\pi}F\Big(\frac{1}{2s}\Big)\frac{\sqrt{\pi}}{4s^3}=\frac{1}{2s^2\sqrt{1+a^2s^2}},$$

which simplifies to $F(s)=\frac{1}{2}\sqrt{\dfrac{\pi}{a^2+s^2}}$. On inverting the $\mathcal{L}_2$-transform (using (D.26)) we obtain the required solution as $f(t)=\dfrac{e^{-a^2t^2}}{t}$. Note that $\operatorname{erf}(x)=\dfrac{2}{\sqrt{\pi}}\int_0^x e^{-t^2}\,dt$.

D.12. Use the Efros theorem to find $\mathcal{L}_2\Big\{\displaystyle\int_0^\infty \operatorname{erfc}\Big(\frac{\tau^2}{2x}\Big)\,d\tau\Big\}$.

SOLUTION. Using (D.28) we get

$$\mathcal{L}_2\Big\{\int_0^\infty \operatorname{erfc}\Big(\frac{\tau^2}{2x}\Big)\,d\tau\Big\}=\mathcal{L}_2\Big\{\int_0^\infty \tau\operatorname{erfc}\Big(\frac{\tau^2}{2x}\Big)\frac{1}{\tau}\,d\tau\Big\}$$
$$=\frac{1}{s^2}\Big[\mathcal{L}_2\Big[\frac{1}{\tau}\Big\}\Big]_{s\to\sqrt{s}}=\frac{\sqrt{\pi}}{4s^{5/4}},$$

which, using (D.26) for $F(\sqrt{s})=\dfrac{\sqrt{\pi}}{4s^{5/4}}$ yields

$$\mathcal{L}_2^{-1}\Big\{\frac{\sqrt{\pi}}{4s^{5/4}}\Big\}=\frac{\sqrt{\pi}}{2\Gamma(\frac{5}{4})}\,x^{1/4}.$$

Table D.1 Some Laplace Transform Pairs

$f(t)$	$F(s)=\int_0^\infty e^{-st} f(t)\,dt$
1. $f(at),\ a>0$	$\frac{1}{a}F\left(\frac{s}{a}\right)$
2. $e^{\pm at} f(t)$	$F(s\mp a)$
3. $H(t-a)f(t-a)$	$e^{-as}F(s)$
4. $e^{b(t-a)}H(t-a)$	$\frac{e^{-as}}{s-b}$
5. $t^n f(t)$	$(-1)^n F^{(n)}(s)$
6. $f'(t)$	$sF(s)-f(0+)$
7. $f''(t)$	$s^2F(s)-sf(0+)-f'(0+)$
8. $f^{(n)}(t)$	$s^nF(s)-\sum\limits_{k=1}^{n} s^{n-k} f^{(k-1)}(0+)$
9. $\int_{0+}^{t} f(y)\,dy$	$\frac{F(s)}{s}$
10. $\int_{0+}^{t} \frac{(t-y)^{n-1}}{(n-1)!} f(y)\,dy$	$\frac{F(s)}{s^n}$
11. $\frac{1}{t}f(t)$	$\int_s^\infty F(y)\,dy$
12. $f(t+T)$, of period T	$(1-e^{-Ts})^{-1}\int_0^T e^{-st} f(t)\,dt$
13. $\frac{1}{\sqrt{t}}$	$\sqrt{\frac{\pi}{s}}$
14. $\delta(t)$	1
15. $\delta(t-T)$	$e^{-Ts},\ T\geqq 0$
16. $\delta^n(t)$	s^n
17. $1,\ H(t)$	$\frac{1}{s}$
18. $H(t-T)$	$\frac{e^{-Ts}}{s},\ T\geqq 0$
19. t^n	$\frac{n!}{s^{n+1}},\ n=0,1,2,\ldots$
20. $e^{\pm at}$	$\frac{1}{s\mp a}$
21. $t^n e^{\pm at}$	$\frac{n!}{(s\mp a)^{n+1}},\ n=0,1,2,\ldots$
22. $(1-at)e^{\pm at}$	$\frac{s}{(s\mp a)^2}$
23. $\sin at$	$\frac{a}{s^2+a^2}$
24. $\cos at$	$\frac{s}{s^2+a^2}$
25. $\sinh at$	$\frac{a}{s^2-a^2}$
26. $\cosh at$	$\frac{s}{s^2-a^2}$

E

Implicit Function Theorem

Let $f : \mathbb{R}n + m \mapsto \mathbb{R}^m$ be a continuously differentiable function, where $\mathbb{R}^{n+m} \equiv \mathbb{R}^n \times \mathbb{R}^m$. Fix a point $(\mathbf{a}, \mathbf{b}) = (a_1, \dots, a_n, b_1, \dots, b_m)$ such that $f(\mathbf{a}, \mathbf{b}) = 0$. Let $U \subset \mathbb{R}^n$ and $V \subset \mathbb{R}^M$, and let $g : U \mapsto V$ such that the graphs of g satisfy the relation $f = 0$ on $U \times V$. Let the Jacobian matrix $J = (Df)(\mathbf{a}, \mathbf{b})$ of f, defined by

$$\begin{aligned}(Df)(\mathbf{a}, \mathbf{b}) &= \begin{bmatrix} \frac{\partial f_1}{\partial x_1}(\mathbf{a}, \mathbf{b}) & \cdots & \frac{\partial f_1}{\partial x_n}(\mathbf{a}, \mathbf{b}) \\ \vdots & \vdots & \vdots \\ \frac{\partial f_n}{\partial x_1}(\mathbf{a}, \mathbf{b}) & \cdots & \frac{\partial f_n}{\partial x_n}(\mathbf{a}, \mathbf{b}) \end{bmatrix} \begin{bmatrix} \frac{\partial f_1}{\partial y_1}(\mathbf{a}, \mathbf{b}) & \cdots & \frac{\partial f_1}{\partial y_n}(\mathbf{a}, \mathbf{b}) \\ \vdots & \vdots & \vdots \\ \frac{\partial f_n}{\partial y_1}(\mathbf{a}, \mathbf{b}) & \cdots & \frac{\partial f_n}{\partial y_n}(\mathbf{a}, \mathbf{b}) \end{bmatrix} \\ &= [\mathbf{x}][\mathbf{y}],\end{aligned}$$

where $\mathbf{x}$ is the matrix of partial derivatives in the variables x_i and $\mathbf{y}$ is the matrix of partial derivatives in the variables y_j. The *implicit function theorem* states that if Y is an invertible matrix, then there are U, V and g such that

$$\{(\mathbf{x}, g(\mathbf{x}) \,|\, \mathbf{x} \in U\} = \{(\mathbf{x}, \mathbf{y}) \in U \times V \,|\, f(\mathbf{x}, \mathbf{y}) = \mathbf{c}\}.$$

Thus, if f is k-times continuously differentiable in $U \times V$, then the same holds for the function g inside U, and

$$\frac{\partial g}{\partial x_j}(x) = -\Big(\frac{\partial f}{\partial y}(x, g(x))\Big)^{-1} \frac{\partial f}{\partial x_j}(x, g(x)).$$

Example E.1. The implicit derivative of y with respect x, and that of x with respect to y, can be found by totally differentiating the implicit function $f(x, y) = x^2 + y2 - 1$ and equating to zero, i.e., $2x\,dx + 2y\,dy = 0$, giving $\frac{dy}{dx} = -\frac{x}{y}$ and $\frac{dx}{dy} = -\frac{y}{x}$. ■

F

Locally Nonsatiated Function

The utility function $u : X \mapsto \mathbb{R}$ represents a binary relation $\succeq$ if $x \preceq y \iff u(x) \geq u(y)$. A preference relation $\succeq$ is locally nonsatiated if for all $x \in X$ and $\varepsilon > 0$, there exists a y such that $\|y - x\| < \varepsilon$ and $y \succ x$. A utility function $u : X \mapsto \mathbb{R}$ is *locally nonsatiated* if it represents a locally nonsatiated preference relation $\succeq$; that is, for every $x \in X$ and $\varepsilon > 0$, there exists a y such that $\|y - x\| < \varepsilon$ and $u(y) \geq u(x)$. This definition leads to the following result:

Theorem F.1. *Suppose $X \in \mathbb{R}^n$. A binary relation $\succeq$ is complete, transitive, and continuous iff it admits a continuous utility representation $u : x \mapsto \mathbb{R}$.*

Theorem F.2. *If $\succeq$ is strictly monotone, then it is locally nonsatiated.*

PROOF. Let x be given, and let $y = x + (\varepsilon/2)\, e$, where $e = \{1, 1, \dots, 1\}$. Then (i) $y_i > x$ for each i; (ii) strict monotonicity implies $y \succ x$; and (iii) $\|y - x\| = \sqrt{\sum_{i=1}^{n} (\varepsilon/n)^2} = \varepsilon/\sqrt{n} < \varepsilon$. Hence, $\succeq$ is locally nonsatiated.

Theorem F.3. *Prove that $\alpha x + (1 - \alpha)y \succeq y$.*

PROOF. Suppose $x \succeq y$, i.e., $x_1 + x_2 \geq y_1 + y_2$. Fix $\alpha \in (0, 1)$. Then $\alpha x + (1-\alpha)y = (\alpha x_1 + (1-\alpha)y_1, \alpha x_2 + (1-\alpha)y_2) = \alpha(x_1 + x_2) + (1-\alpha)(y_1 + y_2) \geq \alpha(y_1 + y_2) + (1 - \alpha)(y_1 + y_2) = y_1 + y_2$, since $x_1 + x_2 \geq y_1 + y_2$. ∎

If $h(u, p)$ is continuous and locally nonsatiated, and $h(u, p)$ is a function, then for all p_1 and p_2, we have

$$(p_2 - p_1)[h(u, p_2) - h(u, p_1)] \leq 0.$$

This is not strictly convex because $(1.0) \succeq (0, 1)$ and $(1, 0) \neq (0, 1)$, but $\frac{1}{2}(1, 0) + \frac{1}{2}(0, 1) = (\frac{1}{2}, \frac{1}{2}) \succeq (0, 1)$.

Finally, if u represents $\succeq$, then (i) $\succeq$ is convex if u is quasi-concave, and (ii) $\succeq$ is strongly convex if u is strictly quasi-concave.

For more details, see Richter [1971].

Bibliography

(Note: First author or single author is cited with last name first.)

Abramowitz, M., and I. A. Stegun (Eds.) 1972. *Handbook of Mathematical Functions.* New York: Dover.

Afriat, S. N. 1967. The construction of utility functions from expenditure data. *International Economic Review.* 8: 67-77.

Aghili A. , A. Ansari, and A. Sedgi. 2007. An inversion technique for the $\mathcal{L}_2$-transform with applications. *Int. J. Contemp. Math. Scs.* 2: 1387-1394.

An, M.Y. 1998. Logconcavity versus logconvexity: A complete characterization. *J. Econ. Theory.* 80: 350-369.

Arrow, K. J., and A. C. Enthoven. 1961. Quasi-concave programming. *Econometrica.* 29: 779-800.

———, L. Hurwicz, and H. Uzawa. 1961. Constraint qualifications in maximization problems. *Naval Res. Log. Quart.* 8: 175-191.

Avriel, M., W. E. Diewert, and I. Zang. 1988. *Generalized Concavity.* New York: Plenum Press.

Bagnoli, Mark, and Ted Bergstrom. 1989. Log-concave Probability and its Applications. Univerisity of Michigan. *Econ. Theory.* 26: 445-469.

———, and T. Bergstrom. 2005. Log-concave probability and its applications. *Econ. Theory.* 26: 445-469.

Barlow, R. E., and F. Proschan. 1981. *Statistical Theory of Reliability and Life Testing: Probability Models.* Silver Springs, MD: McArdale Press.

Barro, R.J., and X. Sala-i-Martin. 2004. *Economic Growth.* Boston: MIT Press.

Barvinok, A. 2002. *A Course in Convexity.* Vol, 54, Graduate Studies in Mathematics. American Mathematical Society.

Beale, E. M. L. 1959. On quadratic programming. *Naval Research Logistic Quart.* 6: 227-243.

Bellman, R. E. 1970. *Introduction to Matrix Algebra.* New York, NY: McGraw-Hill.

Berger, M. 1990. Convexity. *The American Mathematical Monthly.* 97(8): 650-678.

Bergstrom, T., and M. Bagnoli. 2005. Log-concave probability and its applications. *Econom. Theory.* 26: 445-469.

Bertsekas, D. P. 1999. *Nonlinear Programming.* 2nd ed. Athena Scientific, Belmont, MA.

Bertsimas, D.,and J. N. Tsitsiklis. 1997. *Introduction to Linear Optimization.* Athena Scientific, Belmont, MA

Billingsley, Patrick. 1995. *Probability and Measure*, 3rd ed. New York, NY: John Wiley.

Black, F., and M. Scholes. 1973. The pricing of options and corporate liabilities. *Journal of Political Economy.* 81 (3): 637-654.

Boas, Ralph P. 1996. *A Primer of Real Functions.* 4th ed. Washington, DC: Mathematical Association of America.

Borwein, Jonathan M., and Jon D. Vanderwerff. 2010. *Convex Functions: Construction, Characterizations and Counterexamples.* Cambridge University Press.

Boyd, Stephen, and Lieven Vandenberghe. 2004. *Convex Optimization.* Cambridge University Press.

Bazaraa, M. S., H. D. Sherali, and C. M. Shetty. 1993. *Nonlinear Programming: Theory and Algorithms.* 2nd ed. New York, NY: John Wiley.

Chen, Hong-Yi, Cheng-Few Lee, and Weikang Shih. 2010. Derivation and applications of Greek letters: Review and integration, in *Handbook of Quantitative Finance and Risk Management* (Cheng-Few Lee, Alice C. Lee and John Lee, eds.). III: 491-503. New York, Springer-Verlag.

Cormen, Thomas H., Charles E. Leiserson, Ronald L. Rivest, and Clifford Stein. 2001. *Introduction to Algorithms*, 2nd ed. Section 29.3: The simplex algorithm, 790-804. MIT Press and McGraw-Hill.

Crouzeix, J.-P. 2008. Quasi-concavity, in Steven N. Durlauf and Lawrence E. Blume *The New Dictionary of Economics*, 2nd ed. Palgrave: Macmillan.

Davenport, H., and G. Pólya. 1949. On the product of two power series. *Canadian J. Math.* 1: 1-5.

De Finetti, B. 1949. Sulla stratificazioni convesse. *Annali di Mathematica.* 30: 173-183.

D'Esopo, D. A. 1959. A convex programming procedure. *Naval Research Logistic Quarterly*, 6: 33-42.

Dharmadhikari, Sudhakar, and Kumar Joag-Dev. 1988. *Unimodality, Convexity, and Applications.* Boston, MA: Academic Press.

Eggleston, H. G. 1958. *Convexity.* Cambridge University Press.

Fenchel, W. 1953. *Convex Cones, Sets, and Functions.* Lecture Notes, Princeton University.

———. 1983. Convexity through the ages. In *Convexity and Its Applications*, (P. M. Gruber and J. M. Wills, eds.) pages 120-130. Birkhäuser Verlag.

Gill, P. E., W. Murray, M. A. Saunders, and M. H. Wright. 1990. A Schur-complement method for sparse quadratic programming. In M. G. Cox and S. J. Hammarling (eds.) *Reliable Numerical Computation.* Oxford University Press. 113-138.

———, W. Murray, and M. H. Wright. 1991. *Numerical Linear Algebra and Optimization*. Vol. 1. Addison-Wesley Publishing Company, Redwood City.

———, and E. Wong. 2012. Sequential quadratic programming methods. In J. Lee and S. Leyffer (eds.) *Mixed Integer Nonlinear Programming*. Vol. 154 of *The IMA Volumes in Mathematics and its Applications*. Springer New York. 147-224.

Ginsberg, W. 1973. Concavity and quasiconcavity in economics. *Journal of Economic Theory*. 6: 596-605

Golub, G., and C. F. Van Loan. 1989. *Matrix Computations*. 2nd ed. Johns Hopkins University Press.

Halmos, P. R. 1958. *Finite Dimensional Vector Spaces*. Princeton, NJ: Van Nostrand.

Hardy, G. H. 1967. *A Course in Pure Mathematics*. 10th ed. Cambridge University Press.

Hewett, Edwin, and Karl Stromberg. 1965. *Real and Abstract Analysis*. Springer-Verlag, New York.

Hildebrand, F. B. 1974. *Introduction to Numerical Analysis*. 2nd ed. New York, NY: McGraw-Hill.

Hildreth, C. 1957. A quadratic programming procedure. *Naval Research Logistic Quarterly*, 14: 79-85.

Hoffman, K., and R. Kunze. 1961. *Linear Algebra*. Englewood Cliffs, NJ: Prentice-Hall.

Hogg, R., and S. Klugman. 1984. *Loss Distributions*. New York, NY: John Wiley.

Hoggar, S. G. 1974. Chromatic polynomials and logarithmic concavity. *J. Combin. Theory*, Ser. B. 16: 248-254.

Horn, R. A., and C. A. Johnson. 1985. *Matrix Analysis*. Cambridge University Press.

Horst, R., and P. Pardalos. 1994. *Handbook of Global Optimization*. Kluwer.

Hull, John C. 2008. *Options, Futures and Other Derivatives*. 7th ed. Englewood Cliffs, NJ: Prentice-Hall.

Hiriart-Urruty, J.-B., and C. Lemaréchal. 2001. *Fundamental of Convex Analysis*. Springer-Verlag.

Joag-Dev, K., and F. Proschan. 1983. Negative association of random variables with applications. *Ann. Statis.* 11: 286-295.

Johnson, Oliver, and Christina Goldschmidt. 2008. Preservation of log-concavity on summation. Available at *arXiv:math/0502548v2 [math. PR]*, 12 Oct 2005. 1-15.

Kaplan, Wilfred. 1959. *Advanced Calculus*. Addison-Wesley, Reading, MA.

Karush, W. 1939. Minima of functions of several variables with inequalities as side constraints. M. Sc. Dissertation. Dept. of Mathematics, Univ. of Chicago, Chicago, IL.

Klee, V. 1971. What is a convex set? *The American Mathematical Monthly*.

78(6): 616-631.

Krugman, P. 1991. Increasing returns and economic geography. *J. Polit. Econ.* 99: 483-499.

Kuhn, H. W. 1976. Nonlinear Programming. A historical review, in *Nonlinear Programming*, (R. W. Cottle and C. E. Lemke, eds.), Vol 9, *SIAM-AMS Proceedings*, pages 1-26. American Mathematical Society.

———, and A. W. Tucker. 1951. Nonlinear programming. *Proceedings of 2nd Berkeley Symposium*. Berkeley, University of California Press. pp. 481-492.

Kythe, Prem K. 2011. *Green's Functions and Linear Differential Equations: Theory, Applications, and Computation.* Taylor & Francis Group/CRC Press.

Lay, S. R. 1982. *Convex Sets and Their Applications.* New York, NY: John Wiley.

Lekkerkerker, C. G. 1953. A property of logarithmic concave functions, I, II *Indag. Math.* 15: 505-521.

Lipschutz, S. 1968. *Linear Algebra.* New York, NY: McGraw-Hill.

Luenberger, D. G. 1968. Quasi-convex programming. *SIAM Journal on Applied Mathematics.* 16(5).

——— . 1984. *Linear and Nonlinear Programming.* Addison-Wesley.

Mangasarian, O. L. 1969/1994. *Nonlinear Programming.* New York, NY: McGraw-Hill. Reprinted as *Classics in Applied Mathematics.* SIAM, 1994.

Markowitz, H. 1952. Porfolio selection. *The Journal of Finance.* 7(1): 77-91.

Marsden, Jerrold, E., and Anthony J. Tromba. 1976. *Vector Calculus.* W. H. Freemann, San Francisco.

Marshall, A., and I. Olken. 1979. *Inequalities: Theory of Majorization and Its Applications.* New York: Academic Press.

Martin, D. H. 1985. The essence of invexity. *J. Optim. Theory Appl.* 47: 65-76. doi:10.1007/BF00941316.

Martos, B. 1969. Subdefinite matrices and quadratic forms. *SIAM J. Appl. Math.* 17: 1215-1233.

——— . 1971. Quadratic programming with quasi-convex objective function. *Opns. Res.* 19: 87-97.

——— . 1975. *Nonlinear Programming Theory and Methods.* North-Holland.

Mas-Colell, A., M. D. Whinston, and J. R. Green. 1995. *Microeconomic Theory.* Oxford: Oxford University Press.

Merkle, Milan. 1998a. Convolutions of logarithmically concave functions. *Univ. Beograd. Publ. Elektrotehn. Fak. Ser. Math.* 9: 113-117.

——— . 1998b. Logarithmic concavity of distribution functions. *International Memorial Conference "S. S. Mitrinovic"* Niṣ, 1996 collection, in G. V. Milovanovič (ed.) *Recent Progress in Inequalities.* Dordrecht: Kluwer Academic Publishers. pp. 481-484.

Meyer, C.D. 2000. *Matrix Analysis and Applied Linear Algebra.* Society for Industrial and Applied Mathematics.

Michel, Anthony N., and Charles J. Herget. 2007. *Algebra and Analysis.* Boston, MA: Birkhaüser.

Mocedal, J., and S. J. Wright. 1999. *Numerical Optimization.* Springer.

Moyer, Herman. 1969. *Introduction to Modern Calculus.* New York, NY: McGraw-Hill.

Muth, E. 1977. Reliability models with positive memory derived from the mean residual life function, in *Theory and Applications of Reliability*, vol. II, C. Toskos and I. Shimi, eds. New York: Academic Press, pp. 401-436.

Nesterov, Yurii. 2004. *Introductory Lectures on Convex Optimization: A Basic Course.* (Applied Optimization). New York: Springer Science + Business Media.

———, and Lars-Erik Persson. 2006. *Convex Functions and Their Applications.* (CMS Books in Mathematics). Cambridge University Press.

Nicholson, Walter. 1978. *Microeconomic Theory.* 2nd ed. Hinsdale: Dryden Press.

Niculescu, C. P. 2000. A new look at Newton's inequalities. *J. Inequal Pure Appl. Math.* 1, issue 2, article 17; also http://jipam.vu.edu.au/.

———, and Lars-Erik Persson. 2006. *Convex Functions and Their Applications.* (CMS Books in Mathematics).

Nielsen, Lars Tyge. 1993. Understanding $N(d_1)$ and $N(d_2)$: Risk-adjusted probabilities in the Black-Scholes model. *Revue Finance (Journal of the French Finance Association).* 14(1): 95-106.

Osserman, Robert. 1968. *Two-Dimension Calculus.* Harcourt, Bruce & World, New York.

Patel, J. K., C. H. Kapadia, and D. B. Owen. 1976. *Handbook of Statistical Distributions.* New York: Marcel Dekker.

Peajcariaac, Josep E., and Y. L. Tong. 1992. *Convex Functions, Partial Orderings, and Statistical Applications.* Mathematics in Science & Engineering. Boston, MA: Academic Press.

Pečarić, Josep E., Frank Proschan, and Y. L. Tong. 1992. *Convex Functions, Partial Orderings, and Statistical Applications.* Mathematics in Science and Engineering 187. Boston, MA: Academic Press.

Phelps, Robert R. 1993. *Convex Functions, Monotone Operators and Differentiability.* Lecture Notes in Mathematics.

Polyak, B. T. 1987. *Introduction to Optimization.* Optimization Software. Translated from Russian.

Ponstein, J. 1967. Seven kinds of convexity. *SIAM Review.* 9(1): 115-119.

Prékopa, András. 1971. Logarithmic concave measures with applications tom stochastic programming. *Acta Scientiarum Mathematicarum.* 32: 301-316.

Rådo, Lennat, and Bertil Westergren.1995. *Mathematical Handbook for Science and Engineering.* Boston, MA: Birkhäuser.

Rice, John. 1995. *Mathematical Statistics and Data Analysis*, 2nd ed., Duxbury Press.

Richter, M. K., 1971. *Rational Choice*, in J. S, Chipman, I. Hurwitz, M. K. Richter, and H. F. Sonnenschein *Preferences, Utility, and Demand.* A Minnesota Symposium. Chapter 2: 29:58. New York: Harcourt-Bruce-Janovich.

Roberts, A. Wayne, and Dale E. Varberg. 1973. *Convex Functions.* New York: Academic Press.

Rockafellar, Ralph Tyrell. 1970/1997. *Convex Analysis.* Princeton Landmarks in Mathematics and Physics. Princeton University Press.

Ross, S. 1996. *Stochastic Processes.* 2nd ed. New York, NY: Wiley.

Royden, H. L. 1968. *Real Analysis.* 2nd ed. London: Macmillan.

Rudin, W. 1976. *Principles of Mathematical Analysis.* New York: McGraw-Hill.

Ruszczyński, Andrzej. 2006. *Nonlinear Optimization.* Princeton, NJ: Princeton University Press.

Sagan, B. E. 1988. Inductive and injective proofs of log concavity results. *Discrete Math.* 68: 281-292.

———. 1992. Inductive proof of q-log concavity. *Discrete Math.* 99: 289-306.

Schrijver, A. 1986. *Theory of Linear and Integer Programming.* New York, NY: John Wiley.

Schoenberg, I. J. 1951. On Pólya frequency functions, I. The totally positive functions and their Laplace transforms. *J. Analyse Math.* 1: 331-374.

Simon, C., and L. Blume. 1994. *Mathematics for Economists.* New York: W. W. Norton.

Singh, Richa. 2012. *Optimization Methods and Quadratic Programming.* Master's thesis, National Institute of Technology, Rourkela, India.

Skiba, A. K. 1978. Optimal growth with a convex-concave production function. *Econom.* 46(3): 527-539.

Slater, M. 1950. Lagrange multipliers revisited: A contribution to non-Linear programming. Cowles Commission Discussion Papers, Math. 403, November 1950.

Smith, Peter. 1985. *Convexity Methods in Variational Calculus.* Research Studies Press, Letchworth, UK, and John Wiley, New York.

Solow, R. 1956. A contribution to the theory of economic growth. *Q. J. Econ.* 70: 65-94.

Stanley, R. P. 1989. Log-concave and unimodal sequences in algebra, combinatorics, and geometry. In *Graph Theory and its Applications: East and West* (Jinan, 1986); *Ann. New York Acad. Sci.* 576: 500-535.

Strang, G. 1980. *Linear Algebra and its Applications.* New York: Academic Press.

Takayama, Akira. 1993. *Analytical Methods in Economics.* University of Michigan Press.

Tikhomorov, V. M. 1990. Convex analysis. In *Analysis II: Convex Analysis and Approximation Theory.* (R. V. Gamkrelidze, ed.) Vol 14, pages 1-82.

Springer.

Todd, M. J. 2001. Semidefinite optimization. *Acta Numerica.* 10: 515-560.

Valentine, F. A. 1964. *Convex Sets.* New York, NY: McGraw-Hill.

Vandenberghe, L., and S. Boyd. 1995. Semidefinite programming. *SIAM Review*, 49-95.

van Tiel, J. 1984. *Convex Analysis. An Introductory text.* New York, NY: John Wiley.

Varian, Hal R. 1982. The nonparametric approach to demand analysis. *Econometrica.* 50.945-973.

———. 1992. *Microeconomic Analysis.* 3rd ed. New York: Norton.

Veblen, Thorstein B. 1899. *The Theory of the Leisure Class: An Economic Study of Institutions.* London: Macmillan.

von Neumann, J. 1928. Zur Theorie der Gesellschaftsspiele. *Math. Annalen.* 100: 295-320.

———. 1945-46. A model of general economic equilibrium. *Review of Economic Studies.* 13: 1-9.

———, and O. Morgenstern. 1953. *Theory of Games and Economic Behavior.*3rd ed. Princeton University Press; 1st ed. 1944.

Wang, Y. Linear transformations preserving log-concavity. *Linear Algebra Appl.* 359: 161-167.

———, and Y.-N. Yeh. 2005. Log-concavity and LC-positivity. Available at *archiv:math.CO/0504164.*

Webster, R. 1994. *Convexity.* Oxford University Press.

Whittle, P. 1971. *Optimization under Constraints.* New York, NY: John Wiley.

Wilf, H. S. 1994. *Generating Functionology.* 2nd ed. Boston, MA: Academic Press.

Wilmott, P. , S. Howison, and J. Dewynne. 1995. *The Mathematics of Financial Derivatives: A Student Introduction.* Cambridge, U.K.: Cambridge University Press.

Wilson, C. 2012. Concave functions of a single variable. ECON-UA6, University of New York. *http://homepages.nyu.edu/caw1*, Feb 21, 2012.

Yurekli, O., and A. Sadek. 1991. Parseval-Goldstein type theorem on the Widder potential transform and its applications. *Intern. J. Math. Math. Scs.* 14:517-524.

———, and S. Wilson. 2002. A new method of solving Bessel's differential equation using the ℓ_2-transform. *Appl. Math. Comput.* 130: 587-591.

———, and S. Wilson. 2003. A new method of solving Hermite's differential equation using the ℓ_2-transform. *Appl. Math. Comput.* 145: 495-500.

Wolfe, P. 1959. The simplex method for quadratic programming. *Econometrica.* 27: 382-398.

Zalinescu, C. 2002. Convex Analysis in General Vector Spaces. World Scientific.

Index